Handbook of Experimental Pharmacology

Volume 106

Gastrointestinal Regulatory Peptides

Contributors

A. Alemayehu, B. Amiranoff, D.L. Barber, S.R. Bloom
J. Calam, R.E. Carraway, Feng Chen, C.S. Chew,
C.C. Chou, J.M. Conlon, A. Couvineau, E. Ekblad,
T.S. Gaginella, J.J. Galligan, J.R. Grider, R. Håkanson
P. Kitabgi, M. Laburthe, Y.P. Loh, M.S. O'Dorisio
S.M. O'Grady, R.K. Pearson, R.J. Playford
E. Rozengurt, F. Sundler, I. Zachary

Editor

David R. Brown

Springer-Verlag
Berlin Heidelberg New York London Paris
Tokyo Hong Kong Barcelona Budapest

DAVID R. BROWN, Ph.D.
Department of Veterinary Biology
Pharmacology Section
University of Minnesota
1988 Fitch Avenue
Saint Paul, MN 55108
USA

With 50 Figures and 20 Tables

ISBN 3-540-55970-1 Springer-Verlag Berlin Heidelberg New York
ISBN 0-387-55970-1 Springer-Verlag New York Berlin Heidelberg

Library of Congress Cataloging-in-Publication Data. Gastrointestinal regulatory peptides / contributors, A. Alemayehu . . . [et al.]; editor, David R. Brown. p. cm.—(Handbook of experimental pharmacology; v. 106) Includes bibliographical references and index. ISBN 3-540-55970-1 (alk. paper): DM480.00—ISBN 0-387-55970-1 (alk. paper) 1. Gastrointestinal hormones. I. Alemayehu, A. II. Brown, David R. (David Robert), 1954- . III. Series. [DNLM: 1. Gastrointestinal Hormones—physiology. 2. Gastrointestinal System—physiology. 3. Neuropeptides—pharmacology. 4. Neuropeptides—physiology. W1 HA51L v. 106 / WK 170 G2579] QP905.H3 vol. 106 [QP572.G35] 615'.1s—dc20 [612.3'2] DNLM/DLC for Library of Congress 92-49563 CIP

Typesetting: Best-set Typesetter Ltd., Hong Kong
27/3130 – 5 4 3 2 1 0 – Printed on acid-free paper

List of Contributors

Alemayehu, A., Department of Physiology, Michigan State University, East Lansing, MI 48828-1101, USA

Amiranoff, B., INSERM U239, Faculté de Médecine Xavier Bichat, 16 rue H. Huchard, F-75018 Paris, France

Barber, D.L., Department of Stomatology, School of Dentistry, Division of Oral Biology, University of California at San Francisco, San Francisco, CA 94143-0512, USA

Bloom, S.R., Royal Postgraduate Medical School, Hammersmith Hospital, Du Cane Road, London W12 0NN, Great Britain

Calam, J., Royal Postgraduate Medical School, Hammersmith Hospital, DuCane Road, London W12 0NN, Great Britain

Carraway, R.E., Department of Physiology, University of Massachusetts Medical Center, 55 Lake Avenue North, Worcester, MA 01655, USA

Chen, Feng, Departments of Pediatrics and Medical Microbiology and Immunology, The Ohio State University, 700 Children's Drive, Columbus, OH 43205, USA

Chew, C.S., Department of Physiology, Morehouse School of Medicine, 720 Westview Drive, S.W., Atlanta, GA 30310-1495, USA

Chou, C.C., Departments of Physiology and Medicine, Michigan State University, East Lansing, MI 48824-1101, USA

Conlon, J.M., Regulatory Peptide Center, Department of Biomedical Sciences, Creighton University Medical Center, 2500 California Street, Omaha, NB 68178, USA

Couvineau, A., INSERM U239, Faculté de Médecine Xavier Bichat, 16 rue H. Huchard, F-75018, France

Ekblad, E., Department of Medical Cell Research, University of Lund, Biskopsgatan 5, S-223 62 Lund, Sweden

Gaginella, T.S., Gastrointestinal Pharmacology, P.O. Box 1049, Woodland, CA 95695, USA

GALLIGAN, J.J., Department of Pharmacology and Toxicology, Michigan State University, Life Sciences Building, Room B-328, East Lansing, MI 48824, USA

GRIDER, J.R., Departments of Physiology and Medicine, Medical College of Virginia, Box 711, MCV Station, Richmond, VA 23298, USA

HÅKANSON, R., Department of Medical Cell Research, University of Lund, Biskopsgatan 5, S-223 62 Lund, Sweden

KITABGI, P., Institut de Pharmacologie Cellulaire et Moléculaire du CNRS, Sophia Antipolis, F-06560 Valbonne, France

LABURTHE, M., INSERM U239, Faculté de Médecine Xavier Bichat, 16 rue H. Huchard, F-75018 Paris, France

LOH, Y.P., Section on Cellular Neurobiology, Laboratory of Developmental Neurobiology, National Institute of Child Health and Human Development, National Institutes of Health, Bethesda, MD 20892, USA

O'DORISIO, M.S., Departments of Pediatrics and Medical Microbiology and Immunology, The Ohio State University, 700 Children's Drive, Columbus, OH 43205, USA

O'GRADY, S.M., Departments of Physiology and Animal Science, University of Minnesota, 1988 Fitch Avenue, St. Paul, MN 55108, USA

PEARSON, R.K., Center for Basic Research in Digestive Diseases, Mayo Clinic and Mayo Foundation, 200 First Street, S.W., Rochester, MN 55905, USA

PLAYFORD, R.J., Royal Postgraduate Medical School, Hammersmith Hospital, DuCane Road, London W12 0NN, Great Britain

ROZENGURT, E., Imperial Cancer Research Fund, Growth Regulation Laboratory, P.O. Box 123, 44 Lincoln's Inn Fields, London WC2A 3PX, Great Britain

SUNDLER, F., Department of Medical Cell Research, University of Lund, Biskopsgatan 5, S-223 62 Lund, Sweden

ZACHARY, I., Imperial Cancer Research Fund, Growth Regulation Laboratory, P.O. Box 123, 44 Lincoln's Inn Fields, London WC2A 3PX, Great Britain

Preface

Since the discovery of the pancreatic secretagogue secretin by W.M. BAYLISS and E.H. STARLING at the inception of the twentieth century, intense interest has focused on numerous, continually expanding classes of small peptides which appear to serve as regulatory molecules in the gastrointestinal tract, brain, and other organ systems. Initially, many of these substances were, like secretin, discovered in functional assays as "factors" or "activities" extractable in minute quantities from tissues or tissue fluids. By the middle of the century, advances in biochemical and immunological methods for the purification, characterization, and quantification of biologically active peptides in organ systems, tissues, and body fluids provided further impetus to this field. It was readily appreciated that small, biologically active peptides were particularly abundant in the digestive tract. Many peptides such as vasoactive intestinal peptide, gastrin, and more recently peptide YY and galanin were in fact originally discovered in and isolated from gut tissue. Moreover, these peptides were found to have profound actions on the gastroenteropancreatic system in vivo and in vitro. During the past 2 decades, information on regulatory peptides has burgeoned as a result of technological refinements in the synthesis of peptides, improved methods for detecting and visualizing peptides and their precursors in cells and tissues from a variety of species, advances in the functional assessment of peptide activity, and the application of molecular biological techniques to the characterization of peptide gene structure and expression. It has become clear that, as a diverse group of molecules present in a variety of cell types, small peptides are capable of producing a plethora of biological effects and may subserve roles as hormones, neurotransmitters, paracrine mediators, and autocrine factors depending upon their cellular location. In addition, regulatory peptides appear to be associated with aspects of several disease states and have been used both as diagnostic markers and as therapeutic agents.

The purpose of this volume of the *Handbook of Experimental Pharmacology* is to distill, from the vast amount of information collected on gut regulatory peptides, general concepts related to the production, fate, and actions of gut peptides as viewed primarily from cellular and molecular perspectives. Although common aspects of gut peptide biology are highlighted by the use of selected peptides as examples, no attempt was made to exhaustively detail the numerous facts available on specific gut peptides or peptide families. The subject material which follows is at once broad in scope and comprehensive. It is organized into three sections which develop in a logical sequence and is sufficiently balanced to interest pharmacologists and other investigators in a variety of biomedical disciplines. The first six chapters deal with the localization of peptides within cells of the gastrointestinal tract, peptide gene expression, and post-translational processing. In addition, the cellular release of peptides, subsequent interactions of these substances with their receptors on target cells, and the role of peptidase enzymes in terminating peptide action are considered. The next seven chapters discuss the cellular mechanisms by which peptides act to influence the function of specific cell types located in the gastrointestinal tract. Accordingly, the possible roles and pharmacological effects of peptides in such seemingly diverse functions as gastric acid secretion, the activity of enteric neurons, the contractility of gut smooth muscle, active ion transport in intestinal epithelial cells, intestinal blood flow, cellular growth and proliferation, and mucosal immunity are reviewed. The final two chapters focus on the known and suggested roles of peptides in disease states involving the digestive tract as well as the potential importance of these biologically active substances and their antagonists as drugs. It is hoped that this book will serve as an important, one-volume reference source for new and continuing investigations on the biology of regulatory peptides in the gastrointestinal tract and other organ systems well into the next century.

I am exceedingly grateful to all of the distinguished investigators who have authored chapters in this volume. They are to be highly commended, not only for their punctuality, which greatly facilitated the publication process, but for their uniformly excellent contributions which met and in many cases surpassed my expectations. Moreover, I thank the members of the Handbook's Editorial Board for the opportunity to serve as Editor of this prestigious volume, and Ms. DORIS WALKER for her assistance in the publishing process. Finally, I am indebted to my wife, Dr. LAURA J. MAURO, for her encouragement and loving support in the preparation of this book.

DAVID R. BROWN

Contents

CHAPTER 8

Peptides and Enteric Neural Activity

CHAPTER 11

Peptidergic Regulation of Gastrointestinal Blood Flow
C.C. CHOU and A. ALEMAYEHU . . . 325

CHAPTER 12

Peptidergic Regulation of Cell Proliferation Through Multiple Signaling Pathways
I. ZACHARY and E. ROZENGURT. With 1 Figure . . . 343

CHAPTER 13

Peptidergic Regulation of Mucosal Immune Function

CHAPTER 14

Pathophysiological Aspects of Gut Peptide Hormones

CHAPTER 1

Localization and Colocalization of Gastrointestinal Peptides

F. SUNDLER, E. EKBLAD, and R. HÅKANSON

A. Introduction

The digestive tract is the richest source of biologically active peptides outside the brain. The number of identified gut peptides has increased dramatically over the last 2 decades. Methodological advances have made this rapid development possible. Many neurohormonal peptides have C-terminal α-amide groups and the development of a screening method for peptides with amidated C-terminal residues has enabled the isolation of several neurohormonal peptides, such as peptide histidine isoleucine amide (PHI), peptide YY (PYY), neuropeptide Y (NPY) and galanin, all of which are present in the gut (TATEMOTO and MUTT 1978, 1980, 1981; TATEMOTO et al. 1982, 1983). In addition, techniques of molecular biology have been used to identify the precursors of a great number of both known and previously unknown gut peptides (see e.g., LUND et al. 1982; ITOH et al. 1983; ROSENFELD et al. 1983). Many of the precursors of known peptides were found to contain, besides the known peptide, cryptic segments some of which seem to be of biological significance as messenger molecules.

The peptide precursors are subjected to a series of proteolytic cleavages to yield multiple fragments (for reviews, see BROWNSTEIN 1985; EIPPER et al. 1985). The processing of the precursor, which includes modifications such as glycosylation and sulfation, is initiated during its transfer from the endoplasmic reticulum via the Golgi stacks to the secretory granules (ORCI 1986; DOCKRAY et al. 1991b; HALBAN 1991). Proteolytic cleavages and C-terminal amidations are thought to occur mainly within the secretory granules (EIPPER et al. 1985, 1986; DIMALINE 1988). By the time the granule is ready to release its content at the site of exocytosis the precursor has been degraded to fragments. However, the extent of the proteolytic processing may vary, particularly in cases where the precursor has several potential cleavage sites (see Chaps. 2, 3).

Cells designated for chemical communication within the gut include endocrine cells, paracrine cells and enteric neurons (HÅKANSON and SUNDLER 1983, 1986; SUNDLER et al. 1985b, 1989). Not only are they closely integrated in terms of function but they also seem to share a common genetic programming illustrated by the fact that they are capable of producing biologically active peptides, amines, certain types of large proteins such as synaptophysin and chromogranins, and certain types of enzymes, such as DOPA decarboxylase and neuron-specific enolase (SUNDLER et al. 1980; POLAK et al. 1984; SUNDLER and HÅKANSON 1988; WIEDENMANN and HUTTNER 1989).

B. Coexistence of Messengers

It is not uncommon to find two or more chemically related peptide sequences within a precursor molecule. Thus, vasoactive intestinal peptide (VIP) and peptide histidine isoleucine amide (PHI) derive from the same precursor and are structurally quite similar (ITOH et al. 1983). This is also the case with the tachykinins, substance P (SP) and neurokinin A (NKA) (NAWA et al. 1983), and with glucagon and glucagon-like peptides (GLP) I and II (LUND et al. 1982). The biological significance of the tandem arrangement of chemically related peptides within a precursor structure is unknown. It is to be anticipated that the various fragments that arise from a precursor will coexist in the storage granules and this has been amply documented in a number of cases (for reviews see SUNDLER et al. 1985b; VOIGT and MARTIN 1985). During recent years it has become evident that peptides arising from different precursors may coexist. Thus, a major population of enteric neurons of the gut contains both VIP and NPY and some of these neurons contain in addition galanin (for a review, see EKBLAD et al. 1991b). Glicentin and PYY coexist in endocrine cells which are numerous in the distal intestine (ALI-RACHEDI et al. 1984; BÖTTCHER et al. 1984, for a review see SUNDLER and HÅKANSON 1988). Certain granular proteins, such as the chromogranins, have been found to be universally distributed in peptide hormone-producing endocrine cells, including those of the gut, and in enteric neurons (FACER et al. 1985; ROSA et al. 1985; RINDI et al. 1986; LARSSON and SUNDLER 1990; WEIHE et al. 1991; for reviews see WIEDENMANN and HUTTNER 1989; DAYAL 1991). Moreover, for coexisting peptides arising from different precursors there is evidence that they are colocalized in the same granules, both in endocrine cells (ERICSON and SUNDLER 1984; SUNDLER et al. 1985b; BÖTTCHER et al. 1986; JOHNSON et al. 1988) and in neurons (UCHIDA et al. 1985; VOIGT and MARTIN 1985; GULBENKIAN et al. 1986; MERIGHI et al. 1988). As has been repeatedly observed over the years, biologically active peptides may coexist with "classical" neuromessengers both in endocrine cells and in neurons (for reviews, see SUNDLER et al. 1980, 1989; LUNDBERG and HÖKFELT 1986). This is for instance the case in many adrenergic neurons where noradrenaline coexists with NPY (for reviews, see LUNDBERG and HÖKFELT 1986; SUNDLER et al. 1986; HÅKANSON et al. 1990) and in gut endocrine cells where 5-hydroxytryptamine (5-HT) may coexist with enkephalin (ALUMETS et al. 1978) or SP (HEITZ et al. 1976; SUNDLER et al. 1977; for a review see SUNDLER and HÅKANSON 1988).

C. Endocrine and Paracrine Cells

Endocrine and paracrine cells occur scattered in the epithelium throughout the gut (for reviews, see SOLCIA et al. 1987; SUNDLER and HÅKANSON 1988; DAYAL 1991). Their distribution along the human intestine is illustrated in Fig. 1. In the intestine and in the antral portion of the stomach the endocrine/paracrine cells are open, i.e., they reach the lumen via an apical process carrying microvilli. This enables the cells to respond to specific stimuli in the gut lumen, such as pH and various nutrients. In the oxyntic mucosa of the stomach, endocrine/paracrine cells generally fail to reach the glandular lumen (closed cells) and they probably respond to humoral (neuronal and hormonal) or physical (e.g., distension or temperature changes) stimuli.

Endocrine cells deliver their messengers to the blood, while paracrine cells influence neighboring cells by releasing locally acting messengers from cytoplasmic processes. These processes, which usually extend along the base of the epithelium, may be of considerable length. They often end with a club-like swelling, filled with secretory granules. The somatostatin cells in the gastric mucosa have these paracrine features (ALUMETS et al. 1979; LARSSON et al. 1979). Admittedly, the distinction between endocrine and paracrine cells is far from clear and probably cannot be based merely upon

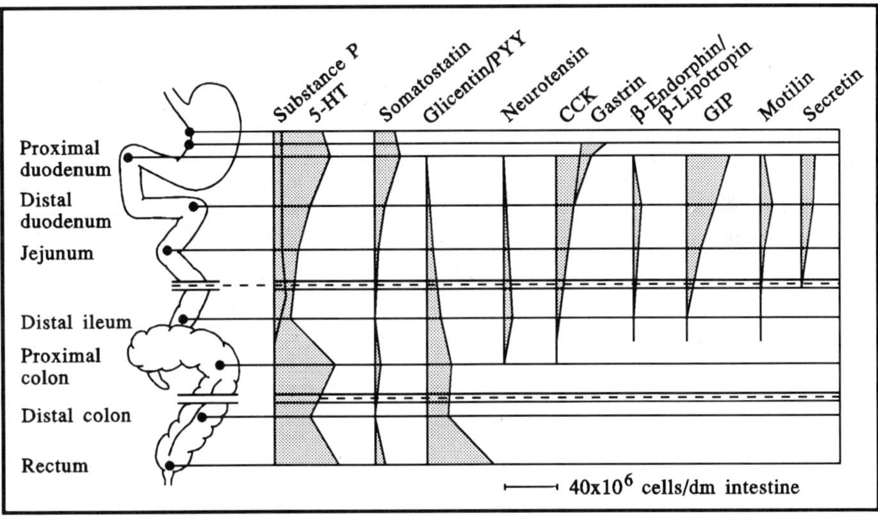

Fig. 1. Schematic diagram illustrating the distribution of the many different populations of endocrine/paracrine cells in the human gut with known peptide hormone and/or serotonin (5-HT) content based on immunocytochemical observations. The *width of the bars* reflects the cell density. Note that a small population of the serotonin-containing enterochromaffin (*EC*) cells contains SP, and that virtually all glicentin cells contain PYY

the presence or absence of basal cytoplasmic processes. Thus, cells which are now considered endocrine may well turn out to have a paracrine function and vice versa and it cannot be ruled out, in fact, that they may serve both endocrine and paracrine functions.

During the last 20 years a great number of different endocrine cell types have been identified in the digestive tract by virtue of the peptide hormones they contain. Each individual cell type has a characteristic distribution pattern (for reviews see Solcia et al. 1987; Sundler and Håkanson 1988; Dayal 1991) (Fig. 1). A few of these cell types, e.g., the enterochromaffin cells and the somatostatin cells, seem to occur throughout the digestive tract. It should be noted, however, that the enterochromaffin cells probably comprise several subpopulations as reflected in the morphology of the secretory granules (Solcia et al. 1976; Sundler and Håkanson 1988; Dayal 1991). Moreover they seem to contain different peptide hormones (Heitz et al. 1976; Sundler et al. 1977; Alumets et al. 1978). The somatostatin cells may be heterogeneous; somatostatin cells in different parts of the digestive tract seem to differ with respect to mechanisms of stimulation and chemical coding (co-existing peptides) (see below).

Several endocrine cell types in the digestive tract, recognized as such by their histochemical and ultrastructural properties, have not yet had identified peptide hormones ascribed to them. Such cells include the two predominating endocrine cell types in the oxyntic mucosa, the so-called ECL cells and the A-like (or X) cells (Sundler and Håkanson 1988, 1991).

As detailed in the succeeding chapters of this handbook, gut hormones and paracrine messengers seem to be involved in the integrated control of digestive processes, such as acid secretion, bicarbonate secretion, enzyme secretion from the pancreas and gut, gallbladder motility, intestinal motility and local blood flow. In addition, they may have indirect effects on these processes by controlling the activity of enteric neurons and other endocrine/paracrine cells (Sundler et al. 1989).

I. Stomach

Although the stomach harbors at least six different endocrine/paracrine cell types as revealed by electron microscopy and various histochemical stainings (Capella et al. 1991; Dayal 1991; Sundler and Håkanson 1991), only two of them, the gastrin cells and the somatostatin cells, have been identified so far on the basis of their peptide content.

Gastrin cells occur in the antrum, where they constitute the predominant endocrine cell population. How the gastrin cells respond to acute and chronic stimulation has been studied in some detail (Håkanson et al. 1982; Dockray et al. 1991a; for reviews, see Dockray et al. 1991b; Sundler et al. 1991c). In the resting cells the majority of granules are mature (old) and the processing of the precursor has generated predominantly small gastrin components. This is associated with large, electron lucent granules.

In the stimulated cells the majority of granules are immature (young) and the processing of the precursor is incomplete. This is associated with a predominance of small and highly electron-dense granules. Upon chronic stimulation the gastrin cells increase in number and in the rat there is a doubling of the cell number within a few weeks with no further increase upon prolonged stimulation as studied for very long periods of time (EISSELE et al. 1991; MATTSSON et al. 1991) (Fig. 2).

The somatostatin cells are distributed throughout the stomach. In the rat, the density of such cells is similar in the acid-producing mucosa and in

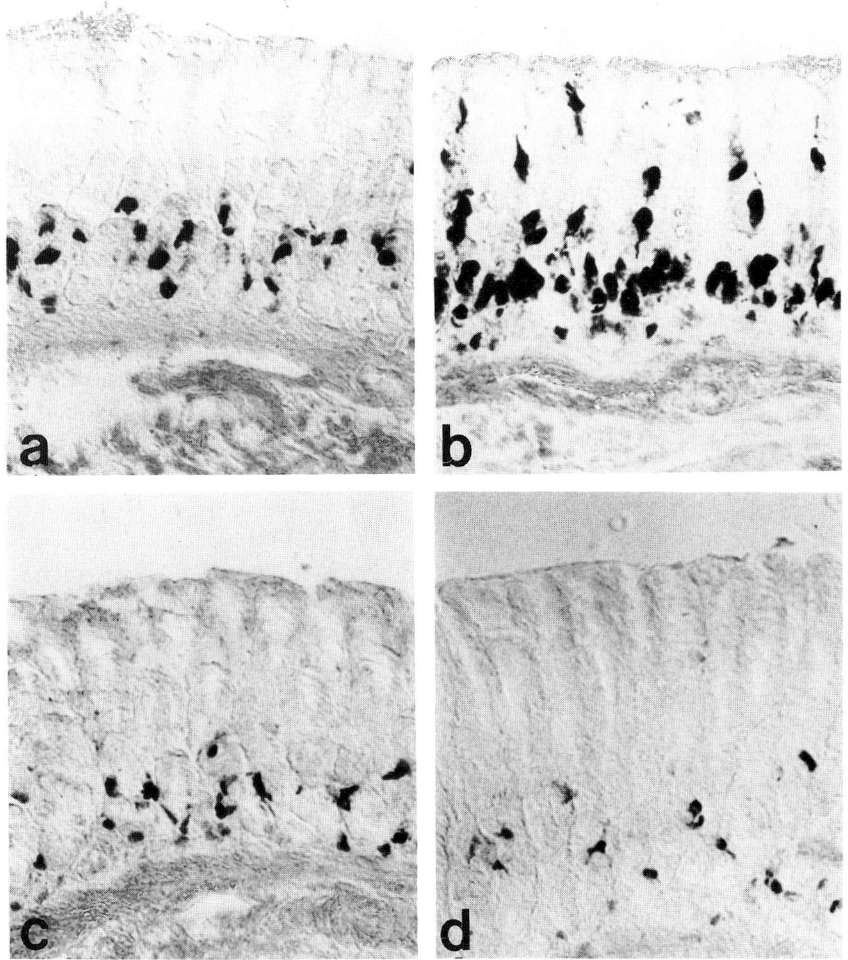

Fig. 2a–d. Rat antrum. Immunostaining for gastrin (**a,b**) and somatostatin (**c,d**). Control (**a,c**). Treatment with large doses of the H^+,K^+-ATPase blocker omeprazole for 10 weeks (**b,d**). The achlorhydria causes hyperplasia (twofold increase) of the gastrin cells, whereas the somatostatin cells are reduced in number, ×200

the antrum (ALUMETS et al. 1979). The cells are of the closed type in the acid-producing mucosa, while being of the open type in the antrum. In both locations they are provided with long cytoplasmic extensions, some of which seem to end on parietal cells in the acid-producing mucosa and on gastrin cells in the antrum (ALUMETS et al. 1979; LARSSON et al. 1979). In the acid-producing mucosa the cells occur scattered all along the glands; in the antrum they predominate – like the gastrin cells – in the basal portion of the glands in rodents and in the mid portion of the glands in carnivores and man. The antral somatostatin cells seem to behave reciprocally with the gastrin cells in that they are reduced in number and activity when the gastrin cells proliferate (KOOP et al. 1987; BRAND and STONE 1988) (Fig. 2).

The ECL cells, which are the major endocrine cell population in the oxyntic mucosa, contain histamine and respond to gastrin (HÅKANSON et al. 1986; for a review see HÅKANSON and SUNDLER 1991). This response is manifested in the release of histamine and the activation of the histamine-forming enzyme, histidine decarboxylase, upon acute gastrin challenge, and in the marked hypertrophy and hyperplasia of the ECL cells that follow a period of sustained hypergastrinemia (Fig. 3).

Interestingly, there is evidence for a gastrin-dependent hormonal role of the stomach in calcium homeostasis (for references, see PERSSON and HÅKANSON 1991), and extracts of the oxyntic mucosa have been found to contain peptide components that enhance calcium uptake into bone (PERSSON et al. 1989). This hypothetical calciotropic peptide, tentatively named gastrocalcin, may turn out to be an ECL cell hormone, keeping in mind the remarkable sensitivity of these cells to gastrin.

Recently, material reacting with certain antisera against endothelin has been demonstrated in the A-like (or X) cells of several mammals (CAPELLA et al. 1991; SUNDLER and HÅKANSON 1991) (Fig. 3). In the rat the A-like cells are the second largest endocrine cell population in the oxyntic mucosa and occur scattered all along the glands. They do not respond with hyperplasia, as do the ECL cells, to sustained hypergastrinemia (Fig. 3). The identity of the amylin- and endothelin-immunoreactive materials has yet to be established.

II. Small Intestine

The peptide hormone-producing cells present in the small intestines display three different types of regional distribution (SJÖLUND et al. 1983; for reviews, see SUNDLER and HÅKANSON 1988; CAPELLA et al. 1991; DAYAL 1991). The cells that contain cholecystokinin (CCK), secretin, motilin, or glucose-dependent insulinotropic peptide (GIP) are numerous in the duodenum. They become gradually fewer distally in the small intestine and are only rarely found in the ileum. In the porcine small intestine, enkephalin-containing endocrine cells, which constitute a subpopulation of the enterochromaffin cells (ALUMETS et al. 1978), also display this pattern of distribu-

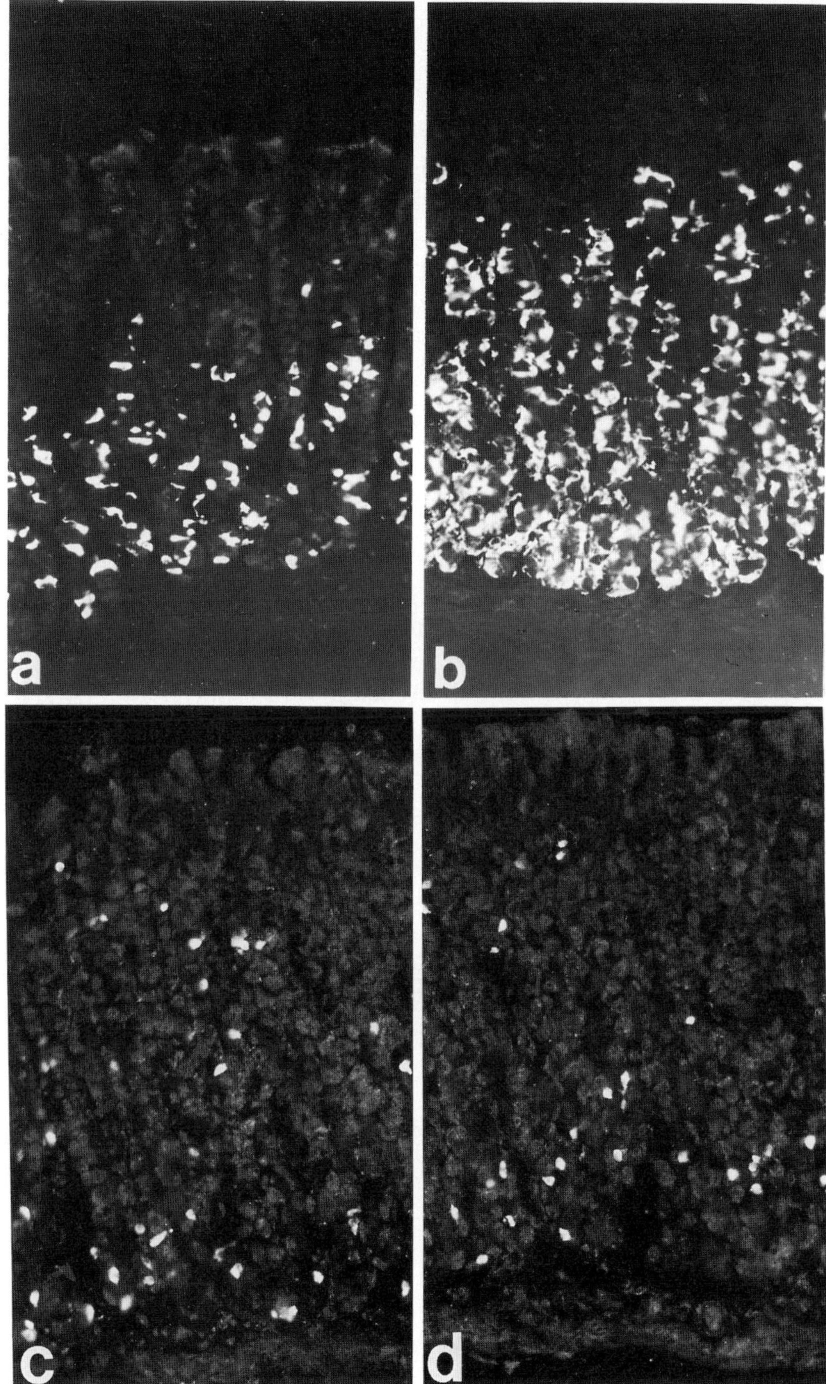

Fig. 3a–d. Rat stomach, oxyntic mucosa. Immunocytochemical demonstration of ECL cells using histidine decarboxylase antiserum (**a,b**) and of A-like cells using endothelin antiserum (**c,d**). Control (**a,c**). Omeprazole treatment as in Fig. 2. (**b,d**) The omeprazole-evoked hypergastrinemia causes a marked hyperplasia of the ECL cells. The A-like cells are unaffected, ×130

tion. The second category comprises somatostatin cells and enterochromaffin cells, which occur throughout the small and large intestines. The third category comprises neurotensin and glicentin/PYY cells, which are absent or rare in the duodenum and upper small intestine, but become gradually more numerous distally in the small intestine; the glicentin/PYY cells remain numerous in the large intestine.

The topographic (crypt/villus) distribution of the various peptide hormone-producing cells differs from one cell type to another (SJÖLUND et al. 1983). Usually, the endocrine cells predominate in the crypts with only occasional cells occurring in the villus epithelium. This distribution pattern applies for CCK, motilin, GIP, and enterochromaffin cells. Secretin cells and neurotensin cells are more numerous on the villi than in the crypts.

III. Large Intestine

Glicentin/PYY cells and enterochromaffin cells predominate in this location. Somatostatin cells are also regularly seen but they are much less frequent. In the proximal portion, occasional neurotensin-immunoreactive cells and gastrin/CCK-immunoreactive cells may occur. In certain species (e.g., the mouse) a subpopulation of the enterochromaffin cells contains SP (HEITZ et al. 1976; SUNDLER et al. 1977).

IV. Coexistence of Peptides in Gut Endocrine Cells

The antral gastrin cells have been reported to display immunoreactivity to a number of peptides that are unrelated to gastrin. Such peptides include PYY, neurotensin and the closely related peptide xenopsin (Fig. 4), ACTH/α-MSH, VIP, thyrotropin-releasing hormone (TRH), and delta sleep-inducing peptide (DSIP)-like immunoreactivity (for references, see Table 1). Some of these immunoreactants are restricted to subpopulations of the gastrin cells and to certain species. This is exemplified by VIP, which occurs in a small subpopulation of the antral gastrin cells in the cat. Most of the data on peptide colocalization in gastrin cells emanate from immunocytochemical studies, and the chemical identity of the immunoreactants has yet to be confirmed. A subpopulation of the somatostatin cells that occur in the oxyntic mucosa of, e.g., rat and guinea pig, contain immunoreactive PYY (own unpublished observations) (Fig. 5). In the intestines the perhaps best-established coexistence of peptides is that of glicentin and PYY. This coexistence has been demonstrated in a great number of species and seems to involve the whole population of glicentin cells in both the small and large intestine.

Much less is known about coexistence of peptide messengers in other intestinal endocrine cells. There is immunocytochemical evidence that DSIP-immunoreactive material is present not only in antral gastrin cells, but

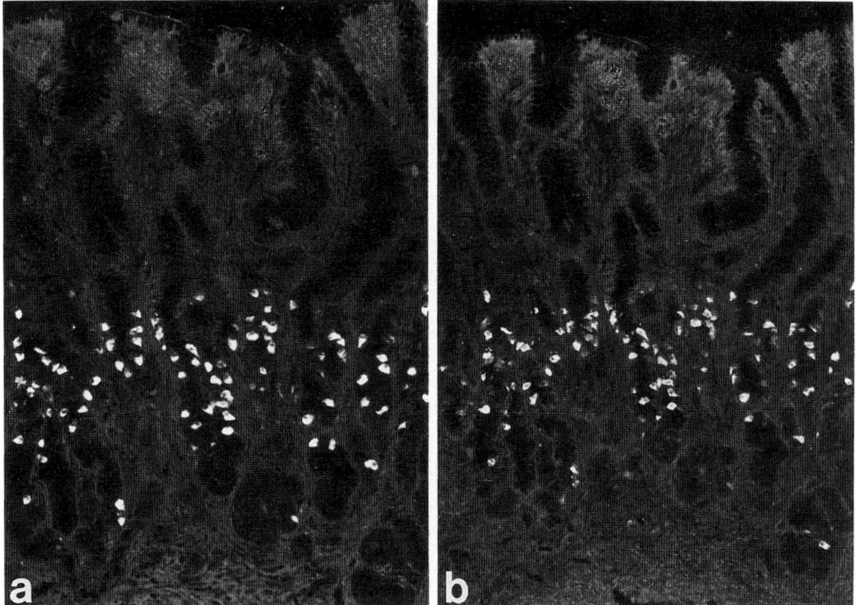

Fig. 4a,b. Human antrum. Double immunostaining for gastrin (**a**) and xenopsin (**b**). The vast majority of gastrin cells store immunoreactive xenopsin, ×100

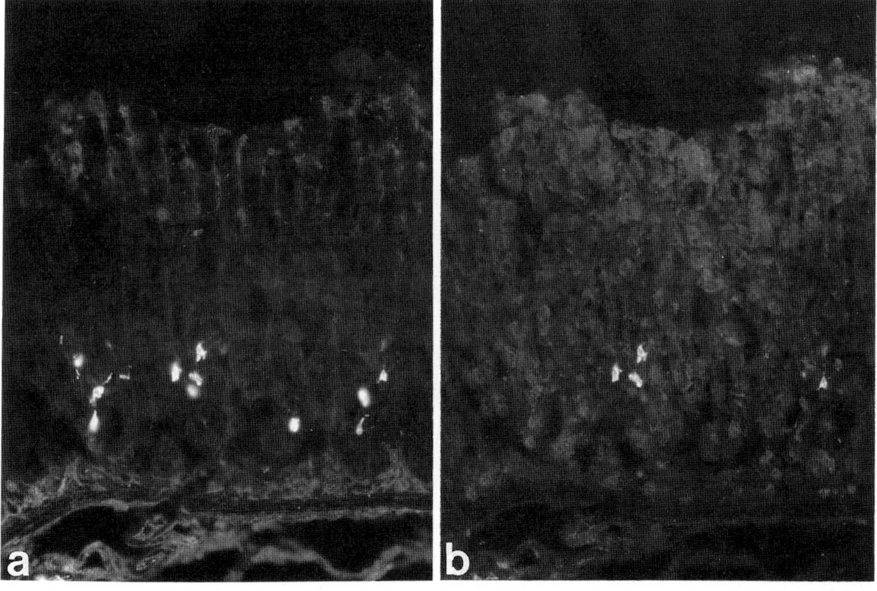

Fig. 5a,b. Rat stomach, oxyntic mucosa. Double immunostaining for somatostatin (**a**) and PYY (**b**). A subpopulation of the somatostatin cells store immunoreactive PYY, ×120

Table 1. Peptides in gut endocrine-paracrine cells

Colocalized peptides in endocrine cells	Species	Location	References
Somatostatin/PYY	Mouse, rat, guinea pig	Oxyntic mucosa	(Own unpublished observations)
Gastrin/VIP	Cat	Antrum	Sundler and Håkanson (1991)
Gastrin/PYY	Several mammals	Antrum	Ondolfo et al. (1989) Sundler and Håkanson (1991) Sundler et al. (1991b) Böttcher et al. (manuscript in preparation)
Gastrin/DSIP	Rat, pig, man	Antrum	Bjartell et al. (1989)
Gastrin/ACTH	Several species	Antrum	Larsson (1978)
Gastrin/xenopsin	Dog, monkey, man	Antrum	Rix et al. (1986) Sundler et al. (1991b)
Gastrin/TRH	Guinea pig	Antrum	Tsuruo et al. (1988)
Gastrin/pancreastatin	Several species	Antrum	Lamberts et al. (1990) Børglum-Jensen et al. (1991)
Secretin/DSIP	Man	Duodenum	Bjartell et al. (1989)
Secretin/SP	Mouse	Small intestine	Roth and Gordon (1990)
Glicentin/PYY	Several species	Small and large intestine	Ali-Rachedi et al. (1984) Böttcher et al. (1984)
Glicentin/PYY/DSIP	Man	Small and large intestine	Bjartell et al. (1989)

also in certain intestinal endocrine cells, notably those storing secretin and those storing glicentin/PYY. Further, a subpopulation of secretin cells in the mouse has been found to store SP.

Chromogranin A can, at least in some cell types, be proteolytically processed to yield fragments, one of which (pancreastatin) has biological effects. Pancreastatin has been localized immunocytochemically to several gut endocrine cells, including the ECL cells and the gastrin cells within the stomach. Also certain intestinal endocrine cells contain pancreastatin (Lamberts et al. 1990; Curry et al. 1990; Børglum Jensen et al. 1991).

D. Enteric Neurons

The nervous control of the gut is exercised primarily by an intrinsic nervous system originating in the two ganglionated plexuses, the submucous (Meissner's) and the myenteric (Auerbach's) plexuses, that occur throughout the gastrointestinal tract. The enteric neurons are as numerous as those in the spinal cord (Furness and Costa 1987) and the enteric nervous system has been referred to as a "little brain." Although characterized by a high degree of autonomy the enteric neurons are controlled by vagosacral parasympathetic nerves, sympathetic nerves emanating in the prevertebral ganglia, and sensory nerves originating in the jugular-nodose ganglionic complex (projecting mainly to the stomach) or in the dorsal root ganglia (projecting both to the stomach and the intestine) (Fig. 6). The parasympathetic fibers seem to terminate in the intramural ganglia. Sympathetic noradrenergic fibers can readily be demonstrated all along the gut and in all layers of the gut, with a predominance in the intramural ganglia, smooth muscle and around blood vessels. A prominent population of SP- and/or calcitonin gene-related peptide (CGRP)-containing, capsaicin-sensitive

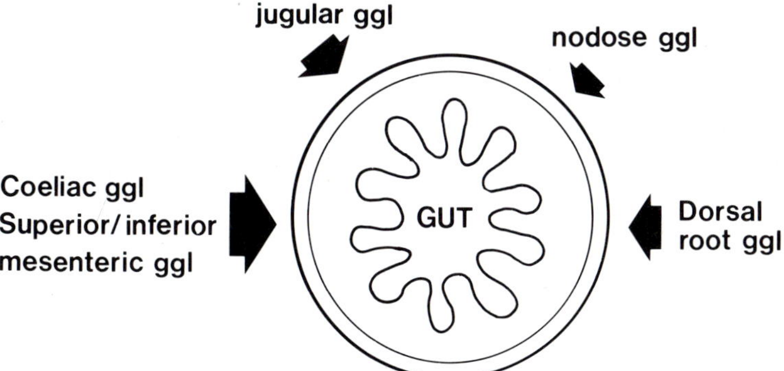

Fig. 6. Schematic outline of the extrinsic nerve supply to the gut. There are several different sources of the extrinsic innervation. The *width of the arrows* reflects the relative size of each contribution, *ggl* = ganglia

fibers occurs in the gastric mucosa and submucosa of many species (for references, see SUNDLER et al. 1991a), but they are less numerous in the intestine. Such fibers are richly distributed around blood vessels and are seen occasionally in the intramural ganglia. They are probably sensory in nature (Figs. 7, 8).

It seems that virtually all enteric neurons contain peptides (Table 2). Examples of the distribution of different peptide-containing fibers in the rat small intestine and colon are given in Figs. 7 and 8.

I. Distribution of Peptide-Containing Nerve Fibers in the Digestive Tract

Numerous populations of enteric peptide-containing nerve fibers can be identified. The predominating one is that containing VIP/PHI (for reviews see FURNESS and COSTA 1987; SUNDLER et al. 1988; EKBLAD et al. 1989b, 1991b). Numerous VIP/PHI-containing nerve fibers are found in all layers of the gut wall throughout the digestive tract. Intramural neurons containing NPY, gastrin-releasing peptide (GRP) and galanin are found to form extensive networks throughout the gastrointestinal tract (for a review see EKBLAD et al. 1991b). Depending on species and location these peptides are sometimes found within the VIP/PHI neurons. In several species VIP-containing neurons in the small intestine contain NPY (EKBLAD et al. 1984b); in some they contain galanin (MELANDER et al. 1985). In rat and humans, the VIP-containing nerve fibers in the gastric mucosa contain GRP but lack NPY (EKBLAD et al. 1991a).

Another population of NPY-containing nerve fibers is identical to the adrenergic nerve fibers that supply the blood vessels (FURNESS et al. 1983; SUNDLER et al. 1983; SU et al. 1987).

CGRP- and SP-containing nerve fibers in the gut derive from two different sources (GIBBINS et al. 1985; SUNDLER et al. 1985a; STERNINI et al. 1987; SU et al. 1987). One source is the enteric neurons that are found mainly in the myenteric ganglia. Another source of CGRP- and SP-containing (perivascular) fibers is the sensory ganglia. Other neuropeptides found in the digestive tract are CCK, neuromedin U (NMU), neurotensin, enkephalin and somatostatin (for a review see EKBLAD et al. 1991b). These populations are not as widespread as the one containing VIP. They occur mainly in the intramural ganglia although species variations in their distribution may exist. Enteric NMU-immunoreactive nerve fibers may exemplify this since this nerve fiber population shows remarkable species variation (EKBLAD et al. 1991b; and manuscript in preparation). NMU-containing neurons are numerous in chicken and rat but very few in man. The topographic distribution of NMU-immunoreactive nerve fibers varies considerably between species. In chicken gut, NMU-containing nerve fibers predominate in myenteric ganglia and smooth muscle while in the pig they occur almost exclusively in the mucosa/submucosa and submucous ganglia.

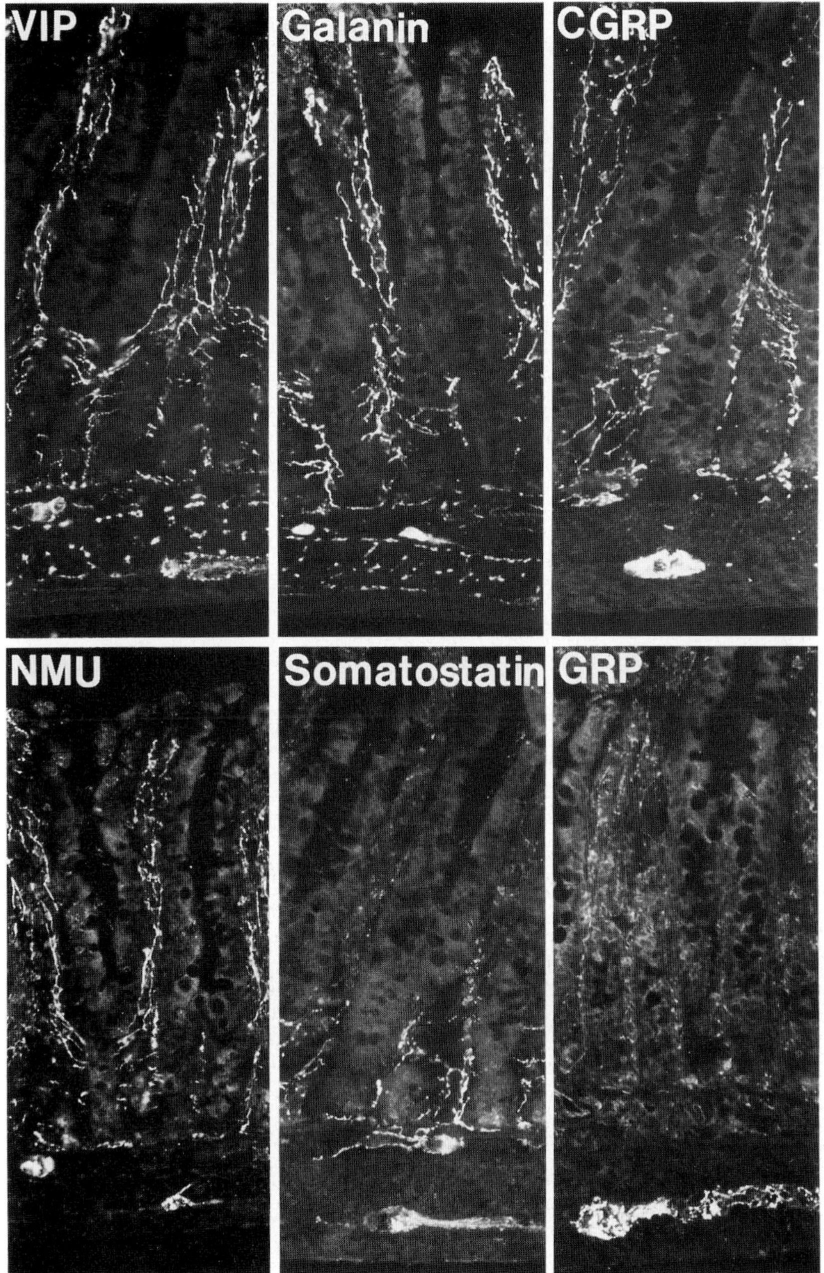

Fig. 7. Rat small intestine. Examples of the distribution pattern of peptide-containing neuronal systems. The illustrations are arranged in order of overall nerve fiber frequency. VIP and galanin fibers are numerous in all layers. CGRP, NMU and somatostatin fibers are moderate in number in the mucosa/submucosa and within the myenteric ganglia and rare in the smooth muscle. Somatostatin fibers predominate in the basal portion of the mucosa and in the submucosa. GRP fibers are numerous in the myenteric ganglia, but are rare in other locations in the gut wall, ×120

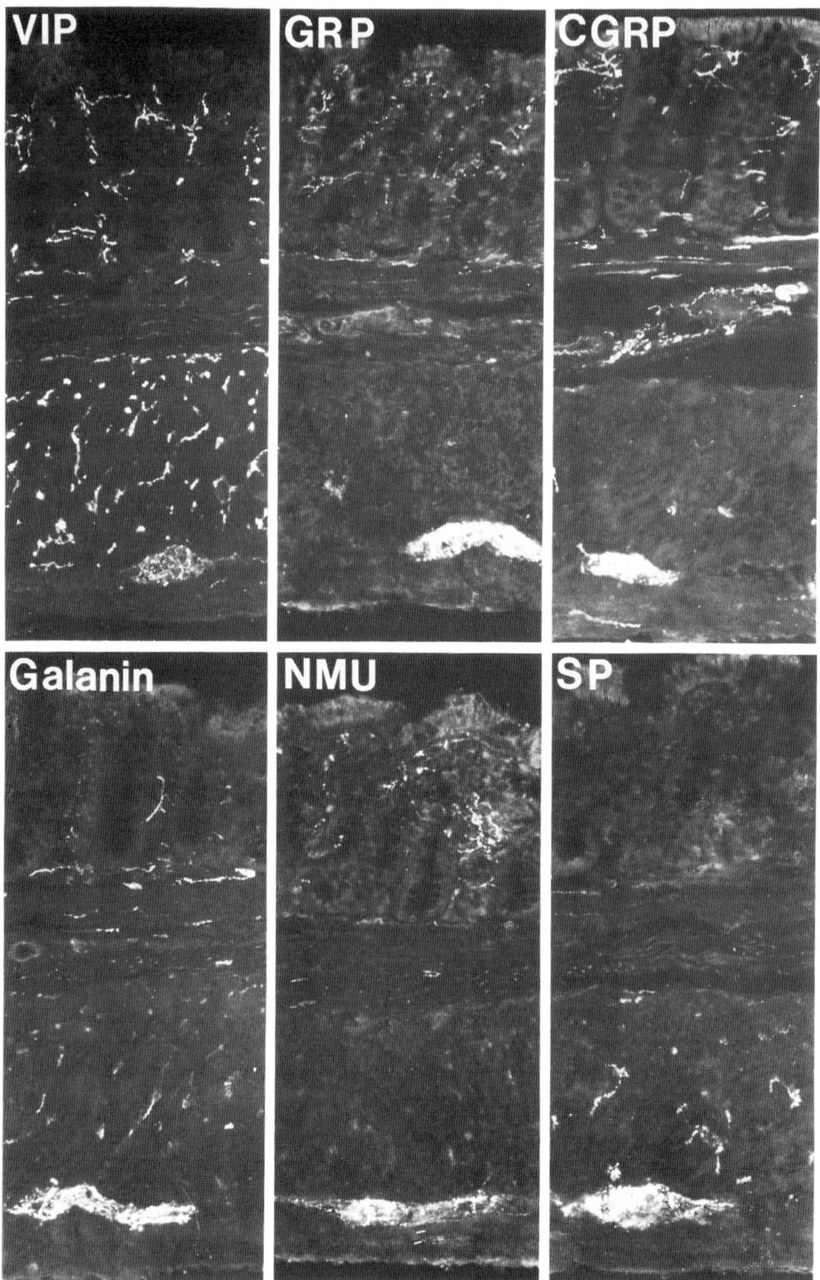

Fig. 8. Rat large intestine. Examples of peptide-containing neuronal systems. Illustrations are arranged in order of overall nerve fiber frequency. VIP fibers are numerous throughout the wall. GRP, CGRP and galanin fibers are moderate in number in the mucosa/submucosa and smooth muscle, but numerous in the intramural ganglia. NMU fibers are moderate in number in the mucosa/submucosa and intramural ganglia, but rare in the smooth muscle. SP fibers are very few in the mucosa/submucosa and submucous ganglia, moderate in number in the smooth muscle, and numerous in the myenteric ganglia, ×120

Table 2. Neuropeptides in the gut

Peptide	Amino acid sequencing	Precursor structure	Demonstration in gut nerves
VIP	MUTT and SAID (1974)	ITOH et al. (1983)	BRYANT et al. (1976)
			LARSSON et al. (1976)
SP	CHANG et al. (1971)	NAWA et al. (1983)	NILSSON et al. (1975)
		KRAUSE et al. (1987)	PEARSE and POLAK (1975)
CCK	MUTT and JORPES (1971)	DESCHENES et al. (1984)	LARSSON and REHFELD (1979)
		GUBLER et al. (1984)	
SOM	BRAZEAU et al. (1973)	SHEN et al. (1982)	HÖKFELT et al. (1975)
GRP	MCDONALD et al. (1979)	SPINDEL et al. (1984)	DOCKRAY et al. (1979)
ENK	HUGHES et al. (1975)	COMB et al. (1982)	ELDE et al. (1976)
		NODA et al. (1982)	
DYN	GOLDSTEIN et al. (1981)	KAKIDANI et al. (1982)	VINCENT et al. (1984)
GAL	TATEMOTO et al. (1983)	RÖKAEUS and BROWNSTEIN (1986)	RÖKAEUS et al. (1984)
NPY	TATEMOTO (1982)	MINTH et al. (1984)	LUNDBERG et al. (1982)
			SUNDLER et al. (1983)
			FURNESS et al. (1983)
α-CGRP	MORRIS et al. (1984)	AMARA et al. (1982)	ROSENFELD et al. (1983)
β-CGRP		AMARA et al. (1985)	MULDERRY et al. (1988)
		STEENBERGH et al. (1985)	
NMU	MINAMINO et al. (1985)	Not known	DOMIN et al. (1987)
PACAP	MIYATA et al. (1989)	KIMURA et al. (1990)	SUNDLER et al. (1992)

SOM, somatostatin.
ENK, enkephalin.
DYN, dynorphin.
GAL, galanin.
PACAP, pituitary adenylate cyclase activating peptide.

The occurrence and topographic distribution of nerve fibers containing VIP, NMU, GRP, galanin, SP, CGRP, somatostatin or enkephalin in the intestines of the rat are illustrated in Figs. 7 and 8.

II. Projections of Enteric Neurons in the Rat

For an understanding of the neuronal circuitry responsible for the regulation of events such as peristalsis, we need to know the projections and pathways of the individual neuronal systems. In order to be able to convey relevant information, enteric neurons engaged in gut motor control have to issue projections that are both descending (preceding the peristalsis wave to induce appropriate relaxation of the circular muscle) and ascending (to initiate or control the contractile wave). Using microsurgical techniques for local denervations of the gut wall and immunocytochemistry for identifying neuronal subpopulations, the projections and polarities of a number of enteric neurons have been described as summarized in Fig. 9 (FURNESS and COSTA 1987; MESSENGER and FURNESS 1990; EKBLAD et al. 1991b).

1. Myenteric Neurons

Myenteric neurons project to other neurons in the myenteric ganglia and to the smooth muscle layers as described in rat gut (EKBLAD et al. 1984a, 1985a,b, 1987, 1988). Except for GRP neurons in the large intestine there is no evidence for projections of myenteric neurons to the mucosa and submucosa. Neurons containing VIP (large intestine), VIP/NPY, GRP, galanin

Myenteric neurons, SI

	Rat		Guinea-pig		Dog	
Enk	←	(7 mm)	←	(6-8 mm)	←	(20-30 mm)
GRP	→	(20 mm)	→	(9-12 mm)	→	(15-30 mm)
NPY	→	(2 mm)	→	(1-2 mm)	←	(30 mm)
Som	→	(6 mm)	→	(8-12 mm)	→	(15-25 mm)
SP	→	(7 mm)	⇄	(1-2 mm)	⇄	(2 mm)
VIP	→	(2 mm)	→	(2-10 mm)	→	(5-20 mm)

Fig. 9. Projections of some myenteric peptide-containing neurons in the small intestine (*SI*) of rat, guinea pig and dog. There are many similarities in the projection pattern between the species. → indicates descending projections; ← indicates ascending projections. The average projection distance for each neuronal population is given in parentheses. (Data from DANIEL et al. 1987; EKBLAD et al. 1987; FURNESS and COSTA 1987) Note that VIP- and NPY-immunoreactive neurons in the rat are identical (see Fig. 10)

or somatostatin (small and large intestine) issue descending anal projections. CGRP neurons issue descending projections in the large intestine, while they project in both oral and anal directions in the small intestine. SP-containing neurons issue descending oral projections in the small intestine, whereas ascending projections are present in the large intestine. Enkephalin neurons give off ascending projections in both small and large intestine. The projection distances for all these neuronal populations have been found to be in the 2–7 mm range, except for galanin and GRP neurons in the small intestine, which issue longer projections, approximately 15–20 mm. The results are summarized in Fig. 10. Similar studies have been performed in the small intestine of guinea pigs (for a review, see FURNESS and COSTA 1987) and dogs (DANIEL et al. 1987), and in the large intestine of guinea pigs (MESSENGER and FURNESS 1990).

2. Submucous Neurons

Submucous neurons project to other neurons in the submucous ganglia and to the submucosa and mucosa (EKBLAD et al. 1984a, 1985b, 1987, 1988). However, neurons storing VIP/NPY, CGRP (small intestine) or VIP (large intestine) seem to issue projections also to subjacent circular muscle. GRP-containing neurons project in both oral and anal directions. Also galanin-, SP- and somatostatin-containing neurons project in both directions in the small intestine. In the large intestine galanin neurons issue no oro-anal

	Myenteric neurons		Submucous neurons	
	SI	LI	SI	LI
CGRP	← → (2 mm)	→ (5 mm)	← (8 mm)	← → (6 mm)
Enk	← (7 mm)	← (5 mm)	0	0
Galanin	→ (15 mm)	→ (6 mm)	← → (2 mm)	↑
GRP	→ (20 mm)	→ (10 mm)	← → (2 mm)	← → (3 mm)
Som	→ (6 mm)	→ (6 mm)	← → (5 mm)	0
Som/CGRP	0	0	0	← → (5 mm)
SP	→ (7 mm)	← (5 mm)	← → (4 mm)	ne
VIP	0	→ (3 mm)	0	← (2 mm)
VIP/NPY	→ (2 mm)	→ (4 mm)	← (4 mm)	↑

Fig. 10. Projection directions (*marked with arrows*) and distances (*in parentheses*) of various peptide-containing neuronal populations in the rat small intestine (*SI*) and large intestine (*LI*). → indicates descending projections; ← indicates ascending projections; ↑ indicates that no oro-anal projections can be demonstrated; *O*, absence of nerve fibers; *ne*, not examined. (Data from EKBLAD et al. 1987, 1988)

projections. SP-containing neurons in the mucosa and submucosa of the large intestine are too rare to allow delineation of the projection pattern. CGRP-containing neurons give off ascending projections in the small intestine but issue both oral and anal projections in the large intestine. VIP/NPY-containing neurons give off ascending projections in the small intestine but issue no oro-anal projections in the large intestine. Neurons storing VIP but not NPY (large intestine) issue ascending projections. Projection distances in the mucosa and submucosa range from 2 to 8 mm, the longest one being that of CGRP neurons. The results are summarized in Fig. 10.

On the whole, the organization of the enteric neurons is similar in the small and large intestine of the same species (rat), and in the different species studied (rat, guinea pig and dog) although minor differences do exist. The myenteric NPY- and VIP-containing neurons in the small intestine may serve as an example. In the rat the two peptides coexist in neurons issuing descending projections (EKBLAD et al. 1987). In the guinea pig (FURNESS and COSTA 1987) and dog (DANIEL et al. 1987), most VIP- and NPY-containing neurons constitute separate populations. In the guinea pig small intestine both these populations of neurons project anally, while in the dog small intestine and in the guinea pig large intestine (MESSENGER and FURNESS 1990) VIP-containing neurons project anally and NPY-containing neurons project orally. Like myenteric NPY-containing neurons in the guinea pig, myenteric SP-containing neurons in the rat may be used to exemplify differences between small and large intestine. These latter neurons issue descending projections in the small intestine (EKBLAD et al. 1987) and ascending projections in the large intestine (EKBLAD et al. 1988). Although these differences are minor in the context of the overall organization of the enteric nervous system, they emphasize the need for a precise mapping of each neuronal population in each species to define the neuronal circuitry involved in the regulation of gut function.

III. Coexistence of Peptides in Enteric Neurons

Numerous neuropeptides have been detected in the enteric nervous system. One neuropeptide is often colocalized with other neuropeptides in the same neuron. A list illustrating the chemical coding, i.e., the combinations of coexisting neuropeptides, is shown in Table 3. However, peptides represent only one type of putative messenger in enteric neurons. Enteric neurons probably also contain conventional transmitters. Thus, neuropeptides may coexist with acetylcholine, noradrenaline, histamine, serotonin, or GABA. Hence, it is to be expected that many (perhaps all) enteric neurons represent multimessenger systems. Sometimes neuropeptide coexistence may be anticipated (when two peptides arise from the same precursor). This is exemplified by VIP/PHI (ITOH et al. 1983) and SP/NKA (NAWA et al. 1983; KRAUSE et al. 1987). Besides the anticipated coexistence there are several examples of nonpredictable coexistence, i.e., coexistence of peptides

Table 3. Peptides in enteric neurons

Colocalized peptides in enteric neurons	Species	Location	References
CGRP/CCK/GAL/NPY/SOM	Guinea pig	Small intestine	FURNESS et al. (1987)
			FURNESS and COSTA (1987)
CGRP/CCK/GAL/NMU/NPY/SOM	Guinea pig	Small intestine	FURNESS et al. (1989b)
CGRP/NMU	Rat	Small intestine	BALLESTRA et al. (1988)
			EKBLAD et al. (1992, manuscript in preparation)
CGRP/SOM	Rat	Large intestine	EKBLAD et al. (1988)
CGRP/VIP	Rat	Large intestine	EKBLAD et al. (1988)
CCK/DYN/ENK/GRP/VIP	Guinea pig	Small intestine	FURNESS and COSTA (1987)
CCK/DYN/GRP/SP/VIP	Guinea pig	Small intestine	FURNESS and COSTA (1987)
CCK/SOM	Guinea pig	Large intestine	SCHULTZBERG et al. (1980)
CCK/SOM/SP	Guinea pig	Small intestine	FURNESS and COSTA (1987)
DYN/ENK/NPY/VIP	Guinea pig	Small intestine	FURNESS and COSTA (1987)
DYN/GAL/NMU/VIP	Guinea pig	Small intestine	FURNESS et al. (1989a)
ENK/SP	Cat	Small intestine	SUNDLER et al. (1987)
	Man	Stomach, small and large intestine	WATTCHOW et al. (1988)
	Guinea pig	Small intestine	FURNESS and COSTA (1987)
			SUNDLER et al. (1987)
			UCHIDA et al. (1985)
GAL/VIP	Dog, pig	Small intestine	MELANDER et al. (1985)
			BISHOP et al. (1986)
			GONDA et al. (1989)
	Guinea pig	Large intestine	MELANDER et al. (1985)
GAL/NPY/VIP	Guinea pig	Small intestine	FURNESS et al. (1989a)
GRP/PACAP	Rat	Small and large intestine	SUNDLER et al. (1991c)
GRP/PACAP/VIP	Rat, man	Stomach	EKBLAD et al. (1991a)
GRP/SP	Rat	Stomach	KUWAHARA et al. (1985)
GRP/VIP	Rat, man	Stomach	EKBLAD et al. (1991a)
NMU/NPY/VIP	Rat, pig	Small intestine	EKBLAD et al. (1992, manuscript in preparation)
NPY/VIP	Rat, man	Stomach, small and large intestine	EKBLAD et al. (1984b, 1985a, 1988, 1989a)
			WATTCHOW et al. (1988)
	Mouse, pig	Small intestine	EKBLAD et al. (1984b)
PACAP/VIP	Rat	Small and large intestine	SUNDLER et al. (1991c)
SP/NMU	Guinea pig	Small intestine	FURNESS et al. (1989b)

that arise from different precursors, e.g., VIP/NPY (Ekblad et al. 1984b), VIP/GAL (Melander et al. 1985) or CCK/CGRP/NPY/SOM (Furness and Costa 1987).

Each population of neurons seems to have a characteristic chemical coding (Furness et al. 1989a), which reflects the programming of that particular neuron or group of neurons. Conceivably the release of multiple messengers may lead to an increased capacity for precise communication with target cells and to an enhanced autocontrol by effective and precise feedback mechanisms. A number of questions still need to be answered. Is the messenger programming of the neuron variable? Are the messengers costored in the same or in different vesicles? Do the messengers cooperate after their release? Do they act on the same or on different target cells? Is there an "optimal" number of coexisting messengers? Studies of these subjects are still at an initial stage. Observations made so far indicate that messenger coexistence is a basic principle and suggest that cotransmission may be functionally important.

Acknowledgements. Grant support from Swedish MRC (proj. no. 4499 and 1007), Påhlsson's Foundation, Bergvall's Foundation and the Medical Faculty, University of Lund.

References

Ali-Rachedi A, Varndell IM, Adrian TE, Gapp DA, van Noorden S, Bloom SR, Polak JM (1984) Peptide YY (PYY) immunoreactivity is co-stored with glucagon-related immunoreactants in endocrine cells of the gut and pancreas. Histochemistry 80:487–489

Alumets J, Håkanson R, Sundler F, Chang KJ (1978) Leu-enkephalin-like material in nerves and enterochromaffin cells in the gut. Histochemistry 56:187–196

Alumets J, Ekelund M, El Munshid HA, Håkanson R, Lorén I, Sundler F (1979) Topography of somatostatin cells in the stomach of the rat; possible functional significance. Cell Tissue Res 202:177–188

Amara SG, Jonas V, Rosenfeld MG, Ong ES, Evans RM (1982) Alternative RNA processing in calcitonin gene expression generates mRNAs encoding different polypeptide products. Nature 298:240–244

Amara S, Arriza JL, Leff SE, Swanson LW, Evans RM, Rosenfeld MG (1985) Expression in brain of a messenger RNA encoding a novel neuropeptide homologous to calcitonin gene-related peptide. Science 229:1094–1097

Ballestra J, Carlei F, Bishop AE, Steel JH, Gibson SJ, Fehey M, Hennessey R, Domin J, Bloom SR, Polak JM (1988) Occurrence and developmental pattern of neuromedin U-immunoreactive nerves in the gastrointestinal tract and brain of the rat. Neuroscience 25:797–816

Bishop AE, Polak JM, Bauer FE, Christofides ND, Carlei F, Bloom SR (1986) Occurrence and distribution of a newly discovered peptide, galanin, in the mammalian enteric nervous system. Gut 27:849–857

Bjartell A, Ekman R, Hedenbro J, Sjölund K, Sundler F (1989) Delta sleep-inducing peptide (DSIP)-like immunoreactivity in gut: coexistence with known peptide hormones. Peptides 10:163–170

Børglum-Jensen T, Fahrenkrug J, Sundler F (1991) Immunocytochemical localisation of pancreastatin and chromogranin A in porcine neuroendocrine tissues. Regul Pept 36:283–298

Böttcher G, Sjölund K, Ekblad E, Håkanson R, Schwartz TW, Sundler F (1984) Coexistence of peptide YY and glicentin immunoreactivity in endocrine cells of the gut. Regul Pept 8:261–266

Böttcher G, Alumets J, Håkanson R, Sundler F (1986) Co-existence of glicentin and peptide YY in colorectal L-cells in cat and man. An electron microscopic study. Regul Pept 13:283–291

Brand SJ, Stone D (1988) Reciprocal regulation of antral gastrin and somatostatin gene expression in omeprazole-induced achlorhydria. J Clin Invest 82:1057–1066

Brazeau P, Vale W, Burgus R, Ling N, Butcher M, Rivier J, Guillemin R (1973) Hypothalamic polypeptide that inhibits the secretion of immunoreactive pituitary growth hormone. Science 179:77–79

Brownstein MJ (1985) Peptide processing: an overview. In: Håkanson R, Thorell J (eds) Biogenetics of neurohormonal peptides. Academic Press, London, pp 105–112

Bryant MG, Bloom SR, Polak JM, Albuquerque RH, Modlin I, Pearse AGE (1976) Possible dual role for vasoactive intestinal peptide as gastrointestinal hormone and neurotransmitter substance. Lancet 1:991–993

Capella C, Finzi G, Cornaggia M, Usellini L, Luinetti O, Buffa R, Solcia E (1991) Ultrastructural typing of gastric endocrine cells. In: Håkanson R, Sundler F (eds) The stomach as an endocrine organ. Elsevier, Amsterdam, pp 27–51 (Fernström symposium, no 15)

Chang MM, Leeman SE, Niall HD (1971) Amino-acid sequence of substance P. Nature [New Biol] 232:86–87

Comb M, Seeburg PH, Adelman J, Eiden L, Herbert E (1982) Primary structure of the human Met- and Leu-enkephalin precursor and its mRNA. Nature 295:663–666

Curry WJ, Johnston CF, Shaw C, Buchanan KD (1990) Distribution and partial characterization of immunoreactivity to the putative C-terminus of rat pancreastatin. Regul Pept 30:207–220

Daniel EE, Furness JB, Costa M, Belbeck L (1987) The projections of chemically identified nerve fibres in canine ileum. Cell Tissue Res 247:377–384

Dayal Y (1991) Neuroendocrine cells of the gastrointestinal tract: introduction and historical perspective. In: Dayal Y (ed) Endocrine pathology of the gut and pancreas. CRC, Boca Raton, pp 1–31

Deschenes RJ, Lorenz LJ, Haun RS, Roos BA, Collier KJ, Dixon JE (1984) Cloning and sequence analysis of a cDNA encoding rat pre-procholecystokinin. Proc Natl Acad Sci USA 81:726–730

Dimaline R (1988) Post-translational modification of peptide messengers in the gut. Q J Exp Physiol 73:873–902

Dockray GJ, Vaillant C, Walsh JH (1979) The neuronal origin of bombesin-like immunoreactivity in the rat gastrointestinal tract. Neuroscience 4:1561–1568

Dockray GJ, Hamer C, Evans D, Karro A, Dimaline R (1991a) The secretory kinetics of the G cell in omeprazole-treated rats. Gastroenterology 100:1187–1194

Dockray GJ, Varro A, Watkinson A, Dimaline R (1991b) Selective processing of peptides in gastric endocrine cells. In: Håkanson R, Sundler F (eds) The stomach as an endocrine organ. Elsevier, Amsterdam, pp 197–210 (Fernström symposium, no 15)

Domin J, Ghatei MA, Chohan P, Bloom SR (1987) Neuromedin U – a study of its distribution in the rat. Peptides 8:779–784

Eipper BA, Mains R, Glembotski CC (1985) Peptide α-amidation: cellular and enzymatic studies. In: Håkanson R, Thorell J (eds) Biogenetics of neurohormonal peptides. Academic Press, London, pp 189–209

Eipper BA, Mains RE, Herbert E (1986) Peptides in the nervous system. Trends Neurosci 9:463–468

Eissele R, Rosskopf B, Koop H, Adler G, Arnold R (1991) Proliferation of endocrine cells in the rat stomach caused by drug-induced achlorhydria. Gastroenterology 101:70–76

Ekblad E, Ekman R, Håkanson R, Sundler F (1984a) GRP neurones in the rat small intestine issue long anal projections. Regul Pept 9:279–287

Ekblad E, Håkanson R, Sundler F (1984b) VIP and PHI coexist with an NPY-like peptide in intramural neurones of the small intestine. Regul Pept 10:47–58

Ekblad E, Ekelund M, Graffner H, Håkanson R, Sundler F (1985a) Peptide-containing nerve fibers in the stomach wall of rat and mouse. Gastroenterology 89:73–85

Ekblad E, Håkanson R, Rökaeus Å, Sundler F (1985b) Galanin nerve fibers in the rat gut: distribution, origin and projections. Neuroscience 16:355–363

Ekblad E, Winther C, Ekman R, Håkanson R, Sundler F (1987) Projections of peptide-containing neurons in rat small intestine. Neuroscience 20:169–188

Ekblad E, Ekman R, Håkanson R, Sundler F (1988) Projections of peptide-containing neurons in rat colon. Neuroscience 27:655–674

Ekblad E, Arnbjörnsson E, Ekman R, Håkanson R, Sundler F (1989a) Neuropeptides in the human appendix: distribution and motor effects. Dig Dis Sci 34:1217–1230

Ekblad E, Håkanson R, Sundler F (1989b) Projections of enteric peptide-containing neurons in the rat. In: Singer MV, Goebell H (eds) Nerves and the gastrointestinal tract. Kluwer, Dordrecht, pp 47–56 (Falk symposium, no 50)

Ekblad E, Håkanson R, Sundler F (1991a) Innervation of the stomach of rat and man with special reference to the endocrine cells. In: Håkanson R, Sundler F (eds) The stomach as an endocrine organ. Elsevier, Amsterdam, pp 79–95 (Fernström symposium, no 15)

Ekblad E, Håkanson R, Sundler F (1991b) Microanatomy and chemical coding of peptide-containing neurons in the digestive tract. In: Daniel EE (ed) Neuropeptide function in the gastrointestinal tract. CRC, Boca Raton, pp 131–191

Elde R, Hökfelt T, Johansson O, Terenius L (1976) Immunohistochemical studies using antibodies to leucine enkephalin: initial observations on the nervous system of the rat. Neuroscience 1:349–351

Ericson LE, Sundler F (1984) Thyroid parofollicular cells. Ultrastructural and functional correlation. In: Motta PM (ed) Ultrastructure of endocrine cells and tissues. Nijhoff, Boston, pp 276–285

Facer P, Bishop AE, Lloyd RV, Wilson BS, Hennesy RJ, Polak JM (1985) Chromogranin: a newly recognized marker for endocrine cells of the human gastrointestinal tract. Gastoenterology 89:1366–1373

Furness JB, Costa M (1987) The enteric nervous system. Churchill Livingstone, London

Furness JB, Costa M, Emson PC, Håkanson R, Moghimzadeh E, Sundler F, Taylor IL, Chance RE (1983) Distribution, pathways and reactions to drug treatment of nerves with neuropeptide Y- and pancreatic polypeptide-like immunoreactivity in the guinea-pig digestive tract. Cell Tissue Res 234:71–92

Furness JB, Costa M, Rökaeus Å, McDonald TJ, Brooks B (1987) Galanin-immunoreactive neurons in the guinea-pig small intestine: their projections and relationships to other enteric neurons. Cell Tissue Res 250:607–615

Furness JB, Morris JL, Gibbins IL, Costa M (1989a) Chemical coding of neurons and plurichemical transmission. Annu Rev Pharmacol Toxicol 29:289–306

Furness JB, Pompolo S, Murphy R, Giraud A (1989b) Projections of neurons with neuromedin U-like immunoreactivity in the small intestine of the guinea pig. Cell Tissue Res 257:415–422

Gibbins IL, Furness JB, Costa M, McIntyre I, Hillyard SJ, Girgis S (1985) Co-localization of calcitonin gene-related peptide-like immunoreactivity with substance P in cutaneous, vascular and visceral sensory neurons of guinea-pigs. Neurosci Lett 57:128–130

Goldstein A, Fischli W, Lowney LI, Hunkapiller M, Hood L (1981) Porcine pituitary dynorphin: complete amino acid sequence of the biologically active heptadecapeptide. Proc Natl Acad Sci USA 78:7219–7223

Gonda T, Daniel EE, McDonald TJ, Fox JET, Brooks BD, Oki M (1989) Distribution and function of enteric GAL-IR nerves in dogs: comparison with VIP Am J Physiol 256:G884–G896

Gubler U, Chua AO, Hoffman BJ, Collier KJ, Eng J (1984) Cloned cDNA to cholecystokinin mRNA predicts an identical preprocholecystokinin in pig brain and gut. Proc Natl Acad Sci USA 81:4307–4310

Gulbenkian S, Merighi A, Wharton J, Varndell IM, Polak JM (1986) Ultrastructural evidence for the coexistence of calcitonin gene-related peptide and substance P in secretory vesicles of peripheral nerves in the guinea pig. J Neurocytol 15:535–542

Håkanson R, Sundler F (1983) The design of the neuroendocrine system: a unifying concept and its consequences. Trends Pharmacol Sci 4:41–44

Håkanson R, Sundler F (1986) The role of peptide messengers in the neuroendocrine system: hormones, neurotransmitters, or neuromodulators. In: Schou J, Greisler A, Norn S (eds) Drug receptors and dynamic processes in cells. Munksgaard, Copenhagen, pp 62–77 (Alfred Benzon symposium, no 22)

Håkanson R, Sundler F (1991) The gastrin concept: the proposed mechanism behind the development of drug-induced gastric carcinoids. In: Håkanson R, Sundler F (eds) The stomach as an endocrine organ. Elsevier, Amsterdam, pp 449–460 (Fernström symposium, no 15)

Håkanson R, Alumets J, Rehfeld JF, Ekelund M, Sundler F (1982) The life cycle of the gastrin granule. Cell Tissue Res 222:479–491

Håkanson R, Böttcher G, Ekblad E, Panula P, Simonsson M, Dohlsten M, Hallberg T, Sundler F (1986) Histamine in endocrine cells in the stomach. A survey of several species using a panel of histamine antibodies. Histochemistry 86:5–17

Håkanson R, Böttcher G, Ekblad E, Grunditz T, Sundler F (1990) Functional implications of messenger coexpression in neurons and endocrine cells. In: Schwartz TW, Hilsted LM, Rehfeld JF (eds) Neuropeptides and their receptors. Munksgaard, Copenhagen, pp 211–232 (Alfred Benzon symposium, no 29)

Halban PA (1991) Structural domains and molecular lifestyles of insulin and its precursors in the pancreatic beta cell. Diabetologia 34:767–778

Heitz P, Polak JM, Timson CM, Pearse AGE (1976) Enterochromaffin cells as the source of gastrointestinal substance P. Histochemistry 49:343–347

Hökfelt T, Johansson O, Efendic S, Luft R, Arimura A. (1975) Are there somatostatin containing nerves in the rat gut? Immunohistochemical evidence for a new type of peripheral nerve. Experientia 31:852–854

Hughes J, Smith TW, Kosterlitz HW, Fothergill LA, Morgan BA, Morris HR (1975) Identification of two related pentapeptides from the brain with potent opiate agonist activity. Nature 258:577–579

Itoh N, Obata K, Yanaihara N, Okamoto H (1983) Human preprovasoactive intestinal polypeptide contains a novel PHI-27-like peptide, PHM-27. Nature 304:547–549

Johnson KH, O'Brien TD, Hayden DW, Jordan K, Ghobrial HKG, Mahoney WC, Westermark P (1988) Immunolocalization of islet amyloid polypeptide (IAPP) in pancreatic beta cells by means of peroxidase-antiperoxidase (PAP) and protein A-gold techniques. Am J Pathol 130:1–8

Kakidani H, Furutani Y, Takahashi H, Noda M, Morimoto Y, Hirose T, Asai M, Inayama S, Nakanishi S, Numa S (1982) Cloning and sequence analysis of cDNA for porcine beta neoendorphin/dynorphin precursor. Nature 298:245–249

Kimura C, Ohkubo S, Ogi K, Hosoya M, Itoh Y, Onda H, Miyata A, Jiang L, Dahl RR, Stibbs HH, Arimura A, Fujino M (1990) A novel peptide which stimulates adenylate cyclase: molecular cloning and characterization of the ovine and human cDNAs. Biochem Biophys Res Commun 166:81–89

Koop H, Willemer S, Steinback F, Eissele R, Tuch K, Arnold R (1987) Influence of chronic drug-induced achlorhydria by substituted benzimidazoles on the endocrine stomach of rats. Gastroenterology 92:406–413

Krause JE, Chirgwin JM, Carter MS, Xu ZS, Hershey AD (1987) Three rat preprotachykinin mRNAs encode the neuropeptides substance P and neurokinin A. Proc Natl Acad Sci USA 84:881–885

Kuwahara A, Mikami S, Yanaihara N (1985) Coexistence of immunoreactive gastrin-releasing peptide and substance P in the myenteric plexus of rat stomach. Biomed Res 6:443–446

Lamberts R, Schmidt WE, Creutzfeldt W (1990) Light and electron microscopic immunocytochemical localization of pancreastatin-like immunoreactivity in porcine tissues. Histochemistry 93:369–380

Larsson L-I (1978) ACTH-like immunoreactivity in the gastrin cell. Independent changes in gastrin and ACTH-like immunoreactivity during ontogeny. Histochemistry 56:245–251

Larsson L-I, Rehfeld JF (1979) Localization and molecular heterogeneity of cholecystokinin in the central and peripheral nervous system. Brain Res 165: 201–218

Larsson L-I, Fahrenkrug J, Schaffalitzky de Muckadell O, Sundler F, Håkanson R, Reheld JF (1976) Localization of vasoactive intestinal polypeptide (VIP) to central and peripheral neurones. Proc Natl Acad Sci USA 73:3197–3200

Larsson L-I, Golterman N, de Magistris L, Rehfeld JF, Schwartz TW (1989) Somatostatin cell processes as pathways for paracrine secretion. Science 205: 1393–1395

Larsson LT, Sundler F (1990) Neuronal markers in Hirschsprung's disease with special reference to neuropeptides. Acta Histochem (Jena) Suppl 38:115–125

Lund PK, Goodman RH, Dee PC, Habener JF (1982) Pancreatic preproglucagon cDNA contains two glucagon-related coding sequences arranged in tandem. Proc Natl Sci USA 79:345–349

Lundberg JM, Terenius L, Hökfelt T, Martling CR, Tatemoto K, Mutt V, Polak J, Bloom S, Goldstein M (1982) Neuropeptide Y (NPY) like immunoreactivity in peripheral noradrenergic neurons and effects of NPY on sympathetic function. Acta Physiol Scand 116:477–480

Lundberg JM, Hökfelt T (1986) Multiple co-existence of peptides and classical transmitters in peripheral autonomic and sensory neurones – functional and pharmacological implications. Prog Brain Res 68:241–262

Mattsson H, Sundler F, Carlsson K, Håkanson R (1991) Antral gastrin and somatostatin cells during long-term hypergastrinemia (Abstr). Gastroenterology 100:A655

McDonald TJ, Jörnvall H, Nilsson G, Vagne M, Ghatei M, Bloom SR, Mutt V (1979) Characterization of a gastrin releasing peptide from porcine non-antral gastric tissue. Biochem Biophys Res Commun 90:227–233

Melander T, Hökfelt T, Rökaeus Å, Fahrenkrug J, Tatemoto K, Mutt V (1985) Distribution of galanin-like immunoreactivity in the gastrointestinal tract of several mammalian species. Cell Tissue Res 239:253–270

Merighi A, Polak JM, Gibson SJ, Gulbenkian S, Valentino KL, Peirone SM (1988) Ultrastructural studies on calcitonin gene related peptide-, tachykinin- and somatostatin- immunoreactive neurones in rat dorsal root ganglia: evidence for the colocalization of different peptides in single secretory granules. Cell Tissue Res 254:101–109

Messenger JP, Furness JB (1990) Projections of chemically-specified neurons in the guinea-pig colon. Arch Histol Cytol 53:467–495

Minamino N, Kangawa K, Matsuo H (1985) Neuromedin U-8 and U-25; novel uterus stimulating and hypertensive peptides identified in porcine spinal cord. Biochem Biophys Res Commun 130:1078–1085

Minth CD, Bloom SR, Polak JM, Dixon JE (1984) Cloning, characterization and DNA sequence of a human cDNA encoding neuropeptide tyrosine. Proc Natl Acad Sci USA 81:4577–4581

Miyata A, Arimura A, Dahl R, Minamino N, Uehara A, Jiang L, Culler MD, Coy DH (1989) Isolation of a novel 38 residue-hypothalamic polypeptide which stimulates adenylate cyclase in pituitary cells. Biochem Biophys Res Commun 164:567–574

Morris HR, Panico M, Etienne T, Tippins J, Girgis SI, MacIntyre I (1984) Isolation and characterization of human calcitonin gene-related peptide. Nature 308:746–748

Mulderry PK, Ghatei MA, Spokes RA, Jones PM, Pierson AM, Hamid QA, Kanse S, Amara SG, Burrin JM, Legon S, Polak JM, Bloom SR (1988) Differential expression of α-CGRP and β-CGRP by primary sensory neurons and enteric autonomic neurons in the rat. Neuroscience 25:195–205

Mutt V, Jorpes E (1971) Hormonal polypeptides of upper intestine. Biochem J 125:57P–58P

Mutt V, Said SI (1974) Structure of the porcine vasoactive intestinal octacosapeptide: the amino-acid sequence. Use of kallikrein in its determination. Eur J Biochem 42:581–589

Nawa H, Hirose T, Takashima H, Inayama S, Nakanishi S (1983) Nucleotide sequences of cloned cDNAs for two types of bovine brain substance P precursors. Nature 306:31–36

Nilsson G, Larsson L-I, Håkanson R, Brodin E, Pernow B, Sundler F (1975) Localization of substance P-like immunoreactivity in mouse gut. Histochemistry 43:97–99

Noda M, Furutani Y, Takahashi H, Toyosato M, Hirose T, Inayama S, Nakanishi S, Numa S (1982) Cloning and sequence analysis of cDNA for bovine adrenal preproenkephalin. Nature 295:202–208

Ondolfo JP, Lehy T, Labeille D, Gres L (1989) Growth pattern of the polypeptide-YY cell population in the upper digestive tract of the rat during the perinatal period and after weaning. Cell Tissue Res 258:569–576

Orci L (1986) The insulin cell: its cellular environment and how it processes (pro)insulin. Diabetes Metab Rev 2:71–106

Pearse AGE, Polak JM (1975) Immunocytochemical localization of substance P in mammalian intestine. Histochemistry 41:373–375

Persson P, Håkanson R (1991) The gastrin-gastrocalcin hypothesis. In: Håkanson R, Sundler F (eds) The stomach as an endocrine organ. Elsevier, Amsterdam, pp 341–350 (Fernström symposium, no 15)

Persson P, Håkanson R, Axelson J, Sundler F (1989) Gastrin releases a blood-calcium lowering peptide from the acid-producing part of the stomach. Proc Natl Acad Sci USA 86:2834–2838

Polak JM, Bloom SR, Marangos PJ (1984) Neuron specific enolase, a marker for neuroendocrine cells. In: Falkmer S, Håkanson R, Sundler F (eds) Evolution and tumour pathology of the neuroendocrine system. Elsevier, Amsterdam, pp 433–452 (Fernström symposium, no 4)

Rindi G, Buffa R, Sessa F, Tortora O, Solcia E (1986) Chromogranin A, B and C immunoreactivities of mammalian endocrine cells. Distribution, distinction from costored hormones/prohormones and relationship with the argyrophil component of secretory granules. Histochemistry 85:19–28

Rix EW, Feurle GE, Carraway RE (1986) Colocalization of xenopsin and gastrin immunoreactivity in gastric antral G-cells. Histochemistry 85:135–138

Rökaeus Å, Brownstein MJ (1986) Construction of a porcine adrenal medullary cDNA library and nucleotide sequence analysis of two clones encoding a galanin precursor. Proc Natl Acad Sci USA 83:6287–6291

Rökaeus A, Melander T, Hökfelt T, Lundberg JM, Tatemoto K, Carlquist M, Mutt V (1984) A galanin-like peptide in the central nervous system and intestine of the rat. Neurosci Lett 47:161–166

Rosa P, Hille A, Lee RWH, Zanini A, de Camilli P, Huttner WB (1985) Secretogranins I and II: two tyrosine-sulfated secretory proteins common to a variety of cells secreting peptides by the regulated pathway. J Cell Biol 101:1991–2011

Rosenfeld MG, Mermod J-J, Amara SG, Swanson LW, Sawchenko PE, Riviera J, Vale WW, Evans RM (1983) Production of a novel neuropeptide encoded by the calcitonin gene via tissue-specific RNA processing. Nature 304:129–135

Roth KA, Gordon JI (1990) Spatial differentiation of the intestinal epithelium: analysis of enteroendocrine cells containing serotonin, secretin and substance P in normal and transgenic mice. Proc Natl Acad Sci USA 87:6408–6412

Schultzberg M, Hökfelt T, Nilsson G, Terenius L, Rehfeld JF, Brown M, Elde R, Goldstein M, Said S (1980) Distribution of peptide- and catecholamine-containing neurons in the gastrointestinal tract of rat and guinea-pig: immunohistochemical studies with antisera to substance P, vasoactive intestinal polypeptide, enkephalins, somatostatin, gastrin/cholecystokinin, neurotensin and dopamine β-hydroxylase. Neuroscience 5:689–744

Shen L-P, Pictet RL, Rutter WJ (1982) Human somatostatin. I. Sequence of the cDNA. Proc Natl Acad Sci USA 79:4575–4579

Sjölund K, Sandén G, Håkanson R, Sundler F (1983) Endocrine cells in human intestine: an immunocytochemical study. Gastroenterology 85:1120–1130

Solcia E, Capella C, Buffa R, Frigerio B (1976) Histochemical and ultrastructural studies on the argentaffin and argyrophil cells of the gut. In: Coupland RE, Fujita T (eds) Chromaffin, enterochromaffin and related cells. Elsevier, Amsterdam, pp 209–255

Solcia E, Capella C, Buffa R, Usellini L, Fiocca R, Sessa F (1987) Endocrine cells of the digestive system. In: Johnson LR (ed) Physiology of the gastointestinal tract, 2nd edn. Raven, New York, pp 111–130

Spindel ER, Chin WW, Price J, Rees LH, Besser GM, Habener JF (1984) Cloning and characterization of cDNA's encoding human gastrin-releasing peptide. Proc Natl Acad Sci USA 81:5699–5703

Steenbergh PH, Höppener JWM, Zandberg J, Lips CJM, Jansz HS (1985) A second human calcitonin/CGRP gene. FEBS Lett 183:403–407

Sternini C, Reeve JR, Brecha N (1987) Distribution and characterization of calcitonin gene-related peptide immunoreactivity in the digestive system of normal and capsaicin-treated rats. Gastroenterology 93:852–862

Su HC, Bishop AE, Power RF, Hamada Y, Polak JM (1987) Dual intrinsic and extrinsic origin of CGRP- and NPY-immunoreactive nerves of rat gut and pancreas. J Neurosci 7:2674–2687

Sundler F, Håkanson R (1988) Peptide hormone-producing endocrine/paracrine cells in the gastro-entero-pancreatic region. In: Björklund A, Hökfelt T, Owman C (eds) The peripheral nervous system. Elsevier, Amsterdam, pp 219–295 (Handbook of chemical neuroanatomy, vol. 6)

Sundler F, Håkanson R (1991) Gastric endocrine cell typing at the light microscopic level. In: Håkanson R, Sundler F (eds) The stomach as an endocrine organ. Elsevier, Amsterdam, pp 9–26 (Fernström symposium, no 15)

Sundler F, Alumets J, Håkanson R (1977) 5-Hydroxytryptamine-containing enterochromaffin cells: storage site of substance P. Acta Physiol Scand Suppl 452:121–123

Sundler F, Håkanson R, Lorén I, Lundquist I (1980) Amine storage and function in peptide hormone-producing cells. Invest Cell Pathol 3:87–103

Sundler F, Moghimzadeh E, Håkanson R, Ekelund M, Emson PC (1983) Nerve fibers in the gut and pancreas of the rat displaying neuropeptide Y immunoreactivity. Intrinsic and extrinsic origin. Cell Tissue Res 230:487–493

Sundler F, Brodin E, Ekblad E, Håkanson R, Uddman R (1985a) Sensory nerve fibers: distribution of substance P, neurokinin A and calcitonin gene-related peptide. In: Håkanson R, Sundler F (eds) Tachykinin antagonists. Elsevier, Amsterdam, pp 3–14 (Fernström symposium, no 6)

Sundler F, Ekblad E, Böttcher G, Alumets J, Håkanson R (1985b) Coexistence of peptides in the neuroendocrine system. In: Håkanson R, Thorell J (eds) Biogenetics of neurohormonal peptides. Academic Press, London, pp 213–243

Sundler F, Håkanson R, Ekblad E, Uddman R, Wahlestedt C (1986) Neuropeptide Y in the peripheral adrenergic and enteric nervous systems. Int Rev Cytol 102:243–269

Sundler F, Bjartell A, Böttcher G, Ekblad E, Håkanson R (1987) Localization of enkephalins and other endogenous opioids in the digestive tract. Gastoenterol Clin Biol 11:14B–26B

Sundler F, Ekblad E, Grunditz T, Håkanson R, Uddman R (1988) Vasoactive intestinal peptide in the peripheral nervous system. Ann NY Acad Sci 527:143–167

Sundler F, Böttcher G, Ekblad E, Håkanson R (1989) The neuroendocrine system of the gut. Acta Oncol 28:303–314

Sundler F, Ekblad E, Håkanson R (1991a) Occurrence and distribution of substance P- and CGRP-containing nerve fibers in gastric mucosa: species differences. Adv Exp Med Biol 298:29–37

Sundler F, Ekblad E, Håkanson R (1991b) The neuroendocrine system of the gut – an update. Acta Oncol 30:419–427

Sundler F, Ekelund M, Håkanson R (1991c) Morphological aspects of gastrin cell activation. In: Håkanson R, Sundler F (eds) The stomach as an endocrine organ. Elsevier, Amsterdam, pp 167–178 (Fernström symposium, no 15)

Sundler F, Ekblad E, Absood A, Håkanson R, Köves K, Arimura A (1992) Pituitary adenylate cyclase activating peptide: A novel vasoactive intestinal peptide-like neuropeptide in the gut. Neuroscience 46:439–454

Tatemoto K (1982) Neuropeptide Y: Complete amino acid sequence of the brain peptide. Proc Natl Acad Sci USA 79:5485–5489

Tatemoto K, Mutt V (1978) Chemical determination of polypeptide hormones. Proc Natl Acad Sci USA 75:4115–4119

Tatemoto K, Mutt V (1980) Isolation of two novel candidate hormones using a chemical method for finding naturally occurring polypeptides. Nature (Lond) 285:417–418

Tatemoto K, Mutt V (1981) Isolation and characterization of the intestinal peptide porcine PHI (PHI-27), a new member of the glucagon-secretin family. Proc Natl Acad Sci USA 78:6603–6607

Tatemoto K, Carlquist M, Mutt V (1982) Neuropeptide Y – a novel brain peptide with structural similarities to peptide YY and pancreatic polypeptide. Nature 296:659–660

Tatemoto K, Rökaeus Å, Jörnvall H, McDonald T, Mutt V (1983) Galanin – a novel biologically active peptide from porcine intestine. FEBS Lett 164:124–128

Tsuruo Y, Hökfelt T, Visser TJ, Kimmel JR, Brown JC, Verhofstadt A, Walsh J (1988) TRH-like immunoreactivity in endocrine cells and neurons in the gastro-intestinal tract of the rat and guinea pig. Cell Tissue Res 253:347–356

Uchida T, Kobayashi S, Yanaihara N (1985) Occurrence and projections of three subclasses of met-enkephalin-Arg^{6}-Gly^{7}-Leu^{8} neurons in the guinea pig duodenum: immunoelectron microscopic study on the co-storage of met-enkephalin-Arg^{6}-Gly^{7}-Leu^{8} with substance P or PHI (1–15). Biomed Res 6:415–422

Vincent SR, Dalsgaard C-J, Schultzberg M, Hökfelt T, Christensson I, Terenius L (1984) Dynorphin-immunoreactive neurons in the autonomic nervous system. Neuroscience 11:973–987

Voigt KH, Martin R (1985) Coexistence of unrelated neuropeptides in nerve terminals. In: Håkanson R, Thorell J (eds) Biogenetics of Neurohormonal Peptides. Academic Press, London, pp 245–272

Wattchow DA, Furness JB and Costa M (1988) Distribution and coexistence of peptides in nerve fibers of the external muscle of the human gastrointestinal tract. Gastroenterology 95:32–41

Weihe E, Hörsch D, Eiden LE, Hartschuh W (1991) Dual presence of chromogranin A-like immunoreactivity in a population of endocrine-like cells and in nerve fibers in the human anal canal. Neurosci Lett 130:190–194

Wiedenmann B, Huttner WB (1989) Synaptophysin and chromogranins/secretogranins – widespread constituents of distinct types of neuroendocrine vesicles and new tools in tumor diagnosis. Virchows Arch B Cell Pathol 58:95–121

CHAPTER 2

Regulation of Gastrointestinal Peptide Hormone Gene Expression

R.K. PEARSON

A. Introduction

Recombinant molecular biology has led to a considerable expansion of our insight into how genes for specific peptide products are regulated. This is of particular importance in the field of endocrinology, where a central theme is the importance of regulation of hormone synthesis, its site as well as relative level. The catastrophic consequences of unregulated hormone synthesis are observed in a variety of clinical syndromes associated with aberrant production of biologically active regulatory peptides; for example, the severe peptic ulceration seen in Zollinger-Ellison syndrome (gastrinoma) and the life-threatening hypoglycemia characteristic of insulin-producing islet cell tumors (see Chap. 14).

In the gastrointestinal (GI) tract, the endocrine cells release peptide hormones, which (with the exception of insulin) are single polypeptide chains and the products of single genes. All somatic cells contain identical genetic material that in humans consists of approximately 10^9 DNA base pairs (bp). The DNA required to code for the predicted 100000 total cellular gene products comprises only a small fraction of this nucleic acid pool and a given cell type is likely to express only a fraction of the possible genes. Thus, an endocrine cell line in the intestinal mucosa differs from its neighboring mucosal enterocyte in the set of genes it is programmed to express during its differentiation and, in particular, the peptide hormone gene to be expressed. This exquisitely regulated cell specificity of expression rests largely, if not exclusively, at the level of transcription or the synthesis of messenger RNA (mRNA) from genomic DNA.

In addition to its specificity of gene expression, the cellular machinery responsible for transcription is also capable of regulating the rate of mRNA synthesis in response to physiological events. While it is clear that the more "distal" steps in peptide hormone biosynthesis, such as translation, processing, vesicular storage, and release are critical in regulation of gut peptides, the era of molecular biology has allowed investigators to begin to understand the importance of transcription in the complex response of GI hormones to a meal and other physiological events.

B. Gene Structure

With the availability of DNA clones representative of the transcribed mRNA (complementary DNA or cDNA), investigators turned towards isolating the genomic DNA elements of genes. When characterized, a surprising twist to gene structure emerged. In contrast to the prokaryote, eukaryotic genes were not linearly represented on the chromosome. Rather, the genetic sequences transcribed to mRNA, including the sequences coding for protein amino acids (exons), are interrupted by intervening DNA sequences (introns). While intronic DNA is initially transcribed to nuclear RNA, it is removed during nuclear splicing and other processing reactions (e.g., polyadenylation) prior to the export of mature mRNA into the cytoplasm for translation (Fig. 1).

This split organization of genes into exons and introns has been viewed as a mechanism of enhancing the rate of evolutionary change. In this scheme, exons are postulated to correspond to specific functional protein domains which could be "shuffled" with other exons to create new proteins. Thus, introns were lost in the prokaryotes in the interest of efficiency of gene expression while eukaryotes exploited the split genes to enhance diversity and the potential for evolutionary change (BLAKE 1983). While some exon gene structures can be clearly related to both structural and functional protein domains (e.g., hemoglobin gene), as more genes were cloned and sequenced the correlation of exons to protein domains became more complex and controversial. While it is quite clear that some domains appear frequently in a variety of otherwise unrelated proteins (i.e., EGF-like domains in protease inhibitors), these domains are often not specified by a single exon. In addition, the functional catalytic domain of many serine proteases is found on two or more exons (BREATHNACH and CHAMBON 1981). For this reason, some investigators favor a structural definition of a domain that represents the smallest peptide fragment capable of autonomously assuming the correct three-dimensional configuration. A resolution of this issue awaits the X-ray crystallography of structures of additional proteins for which the gene structure is known.

The genes for several gut hormones have been cloned, sequenced, and characterized. They all contain at least one intron and are shown schematically in Fig. 2. There is clearly conservation in gene structure with at least some consideration to protein domains contained in exons with functional importance. For example, all genes except secretin contain an intron separating the start of transcription from the start of translation (initiator methionine). In addition, generally a second intron separates the amino terminal signal sequence from the exon coding for the biologically active, mature, secreted hormone. Further, genes coding for more than one released peptide product with potential biological activity often have these peptides segregated into specific exons (e.g., peptide histidine isoleucine on exon 4 and vasoactive intestinal peptide (VIP) on exon 5 of the VIP gene (see Fig. 2).

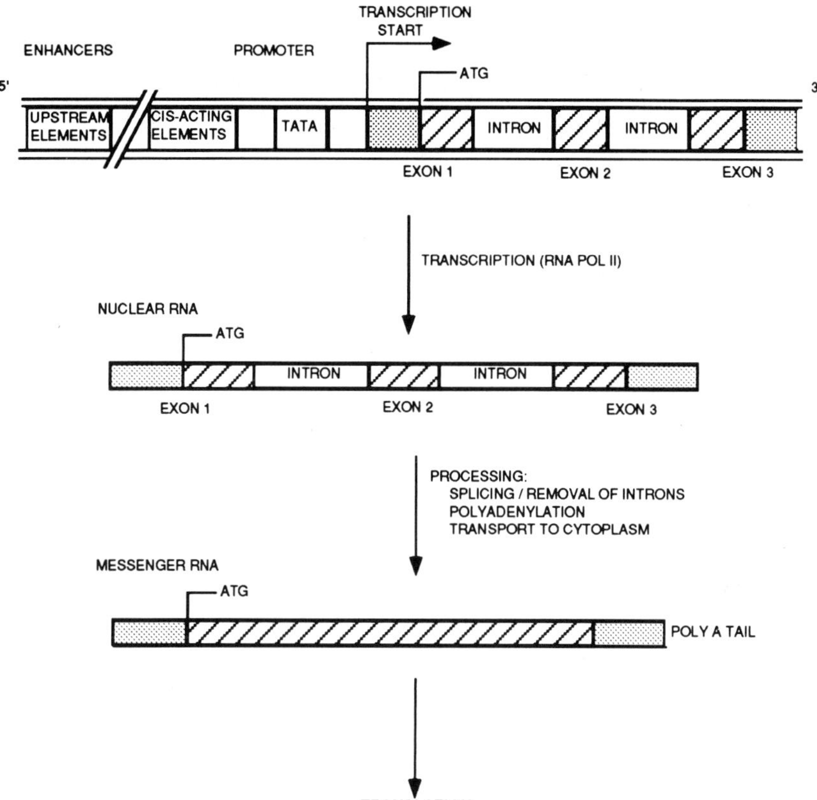

Fig. 1. Schematic representation of a typical eukaryotic gene with the position of the transcriptional start site and initiation of translation methionine *indicated by arrows*. The exons are DNA segments represented in the mature, processed mRNA while introns are DNA sequences removed from the RNA transcript during processing in the nucleus. The *cross-hatched* region represents sequences translated into protein, with the 5′ and 3′ transcribed but nontranslated regions *shaded*. The regulatory elements of a gene are generally located 5′ to the start of transcription. The promoter consists of sequences which direct the correct initiation point of transcription, including an AT-rich region important in the early assembly of the transcriptional apparatus. Other regulatory DNA or *cis*-acting elements are located in close proximity (CAAT motif) to the promoter or act at great distances ("enhancers"). These DNA elements all function by binding protein or *trans*-acting factors which initiate binding or influence the rate of initiation of RNA polymerase II activity

C. RNA Polymerase II

Eukaryotic organisms employ three distinct nuclear enzymes capable of polymerizing ribonucleotides into RNA using DNA as a template to transcribe different sets of genes. A variety of in vitro studies in simple and complex organisms have clearly shown that RNA polymerase I synthesizes

1. CCK / GASTRIN

Human Gastrin
(Wilborg, et. al., 1984)

EGF-RE
IEE
TATA
ATG
G-17
3500 bp
129 bp

Rat CCK
(Deschenes, et. al., 1985)

SP1
AP-1
AP-4
CRE
TATA
ATG
CCK-33
1100 bp
5300 bp

2. SECRETIN / GLUCAGON

Rat Secretin
(Kopin, et. al., 1991)

AP-2
SP1
TATA
ATG
SECRETIN
81 bp
105 bp
104 bp

Human VIP
(Tsukada, et. al., 1985)

CRE
TATA
ATG
PHM
VIP
1300 bp
1700 bp
1000 bp
700 bp
800 bp
1700 bp

Rat Glucagon
(Heinrich, et. al., 1984)

CRE
TATA
ATG
GLUCAGON
GLP-1
GLP-2
3000 bp
1600 bp
1600 bp
1600 bp
600 bp

3. PANCREATIC POLYPEPTIDE

Human P P
(Leiter, et. al., 1985)

TATA
ATG
PP
750 bp
245 bp
190 bp

Human Neuropeptide Y
(Minth, et. al., 1986)

SP1
AP-1
TATA
ATG
NPY
965 bp
4300 bp
2300 bp

4. MOTILIN

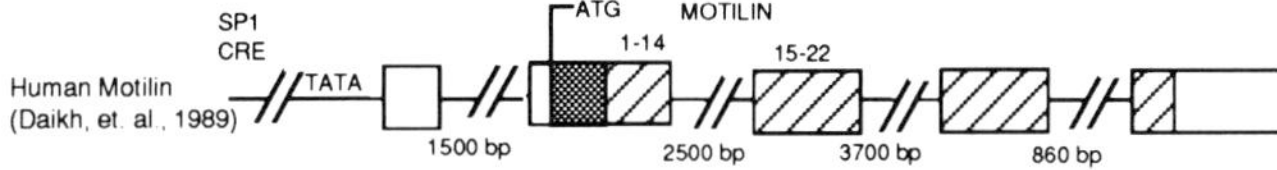

5. SOMATOSTATIN

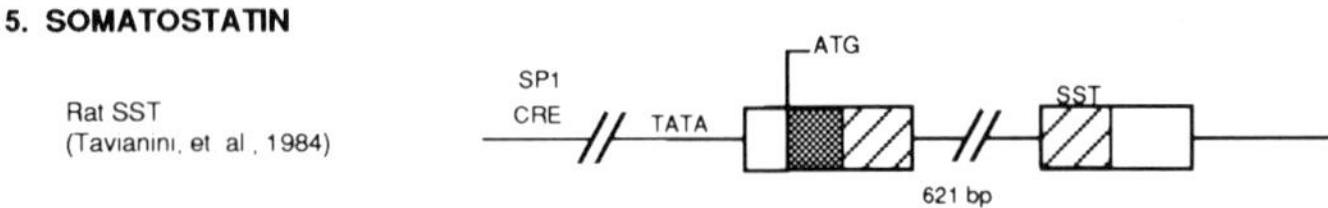

Fig. 2. Representation of the structural features of GI regulatory peptide genes. Exons (not drawn to scale) are *indicated by the boxes* and the intervening sequences (introns) *represented by the broken line*, with their approximate size in base pairs below. Nontranslated sequences are *indicated by open boxes*, the signal peptide *by shading* and the prohormone region *in cross-hatches*. The exon containing the mature, secreted peptide is identified. All genes contain a TATA element from 20 to 30 bp from the transcriptional start site. Consensus sequences for transcription factor binding sites found in the 5′ regulatory region are indicated. *IEE*, insulin enhancer element; *PP*, pancreatic polypeptide; *SST*, somatostatin

ribosomal precursors, RNA polymerase II (pol II) transcribes mRNA for protein-encoding genes, and RNA polymerase III transcribes 5S and transfer RNA (tRNA) genes. Because of its pivotal role in gene expression, pol II is the best characterized of the polymerases and has been purified to near homogeneity from more than 20 different organisms (SAWADOGO and SENETAC 1990). Pol II is a large, highly conserved multi-subunit enzyme with a molecular mass approaching 600000 daltons. It contains two large subunits ($M_r > 140000$) and a collection of eight to ten smaller protein components. Despite intense study at both the biochemical and genetic level, the functional importance of these smaller subunits remains uncertain.

Initiation of mRNA synthesis by pol II is a pivotal and primary point of regulation of tissue-specific gene expression. In addition, the rate at which pol II initiates transcription is also a key mechanism by which cells modulate the level of active gene expression (and thus protein product) in response to a variety of intra- and extracellular signals. However, the amount of mRNA available for translation to protein is dependent on the rate of pol II initiation and the rate of elongation or synthesis, balanced against degradation of mRNA by cellular RNases. Because of the scope of this review, I will focus primarily on mechanisms of regulation of pol II initiation. While the other processes have been increasingly recognized as important in regulating some genes (e.g., transferrin receptor mRNA degradation increased in the presence of elemental iron; THEIL 1990), RNA stability has not been demonstrated to influence directly GI peptide hormone mRNA levels.

D. Transcriptional Regulation Overview

I. DNA Elements

Given the complexity of the eukaryotic cell and its genome with its long stretches of transcriptionally inactive DNA and complex chromatin structure, it is not surprising that the mechanisms of initiation and regulation of transcription are complex. In contrast to prokaryotes, a wide assortment of DNA elements responsible for the accurate and efficient initiation of pol II transcription are located both upstream and downstream from the origin of transcription or cap site. These DNA or *cis*-acting elements are modules or DNA sequences which bind proteins or transcription factors in a sequence-specific fashion. They are relatively short (6–20 bp in length) linear sequence motifs which have been demonstrated by a variety of biochemical techniques to be important in the activity of a promoter. Deletional and site-directed mutational analysis identify the crucial nucleotides required for mediating the protein binding – similar to the structural requirements of a classical ligand/receptor interaction. These proteins then act by either activating or repressing transcription presumably by affecting the stability of the pol II transcription complex. Thus, each gene contains an array of DNA sequences with positive and negative effects on transcription. The level of transcription in a given cell would be dependent on the presence, absence, or physical state of specific transcription factors which bind to these DNA elements (MANIATIS et al. 1987; MITCHELL and TJIAN 1989; SALTZMAN and WEINMAN 1989).

The control region immediately adjacent to the cap site of a gene to which pol II binds is called a promoter. This region includes DNA elements absolutely required for accurate initiation of transcription. These "core" promoter elements include the site of RNA pol II binding and direct the site of mRNA synthesis. Many gene core promoters, including all of the GI peptide hormone genes sequenced to date, contain an adenine-thymine (AT)-rich region designated as the TATA box (BREATHNACH and CHAMBON 1981). Found within 30 bp of the cap site, the TATA box binds a transcription factor known as TFIID (GREENBLATT 1991). In genes containing a TATA box, mutational analysis of this motif demonstrates that it is absolutely required for accurate transcriptional initiation of these genes. The promoter also contains other DNA modules upstream (or 5′) of the TATA box which bind transcription factors in a sequence-specific fashion. These upstream promoter elements are responsible for regulating the rate of transcriptional initiation from the core promoter. Common 5′ promoter elements found in many genes include the guanine-cytosine (GC) sequence motif, which binds the transcriptional activator, SP1 (DYNAN et al. 1986), and the CCAAT box (BENOIST et al. 1980), which binds many different cellular DNA binding proteins. These elements are generally found within 100 bp of the transcriptional start site and greatly enhance the rate of transcription.

Traditionally, activating regulatory DNA elements were categorized as promoters or enhancers. Enhancers were distinguished by being active at great distances from the origin of transcription in an orientation-independent fashion. Enhancers can also modulate transcription downstream (3′) as well as upstream (5′) from the promoter. However, promoter and enhancer elements are similarly organized into short DNA sequences which function by binding specific protein or *trans*-acting factors which activate transcription. In addition, some of these sequence elements are interchangeable; an enhancer element found in the immunoglobulin gene (*Oct* 1) is also found in the promoter region of a number of other genes (MITCHELL and TJIAN 1989). Thus, the distinction between promoter and enhancer elements may be quite arbitrary and dependent upon their orientation or number in a given gene. For example, a single copy of a heat shock element requires close proximity to the promoter to be active while duplication of this element allows it to activate transcription from a distance, meeting the criterion of an enhancer (BIENZ and PELHAM 1986).

II. Transcription Factors

Specific binding of regulatory proteins to DNA in a sequence- and site-specific fashion is pivotal in transcriptional regulation. Eukaryotic pol II does not bind efficiently or accurately to promoters. Rather, transcription is dependent on a series of specific protein-DNA interactions. While the distinction between the *cis*-acting promoters, upstream elements and enhancers is blurred, *trans*-acting factors can be neatly divided into general initiation and regulatory factors.

1. General Transcription Factors

General transcription factors act by binding to the core promoter elements (including the TATA box) and are required for basal levels of transcription (CONAWAY and CONAWAY 1991; SAWADOGO and SENTENAC 1990). The critical binding event is TFIID binding to the TATA region, which "commits" this gene to transcription. The TFIID-promoter complex then initiates an ordered assembly of a series of proteins, including pol II and at least four other subunits (TFIIA, TFIIB, TFIIE, and TFIIF) on the promoter element. This assembly is required for a stable multi-protein complex which is then capable of basal gene transcription illustrated in Fig. 3. Binding of the general transcription assembly to the template is sufficient for low levels of basal transcription and requires only core promoter elements including the TATA box and nearby initiation sequences.

This complex serves to "interrupt" the chromatin structure such that the repression by chromosomal histone proteins is prevented. Studies in vivo and in vitro demonstrate that a functional competition exists between the nucleosome structure and the transcriptional apparatus for the TATA box (MEISTERERNST et al. 1990). The DNA template transcriptional initiation complex does not require adenosine triphosphate (ATP). Once bound, initiation of transcription occurs rapidly in an enzymatic reaction which requires ribonucleoside triphosphates and hydrolysis of ATP. Elongation of the RNA transcript is also mediated by specific protein factors. Regulation of expression by altered rates of elongation appears to be important for some genes. For example, transcription of the c-*myc* gene results in prematurely truncated and, thus, inactive RNA species, presumably due to pauses or blocks in elongation of mRNA which can be overcome by specific factors under appropriate conditions (BENTLEY and GROUDINE 1986).

Therefore, the assembly of the transcriptional complex on the template is a slow, rate-limiting step. Once transcription is terminated, the complex dissociates. For efficient and regulated transcription to occur, other activating proteins are required.

2. Regulated Transcription Factors

The regulated transcription factors bind to upstream promoter elements and enhancers and mediate tissue-specific gene expression as well as the regula-

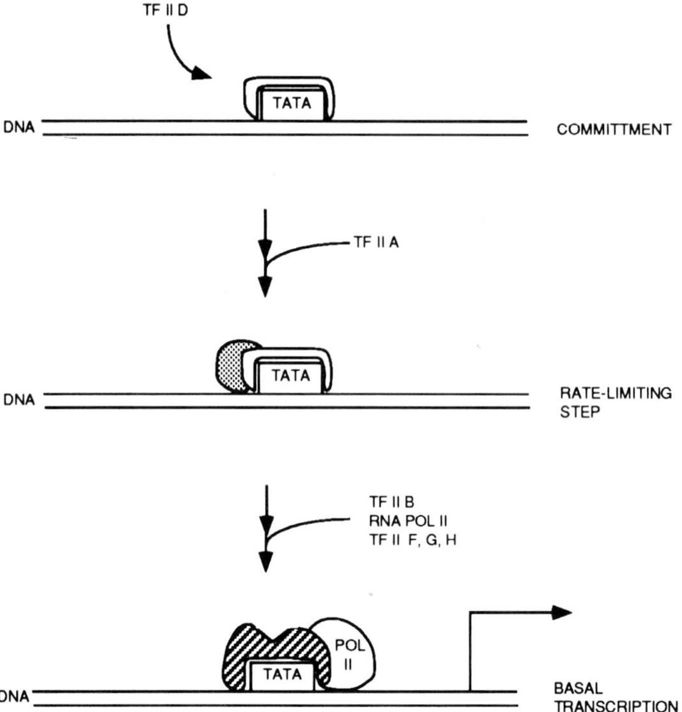

Fig. 3. Model for the assembly of the transcriptional complex. The initial or commitment step is binding of the general transcription factor TFIID to the core promoter including the TATA element. TFIID acts to "derepress" the promoter by blocking nucleosome formation and provides a target for other members of the apparatus to assemble. Binding of TFIIA (at least in some promoters) stabilizes this initial complex. The remainder of the factors assemble in a rapid, but ordered fashion. TFIIB and perhaps TFIIF are required for the selective binding of RNA polymerase II to the complex. Once assembled, this apparatus is capable of initiating transcription at a basal level. (Adapted from CONAWAY and CONAWAY 1991; LEWIN 1990; GREENBLATT 1991)

tion of mRNA synthesis by intracellular messengers such as adenosine-3′,5′-cyclic monophosphate (cAMP) (JOHNSON and MCKNIGHT 1989). These proteins bind to sequence-specific sites in the gene and, through interactions with the basal transcription complex, activate the rate of initiation of transcription. The exact mechanism by which this activation occurs has not been definitely established and is the subject of some controversy. Most investigators agree that for activation to occur at a distance, intervening DNA sequences must be bent or looped out (MASTRANGELO et al. 1991). However, whether the activating domains of enhancer binding proteins interact directly with the basal transcriptional apparatus (LIN and GREEN 1991) or act indirectly by binding a third class of regulatory proteins (coactivators or adapters), which in turn bind to the transcriptional apparatus (PUGH and TJIAN 1990; BERGER et al. 1990; DYNLACHT et al. 1991), is not certain (Fig. 4).

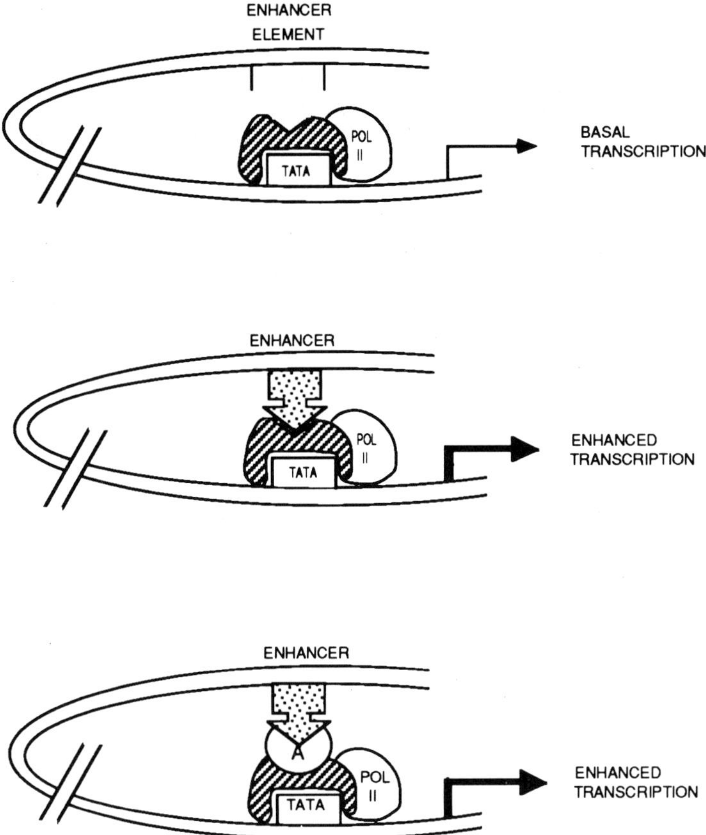

Fig. 4. A model for activated transcription. To enhance the efficiency of pol II directed transcription, DNA promoter or enhancer elements bind regulatory transcription factors, which influence the rate at which pol II initiates new rounds of transcription. These factors are composed of separate domains responsible for mediating the specificity of DNA binding (represented by the box structure) and mediating the transcriptional activation (*arrow*). In order for these proteins to act at a distance, DNA must be bent or looped out. Regulatory factors may either interact directly with the general transcriptional apparatus (*top*) or act through, or in concert with, a coactivator or adaptor (*A*), which in turn interacts with a basal transcription factor. Basal factors identified as targets for interaction with regulatory proteins include TFIID and B. One likely mechanism of activation is stabilization of TFIID binding to the core promoter in support of multiple rounds of transcription. (From LEWIN 1990; GREENBLATT 1991)

Experimental evidence exists to support both mechanisms, aided by the recent cloning of two human basal transcription factors, TFIID (PETERSON et al. 1990; KAO et al. 1990; HOFFMANN et al. 1990) and TFIIB (HA et al. 1991). The direct interaction and coactivation hypotheses are not mutually exclusive and are likely to operate together in regulating gene expression.

One mechanism which likely exists is that, by binding to *cis*-acting promoter elements, these transcription factors stabilize the binding of some of the general initiation factors by way of protein-protein interactions.

In recent years, many transcription factors have been purified and their cDNAs cloned. Similar to the modular nature of the *cis*-acting DNA elements, these proteins have structural motifs which correspond to a functional activity. These include DNA binding regions which mediate the specificity of the sequences required for binding, and a domain involved in protein-protein interactions which are required to activate the transcriptional apparatus (LEWIN 1990; PTASHNE and GANN 1990). Thus, transcription factors may be divided into functional families by their shared structural motifs responsible for DNA binding. These include: (1) zinc finger motifs; (2) helix-turn-helix domains; and (3) basic leucine (Leu) zipper. It is important to highlight that the factors within each family have divergent DNA sequence recognition sites. Although binding to DNA in a sequence-specific fashion is necessary, it is not sufficient for transcriptional activation, which is dependent on other domains involved in protein-protein interactions. While a detailed discussion of the mechanism of transcriptional activation by specific proteins is beyond the scope of this review, a list of important and well-characterized transcription factors along with their sequence recognition sites is provided in Table 1. Interested readers are referred to excellent recent reviews on this topic (HARRISON 1991; JOHNSON and MCKNIGHT 1989).

3. Modulation of Transcription Factor Activity

The regulation of transcription in a particular cell involves the interactions of specific, regulatory, *trans*-activating proteins with DNA sequence modules. The number and position of these motifs in a given gene determine its pattern of expression. The conservation and efficiency of this complex process are further illustrated by the observation that transcription factors dimerize prior to DNA binding. Indeed, members of the Leu zipper family of nuclear proteins require dimerization to bind DNA (JONES 1990). These factors may dimerize with themselves or to a related protein, and the consequences of homo- or heterodimerization include changes in the DNA sequence motif required for recognition, the relative affinity of binding, and the functional consequences of DNA binding, ranging from efficient activation to complete repression. An example of this phenomenon is the *jun-fos* family of *trans*-activators. These factors recognize the AP-1 binding site, a *cis*-acting element (TGA$^{C}/_{G}$TCA) which confers responsiveness to phorbol esters (BOHMANN et al. 1987). The *jun* family of proteins contains at least six members which can bind to the AP-1 site as a homo- or heterodimer with *fos* protein (CHIU et al. 1989). However, the consequences of binding are markedly different, with *jun* A-*jun* A or *jun* A–*fos* dimers activating a promoter containing a single AP-1 site, while *jun* B homo- or heterodimers

Table 1. Mammalian transcriptional regulatory proteins

Regulatory protein	DNA binding element	Comment	References
Group I: Helix-turn-helix			
Oct-1	ATTTGCAT	Ubiquitous; often proximal promoter element	LEVINE and HOEY (1988)
Oct-2	ATTTGCAT	Lymphocyte specific, activates immunoglobulin genes	
Pit-1	$^{A}/_{T}{}^{A}/_{T}$TATNCAT	Tissue-specific activator in pituitary cells (e.g., prolactin gene)	PEERS et al. (1991)
Group II: Zinc finger			
Sp1	GGGCGG GCGGGGGCG	Ubiquitous activator; generally proximal promoter element	KADONAGA et al. (1987)
Egr-1		Mediates transcriptional activation by growth factors	CHRISTY and NATHANS (1988)
Steroid receptors:			
Glucocorticoid, progesterone, androgen, mineralocorticoid	GGTACANNNTGTTCT	Superfamily of hormone receptors; activate and repress transcription. Activity dependent on hormone binding	BEATO (1989)
Estrogen	AGGTCANNNTGACCT	Domains: DNA-binding, hormone binding and *trans*-activating	
Thyroid hormone (c-*erb*A)	TCAGGTCANNNTGACC	Viral oncogene v-*erb*A lacks hormone binding	

Table 1. *Continued*

Regulatory protein	DNA binding element	Comment	References
Group III: Basic, leucine zipper		Large family (at least six genes); dimerize via Leu zipper	CHIU et al. (1989); RAUSCHER et al. (1988a,b); BOHMANN et al. (1987)
Jun A *Jun* B *Fos*	NTGA$^{C}/_{G}$TCANN	Confers responsiveness to phorbol esters. *Fos* forms dimer with *jun* and binds to AP-1 site	
CREB ATF	NTGACGTCAN GTGACGT$^{A}/_{C}{}^{A}/_{G}$N	Confers responsiveness to cAMP. Phosphorylation promtes dimerization, DNA binding and *trans*-activation. Multiple proteins bind CRE; phosphorylated by PKA	MONTMINY and BILEZIKJIAN (1987) REHFUSS et al. (1991)
C/EBP	CCAAT	One of many factors binding to proximal promoter element	JOHNSON et al. (1987)
Group IV: Basic, helix-loop-helix			
Myc	CACGTG	Specific DNA binding dependent on dimers (with Max); tissue specific; induces differentiation to muscle cells	BLACKWOOD and EISENMAN (1991)
MyoD	CACCTG		DAVIS et al. (1987)
Group V: Unclassified			
AP-2	TCCCCAGGCG	Confers responsiveness to cAMP and PKC	IMAGAWA et al. (1987)
CTF/NF-I	GCCAAT	Proximal CAAT box binding factor	CHODOSH et al. (1988)
SRF	GATGTCCATATTAGG-ACATC	Responsive to serum, EGF, TPA and insulin; dimerizes	NORMAN et al. (1988)
Pan-1	CACCTGTC	*Trans*-activator for pancreatic acinar genes	NELSON et al. (1990)

do not activate transcription. Furthermore, *jun* B dimers can compete for binding to a single AP-1 site, thus "repressing" *jun* A activity for a given promoter. However, if multiple AP-1 sites are present, *jun* B functions as an efficient activator (CHIU et al. 1989). This demonstrates how important competition between transcription factor family members is for DNA binding and dimerization and its impact on the rate of transcription. Furthermore, the number and position of the DNA elements is also critical. This combinatorial and competitive mechanism between multiple protein complexes for similar DNA elements increases the possible precision and flexibility of regulation while efficiently minimizing the number of proteins required. To understand the regulation of a specific gene, the critical DNA elements must be identified along with all of the proteins capable of interacting with these elements. The number and spatial array of these DNA elements on the template, along with the relative abundance and physical state of a number of interacting transcription factors, will determine the ultimate rate of transcription.

This process must also be responsive to signaling from extracellular stimuli and during differentiation. The AP-1 family of regulatory proteins is tightly regulated. The *jun/fos* factors are important in the normal growth and proliferation response, including the transcription of many other cellular genes during exposure to growth factors, mitogens, and oncogenes (CURRAN and FRANZA 1988). *Jun-fos* heterodimers are the most potent transcriptional activators, and their respective genes are regulated by different mechanisms. *Jun* induces transcription of its own gene, and its activity is primarily regulated by a cellular inhibitor which interacts with a specific domain of *jun* and blocks its *trans*-activating activity. The importance of this mechanism is underscored by the transforming capacity of the retroviral form of *jun*, which lacks this regulatory domain (BAICHWAL and TJIAN 1990). By contrast, *fos* represses the activity of its own gene, which is dependent on serine phosphorylation of its carboxyl terminus (OFIR et al. 1990). Signal-transduction pathways which induce AP-1 activity (e.g., phorbol esters) likely operate by interrupting the *jun*-inhibitor complex and dephosphorylating *fos*. A more direct example of mechanisms of signaling transcription from extracellular cues is the phosphorylation by protein kinase A of a family of transcription factors which mediate enhanced transcription in response to cAMP discussed below.

III. Transcriptional Repression

The overview of transcription above has focused on recent advances in the understanding of the selective activation of gene expression. Considerably less is known about the mechanisms by which eukaryotes repress transcription, but they are almost certainly important for transcriptional regulation and may be as widely used as activation in the selective regulation of promoters. When one considers the importance of having some critical genes completely inactivated in the appropriate cell type and the need to turn off transcription of inducible gene products rapidly under the appropriate stimuli, then the advantage of a system of repression as well as activation of gene activity is readily apparent. To date, the number of proteins isolated from eukaryotes with gene repressive activity are few when compared with the *trans*-activators. At least four mechanisms of transcriptional repression have emerged (LEVINE and MANLEY 1989; RENKAWITZ

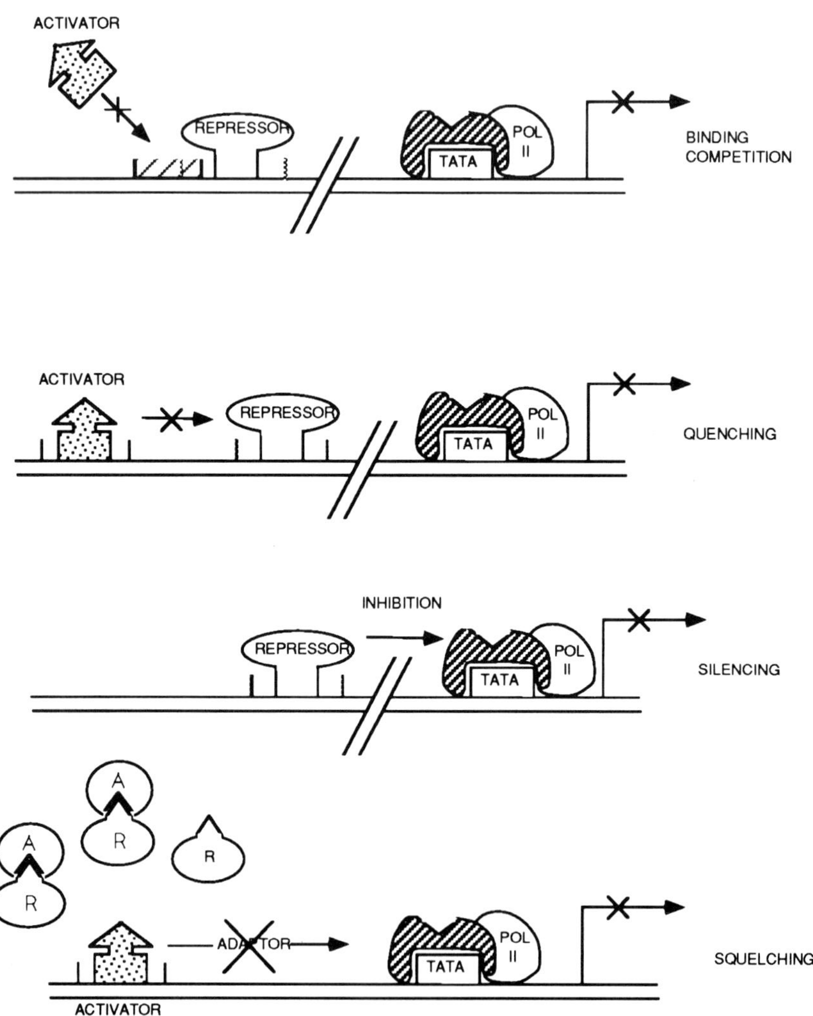

Fig. 5. Possible models for transcriptional repression. Details are provided in the text. (Adapted from LEVINE and MANLEY 1989)

1990) (Fig. 5). Direct competition for DNA binding sites of either the general transcriptional apparatus or upstream activating transcriptional factors by repressors may be the most common mechanism of repression in prokaryotes and eukaryotes. Secondly, the repressor could act by blocking the *trans*-activating function of a DNA bound factor by impeding its interaction with the basal transcriptional apparatus, a mechanism termed "quenching." A bound repressive factor could more directly repress transcription by directly interacting and blocking the activity of the basal transcriptional apparatus. DNA elements binding to such factors would be termed silencers. Repression may not involve DNA binding directly but rather involve dimerization schemes of an inhibitory protein with a transcriptional enhancer binding protein. This could change either the DNA binding affinity or transcriptional activity of the transactivator.

The best characterized proteins capable of repressing transcription in mammalian systems are the steroid hormone receptors (BEATO 1989, 1991). These nuclear receptors/transcription factors belong to the functional family employing a zinc finger DNA binding motif. These receptors mediate their biological effects by binding to DNA in the presence of hormones and modulating transcription. Examples of these are receptors for estrogen, progesterone, glucocorticoids, mineralocorticoids, thyroid and androgen hormones. It has been known for some time that these proteins, when bound with ligand, are capable of both activating and repressing transcription. Both the positive and negative effects are mediated through DNA binding by the same class of receptor molecules to a hormone-responsive element. There are no consistent sequence differences in the *cis* elements which repress or activate transcription. In fact, under different conditions and in different genes the identical element can have opposite effects (DIAMOND et al. 1990). Thus, the effect of steroid receptor binding likely reflects its position within the promoter, or in very subtle differences in the sequence of the element. The former mechanism is illustrated by the glucocorticoid-induced repression of the gene for the α-subunit of the glycoprotein family of hormones. This gene is induced by cAMP and inhibited by glucocorticoids. However, this inhibition is noted only in cells in which the cAMP element is active, suggesting that the glucocorticoid receptor (GR) counteracts the effects of cAMP (AKERBLOM et al. 1988). This has been largely confirmed in experiments demonstrating GR binding to DNA that overlaps two binding sites containing cAMP responsive elements (CRE), and that repression requires only DNA binding. This suggests that there is a direct competition for DNA elements between cAMP responsive factors and GR. In addition, other models of steroid hormone-induced repression suggest roles for heterodimerization of GR with the *jun-fos* family of transcription factors as a mechanism of repression (DIAMOND et al. 1990). Thus, a repressive transcription factor may act by multiple mechanisms.

IV. Genes Lacking TATA Elements

Generally, promoters for genes encoding messenger RNA contain the core promoter elements, the TATA box, and, less frequently, the CCAAT motif. However, "TATA-less" genes are becoming increasingly recognized. The first sequenced were the so-called housekeeping genes, which encoded enzymes which are widely distributed and perform an essential metabolic function (e.g., dihydrofolate reductase; MELTON et al. 1986). While these genes are regulated, early observations suggested that these TATA-less promoters directed RNA synthesis at a relatively low level with less regulation and minimal tissue specificity. Interestingly, recent studies have confirmed that the same basal transcriptional apparatus, including the binding of the TATA factor TFIID, is operative in these genes (GREENBLATT 1991). However, the site of initiation of mRNA synthesis may not be as accurately defined, and these genes often demonstrate multiple minor transcriptional start sites.

The number of genes cloned lacking a TATA box has been expanded recently and includes the genes for the EGF receptor (ISHII et al. 1985b), the human HA-*ras* oncogene product (ISHII et al. 1985a), and the recently cloned cystic fibrosis gene (YOSHIMURA et al. 1991). While the genes for the peptide hormones characterized to date all contain typical TATA sites, the first processing enzyme gene cloned, carboxypeptidase E (CPE), which removes the C-terminal basic amino acids from the endoproteolytic cleavage products of peptide prohormones (see Chap. 3), has a promoter lacking a TATA motif (JUNG et al. 1991). Distinct from the housekeeping genes, CPE exhibits marked tissue-specific expression and appears to be exclusively expressed in hormone-producing cells. Indeed, it appears that CPE expression is an excellent marker for the expression of peptide hormones and neuropeptide transmitters. Furthermore, several in vivo and in vitro studies have demonstrated significant regulation of CPE mRNA by a variety of stimuli and factors, which is often coordinated with the regulation of the prohormone gene (RODRIGUEZ et al. 1989; BONDY et al. 1989).

Besides lacking a TATA box, these promoters all have GC-rich regions (up to 70%) in the 5′ flanking region, 100–150 nucleotides from the cap site. These genes also contain GC boxes further upstream which presumably bind transcription factors such as SP1. The CPE gene additionally contains potential sites for the regulatory elements, AP-2, NF-1 and Pan-1 in its 5′ flanking region (JUNG et al. 1991). Because this gene is expressed in all cells which process and secrete peptide hormones, identifying the tissue-specific *cis* elements that regulate transcription will be important.

E. Analysis of Gene Expression

I. RNA Quantitation

It is clear from the brief outline above that the regulation of a given gene will likely be extraordinarily complex with multiple DNA elements binding a large menu of possible transcription factors. Fortunately, molecular tools do exist which enable the researcher to begin unraveling some of the mechanisms underlying this conundrum.

The simplest method to study the transcriptional activity is to measure the level of mRNA from the gene of interest. Northern analysis consists of isolating RNA from a tissue or cultured cell after experimental manipulations and identifying specific mRNA by hybridization with a labeled cDNA or cRNA probe. This usually involves separating the mRNA species by size with denaturing agarose gels and subsequently transferring and immobilizing the nucleic acid on nitrocellulose filters. The resulting signal from the hybridization of the labeled probe and specific RNA can be quantitated by autoradiography and densitometry. These studies always require an internal control by blotting the RNA with a probe for an expressed gene with a constant level of expression such as β-actin or glyceraldehyde 3-phosphate dehydrogenase (GAPD) to control for experimental artifacts due to the quantity or quality of the RNA isolated. Similar types of estimation of mRNA quantity may be done to even higher sensitivity using polymerase chain reaction (PCR) technology. An important extension of this methodology is in situ hybridization (GIBSON and POLAK 1990), which detects specific DNA or RNA sequences on tissue sections or intact cell preparations and allows cellular localization of specific sequences in heterogeneous populations.

Hybridization methods do not distinguish whether changes observed in mRNA content are due to alterations in the rate of transcription initiation, RNA turnover/degradation, transport, or synthesis. The rate of transcription initiation can be more directly determined by performing nuclear run-on transcription assays. In these studies, nuclei from cells under various experimental conditions are isolated, and newly synthesized nascent RNA is specifically labeled with uracil-5′-triphosphate (UTP). Specific RNA transcripts are detected by hybridization of the labeled RNA mix with a specific cDNA immobilized on nitrocellulose. This method basically determines the number of RNA pol II complexes on a given gene at a given time and is the most sensitive assay for determining specific gene transcription as a function of cell state.

II. Promoter Studies

While quantitative RNA measurements give important insights into regulators of gene expression, the characterization of the DNA elements and protein factors involved requires the sequence and characterization of the gene. Because most cDNAs are not full length, especially on their 5′ end, and genes can contain introns in the 5′ nontranslated region, it is important to unequivocally establish the site of initiation of transcription. This can be done by isolating full-length RNA from an expressing cell line and hybridizing it to DNA probes based on the genomic sequence. The resulting DNA/RNA heteroduplex provides a template for primary extension or protection from single-stranded S1 nuclease digestion, two methods which determine the site of transcriptional initiation with precision. To begin studies defining the elements responsible for controlling gene expression, one compares the genomic sequence 5′ to the defined cap site with known regulatory DNA elements such as the TATA, CAT, and GC boxes common to many promoters, as well as other well characterized upstream elements (Table 1).

To definitively establish a fragment of DNA as a promoter or as an element important in regulating a gene, the ability to direct or regulate transcription must be proven experimentally. These studies have been greatly simplified by two methodologic developments: eukaryotic transfection and reporter genes (Alam and Cook 1990).

Transfection is the introduction of functional foreign DNA into eukaryotic cells. This is accomplished by either creating a chemical environment that enhances DNA attachment to the cell surface [calcium phosphate precipitation, diethylaminoethyl (DEAE)-dextran, lipid fusion] for uptake by an ill-defined endocytic mechanism. Alternatively, electroporation uses an electrical field to open pores in the cell membrane through which DNA enters.

Once introduced, the transfected DNA need not be integrated into the cellular genome to be expressed. *Transient* transfection assays quantitate expression as early as 12 h and as late as 72 h after transfection. Measurement of the transfected gene activity has been enhanced by the use of reporter genes which obviate the need to directly quantitate mRNA, a laborious and insensitive method. Reporter genes code for proteins with an easily assayable, sensitive and unique activity (generally enzymatic). Studies of gene expression employ fusion genes with the putative regulatory elements from the gene to be studied aligned to transcribe the reporter gene protein. For example, the most popular reporter gene is chloramphenicol acetyltransferase (CAT), an exclusively prokaryotic enzyme which transfers acetyl groups to chloramphenicol, both of which are readily available in a radiolabeled form. The product, by virtue of its different solubility in organic solvents, is readily distinguished from the precursor by its migration position by thin-layer chromatography (TLC). Because CAT activity is not endogenous to mammalian cells, the enzymatic activity measured after transient transfection reflects the activity of the cloned regulatory DNA elements which "drive" the expression of the fusion gene. An appropriate positive control is a strong, ubiquitous promoter aligned to transcribe a different reporter gene which allows for comparisons between experiments by controlling for the efficiency of transfection. Other reporter genes which are employed include firefly luciferase (enzymatic), human growth hormone (radioimmunoassay), and β-galactosidase (enzymatic).

Once a DNA fragment has been shown to function as a promoter, the transient transfection method allows further functional and structural characterization. Employing nucleases or convenient restriction enzyme sites, deletions of the promoter/CAT gene construct are performed. Comparing the measureable CAT activity of the truncated constructs with the full-length promoter aids in isolating stretches of promoter DNA responsible for activating (decreased CAT activity after deletion) or repressing transcription (increased CAT activity after deletion). This method can be further refined by site-directed mutagenesis of residues within a DNA element to establish its importance in regulating transcription. Finally, by comparing the CAT activity of the full length, truncated, and mutated promoter elements in cells which do and do not express the gene under study, elements important in mediating tissue-specific expression can be identified. An important and powerful adjunct to transient transfection assays in vitro is the use of promoter/reporter gene fusions in transgenic mice. This allows the determination of tissue-specific patterns of expression during development and differentiation (EFRAT et al. 1988). In addition, expression is characterized in vivo (i.e., in tissue), rather than in a (generally) transformed cell line in vitro using extrachromosomal transcription from a bacterially derived vector.

With the important regulatory DNA elements identified, the affinity and specificity of the DNA-protein binding interaction is exploited to biochemically characterize the transcription factors involved. Confirmation of protein binding to a specific element is provided by demonstrating protection of the protein-bound DNA from nuclease digestion (DNase protection assay; GALAS and SCHMITZ 1978) or a shift in the migration of a labeled sequence-specific oligonucleotide by protein binding on gel electrophoresis (band shift assays; MISKIMINS et al. 1985). In these studies, nuclear extracts from an appropriate cell are used to provide the source of transcription factors.

Specific factors can be further isolated and cloned using strategies based on DNA/protein binding. For example, the transactivator which binds to a DNA element conferring cAMP responsiveness (CRE) was purified. Furthermore, it was shown that the flow-through from the affinity column, depleted of CRE binding activity, would not support transcription of the somatostatin (SS) gene in vitro. Transcription was reinitiated by the addition of the purified cAMP response element binding protein, or CREB (ANDRISANI et al. 1989). Furthermore, CREB was cloned using a variation of standard cloning strategies. An oligonucleotide corresponding to the CRE was labeled and bound specifically to a CREB cDNA clone expressed as a β-galactosidase fusion protein in a bacteriophage expression library

(HOEFFLER et al. 1988). This method, known as "Southwestern" hybridization, has been employed in the biochemical characterization and cloning strategies for other DNA binding proteins (SINGH et al. 1988; VINSON et al. 1988).

F. Regulation of Specific Hormone Gene Expression

I. Preface

This section will detail what is known about the regulation of specific gut hormone genes. As shown in Fig. 2, the structures of many genes have been characterized. Unfortunately, biochemical details on the mechanisms of transcriptional regulation and specificity are mostly lacking from cells of GI endocrine origin due to a lack of cell lines for in vitro studies. Many of the studies outlined above are dependent on maintaining transfected cell lines in culture or isolating nuclear extracts from plentiful supplies of cells. GI endocrine cells are neither plentiful nor easily isolated and, with the exception of islet cells, are not available as transformed cell lines. Thus, studies of regulation of GI hormone genes have often depended on using transformed cell lines, which perhaps represent aberrant expression (e.g., cholecystokinin expression in rat medullary thyroid carchinomas). Generally, common mechanisms of regulation are assumed to exist, but studying the effects of luminal events on gene regulation is not possible with these models. Fortunately, the methods of detecting mRNA in tissue are sufficiently sensitive that these relatively rare messages can be detected and quantitated under various physiological conditions. Successfully enriched specific GI endocrine cells for study in vitro have been described (SOLL et al. 1984 and see Chap. 4). The successful maintenance of these cells in culture will allow advances in our understanding of GI hormone gene regulation.

Since the advent of amino acid sequence data for gut peptides, these substances have been grouped into families based on sequence and structure and function similarities. Both cDNA and genomic clones have supported the prediction that these peptide families arose from common ancestral genes from which the various members diverged.

II. Gastrin/Cholecystokinin Family

These peptide hormones are clearly related structurally and functionally, including an identical C-terminal peptide sequence (Gly-Trp-Met-Asp-Phe-NH_2), which is crucial to the biological activity of both peptides. Furthermore, a peptide is CCK- or gastrin-like depending on the position of the immediate HN_2^- terminal tyrosine (Tyr) residue:

Gastrin	-Ala-Tyr-Gly-Trp-Met-Asp-Phe-NH_2
CCK	-Tyr-Met-Gly-Trp-Met-Asp-Phe-NH_2

Sulfation of the Tyr is crucial for CCK activity at its classical peripheral targets.

A survey by LARSSON and REHFELD (1977) using immunocytochemical techniques in different vertebrate species demonstrated that CCK- and gastrin-like cells were separable in mammals, birds, and reptiles, while amphibians and teleosts (bony fish) had no separable CCK/gastrin components or cells. Rather, CCK and gastrin activities were represented by common caerulein-like peptides. The recent purification of a neuropeptide from the coelenterate *Ciona intestinalis* demonstrates complete conservation of the C-terminal peptide with a surprising structure: two adjacent sulfated Tyr residues (JOHNSEN and REHFELD 1990). This structure resembles CCK and gastrin equally in terms of the position of the critical tyrosine residue and is a good candidate as a common ancestral gene. Coelenterates diverged from the lineage giving rise to mammals more than a billion years age and the evolutionary distance of the divergence of the CCK and gastrin genes is supported by their location on different human chromosomes (CCK: chromosome 3q and gastrin: chromosome 17q; LUND et al. 1986).

1. Gastrin: Background

Gastrin has a dual function in the GI tract as a major regulator of gastric acid secretion and as a stimulator of mucosal proliferation (see Chaps. 7 and 12). In adults, gastrin is expressed in the GI endocrine cells, primarily found in the gastric antrum and in the nervous system as a putative neurotransmitter. While cDNAs for gastrin have been cloned from a number of species (YOO et al. 1982; FULLER et al. 1987; DIMALINE and LEE 1990), the only gastrin gene fully characterized to date is from humans. Attempts at cloning the gastrin gene illustrate the pitfalls in assigning full-length status to a genomic clone. Initially the human gastrin gene was reported to be less than 1 kilobase (kb) in length and contained only a single intron. However, it was subsequently shown by WIBORG et al. (1984) that the human gene spans greater than 4 kb and contains an additional large intron (>3.5 kb) which begins 5 bp 5′ to the translation initiator ATG codon. Thus, the first exon contains only noncoding sequences, a common structure in GI peptide genes (Fig. 2).

Because of its importance in regulating gastric acid secretion, considerable experimental effort has been directed towards understanding the regulation of gastrin secretion. Gastric acid inhibits gastrin secretion, providing a negative feedback control to prevent excessive secretion (WALSH and GROSSMAN 1975). Substantial evidence supports somatostatin (SS) release from neighboring D cells in the gastric antrum in response to acidification as a major inhibitor of gastrin release (LARSSON 1980; BLOOM et al. 1974).

2. Luminal Regulation of Gastrin Gene Expression

To determine if these mechanisms operate at the level of gene transcription, the effect of omeprazole-induced achlorhydria on antral gastrin and SS gene expression has been examined. Brand and Stone (1988) first demonstrated a reciprocal change in SS and gastrin gene expression in response to abolition of parietal cell acid secretion by the H^+,K^+-ATPase inhibitor omeprazole in rats. Quantitative Northern blots of antral RNA with specific gastrin and SS cDNA probes demonstrated a fourfold increase in gastrin mRNA and a threefold decrease in SS mRNA. Omeprazole had no effect on SS mRNA in duodenum, excluding a direct effect of the drug on D cells. A direct effect of SS in regulation of gastrin gene expression was confirmed by administration of an SS analog that prevented the hypergastrinemia and blocked the increase in gastrin mRNA induced by achlorhydria. These changes were observed after 24 h of achlorhydria. This delay in alteration of RNA levels largely reflects the large pool of SS and gastric mRNA in the antral mucosa and their relatively long half-life (>12 h in islet cell lines). Thus, the early (within minutes to hours) changes in hormone secretion reflect regulation of release of stored, synthesized peptides.

Wu et al. (1990b) extended these observations in rats by including a surgical fundectomy model for achlorhydria, in addition to omeprazole treatment. A similar reciprocal change in gastrin and SS mRNA levels was observed, confirming the importance of gastric acid neutralization. Larger increases in gastrin mRNA were observed in the surgical model relative to omeprazole treatment (6.5-fold induction after 4 weeks), while the SS mRNA changes were less striking (77% vs. 200% decrease). This suggested the presence of at least a second, undefined mechanism other than decreased SS expression mediating the gastric gene induction by achlorhydria. In this same study, direct measurements of transcriptional initiation were done using nuclear run-on assays. These demonstrated a 65% increase in nuclear gastrin RNA precursors after omeprazole treatment. Thus, an increased rate of transcriptional initiation at least participates in the increased gastrin mRNA. In addition, the existing gastrin mRNA may have been selectively stabilized. These studies confirm that regulation of gut hormones by luminal, physiological events (e.g., acid secretion) as well as by other regulatory factors (hormones, neurotransmitters) occurs at the level of the gene expression.

3. Islet Cell Gastrin Expression

Gastrin undergoes a complex pattern of expression during development, including expression in the neonatal pancreatic islets during the late fetal period (Brand et al. 1984). This corresponds to a critical period of pancreatic development with both dramatic growth and differentiation of the exocrine and the endocrine elements. Gastrin's potential growth-promoting targets are wide and include islet and acinar cells, and it has been advanced as a potential mediator of some of these developmental changes (Brand and Fuller 1988).

Using quantitative Northern analysis, BRAND and FULLER (1988) detected pancreatic gastrin levels in the 18-day rat fetus, which represented 1%–3% of the levels found in the adult antrum. There was a steady decline in gastrin expression, being undetectable by postnatal day 10. Reciprocal changes in expression were observed for the traditional islet cell products of insulin, SS, and glucagon. Thus, gastrin expression precedes the onset of differentiation of the individual islet cell populations. In contrast, antral SS and gastrin expression are parallel, being low in the fetus and rapidly rising after birth to reach adult levels between postnatal day 10 and 20. By contrast again, duodenal gastrin, SS, and glucagon have stable pre- and postnatal expression, while duodenal CCK reciprocates pancreatic gastrin activity, rising from undetectable fetal levels to detectable levels by postnatal day 3. The observation that pancreatic gastrin is 100% sulfated (as opposed to 50% sulfation of antral gastrin) supports the hypothesis that pancreatic gastrin may act as a paracrine CCK equivalent in the developing fetal pancreas (BRAND and FULLER 1988). Furthermore, selective regulation of fetal islet gastrin expression by calcium was observed (BRAND and WANG 1988). As cDNA clones are now available for the CCK/gastrin family of receptors, it will be interesting to observe the pattern of expression of the receptor subtypes in the developing acinar and islet cells.

4. Gastrin Promoter Analysis

Gastrin promoter analyses highlight a distinctive, developmental profile of regulatory peptide gene expression in pancreas, antrum, and duodenum. Furthermore, there is a single gastrin gene with identical transcriptional initiation start sites in all tissues (BRAND and FULLER 1988). Therefore, the same promoter elements are responsible for this complex pattern of tissue-specific expression. The DNA regulatory elements of the gastrin gene have been characterized in some detail, owing at least in part to the availability of stable gastrin-expressing cells in culture from gastrinomas, neuroblastomas, and pituitary cell lines. The neoplastic transformation of islets, therefore, results in the re-expression of the gastrin gene which is normally only transiently expressed early in development.

To characterize the DNA elements responsible for this exquisite and complex regulation, BRAND and coworkers have employed 1300 bp of the human gastrin promoter aligned to transcribe the reporter gene CAT (GASCAT) in transient transfection studies in a variety of gastrin-expressing cell lines. Antral G cells would be ideal for gastrin gene expression studies, but the limited viability of these cells when isolated preclude their use in transient expression assays. The pituitary cell line, GH_4, provides an attractive alternative because it possesses receptors for epidermal growth factor (EGF), gastrin-releasing peptide (GRP), thyrotropin-releasing hormone (TRH), VIP, and SS, all putative regulators of antral G cell secretion. Their efforts have led to important new models of gastrin regulation as well as provided the most carefully characterized GI hormone regulatory elements.

a) Studies in Pituitary Cell Lines

GASCAT confers tissue specificity with greater than tenfold more CAT activity in expressing (e.g., GH_4 cells) than nonexpressing cell lines (fibroblasts) (Godley and Brand 1989). In GH_4 cells, GASCAT activity was equal to the activity of a reporter gene employing the endogenously expressed growth hormone promoter, suggesting the absence of some repressive activity or sequences from the transiently expressed GASCAT plasmid. Epidermal growth factor (EGF) was the most potent inducer of GASCAT expression (threefold), while TRH (twofold) and VIP (1.5-fold) also enhanced GASCAT activity. Interestingly, GRP, the most potent stimulant of gastrin release, induced a small increase in gastrin transcription (1.2-fold). SS antagonized the induction of gastrin gene activity by all of these agents but had little effect on basal promoter activity.

Godley and Brand (1989) used deletional analysis and identified a critical DNA fragment at −82 to −40 bp upstream (5′ relative to the transcriptional start site). Deletion of this fragment abolished induction by EGF and TRH with 50% of the basal activity of the full-length GASCAT. This fragment contained sites for interactions with the *trans*-activators SP1 and AP-2. While SP1 is a *trans*-activator thought not to be inducible, AP-2 confers responsiveness to both cAMP and protein kinase C (PKC). This latter activity may explain the induction by TRH and VIP, both of which can act through cAMP induction. However, EGF-induced transcription was not easily explained.

Merchant et al. (1991) recently extended these observations and identified a novel EGF response element (ERE) in this −82 to −40 region of the promoter. GC-rich, competition and site-directed mutation studies confirmed that the element GGGGCGGGGTGGGGGG confers responsiveness to EGF, even when cloned into unresponsive, heterologous promoters. Its sequence is distinct from EREs from other genes and, in spite of a high affinity binding site (GGCGGG), does not bind SP1. Furthermore, a variety of biochemical techniques (gel mobility shift assays, DNase footprinting) demonstrate specific interactions of the element with a nuclear DNA binding protein which is distinct from previously described factors which bind GC-rich regions and mediate transcriptional effects of growth factors (e.g., EGR-1, AP-2) (Fisch et al. 1989).

These observations led Merchant et al. (1991) to propose a "feed forward" mechanism of gastrin gene expression, acid secretion, and mucosal growth and protection mediated by intraluminal EGF/transforming growth factor-α (TGF-α). These factors are well-known mucosal growth stimulants and TGF-α is expressed locally in the antral mucosa (Beauchamp et al. 1989). The striking dependence of gastrin expression on EGF receptor stimulation by the ERE ensures coupling of acid secretion with mucosal proliferation and integrity. EGF/TGF-α may act either as a paracrine mediator or as a luminal messenger acting at the apical surface of G cells to link gastrin expression with adequate capacity for mucosal growth. In this scheme SS, in response to acidification, is the major inhibitor of gastrin expression. A fall in gastrin would presumably result from decreased secretion of TGF-α/EGF.

b) Studies in Islet Cells

WANG and BRAND (1990) used the same GASCAT construct to examine the DNA elements regulating gastrin expression in pancreatic islets using a series of rat insulinoma cell lines. Deletional analysis identified different DNA fragments important in transcriptional regulation. They identified a negative regulatory element with a fivefold increase in CAT activity when a DNA fragment from −194 to −82 (*Hin*cII/*Nde*I restriction sites) was deleted. Further analysis refined the identification of this element to be an 8-bp sequence (ATTCCTCT), which is identical to a negative element in the β-interferon promoter. Point mutations in this 8-bp motif disrupted the transcriptional repression.

Further deletions identified an immediately adjacent positive element. CAT activity fell 40-fold with deletions of the −82 to −76 sequences (CATATGG) which are homologous to enhancer- and tissue-specific elements in the rat insulin promoter (Moss et al. 1988). To confirm transcription factor binding to these elements, elegant competition experiments were performed. GASCAT was co-transfected with an excess of plasmid containing the negative element (−194 to −82), which enhanced transcription. In separate transfections with a plasmid containing the insulin promoter, CAT activity decreased. Thus, the co-transfected competing *cis*-acting elements effectively bound the limited nuclear proteins which would normally regulate GASCAT expression. This confirms that these overlapping elements bind *trans*-activators and repressors. Distinct nuclear protein binding to both elements was confirmed directly with gel mobility shift and DNase protection assays. Multiple potential repressor proteins were identified, while the positive gastrin element bound a protein with similar properties as the insulin *trans*-activator. This tandem array of *cis* elements may function as a switch during development with insulin and gastrin expression activated in fetal islets by the same transactivator (WANG and BRAND 1990). During differentiation into a β-cell, the expression of gastrin is turned off selectively by repressor binding, while insulin transcription continues because its promoter does not contain the negative element. By having these *cis*-acting elements overlap in the gastrin promoter, the repressor may act by blocking the binding of the insulin transactivator to its element (see Fig. 4).

5. Cholecystokinin: Background

Like many GI regulatory peptides, CCK has a dual role as a classical hormone released from endocrine cells in the proximal small bowel in response to nutrients and as a neuropeptide, where it is found in very high concentrations in the central nervous system (CNS). Developmentally, differential CCK gene expression between gut and brain has been demonstrated in the mouse by FRIEDMAN et al. (1985) and VITALE et al. (1990) with adult levels of CCK mRNA present from birth in the gut, while brain expression remains undetectable until a dramatic rise by postpartum day 9.

6. Cholecystokinin Gene Regulation: RNA Quantitation

Intraluminal nutrients are the major regulators of hormonal CCK release. Studies examining the relationship between secretion and intestinal CCK gene expression have been initiated by LIDDLE et al. (1988) in the rat. Unlike other species which release CCK in response to fat and amino acids, the rat's intake of dietary protein is the major stimulant (LIDDLE et al. 1986). The mechanism is a well-established feedback loop involving protease inhibition of CCK release. Thus, protein substrate, or trypsin inhibitors, inactivate trypsin, which in turn leads to enhanced CCK release (GREEN and LYMAN 1972). A model has been developed in which rats receive full nourishment by an elemental diet that does not stimulate CCK release. With the addition of a small amount of soybean trypsin inhibitor (SBTI), CCK is released without significant changes in the intestinal, nutritional milieu.

In their studies, LIDDLE et al. (1986) found that trypsin inhibition induced a fivefold elevation in circulating CCK with a concomitant, rapid (within 4h), 330% increase in CCK mRNA. Nuclear run-on assays at 4h showed a similar 300% increase in signal in the trypsin-inhibited intestine, indicating that the CCK gene transcriptional rate was increased. As a control, no change in the rate of β-actin transcription was observed. A similar study demonstrated no change in tissue levels of CCK under these conditions in the rat, suggesting a close coupling of secretion with gene expression.

Another example of coupling secretion with gene expression has been demonstrated by two studies measuring CCK secretion and mRNA levels in response to fasting (GREENSTEIN et al. 1990; KANAYAMA and LIDDLE 1991a). In the rat, a twofold fall in duodenal CCK mRNA after 3 days of fasting followed a more rapid decline in circulating CCK levels. Refeeding resulted in a prompt (within 24h) rebound of both mRNA and circulating CCK. The specificity of this response was demonstrated by no change in duodenal SS or β-actin mRNA. Similar to the relationship of SS to gastrin in the antrum, D-cells are found throughout the GI tract, and SS inhibits the release of CCK (SCHLEGEL et al. 1977). In the rat SBTI-CCK release model, exogenous SS was able to partially inhibit the CCK release and block the increase in mRNA (KANAYAMA and LIDDLE 1990). Thus, the induction of secretion and gene expression is mediated, at least in part, by mechanisms responsive to inhibition by SS. CCK release is also modulated by other neurohormonal regulatory peptides, including bombesin. KANAYAMA and LIDDLE (1991b) confirmed an elevation in circulating CCK in response to bombesin in rats, which peaked at 1 h with a return to basal levels at 4 and 24 h in spite of continuous infusion. However, unlike dietary manipulation, bombesin had no effect on CCK mRNA demonstrating an uncoupling of hormone secretion and gene expression. One possible mechanism includes differences in the signaling mechanisms employed by a luminal factor versus a neuropeptide and the different plasma membrane domains with which these factors interact. Alternatively, prolonged stimulation of CCK release may be needed before transcription is activated.

7. Cholecystokinin: Promoter Analysis

Similar to gastrin, CCK is encoded by a single gene with an identical transcriptional start in all cell types examined. DESCHENES et al. (1985) cloned the rat gene and have provided the most detailed analysis of its promoter function. Employing 800 bp of 5′ regulatory DNA elements used to transcribe the CAT gene (CCK CAT), a high level of enzyme activity was observed in a wide variety of cells (Chinese hamster ovary, HeLa, MCF7) which do not normally express CCK. Thus, in a fusion gene plasmid construct, these DNA elements do not confer tissue specificity.

Deletional analysis of this promoter construct resulted in a complex pattern of transcriptional activity (HAUN and DIXON 1990). Elimination of sequences from −780 to −331 resulted in a sixfold increase in CAT activity. Basal levels of expression (compared to full-length CCK CAT) were obtained with further deletions to −155. While these results suggest the presence of both positive and negative elements, this region was not further studied. Instead, the "basal" transcriptional activity from −155 was examined in detail. Further deletion of sequences to −119 resulted in an 11-fold increase in CAT activity which was preserved until a precipitous drop is observed with removal of the sequence between −102 and −81. This sequence (−102 to −81) was further tested extensively and found to have the properties of an enhancer element, activating a heterologous promoter at a distance in either orientation. DNase protection and gel mobility shift assays confirmed nuclear protein binding to this region. Contained in the −102 to −81 segment is a sequence of 11 nucleotides (CTGCGTCAGCA) identical to a region of the c-*fos* gene which is a target for binding of the transcription factor dimers, *fos*/*jun*. In the c-*fos* gene, this element acts to repress basal transcription unless it is displaced from its surrounding elements when it activates the same gene (SHAW et al. 1989). This provides an illustration of the importance of the adjacent DNA sequences in the determination of the activity of a control element.

In the CCK promoter, on both sides of this element, are degenerate sequences (CACGTG and CAGATG) which also specifically bind nuclear proteins. By cross-competition in gel mobility shift assays, these sequences bind the same or very similar factors. It is likely that it is the interaction of these promoter-specific factors with protein binding to the *fos*-like enhancer element that is ultimately responsible for modulation of promoter activity (HAUN and DIXON 1990).

Yet to be characterized are the location of the elements of the CCK gene responsible for tissue specificity and the mechanism of the apparent repression by the upstream sequences in the CCK CAT construct. Several cell lines derived from rat medullary thyroid carcinomas express, process, and secrete CCK (HAUN et al. 1989; ODUM and REHFELD 1990). CCK CAT has been demonstrated to be active in one of the cell lines, but detailed studies of the CCK promoter have not been published from a CCK-expressing cell line.

III. Somatostatin

SS is widely expressed in several CNS loci, pancreatic islets, thyroid C cells and throughout the GI tract. In the gut, SS is a paradigm of a paracrine mediator that exerts a consistently inhibitory action on physiological events, primarily by inhibiting the release of other hormones. The ability to quantify specific mRNA has provided an important tool in studies of SS regulation because circulating peptide levels may not reflect the activity of the paracrine-delivered messenger.

1. Somatostatin Gene Regulation

Published studies summarized above have indicated that SS exerts some of its effects on hormones at the level of gene expression and that SS gene activity is responsive to luminal events. The reciprocal decrease in SS antral mRNA and increase in gastrin expression in response to achlorhydria is compatible with a direct effect of luminal acid on SS gene expression which, in turn, regulates gastrin gene transcription via paracrine mechanisms (BRAND and STONE 1988). Similarly, duodenal SS mRNA in the rat is increased in response to luminal protease inhibition, and exogenous SS blocks the CCK and SS mRNA induction in this model (KANAYAMA and LIDDLE 1990). This capacity for autoregulation of SS gene expression is consistent with findings that SS inhibits its own release in vitro from isolated canine D cells (PARK et al. 1989). SS is also controlled by other neurohormonal regulators. KANAYAMA and LIDDLE (1991b) demonstrated a transient 1.5-fold induction of duodenal SS mRNA levels in rats infused with bombesin.

Because it is expressed at many different sites, SS provides an interesting model to study tissue-specific regulation. PAPACHISTOU et al. (1989) found that in a streptozotocin-induced, diabetic rat model, SS mRNA increased in pancreatic islets and stomach and was partially reversed by insulin. No change in mRNA was noted at other sites of expression such as jejunum, CNS, and thyroid. Dissociation of small intestinal and antral SS gene expression has also been observed in response to fasting in rats. WU et al. (1990a) measured significant increases in antral SS mRNA, while

Kanayama and Liddle (1991a) showed no change in duodenal SS expression in response to fasting.

2. Model of cAMP Regulation

The rat SS gene was one of the first neuropeptide genes to be cloned and characterized and has been studied extensively. This is due, in part, to the original observation by Montminy et al. (1986) that SS expression is regulated by cAMP. Fusion gene constructs containing deletions of the SS regulatory 5′ elements aligned to transcribe CAT identified a sequence from −29 to −60 bp upstream from the transcriptional start site that confers cAMP responsiveness, when cloned onto a heterologous, unresponsive promoter. Further deletional and site-directed mutation studies identified the cAMP response element (CRE) as an 8-bp palindrome (TGACGTCA) which is conserved in other cAMP-regulated genes (including VIP and proenkephalin) (Andrisani and Dixon 1990; Goodman et al. 1990).

Transfection of the SS CAT fusion gene into protein kinase A (PKA) deficient cell lines abolished its responsiveness to cAMP (Montminy et al. 1986). Two other approaches confirmed the necessity of PKA activity in CRE induction of gene expression by overexpressing or microinjecting PKA inhibitors (inactivates CRE) or the catalytic subunit of PKA (recovery of cAMP induction). This suggested that the SS CRE is the target of a nuclear PKA substrate (Grove et al. 1987; Riabowol et al. 1988).

The high affinity and specific interaction of the CRE for its DNA protein was exploited in the purification and eventual cloning of the CRE binding protein, or CREB (Gonzalez et al. 1989). A 43-kD protein, it shares the structural characteristics of the basic, Leu zipper family with a basic DNA binding domain and a region with regularly repeating Leu residues for dimerization, a step necessary for specific DNA binding. The amino (N-) terminal region contains consensus sequences for phosphorylation by PKA as well as a number of other kinases, including PKC and calmodulin-dependent kinases (Gonzalez and Montminy 1989). Site-directed mutations in the PKA consensus sequence abolished the transcriptional activity of the CREB mutants, indicating that phosphorylation is required to render CREB transcriptionally active (Montminy et al. 1990). In addition, some experimental evidence supports a role for PKC in enhancing dimer formation and DNA binding (Yamamoto et al. 1988). Finally, the same serine residue phosphorylated by PKA is also a substrate for calmodulin-dependent kinases. Thus, CREB is also responsive to changes in intracellular calcium and may function as an integration point for multiple signaling pathways (Sheng et al. 1991). The N-terminal region is also very acidic (primarily glutamic acid), a well-established motif important in *trans*-activation. Phosphorylation may act by altering the secondary structure of this domain and allow it to interact with other components of the transcriptional machinery.

The CRE is important in mediating basal SS gene expression, its induction by cAMP and, in at least one cell line (rat medullary thyroid carcinoma, CA-77), its tissue-specific enhanced expression (ANDRISANI et al. 1987; DIXON et al. 1990). In other genes, the CRE is required but not sufficient for tissue specificity (DELEGEANE et al. 1987). By contrast, the glucagon gene contains a CRE, but the gene is not responsive to cAMP in pancreatic A cells (PHILIPPE et al. 1988). The issue of how a single 8-bp element can mediate this degree of regulatory versatility has not been answered but likely resides in the heterogeneity of the surrounding DNA sequences and the proteins they bind. DEUTSCH et al. (1988) performed a detailed structure/function analysis of the CRE from a number of genes. While the 8-bp element is absolutely conserved, the adjacent sequences diverge. The activity of the palindrome in a gene is heavily dependent on the adjacent DNA "context," which does not appear to follow any simple sequence rules.

Furthermore, recent studies support the existence of multiple CREB-related factors, including the factor ATF-1 (also responsive to phosphorylation by PKA) and the recently cloned CREB-BP1 (REHFUSS et al. 1991; MAEKAWA et al. 1988). The AP-1 element, binding site for the *jun/fos* transcription factors, differs in only a single base from the CRE. *Jun/fos* dimers will also bind, albeit less efficiently, to the CRE. Responsiveness to both cAMP and phorbol esters has been demonstrated from the same CRE in the VIP gene (FINK et al. 1991). The possibility of competition for homodimerization (CREB-CREB) vs heterodimerization (CREB-*jun*) would dramatically increase the combinations for regulation. These dimers would likely dramatically differ in their affinity for a given CRE, transcriptional activity, and response to phosphorylation. While complicating the analysis of CRE activity, this combinatorial hypothesis may provide the level of complexity and specificity necessary to explain the diverse activity of this single element which binds ubiquitous regulatory proteins.

3. Tissue Specific Expression

It remains unknown whether the elements and binding factors responsible for tissue-specific expression of the SS gene at its various sites are similar. Taking advantage of transformed rat islet cell lines preferentially expressing SS, POWERS et al. (1989) found specificity was directed in part by the sequences surrounding the CRE. For example, substituting the nine bases immediately 3′ from the glucagon gene CRE into the same position of SS-CAT construct dramatically reduced expression. The importance of the CRE in activated SS gene transcription is in agreement with findings in thyroid C cell (ANDRISANI et al. 1987). However, the basal transcription of the truncated SS-CAT constructs was considerable in non-SS-expressing islet cells. This was dramatically reduced by including segments of the rat SS gene from −65 to −250, suggesting that the SS gene may be under negative control by repressors binding to this region (Powers et al. 1989). To what degree these mechanisms are conserved in the regulation of SS expression in intestinal D cells is unknown. Results from studies of the gastrin and glucagon genes suggest that different *cis*-acting elements will be important at different sites.

IV. Glucagon

Another gene with diverse but highly tissue specific expression is the proglucagon gene, transcribed in the pancreatic islet A cells, endocrine L cells of the gut, and specific CNS nuclei. The proglucagon gene encodes for multiple, potential biologically active peptides which result from tissue-specific post-translational processing events (see Chap. 3).

Employing short-term fetal rat intestinal cell cultures (FRIC), Drucker and Brubaker (1989) demonstrated that transcription of the rat glucagon gene in intestine generates the same mRNA transcript as pancreatic islets. This confirms the importance of post-translational processing and the generation of different peptide products from the proglucagon precursor rather than changes in the processing of the mRNA. The FRIC system stores and secretes glucagon peptides identically to normal intact adult and fetal ileum. In addition, glucagon, SS, peptide YY (PYY), and CCK mRNAs from FRIC are identical in size to normal rat intestinal transcripts.

Previous studies by Philippe et al. (1987) had shown a lack of responsiveness of the glucagon gene to cAMP in islet cells despite the presence of a CRE. Signaling by phorbol esters seems to predominantly regulate glucagon gene expression in islet cells. In FRIC, regulation of glucagon gene expression, as assessed by quantitative Northern blot analysis, was quite the opposite. Phorbol esters resulted in no change in glucagon mRNA levels, while several agents exerting their effects via cAMP-dependent pathways (i.e., dibutryl cAMP, forskolin, cholera toxin) induced two- to threefold increases in glucagon gene expression. While this study did not address how much enhanced transcriptional activity versus stabilization of mRNA contributed to the observed changes, the results provide strong evidence for the importance of the cAMP pathway in intestinal glucagon gene expression and provide another example of distinct, tissue-specific responsiveness to the same regulatory element.

Further evidence that the DNA sequence responsible for directing glucagon gene expression in intestine differ from those at other targets come from transgenic mouse studies by Efrat et al. (1988). A construct containing 850 bp rat glucagon 5′ regulating elements aligned to transcribe the simian virus 40 large tumor antigen was expressed in brain and pancreas but not in intestine. This suggested that intestinal expression required additional regulatory genetic elements for expression at this site.

V. Secretin

Kopin et al. (1990) have recently described the cloning and sequencing of the cDNAs for rat and porcine secretin. Two unusual features were noted. First, preprosecretin contained an unusually long (72 amino acids) C-terminal peptide. Poor conservation between the two species (39%) in this region compared with the greater than 95% homology in the mature secretin sequence casts doubt on the physiological relevance of this flanking peptide. Secondly, the tissue distribution of secretin expression revealed higher levels of mRNA in ileum than duodenum. The physiological significance of high secretin expression in the rat ileum as well as low level expression in the colon and CNS awaits further study.

These same investigators (Kopin et al. 1991) also cloned the rat secretin gene. As shown in Fig. 2, it is the only member of the secretin/glucagon family that does not have an intron separating the transcriptional and translational start sites. Shorter secretin transcripts isolated from intestinal RNA were identical to full-length transcripts except for the exclusion of the segment corresponding to exon 3. This exon codes for a portion of the long C-terminal flanking peptide. The difference in the sizes of the mRNA transcript likely occurs by differential splicing of a single preprosecretin RNA precursor. The biological significance of this regulatory event is unknown. The abundance of secretin mRNA was also quantitated in fetal, neonatal, and adult rat intestine by Northern blot analysis. Secretin transcript levels peaked on day 20 in the fetus and gradually fell to adult levels by day 10 postpartum. The appearance of secretin mRNA on day 17 of gestation antedates the onset of gastric acid secretion by at least several days. Thus, developmental regulation of secretin expression is independent of its major release stimulant in adult animals. As noted above, this period of enhanced fetal secretin expression occurs during a period important in pancreatic growth and differentiation, inviting speculation for a potentiation of the growth-promoting effects by secretin of the CCK/gastrin peptide family during development.

VI. Motilin

Motilin is a 22 amino acid peptide found in highest concentration in the proximal small bowel. It undergoes regular, cyclical release and has been proposed to be involved in the initiation of interdigestive gut motor activity (see Chap. 9). Daikh et al. (1989) have cloned the human motilin gene, which has an unusual structure because the sequence for the mature secreted form of motilin is split between exons 2 and 3 (see Fig. 2). This may suggest functional differences in these regions of the peptide. Motilin expression was identified in the small intestine and colon. The significance of widespread expression in non-GI tissue is not known.

G. Conclusion

The past decade has witnessed remarkable advances in the understanding of the basic mechanisms of gene expression. The cornerstone of transcriptional activity regulation is the specific interaction of transcription factors with control DNA elements, simplistically outlined in cartoon form in this chapter. However, the growing appreciation of the importance of the interplay of protein/protein interactions of regulatory transcription factors with the transcriptional apparatus has added considerable complexity and challenge to understanding how a cell regulates a given gene. In the GI endocrine system, regulation of regulatory peptides is the key to integrating responses of the digestive tract to a meal. The successful application of molecular biological techniques to the study of GI hormones has shown that regulation of transcriptional activity is important in this regulation and that luminal events can influence hormone gene activity. Furthermore, the DNA elements and transcription factors mediating the exquisite tissue specificity of expression of these regulatory peptides are likely to be different from site to site (CNS vs GI endocrine cells). Determination of these tissue specificity and regulatory mechanisms in the GI endocrine cell awaits novel and creative methods to isolate, maintain, and eventually propagate these cells in vitro.

References

Akerblom IW, Slater EP, Beato M, Baxter JD, Mellon PL (1988) Negative regulation by glucocorticoids through interference with a cAMP responsive enhancer. Science 241:350–353

Alam J, Cook JL (1990) Reporter genes: application to the study of mammalian gene transcription. Anal Biochem 188:245–254

Andrisani OM, Dixon JE (1990) Somatostatin gene regulation. Annu Rev Physiol 52:793–806

Andrisani OM, Zhu Z, Pot DA, Dixon JE (1989) In vitro transcription directed from the somatostatin promoter is dependent upon a purified 43 kDa DNA-binding protein. Proc Natl Acad Sci USA 86:2181–2185

Andrisani OM, Hayes TE, Roos B, Dixon JE (1987) Identification of the promoter sequences involved in the cell specific expression of the rat somatostatin gene. Nucleic Acids Res 15:5715–5728

Baichwal VR, Tjian R (1990) Control of c-*Jun* activity by interaction of a cell-specific inhibitor with regulatory domain δ: differences between v- and c-*Jun*. Cell 272:815–825

Beato M (1989) Gene regulation by steroid hormones. Cell 56:335–344

Beato M (1991) Transcriptional control by nuclear receptors. FASEB J 5:2044–2051

Beauchamp RD, Bernard JA, McCutchen CM, Cherner JA, Coffey RJ Jr (1989) Localization of transforming growth factor α and its receptor in gastric mucosal cells. J Clin Invest 84:1017–1023

Benoist C, O'Hare O, Breathnach R, Chambon P (1980) The ovalbumin gene – sequence of putative control regions. Nucleic Acids Res 8:127–142

Bentley DL, Groudine M (1986) A block to elongation is largely responsible for decreased transcription of c-*myc* in differentiated HL60 cells. Nature 321:702–706

Berger SL, Cress WD, Cress A, Triezenberg SJ, Guarente L (1990) Selective inhibition of activated but not basal transcription by the acidic activation domain of VP16: evidence for transcriptional adaptors. Cell 61:1199–1208

Bienz M, Pelham HRB (1986) Heat shock regulatory elements function as an inducible enhancer in the *Xenopus* hsp 70 gene and when linked to a heterogenous promoter. Cell 45:753–760

Blackwood EM, Eisenman RN (1991) Max: a helix-loop-helix zipper protein that forms a sequence-specific DNA-binding complex with *myc*. Science 251:1211–1217

Blake C (1983) Exons – present from the beginning? Nature 306:535–537

Bloom SR, Mortimer CH, Thorner MD (1974) Inhibition of gastrin and gastric acid secretion by growth hormone release inhibiting hormone. Lancet 2:1106–1109

Bohmann D, Bos T, Admon A, Nishimura T, Vogt P, Tjian R (1987) Human proto-oncogene c-*jun* encodes a DNA binding protein with structural and functional properties of transcription factor AP-1. Science 238:1386–1392

Bondy CA, Whitnall MH, Brady LS (1989) Regulation of carboxypeptidase gene expression in magnocellular neurons: response to osmotic stimulation. Mol Endocrinol 3:2086–2092

Brand SJ, Fuller PJ (1988) Differential gastrin gene expression in rat gastrointestinal tract and pancreas during neonatal development. J Biol Chem 263:5341–5347

Brand SJ, Stone D (1988) Reciprocal regulation of antral gastrin and somatostatin gene expression by omeprazole-induced achlorhydria. J Clin Invest 82:1059–1066

Brand SJ, Wang TC (1988) Gastrin gene expression and regulation in rat islet cell lines. J Biol Chem 263:16597–16603

Brand SJ, Andersen BM, Rehfeld JF (1984) Complete tyrosine-*O*-sulphation of gastrin in neonatal rat pancreas. Nature 309:456–458

Breathnach R, Chambon P (1981) Organization and expression of eucaryotic split genes coding for proteins. Annu Rev Biochem 50:349–383

Chiu R, Angel P, Karin M (1989) *Jun*-B differs in its biological properties from, and is a negative regulator of, c-*Jun*. Cell 59:979–986

Chodosh LA, Olesen J, Hahn S, Baldwin AS, Guarente L, Sharp PA (1988) A yeast and a human CCAAT-binding protein have heterologous subunits that are functionally interchangeable. Cell 53:25–35

Christy B, Nathans D (1988) A gene activated in mouse 3T3 cells by serum growth factors encodes a protein with zinc finger sequences. Proc Natl Acad Sci USA 85:7857–7861

Conaway JW, Conaway RC (1991) Initiation of eukaryotic messenger RNA synthesis. J Biol Chem 266:17721–17724

Curran T, Franza BR Jr (1988) *Fos* and *Jun*: the AP-1 connection. Cell 55:395–397

Daikh DI, Douglass JO, Adelman JP (1989) Structure and expression of the human motilin gene. DNA 8:615–621

Davis RL, Weintraub H, Lassar AB (1987) Expression of a single transfected cDNA converts fibroblasts to myoblasts. Cell 51:987–1000

Delegeane AM, Ferland LH, Mellon PL (1987) Tissue-specific enhancer of the human glycoprotein hormone α-subunit gene: dependence on cyclic AMP-inducible elements. Mol Cell Biol 7:3994–4002

Deschenes RJ, Haun RS, Funckes CL, Dixon JE (1985) A gene encoding rat cholecystokinin. J Biol Chem 260:1280–1286

Deutsch PJ, Hoeffler JP, Jameson JL, Lin JC, Habener JF (1988) Structural determinants for transcriptional activation by cAMP-responsive DNA elements. J Biol Chem 263:18466–18472

Diamond MI, Miner JN, Yoshinaga SK, Yamamoto KR (1990) Transcription factor interactions: selectors of positive or negative regulation from a single DNA element. Science 249:1266–1272

Dimaline R, Lee CM (1990) Chicken gastrin: a member of the gastrin/CCK family with novel structure-activity relationships. Am J Physiol 259:G882–G888

Dixon JE, Zhu A, Andrisani O (1990) Efforts directed at understanding cell-specific somatostatin gene expression. Metabolism 39:17–19

Drucker DJ, Brubaker PL (1989) Proglucagon gene expression is regulated by a cyclic AMP-dependent pathway in rat intestine. Proc Natl Acad Sci USA 86:3953–3957

Dynan WS, Sazer S, Tjian R, Schimke RT (1986) Transcription factor Sp1 recognizes a DNA sequence in the mouse dihydrofolate reductase promoter. Nature 319:246–248

Dynlacht BD, Hoey T, Tjian R (1991) Isolation of coactivators associated with the TATA-binding protein that mediate transcriptional activation. Cell 66:563–576

Efrat S, Teitelman G, Anwar M, Ruggiero D, Hanahan D (1988) Glucagon gene regulatory region directs oncoprotein expression to neurons and pancreatic α cells. Neuron 1:605–613

Fink JS, Verhave M, Walton K, Mandel G, Goodman RH (1991) Cyclic AMP- and phorbol ester-induced transcriptional activation are mediated by the same enhancer element in the human vasoactive intestinal peptide gene. J Biol Chem 266:3882–3887

Fisch TM, Prywes R, Roeder RG (1989) An AP1-binding site in the c-*fos* gene can mediate induction by epidermal growth factor and 12-*O*-tetradecanoyl phorbol-13-acetate. Mol Cell Biol 9:1327–1331

Friedman J, Schneider BS, Powell D (1985) Differential expression of the mouse cholecystokinin gene during brain and gut development. Proc Natl Acad Sci 82:5593–5597

Fuller PJ, Stone DL, Brand SJ (1987) Molecular cloning and sequencing of a rat preprogastrin complementary deoxyribonucleic acid. Mol Endocrinol 1:216–223

Galas D, Schmitz A (1978) DNase footprinting: a simple method for the detection of protein-DNA binding specificity. Nucleic Acids Res 5:3157–3170

Gibson SJ, Polak JM (1990) Principles and applications of complementary RNA probes. In: Polak JM, McGee J (eds) In situ hybridization: principles and practice. Oxford University Press, New York, pp 81–94

Godley JM, Brand SJ (1989) Regulation of the gastrin promoter by epidermal growth factor and neuropeptides. Proc Natl Acad Sci USA 86:3036–3040

Gonzalez GA, Montminy MR (1989) Cyclic AMP stimulates somatostatin gene transcription by phosphorylation of CREB at serine 133. Cell 59:675–680

Gonzalez GA, Yamamoto KK, Fischer WH et al. (1989) A cluster of phosphorylation sites on the cyclic AMP-regulated nuclear factor CREB predicted by its sequence. Nature 337:749–752

Goodman RH, Rehfuss RP, Verhave M, Ventimiglia R, Low MJ (1990) Gene regulation, biosynthesis, and processing. I. Gene regulation. Somatostatin gene regulation: an overview. Metabolism 39:2–5

Green GM, Lyman RL (1972) Feedback regulation of pancreatic enzyme secretion as a mechanism for trypsin inhibitor-induced hypersecretion in rats. Proc Soc Exp Biol Med 140:6–12

Greenblatt J (1991) Roles of TFIID in transcriptional initiation by RNA polymerase II. Cell 66:1067–1070

Greenstein RJ, Isola L, Gordon J (1990) Differential cholecystokinin gene expression in brain and gut of the fasted rat. Am J Med Sci 299:32–37

Grove JR, Price DJ, Goodman HM et al. (1987) Recombinant fragment of protein kinase inhibitor blocks cyclic AMP-dependent gene transcription. Science 238: 530–533

Ha I, Lane WS, Reinberg D (1991) Cloning of a human gene encoding the general transcription initiation factor IIB. Nature 352:689–695

Harrison SC (1991) A structural taxonomy of DNA-binding domains. Nature 353: 715–719

Haun RS, Dixon JE (1990) A transcriptional enhancer essential for the expression of the rat cholecystokinin gene contains a sequence identical to the −296 element of the human c-*fos* gene. J Biol Chem 265:15455–15463

Haun RS, Beinfeld MC, Roos BA, Dixon JE (1989) Establishment of a cholecystokinin-producing rat medullary thyroid carcinoma cell line. Endocrinology 125:850–856

Heinrich G, Gros P, Habener JF (1984) Glucagon gene sequence: four of six exons encode separate functional domains of rat pre-proglucagon. J Biol Chem 259: 14082–14087

Hoeffler JP, Meyer TE, Yun Y, Jameson LJ, Habener JF (1988) Cyclic AMP-responsive DNA-binding protein: structure based on a cloned placental cDNA. Science 242:1430–1433

Hoffmann AE, Sinn E, Yamamoto T, Wang J, Roy A, Horikoshi M, Roeder RG (1990) Highly conserved core domain and unique N terminus with presumptive regulatory motifs in a human TATA factor (TFIID). Nature 346:387–390

Imagawa M, Chiu R, Karin M (1987) Transcription factor AP-2 mediates induction by two different signal-transduction pathways: protein kinase C and cAMP. Cell 51:251–260

Ishii S, Merlino GT, Pastan I (1985a) Promoter region of the human Harvey *ras* proto-oncogene: similarity to the EGF receptor proto-oncogene promoter. Science 230:1378–1380

Ishii S, Xuy Y, Stratton RH, Roe BA, Merlino GT, Pastan I (1985b) Characterization and sequence of the promoter region of the human epidermal growth factor receptor gene. Proc Natl Acad Sci USA 82:4920–4924

Johnsen AH, Rehfeld JF (1990) Cionin: a disulfotyrosyl hybrid of cholecystokinin and gastrin from the neural ganglion of the protochordate *Ciona intestinalis*. J Biol Chem 265:3054–3058

Johnson PF, McKnight SL (1989) Eukaryotic transcriptional regulatory proteins. Annu Rev Biochem 58:799–839

Johnson PF, Landschulz WH, Graves BJ, McKnight SL (1987) Identification of a rat liver nuclear protein that binds to the enhancer core element of three animal viruses. Genes Dev 1:133–146

Jones N (1990) Transcriptional regulation by dimerization: two sides to an incestuous relationship. Cell 61:9–11

Jung Y-K, Kuncz CJ, Pearson RK, Dixon JE, Fricker LD (1991) Structural characterization of the rat carboxypeptidase-E gene. Mol Endocrinol 5:1257–1268

Kadonaga JT, Carner KR, Masiarz FR, Tjian R (1987) Isolation of cDNA encoding transcription factor Sp1 and functional analysis of the DNA binding domain. Cell 51:1079–1090

Kanayama S, Liddle RA (1990) Somatostatin regulates duodenal cholecystokinin and somatostatin messenger RNA. Am J Physiol 258:G358–G364

Kanayama S, Liddle RA (1991a) Influence of food deprivation on intestinal cholecystokinin and somatostatin. Gastroenterology 100:909–915

Kanayama S, Liddle RA (1991b) Regulation of intestinal cholecystokinin and somatostatin mRNA by bombesin in rats. Am J Physiol 261:G71–G77

Kao CC, Lieberman PM, Schmidt MC, Zhou Q, Pei R, Berk AJ (1990) Cloning of a transcriptionally active human TATA binding factor. Science 248:1646–1649

Kopin AS, Wheeler MB, Leiter AB (1990) Secretin: structure of the precursor and tissue distribution of the mRNA. Proc Natl Acad Sci USA 87:2299–2303

Kopin AS, Wheeler MB, Nishitani J, McBride EW, Chang T-M, Chey WY, Leiter AB (1991) The secretin gene: evolutionary history, alternative splicing, and developmental regulation. Proc Natl Acad Sci USA 88:5335–5339

Larsson L-I (1980) Peptide secretory pathways in GI tract: cytochemical contributions to regulatory physiology of the gut. Am J Physiol 239:237–246

Larsson L-I, Rehfeld JF (1977) Evidence for a common evolutionary origin of gastrin and cholecystokinin. Nature 269:335–338

Leiter AB, Montminy MR, Jamieson E, Goodman RH (1985) Exons of the human pancreatic polypeptide gene define functional domains of the precursor. J Biol Chem 260:13013–13017

Levine M, Hoey T (1988) Homeobox proteins as sequence-specific transcription factors. Cell 55:537–540

Levine M, Manley JL (1989) Transcriptional repression of eukaryotic promoters. Cell 59:405–408

Lewin B (1990) Commitment and activation at Pol II promoters: a tail of protein-protein interactions. Cell 61:1161–1164

Liddle RA, Green GM, Conrad CK, Williams JA (1986) Proteins but not amino acids, carbohydrates, or fats stimulate cholecystokinin secretion in the rat. Am J Physiol 251:G243–G248

Liddle RA, Carter JD, McDonald AR (1988) Dietary regulation of rat intestinal cholecystokinin gene expression. J Clin Invest 81:2015–2019

Lin Y-S, Green MR (1991) Mechanism of action of an acidic transcriptional activator in vitro. Cell 64:971–981

Lund T, van Kessel AHMG, Haun S, Dixon JE (1986) The genes for human gastrin and cholecystokinin are located on different chromosomes. Hum Genet 73: 77–80

Maekawa T, Dakura H, Kanei-Ishii C et al. (1988) Leucine zipper structure of the protein CRE-BPI binding to the cyclic AMP response element in brain. EMBO J 8:2023–2028

Maniatis T, Goodbourn S, Fischer JA (1987) Regulation of inducible and tissue-specific gene expression. Science 236:1237–1245

Mastrangelo IA, Courey AJ, Wall JS, Jackson SP, Hough PVC (1991) DNA looping and Sp1 multimer links: a mechanism for transcriptional synergism and enhancement. Proc Natl Acad Sci USA 88:5670–5674

Meisterernst M, Horikoshi M, Roeder RG (1990) Recombinant yeast TFIID, a general transcription factor, mediates activation by the gene-specific factor USF in a chromatin assembly assay. Proc Natl Acad Sci USA 87:9153–9157

Melton DW, McEwan C, McKie AB, Reid AM (1986) Expression of the mouse HPRT gene: deletional analysis of the promoter region of an X-chromosome linked housekeeping gene. Cell 44:319–328

Merchant JL, Demediuk B, Brand SJ (1991) A GC-rich element confers epidermal growth factor responsiveness to transcription from the gastrin promoter. Mol Cell Biol 11:2686–2696

Minth CD, Andrews PC, Dixon JE (1986) Characterization, sequence, and expression of the cloned human neuropeptide Y gene. J Biol Chem 261:11974–11979

Miskimins WK, Roberts MP, McClelland A, Ruddle FH (1985) Use of a protein-blotting procedure and a specific DNA probe to identify nuclear proteins that recognize the promoter region of the transferrin receptor gene. Proc Natl Acad Sci USA 82:6741–6744

Mitchell PJ, Tjian R (1989) Transcriptional regulation in mammalian cells by sequence-specific DNA binding proteins. Science 245:371–378

Montminy MR, Bilezikjian LM (1987) Binding of a nuclear protein to the cyclic-AMP response element of the somatostatin gene. Nature 328:175–178

Montminy MR, Sevarino KA, Wagner JA, Mandel G, Goodman RH (1986) Identification of a cyclic-AMP-responsive element within the rat somatostatin gene. Proc Natl Acad Sci 83:6682–6686

Montminy MR, Gonzalez GA, Yamamoto KK (1990) Gene regulation, biosynthesis, and processing. I. Characteristics of the cAMP response unit. Metabolism 39: 6–12

Moss LG, Moss JB, Rutter WJ (1988) Systematic binding analysis of the insulin gene transcription control region: insulin and immunoglobulin enhancers utilize similar transactivators. Mol Cell Biol 8:2620–2627

Nelson C, Shen L-P, Meister A, Fodor E, Rutter WJ (1990) Pan: a transcriptional regulator that binds chymotrypsin, insulin, and AP-4 enhancer motifs. Genes Dev 4:1035–1043

Norman C, Runswick M, Pollock R, Treisman R (1988) Isolation and properties of cDNA SRF, a transcription factor that binds to the c-*fos* serum response element. Cell 55:989–1003

Odum L, Rehfeld JF (1990) Expression and processing of procholecystokinin in rat medullary thyroid carcinoma cell line. Biochem J 271:31–36

Ofir R, Dwarki VJ, Rashid D, Verma IM (1990) Phosphorylation of the C terminus of *Fos* protein is required for transcriptional transrepression of the c-*fos* promoter. Nature 348:80–82

Papachristou DN, Pham K, Zingg HH et al. (1989) Tissue-specific alterations in somatostatin mRNA accumulation in streptozocin-induced diabetes. Diabetes 38:752–757

Park J, Chiba T, Yokotani K, Delvalle J, Yamada T (1989) Somatostatin receptors on canine fundic D-cells: evidence for autocrine regulation of gastric somatostatin. Am J Physiol 257:G235–G241

Peers B, Monget P, Nalda MA, Voz ML, Berwaer M, Belayew A, Martial JA (1991) Transcriptional induction of the human prolactin gene by cAMP requires two *cis*-acting elements and at least the pituitary-specific factor *pit*-1. J Biol Chem 266:18127–18134

Peterson MG, Tanese N, Pugh BF, Tjian R (1990) Functional domains and upstream activation properties of cloned human TATA binding protein. Science 248:1625–1630

Philippe J, Drucker DJ, Knepel W, Jepeal L, Misulovin Z, Habener JF (1988) Alpha-cell-specific expression of the glucagon gene is conferred to the glucagon promoter element by the interactions of DNA-binding proteins. Mol Cell Biol 8:4877–4888

Powers AC, Tedeschi F, Wright KE, Chan JS, Habener JF (1989) Somatostatin gene expression in pancreatic islet cells is directed by cell-specific DNA control elements and DNA-binding proteins. J Biol Chem 264:10048–10056

Ptashne M, Gann AAF (1990) Activators and targets. Nature 346:329–331

Pugh BF, Tjian R (1990) Mechanism of transcriptional activation by Sp1: evidence for coactivators. Cell 61:1187–1197

Rauscher FJ III, Cohen DR, Curran T, Bos TJ, Vogt PK, Bohmann D, Tjian R, Franza BR Jr (1988a) *Fos*-associated protein p39 is the product of the *jun* proto-oncogene. Science 240:1010–1016

Rauscher FJ III, Voulalas PJ, Franza BR Jr, Curran T (1988b) *Fos* and *Jun* bind cooperatively to the AP-1 site: reconstitution in vitro. Genes Dev 2:1687–2699

Rehfuss RP, Walton KM, Loriaux MM, Goodman RH (1991) The cAMP-regulated enhancer-binding protein ATF-1 activates transcription in response to cAMP-dependent protein kinase A. J Biol Chem 266:18431–18434

Renkawitz R (1990) Transcriptional repression in eukaryotes. Trends Genet 6:192–197

Riabowol KT, Fink JS, Gilman MZ et al. (1988) The catalytic subunit of cAMP-dependent protein kinase induces expression of genes containing cAMP-responsive enhancer elements. Nature 238:83–86

Rodriguez C, Brayton KA, Brownstein M, Dixon JE (1989) Rat preprocarboxy-peptidase H: cloning, characterization, and sequence of the cDNA and regulation of the mRNA by corticotropin-releasing factor. J Biol Chem 264:5988–5995

Saltzman AG, Weinmann R (1989) Promoter specificity and modulation of RNA polymerase II transcription. FASEB J 3:1723–1733

Sawadogo M, Sentenac A (1990) RNA polymerase B (II) and general transcription factors. Annu Rev Biochem 59:711–754

Schlegel WS, Raptis RF, Harvey JM, Pfeiffer EF (1977) Inhibition of cholecystokinin-pancreozymin release by somatostatin. Lancet 2:166–168

Shaw PE, Frasch S, Nordheim Å (1989) Repression of c-*fos* transcription is mediated through $p67^{SRF}$ bound to the SRE. EMBO J 8:2567–2574

Sheng M, Thompson MA, Greenberg ME (1991) CREB: a Ca^{2+}-regulated transcription factor phosphorylated by calmodulin-dependent kinases. Science 252: 1427–1430

Singh H, LeBowitz JH, Baldwin AS Jr, Sharp PA (1988) Molecular cloning of an enhancer binding protein: isolation by screening of an expression library with a recognition site DNA. Cell 52:415–423

Soll AH, Yamada T, Park J, Thomas LP (1984) Release of somatostatin-like immunoreactivity from canine fundic mucosal cells in primary culture. Am J Physiol 247:G558–G566

Tavianini MA, Hayes TE, Magazin MD, Minth CD, Dixon JE (1984) Isolation, characterization, and DNA sequence of the rat somatostatin gene. J Biol Chem 259:11798–11803

Theil EC (1990) Regulation of ferritin and transferrin receptor mRNAs. J Biol Chem 265:4771–4774

Tsukada T, Horovitch SJ, Montminy MR, Mandel G, Goodman RH (1985) Structure of the human vasoactive intestinal polypeptide gene. DNA 4:293–300

Vinson CR, LaMarco KL, Johnson PF, Landschulz WH, McKnight SL (1988) In situ detection of sequence-specific DNA binding activity specified by a recombinant bacteriophage. Genes Dev 2:801–806

Vitale M, Vashishtha A, Linzer E, Powell DJ, Friedman JM (1990) Molecular cloning of the mouse CCK gene: expression in different brain regions and during cortical development. Nucleic Acids Res 19:169–177

Walsh JH, Grossman MI (1975) Gastrin. N Engl J Med 292:1324–1332

Wang TC, Brand SJ (1990) Islet cell-specific regulatory domain in the gastrin promoter contains adjacent positive and negative DNA elements. J Biol Chem 265:8908–8914

Wiborg O, Berglund L, Boel E, Norris F, Norris K, Rehfeld JF, Marcker KA, Vuust J (1984) Structure of a human gastrin gene. Proc Natl Acad Sci USA 81:1067–1069

Wu SV, Sumii K, Tari A, Mogard M, Walsh JH (1990a) Regulation of gastric somatostatin gene expression. Metabolism 39:125–130

Wu SV, Giraud A, Mogard M, Sumii K, Walsh JH (1990b) Effects of inhibition of gastric secretion on antral gastrin and somatostatin gene expression in rats. Am J Physiol 258:G788–G793

Yamamoto KK, Gonzalez GA, Biggs WH III et al. (1988) Phosphorylation-induced binding and transcriptional efficacy of nuclear factor CREB. Nature 334:494–498

Yoo OJ, Powell CT, Agarwal KL (1982) Molecular cloning and nucleotide sequence of full-length cDNA coding for porcine gastrin. Proc Natl Acad Sci USA 79:1049–1053

Yoshimura K, Nakamura H, Trapnell BC, Dalemans W, Pavirani A, Lecocq J-P, Crystal RG (1991) The cystic fibrosis gene has a "housekeeping"-type promoter and is expressed at low levels in cells of epithelial origin. J Biol Chem 266:9140–9144

CHAPTER 3

Post-Translational Processing of Regulatory Peptides

R.E. CARRAWAY and Y.P. LOH

A. Introduction

Many biologically active peptides are biosynthesized as parts of precursor proteins which undergo multiple post-translational processing steps, including "limited proteolysis" to excise specific sequences as well as modifications such as amidation, acetylation, pyroglutamate formation, sulfation, phosphorylation and glycosylation to produce the final product(s).

The primary intent of this review is to identify common features in the design of these regulatory peptide systems in hopes of simplifying the information and stimulating the reader to consider new concepts in this field. It is also our aim to present a novel perspective which we believe gives the reader a broader, more comprehensive viewpoint than has been presented in the excellent reviews already available (ANDREWS et al. 1989; DARBY and SMYTH 1990; FISHER and SCHELLER 1988; GLUSCHANKOF and COHEN 1987; HARRIS 1989; LOH et al. 1984; SCHWARTZ 1990). Rather than attempt to cover the entire literature for the nearly 100 currently known peptides, we have chosen to focus on selected examples to illustrate our points and hope that those whose work is not covered will be forgiving.

B. General Design of Processing Systems

The various biologic signals employed by animal cells for communication might be broadly classified into those which are stored in "active" form and those which are stored as "inactive" precursors. While the former category includes substances such as the monoamine and amino acid transmitters and the nucleotide phosphates, examples for the latter group are steroids (stored as cholesterol), adenosine-3′,5′-cyclic monophosphate (cAMP), stored as adenosine triphosphate (ATP) and prostaglandins (stored as arachidonic acid esters). It can be seen in Table 1 that a similar classification scheme can be a useful way of organizing information about regulatory peptides.

Regulatory peptides which are primarily found in "active" form (Table 1, class A) can be divided into those which are stored within cellular vesicles and those which are released immediately. Examples of the former category include hormones [e.g., adrenocorticotrophic hormone (ACTH) and insulin], as well as peptide transmitters and modulators (e.g., substance P and

Table 1. General designs for processing systems

A. Found primarily as "active" peptide	B. Found primarily as "inactive" precursor
1. "Active peptide stored in secretory granules	1. Precursor activated in or near the cell of origin
2. "Active" peptide released constitutively	2. Precursor activated in interstitial space or blood system
	3. Precursor activated at the target or by other cells

neurotensin). Processing of these peptides is, for the most part, performed within the cells of origin and they are fully active upon release from storage granules. Constitutively secreted peptides (e.g., interleukins) can also be fully processed when they leave the cell but peptide output is primarily regulated by transcriptional controls.

Peptides which are stored as "inactive" precursors (Table 1, class B) fall into several subcategories, depending upon whether the precursor is activated inside the cell of origin, in the extracellular space or at the target cell. Each of these categories can be further subdivided according to where the precursor is stored. For example, in some cases (e.g., leukotrienes), the precursor is both stored and activated in the same cells, while, in others, activation occurs during release (e.g., atrial natriuretic peptide). Another common arrangement entails storage and activation of precursors in the extracellular compartment (e.g., bradykinin); however, recent studies indicate that some of these precursors are found anchored to cell surfaces [e.g., transforming growth factor α (TGFα)]. Finally, there is the possibility (although the evidence is less clear) that activation of precursors occurs at the target [e.g., epidermal growth factor (EGF)] and in some cases even involves storage of precursor within the target itself (e.g., steroid receptors).

Figure 1 illustrates this simple method of classifying peptide-messenger systems. All precursor proteins are pictured as undergoing the same eventual fate – cleavage to form peptides – as a necessary part of information transfer and the locations for precursor storage and processing vary. It should also be noted that these various arrangements are not mutually exclusive and that some peptides may belong to several groups. A few examples for each of the categories listed in Table 1 are discussed in some detail below.

I. Active Peptide Stored in Secretory Vesicles (Class A1)

Most of the known bioactive peptides fall into this category, whereby signaling is initiated by exocytosis of the fully processed peptide from storage granules. Although steady state levels of precursor are usually <5% of total peptide, there are some exceptions to this rule. Tissue levels of the

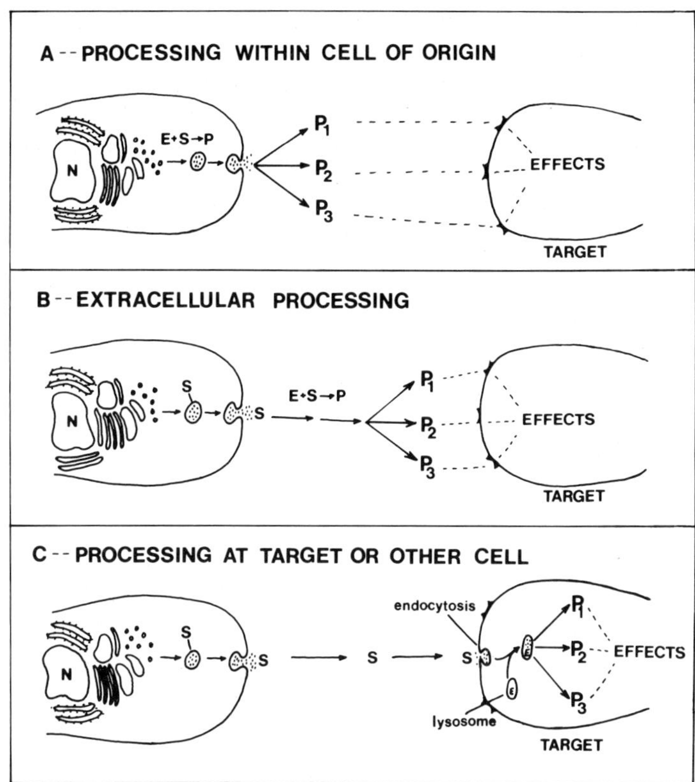

Fig. 1A–C. Diagrammatic representation of three basic designs for procesessing of precursors to bioactive peptides. **A** Some peptides (e.g., neurotensin) appear to be fully processed before secretion. **B** Other peptides (e.g., angiotensin) are formed primarily via extracellular cleavage of the secreted precursors. **C** Although speculative at this point, a third possibility involves processing after endocytosis by the target or other cell. *E*, enzyme; *S*, substrate; *P*, product; *N*, nucleus

neurotensin (NT) precursor, for example, are ~3% of total in intestine, 15%–30% in adrenal and >75% in induced PC-12 cells. Thus, the release rate is the primary determinant of activity for these systems but processing steps might become rate-limiting in some situations.

One of the most extensively studied precursors is proopiomelanocortin (POMC), which is cleaved to yield at least seven different biologically active peptides and is differentially processed in anterior and intermediate lobes of the pituitary and in brain (EIPPER and MAINS 1980). The principal products in the anterior lobe are an N-terminal glycopeptide, ACTH and β-lipotrophin (β-LPH), while in the intermediate lobe further processing gives rise to γ- and α-melanocyte-stimulating hormone (MSH) and β-endorphin (Fig. 4). Several of these peptides are variably N-acetylated and this has profound effects on biologic potency, enhancing activity for ACTH α-MSH and reducing activity for endorphins. The precursor is also glycosylated and phosphorylated to some extent, which may provide a means of regulation.

BLALOCK (1989) has investigated the expression of POMC-derived peptides in leukocytes and found it to differ depending upon the stimulus. While treatment of cells with CRF (corticotrophin releasing factor) stimulated production of $ACTH_{1-39}$ and β-endorphin, endotoxin treatment yielded $ACTH_{1-24}$ and α- or γ-endorphin. Since these products exhibit different biologic activities on the immune system, this appears to be an example whereby the same precursor is utilized to elicit a variety of responses depending upon the stimulus.

Another well-studied system is the *Aplysia* egg-laying hormone (ELH) complex, which is an array of peptides regulating this stereotypic behavior in marine snails. Elegant studies concerning the processing of the precursor and the fates of its nine peptide products support the absolutely novel concept that different portions of the same precursor can be packaged in separate vesicles and routed to different cellular locations (FISHER et al. 1988). Thus, the 271-amino acid ELH precursor undergoes an initial cleavage within the Golgi apparatus of bag cell neurons to yield two fragments which are then sorted into separate vesicle classes (Fig. 4). Product peptides from the larger, 18-kDa fragment were localized to the cell soma, while those derived from the 8-kDa piece were transported to the nerve endings. Furthermore, the kinetics of processing and the steady-state levels of the peptides generated from the two fragments differed greatly. These exciting findings suggest that portions of the same precursor can be exposed to different processing enzymes, depending upon how they are sorted and then they can be targeted for different needs, e.g., exocrine vs. endocrine vs. paracrine release sites.

II. Active Peptide Released Constitutively (Class A2)

Some bioactive peptides do not accumulate within the producing cells and we think of their secretion as being constitutive or nonregulated, such as occurs for albumin. The interleukins and interferons may be examples of this concept since some of them are not packaged in dense secretory granules and their release into culture media is relatively rapid and not known to be effected by secretagogues (WHITACRE 1990). Interleukin (IL)-1β, an inflammatory mediator produced by activated monocytes, is synthesized as a 33-kDa precursor which upon secretion undergoes processing to yield the 17-kDa bioactive form. The actual pathway for transport of IL-1β to the cell surface is as yet undefined, since the cDNA for its precursor does not have a normal signal sequence and since its secretion is not blocked by drugs which block the normal endoplasmic reticulum (ER)/Golgi route.

Although secretory cells are classically divided into those exhibiting regulated secretion and those which release products constitutively, some hormone-producing cells have been shown to utilize both modes (MOORE

1987). In AtT-20 cells, for example, two classes of secretory proteins could be distinguished. ACTH and four other proteins were stored in granules and released in response to cAMP, while another set of proteins was secreted in a nonregulated, rapid manner rather than being stored. Constitutive release of ACTH was increased when chloroquine was used to block the regulated pathway; however, in this case, processing of the precursor did not take place (MOORE et al. 1983). Thus, it may be possible to regulate the extent of processing simply by shifting the secretory protein from one pathway to another.

III. Precursor Activated in or Near Cell of Origin (Class B1)

1. Precursor Stored Intracellularly

Atrial natriuretic peptide (ANP), a cardiac hormone affecting blood pressure and electrolyte balance, is stored in atrial granules as the 126-amino acid prohormone (THIBAULT et al. 1987). The bioactive species, a 28-residue peptide derived from the C-terminal region of the precursor, appears to be generated during secretion with atrial stretch as the stimulus. Processing occurs extracellularly via a Ca^{2+}- and aprotinin-sensitive protease which cleaves primarily at the Arg^{98}-Ser^{99} bond (WYPIJ and HARRIS 1988). The enzyme, which may be costored with the ANP precursor in an inhibited state, is apparently activated upon dilution into the extracellular compartment.

Endothelin (ET-1), a vasoactive peptide of 21 residues, is synthesized in endothelial cells as a 203-amino acid prepro form which undergoes cleavage at a dibasic pair to form a 39-residue intermediate, big ET-1. Conversion of big ET-1 to ET-1 occurs via an unusual processing of a Trp^{21}-Val^{22} bond during or shortly after secretion. Of the several potential converting enzymes, phosphoramidon-sensitive metalloproteinases in endothelial cells and leukocytes (SESSA et al. 1991) appear to be the best candidates for conversion after secretion. It is also possible that some peptide conversion occurs within the acidic secretory granules via an enzyme similar to lysosomal cathepsin E since production of ET-1 is inhibited by pepstatin (MATSUMURA et al. 1990).

The peptidyl leukotriene LTC_4 is another example of a bioactive peptide generated from cell-stored precursors, in this case arachidonic acid and glutathione. The generation of this inflammatory mediator by leukocytes can be initiated by the release of arachidonate from membranes, which is initially converted to LTA_4 by 5-lipoxygenase. The second step, conjugation to glutathione, can either take place within the same cell or in other vascular cells after release and uptake of LTA_4. Thus, endothelial cells are thought to be important contributors to LTC_4 production; however, they can only do so by way of a cooperative synthesis involving blood cells (FEINMARK 1988).

2. Precursor Anchored to Cell Surface

A new development has been the finding that mature forms of the growth factors EGF, TGFα, and colony-stimulating factor (CSF) as well as the cytotoxin tumor necrosis factor (TNF) are cleaved from precursors anchored to cell surfaces (Fig. 2). This novel mode of secretion was suggested when it was noticed that the complementary DNA (cDNA) structures have, in addition to a signal peptide, a membrane-spanning segment followed by a cluster of basic residues typical of "stop transfer" signals (MASSAGUE 1990). Some of these precursors are bioactive, mediating cell-cell stimulation by contact, a process called "juxtacrine" regulation. Precursor processing, in this case, might turn off cell-cell communication and turn on paracrine signaling.

In regards to wound healing, data support the idea that leukocyte elastase could process pro-TGFα on the surface of vascular smooth muscle cells to signal for growth and repair (MUELLER et al. 1990). In cells trans-

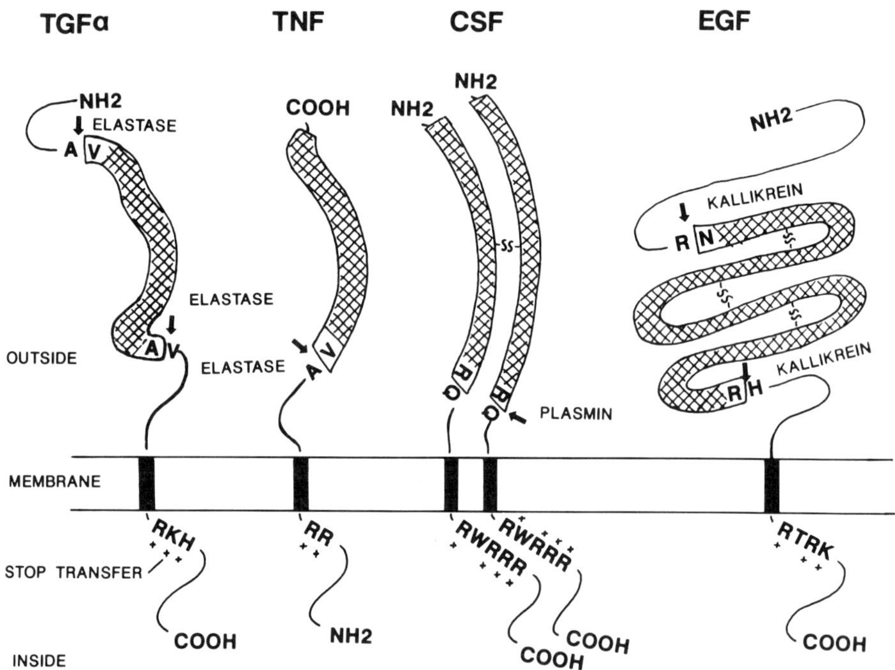

Fig. 2. Diagrammatic representation of membrane-anchored precursors to several growth factors. For each precursor (*named at top*), the NH_2 to COOH orientation in the plasma membrane is shown, with the transmembrane segment indicated *in black* and the stop transfer signal just inside the cell. Soluble forms (*cross-hatched*) can be released by the actions of the enzymes, elastase, plasmin and kallikrein (as shown) at cleavage sites *indicated by single-letter code* (*A*, alanine; *R*, arginine; *Q*, glutamine; *N*, asparagine; *H*, histidine; *K*, lysine; *W*, tryptophan; *T*, threonine; *G*, glycine; *L*, leucine; *V*, valine)

fected with pro-TGFα cDNA, conversion to soluble TGFα can be stimulated by elevation of cellular Ca^{2+} and activation of protein kinase C (PANDIELLA and MASSAGUE 1991). A similar scheme may explain the appearance of a cell surface form of TNF in lipopolysaccharide (LPS)-activated monocytes which is thought to provide a local cytotoxic signal (KRIEGLER et al. 1988). Here again leukocytic elastase would seem to be an ideal protease to cleave the Ala-Val bond, liberating active TNF (Fig. 2).

The macrophage colony-stimulating factor, CSF-1, stimulates the proliferation of progenitor cells committed to the mononuclear phagocytic lineage. Analysis of CSF-1 messenger RNA (mRNA) indicated that it contained a transmembrane hydrophobic domain, and transfection of 3T3 cells led to expression of the precursor at the cell surface (RETTENMIER et al. 1987). Since trypsin-like enzymes can liberate cell surface-bound CSF-1, it was suggested that during inflammation plasminogen activator, released from macrophages, could lead to plasmin-mediated CSF-1 release from fibroblasts. This feed-forward system would ensure macrophage function during an inflammatory crisis.

A variation on this theme is seen with TGF-β, a 25-kDa homodimer which may function in wound healing by inducing matrix synthesis and decreasing levels of collagenase and plasminogen activator. The 390-residue precursor undergoes processing within lysosomes and an inactive complex of TGF-β associated with part of its precursor appears on the cell surface (MIYAZONO et al. 1988). In endothelial/smooth muscle cocultures, activation of latent TGF-β appears to be mediated by plasmin and part of a feedback loop (SATO et al. 1990). In this example, however, the inhibitory segment is cleaved rather than TGF-β itself.

Epidermal growth factor is another hormone which can be present as a transmembrane protein. Its 1217-residue precursor contains EGF (53 residues), multiple EGF-related structures and a hydrophobic segment followed by a "stop transfer" signal (Fig. 2). Processing varies greatly from tissue to tissue (FISHER and LAKSHMANAN 1990). While full processing to EGF occurs in mouse submaxillary gland where EGF is stored in granules and secreted into saliva and blood, pro-EGF in the kidney is processed constitutively to a membrane-bound 200-kDa form. Although the enzymes which liberate EGF from cell surfaces have not been identified, the single basic cleavages suggest tissue kallikrein as a candidate.

IV. Precursor Activated in the Interstitial Space or Blood (Class B2)

Bradykinin (BK) and angiotensin (ANG) are classic examples of peptides liberated from precursors in the blood (REGOLI and BARABE 1980; PEACH 1977). Since the precursor levels are very high (>200 n*M*), bioactive amounts of the peptides (~1 n*M*) can easily be produced by release of the processing enzymes, kallikrein and renin. Interestingly, both enzymes circulate in zymogen forms which can be activated via a cascade of steps

(SEALEY et al. 1983; CLEMENTS 1989). Some advantages of this design are: (a) the power to generate activity is given to any cell which can trigger an input, (b) the catalytic nature of the trigger yields rapid signaling with minimal energy expenditure and (c) the cascade integrates complex inputs and produces an amplified signal.

The kallikrein-kinin system is made up of two substrates, high and low molecular weight kininogen, and two enzymes, plasma and tissue kallikrein, which can liberate BK or Lys-BK through a cascade of proteolytic activation steps. While subsequent trimming with carboxypeptidase N yields des-Arg^9BK, an active peptide with BK-2 receptor selectivity, exposure to angiotensin-converting enzyme (ACE) yields inactive peptides (Fig. 3). Inhibitors of ACE, which are thought to lower blood pressure via effects on angiotensin, also enhance the depressor response to BK and increase urinary excretion of kinin. Thus, the lowered blood pressure could be partly

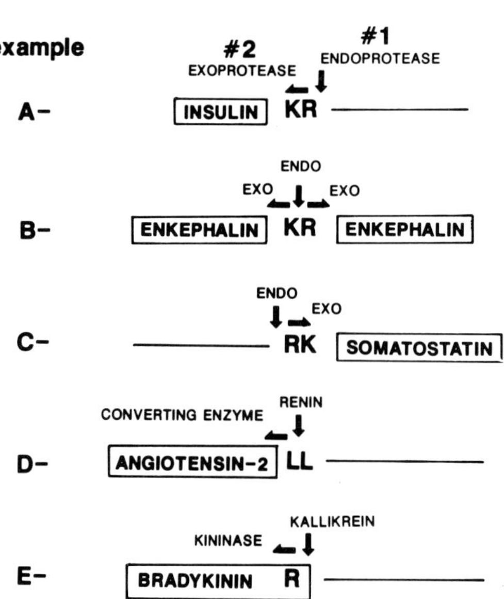

Fig. 3. Examples of two-step processing for regulatory peptides. For *A–C*, the first step entails endoproteolytic cleavage to the right of the dibasic pair (*A*), in the middle of the dibasic (*B*) and to the left of the dibasic (*C*). This is followed by exoproteolytic trimming by carboxypeptidase B-like and aminopeptidase B-like enzymes. For (*D*), renin cleaves a Leu-Leu bond and converting enzyme removes two C-terminal amino acids to generate angiotensin-2. For (*E*), endoproteolytic cleavages by kallikrein, liberating bradykinin, can be followed by removal of one residue (kininase I) to give an active peptide or removal of two residues (kininase II) to inactivate. Single-letter designations for amino acids appear as listed in the legend to Fig. 2

due to an enhancement of BK-mediated vasodilation and diuresis. A wealth of studies measuring both substrate and product support roles for the kinin system in the pathogenesis of conditions such as angioedema and arthritis (Worthy et al. 1990).

Recently, a third kinin system was discovered involving the liberation of T-kinin (Ile-Ser-BK) from T-kininogen upon cleavage by a thiol protease (Greenbaum et al. 1989). Levels of T-kininogen and T-kinin are elevated by inflammatory stimuli and decreased by antiinflammatory drugs. There are several pathways by which this system might act since T-kininogen is a protease inhibitor and T-kinin can produce vasodilation and edema.

While there is much known about the blood clotting and complement cascades, the possibility that other peptide families might be represented in this fashion has hardly been tested and it is probable that the importance of extracellular processing is underestimated. Some enticing information is available, however, regarding the generation of opioid peptides in spinal fluid, which has been shown to contain large molecular opiates derived from pro-enkephalin (pro-ENK) and pro-dynorphin (Nyberg et al. 1986) as well as various converting-enzyme activities (Terenius and Nyberg 1988). Although it is difficult to estimate the extent of conversion, the fact that the large forms are in much higher concentration than the small peptides indicates a potential for extracellular processing. Thus, the larger peptides, which have unique activities of their own, could serve both as long-term modulators and as reservoirs of potential opioids exhibiting various receptor selectivities.

A number of enzymes could have important extracellular roles with regard to peptide action within the central nervous system (CNS). Spinal fluid contains α-amidating enzyme, ACE, dynorphin-converting enzyme, as well as degradative activities such as enkephalinase, aminopeptidase and carboxypeptidase (Terenius and Nyberg 1988). These enzymes could be acting locally to determine peptide content and their concentrations may be a reflection of neuronal activity. For example, ACE activity, which is nonuniformly distributed with high levels in basal ganglia, is lower than normal in spinal fluid from patients with diseases involving the dopaminergic system. Lowered ACE activity might be expected to increase the effective levels of BK and ENK while decreasing those of ANG-2.

V. Precursor Activated at Target or by Other Cells (Class B3)

There are many examples of "hormones" which become activated after secretion by way of interactions with other cells, including their targets. Thyroxines, for example, are cleaved from their precursor, thyroglobulin, by lysosomal proteases after endocytosis of extracellular colloid by the follicular cells of the thyroid. In addition, circulating thyroxine might be viewed as a precursor to 3,5,3′-triiodothyroxine, because it is activated after uptake by the liver and other organs.

In regard to peptides, however, the evidence supporting the existence of such mechanisms is less firm. Common sense suggests that large (50–200 residue) protein hormones might be further processed to generate additional signals, since most biologically active peptides are much smaller than this. Furthermore, endocytosis of hormones at their targets and entry into lysosomes is a well-known phenomenon. Some attractive features of target cell processing are that a single hormone could provide its targets with the opportunity to generate a variety of peptides in a concerted fashion and, depending upon the enzymes used, different arrays of peptides might be formed.

1. Precursor Stored Outside Target

Epidermal growth factor is a peptide which undergoes processing subsequent to receptor binding and internalization (SCHAUDIES et al. 1987). Initial processing occurs at the plasma membrane and in the early endosome, resulting in loss of the C-terminal arginine. As EGF moves through the endosomal compartments, it is completely converted to the des-48-53 species. These conversions involve the actions of a carboxypeptidase B-like protease in early endosomes combined with a trypsin-like protease in late endosomes (RENFREW and HUBBARD 1991). It is not clear whether internalized EGF participates in the mitogenic signal; however, the majority of EGF associated with fibroblasts during induction of mitogenesis is intracellular. Furthermore, compounds which inhibit endocytosis and processing also inhibit stimulation of DNA synthesis. The fact that regenerating liver handles EGF differently from normal liver also suggests that target cell processing may contribute to the mitogenic signal (MARTI et al. 1989). The uptake of [^{125}I]EGF was threefold higher in regenerating liver and 27% of the endocytosed radioactivity was found in nuclei as compared to <0.5% in controls. Although a nuclear EGF-binding receptor has been demonstrated, it is not yet clear whether EGF or processed form(s) of this peptide act within the nucleus.

Target cell processing may also be important for the lipolytic effect of β-lipotropin in adipocytes since tissue response and hormone degradation were related (RICHTER et al. 1990). Thus, all substances which inhibited degradation – lysosomotropic agents, protease inhibitors, cytoskeletal disruptors, etc. – also inhibited the lipolytic response and, in addition, the two correlated in different physiologic states. Parathyroid hormone (PTH) may provide another example since macrophages which endocytose PTH_{1-84} process it to PTH_{1-34} via endosomal/lysosomal cathepsin D and return this bioactive fragment to the medium (DIMENT et al. 1989).

2. Precursor Stored Inside Target

Another variation on this theme, processing of a precursor which is stored at the target, might be illustrated by steroid receptors, the transformation of which appears to involve a limited proteolysis induced by the specific steroid (Puca et al. 1986). Thus, in the case of the estrogen receptor, a chymotrypsin-like activity could be expressed by the immunoaffinity-purified hormone-binding subunit of the receptor in the presence but not in the absence of estradiol (Molinari et al. 1991). Furthermore, serine protease inhibitors prevented the structural changes in the receptor necessary for DNA binding. Since transformation of steroid receptors is generally an irreversible reaction, it seems likely that the activation process involves proteolysis of the receptor itself. It is also possible that the proteolytic activity of the activated receptor might be utilized to generate other bioactive peptides.

VI. Mixed Systems (Several Classes)

Signaling by the renin-angiotensin (ANG) system may be accomplished in a number of ways, since ANG-2 is stored both as the prohormone in blood and as the active peptide within neurons and other tissues (Cambell 1987). An added complexity is that renin is found in an inactive pro-form which can account for most of the circulating potential activity. Furthermore, renin is not only localized to the juxtaglomerular (JG) cells of the kidney and angiotensinogen is not only secreted by the liver as originally thought; local generation also takes place within tissues like brain, heart and adrenal (Goldblatt 1986). Some tissues contain mRNA for renin and angiotensinogen, while others may take up these proteins from blood. Proteases other than renin may also be involved, since cathepsin D can generate ANG-1 and tonin can form ANG-2 from the precursor.

Operation of the classical renin-ANG system is triggered by release of renin from the JG cells of the kidney in response to lowered blood pressure or sodium concentration. Renin cleaves a Leu-Leu bond in angiotensinogen, a 58-kDa protein secreted by the liver, to generate the decapeptide ANG-1. On passage through the pulmonary vasculature, ANG-1 is converted to the bioactive peptide, ANG-2, by removal of His-Leu from its C terminus by converting enzyme. In addition to secreting large amounts of active renin, however, JG cells also make and store small amounts of ANG-1 and ANG-2 (Brooks et al. 1982). Thus, it might be suggested that the pre-formed peptides are used for "paracrine" signaling and renin, which can generate far more peptide, could be reserved for long-distance "hormonal" effects.

Work with a tumor cell-line of testicular Leydig cells has shown synthesis and release of renin-ANG in a gonadotropin-dependent manner (Pandey and Inagami 1986). Treatments with luteinizing hormone (LH) increased the content of ANG-1 and ANG-2 by 150- and 40-fold. Renin, which increased 40-fold, remained inside the cells. While ANG-1 was also not secreted, ANG-2 was mostly in the medium. Thus, it appears that renin functions intracellularly in Leydig cells to generate ANG-1 and that formation of ANG-2 could occur before or after secretion. These examples illustrate that multiple formats are possible and future work will probably uncover similar complexities for many peptides.

C. Intracellular Trafficking and Processing Pathways

Peptide hormone precursors (pre-pro-peptides) are specifically processed in a temporal order through different intracellular compartments of an endocrine or neuroendocrine cell (Fig. 1) These peptide prohormone precursors are first synthesized on ribosomes with a 16–22 amino acid hydrophobic sequence at the N terminus known as the signal peptide. This signal peptide serves as a recognition site for routing the protein into the rough endoplasmic reticulum (RER) cisternae. Its hydrophobic nature facilitates the threading of the protein across the RER membrane. Subsequent to its insertion into the RER cisternae, which involves the participation of a signal recognition particle and receptor, the signal peptide is cleaved off by a signal peptidase, an enzyme with neutral pH optimum (EVANS et al. 1986). Protein folding occurs in the RER cisternae and in some cases the pro-peptides also undergo N-glycosylation and/or disulfide bond formation (ELBEIN 1987; PFEFFER and ROTHMAN 1987). The pro-peptides are then transported to the Golgi where they undergo further modifications such as oligosaccharide maturation, O-glycosylation, phosphorylation and sulfation. Endoproteolytic processing of the pro-peptides may begin at the *trans*-most compartment of the Golgi apparatus, just prior to packaging into clathrin-coated immature secretory granules (SCHNABEL et al. 1989). However, for the most part, proteolytic processing as well as N-acetylation, amidation and pyroglutamate production occur within the secretory granules (FISHER and SCHELLER 1988).

Large regions of many precursors have no apparent function and they may simply serve to meet a size requirement or to confer a suitable conformation for processing enzymes to act. Pro-insulin was shown to be an example of the latter when a cDNA construct missing the C-peptide was transfected into yeast and cleavage at the paired basic residues did not occur (THIM et al. 1986). Recent studies suggest the presence of a motif in the N-terminal part of some precursors which promotes sorting into secretory granules (SEVARINO et al. 1989; STROLLER and SHIELDS 1989; KIZER and TROPSHA 1991). For the vitamin K-dependent proteins (e.g., prothrombin) the propeptide serves as a signal to direct γ-carboxylation of specific glutamic acids (FURIE and FURIE 1990). Thus, regions of precursors may have subtle roles as signals for sorting, packaging and processing.

The list of post-translational modifications should also include intracellular degradation since this can be an important means of regulating the storehouse of hormone. For example, while degradation of $ACTH_{1-39}$ via specific cleavage at its Lys-Lys-Arg-Arg site is a major occurrence in newborn rat corticotropes, this effect largely disappears by adulthood. Since degradation can be suppressed by physiologic levels of glucocorticoids, it may be part of a positive feedback loop for pituitary-adrenal function (SATO and MAINS 1988). Degradation also appears to be a determinant of PTH production in response to signals such as calcium (BIENKOWSKI 1983). Although lysosomal proteases have generally been implicated in the breakdown of hormones via an autophagia of granules when secretion is inhibited, some findings suggest the involvement of other proteases (MACGREGOR et al. 1986).

D. Processing Sites and General Mechanisms

I. Two-Step Processing

There are many examples of prohormone maturation which involve a two-step process, first cleavage by an endopeptidase and then trimming with an exopeptidase (Fig. 3). Since various combinations might be utilized, this arrangement lends versatility and affords more control. In some cases (e.g., ANG-2), trimming is an activating step, while in others (e.g., BK), it can alter receptor selectivity (des-Arg^9-BK) or lead to degradation.

II. Dibasic, Monobasic and Nonbasic Sites

By far the majority of prohormones known to date undergo endoproteolytic processing at dibasic cleavage sites, most commonly at Lys-Arg but also at Arg-Arg, Arg-Lys and Lys-Lys (DICKERSON et al. 1990; STROLLER and SHIELDS 1989; THORNE and THOMAS 1990). Recently, however, a second mechanism has been recognized which involves processing at single basic residues, primarily Arg (SCHWARTZ 1986). This has led to the idea that there are, for the most part, two major cleavage types, both involving enzymes with trypsin-like specificity. However, there are an increasing number of peptides which do not fit into either class and for which the processing might be generally thought of as occurring at hydrophobic residues. A few examples and their cleavage sites (in parentheses) include endothelin (Val-Trp), ANG (Leu-Leu), relaxin (Leu-Ser), γ-endorphin (Leu-Phe) and TNF (Val-Ala). It has been suggested that peptide processing may have evolved from mechanisms used initially to digest food and, if so, then one might expect the participation of multiple endoproteases similar to the known digestive enzymes (CARRAWAY and REINECKE 1984).

III. Conformation and Consensus Features

Certain dibasic and monobasic sites within prohormones are not cleaved, perhaps because other features are important for recognition. Although there is no consensus sequence, dibasic cleavage sites are often preceded on the N terminal by a basic or hydrophilic region (BRAKCH et al. 1989). Monobasic sites are generally found in regions rich in Ala or Leu and sometimes a second basic site is 3, 5 or 7 residues from the N terminus. Another class of monobasic sites have adjacent prolyl residues which can direct the cleavage to the N- or C-terminal side (SCHWARTZ 1986). In addition, the predicted conformation near the site can be a useful indicator, since processing sites are commonly associated with β-turns flanked by α-helixes (RHOLAM et al. 1986).

IV. Sequentially Ordered Reactions

Some evidence indicates that processing events can occur in distinct stages which are temporally ordered. In the processing of POMC in pituitary, for example, the prohormone is first cleaved at the dibasic sequence which links the C terminal of ACTH-CLIP to the N terminal of lipotropin. This reaction goes to completion before proteolysis occurs at a C-terminal dibasic site in β-LPH to generate β-endorphin (SMYTH et al. 1988). It is possible that a conformational change occurs with the first excision to expose the second site.

Another example is proinsulin, for which processing at Lys^{64}-Arg^{65} occurs before cleavage at Arg^{31}-Arg^{32}. In this case, the ordering of the reactions is nicely explained by sequential activation of two processing enzymes with differing pH and Ca^{2+} requirements. Thus, the type II protease, which has a broader pH range and requires less Ca^{2+}, becomes activated in the Golgi apparatus, while the type I protease can only act in the more mature granules having low pH and high $[Ca^{2+}]$ (DAVIDSON et al. 1988).

In some systems (e.g., *Aplysia* ELH), a tetrabasic sequence is a signal for rapid cleavage, which is followed by sorting of the fragments into different vesicles and further processing (FISHER et al. 1988). Whether this is regulated by specific enzyme(s) or conditions in the Golgi apparatus is not clear.

V. Tissue-Specific Processing

Precursors have been found to contain only one (e.g., insulin) or several (e.g., POMC) bioactive peptides and they can even have multiple copies of the same peptide (e.g., pro-ENK). In some cases, alternate gene-splicing leads to the formation of several different versions of the same precursor (e.g., pro-tachykinins) which can lead to differential expression of a set of peptides. Another scenario involves tissue-specific processing, such that the same precursor yields different products depending upon the processing enzymes it encounters. For example, pro-enkephalin, which contains 7 repeat sequences of methionine-enkephalin and 1 of leucine-enkephalin, is processed to large forms in the adrenal medulla, whereas it yields primarily the pentapeptides in brain (GIRAUD et al. 1984). Likewise, there is differential processing of prodynorphin in regions of brain, yielding either Leu-enkephalin or extended forms (ZAMIR et al. 1984). A number of mechanisms might account for this:

1. There may be different enzymes specific for certain cleavage sites and their distributions may be tissue specific.
2. The ability of the processing enzyme to cleave might be influenced by nearby glycosylation or phosphorylation in a tissue-specific manner.
3. Ca^{2+} concentration, pH or specific inhibitors might vary within each tissue and serve to regulate the extent of processing. Chromogranin may be such an inhibitor (SEIDAH et al. 1987).

E. Enzymes Involved in the Processing of Prohormones

I. Dibasic Residue-Specific Endoproteases

An ability to process at paired basic sites is a feature common to organisms throughout the animal kingdom. Thus, the same processing signals appear to be used for yeast α-mating factor (KURJAN and HERSHOWITZ 1984), *Aplysia* egg-laying hormone (FISHER et al. 1988), anglerfish somatostatin (MACKIN and NOE 1987) and numerous mammalian prohormones (LOH et al. 1984). The most common paired basic residues appear to be Lys-Arg followed by Arg-Arg and Arg-Lys and only a few cleavages occur at Lys-Lys. Various cell lines have been used in combination with site-directed mutagenesis to investigate the effects of altered residues on processing specificity and kinetics. Mutation of the Arg-Lys cleavage site in pro-somatostatin (SS) to other basic residues had little effect on the efficiency of SS-14 production in GH_3 cells (STOLLER and SHIELDS 1989). In contrast, processing of pro-neuropeptide Y (NPY) in AtT-20 cells was greatly slowed by mutation of the native Lys-Arg site to Lys-Lys or Arg-Lys (DICKERSON et al. 1990). Findings of this sort suggest that a number of processing enzymes having differing selectivities can be utilized.

Several endoproteases which cleave at paired basic residues have been identified and they are summarized in Table 2. The list is restricted to endoproteases which have been shown to properly cleave a pro-hormone or a large fragment of a prohormone and which have been localized to secretory vesicles. Many of these enzymes have been found to cleave short synthetic substrates poorly or not at all. Those which have been cloned include yeast KEX-2 and its three mammalian homologues, known as PC1 (PC3), PC2 and furin (SEIDAH et al. 1991; SMEEKENS et al. 1991; VAN DEN OUWELAND et al. 1990). These enzymes belong to a subtilisin-like family of Ca^{2+}-activated serine proteases and appear to be highly specific for paired basic residues. PC1 and PC2 have been shown to properly cleave dibasic

Table 2. Paired basic residue-specific endoproteases

Source	Substrate	Product(s)	pH optimum	Size	Specificity[a]	Class	Reference
Yeast	Pro-α-mating factor	α-mating factor	7.0	125–135 kDa	Lys-Arg	Ca^{2+}-activated subtilisin-like serine protease KEX-2	FULLER et al. (1989) MIZUNO et al. (1989)
Yeast	Pro-α-mating factor	α-mating factor	4.0–5.0	65 kDa	Lys-Arg	Aspartic protease YAP3	EGEL-MITANI et al. (1990)
Rat, mouse pituitary, bovine adrenal medulla chromaffin granules, AtT-20 cells	Proopiomelanocortin	ACTH β-LPH	?	70 kDa (deduced from cDNA)	Lys-Arg	Ca^{2+}-activated subtilisin-like serine protease PC1 (PC3)	SEIDAH et al. (1991) SMEEKENS et al. (1991) CHRISTIE et al. (1991)
Rat, mouse pituitary, human insuloma, bovine adrenal medulla chromaffin granules	Proopiomelanocortin	α-MSH β-Endorphin	?	64 kDa (deduced from cDNA)	Lys-Arg Lys-Lys Arg-Arg	Ca^{2+}-activated subtilisin-like serine protease PC2	SMEEKENS and STEINER (1990) SEIDAH et al. (1990) CHRISTIE et al. (1991)
Bovine pituitary intermediate lobe secretory granules	Proopiomelanocortin ACTH	21–23K ACTH β-LPH β-Endorphin α-MSH	6.0–8.0	–	Lys-Arg Lys-Lys Arg-Arg	Ca^{2+}-activated PC2-like protease	ESTIVARIZ et al. (1992)
Bovine pituitary intermediate lobe secretory granules	Proopiomelanocortin	ACTH β-LPH β-Endorphin α-MSH	4.0–5.0	70 kDa	Lys ↓ Arg ↓ Arg ↓ Lys	Ca^{2+}-activated aspartic protease	LOH et al. (1985) LOH et al. (1986) ESTIVARIZ et al. (1989); BIRCH et al. (1991b)
Rat insuloma secretory granules	Pro-insulin	Insulin	5.5	–	Lys-Arg Arg-Arg	Ca^{2+}-activated protease	DAVIDSON et al. (1988)
Golgi/secretory granules	Pro-insulin	C-Peptide A Chain	5.5–6.5	–	LysArg	Ca^{2+}-activated protease	DAVIDSON et al. (1988)
Anglerfish pancreas secretory granules	Pro-somatostatin I	SS-14	5.0	57 kDa	Arg-Lys	Ca^{2+}-activated subtilisin-like serine protease (PC2-like)	MACKIN et al. (1991a)

Bovine hypothalamic secretory granules	Pro-GnRH/GAP	Gonadotropin associated peptide (GAP)	7.5	64.5 kDa	Lys-Arg	Ca^{2+}-activated serine protease	RANGARAJU and HARRIS (1991)
Bovine pituitary intermediate lobe secretory granules	Pro-vasopressin	Vasopressin	4.0–5.0	70 kDa	Lys-Arg	Aspartic protease	PARISH et al. (1986)
Rat brain secretory granules	Somatostatin 28 synthetic fragment	Somatostatin	7.4	90 kDa	↓ Arg-Lys	Sensitive to thiol reagents	GLUSCHANKOF et al. (1984, 1987)
Bovine adrenal medulla chromaffin granules	Pro-enkephalin peptide F	Intermediates Met-enkephalin	5.5	33 kDa	Lys-Arg Lys-Lys	Cysteine (thiol) protease	KRIEGER and HOOK (1991)
Bovine adrenal medulla	Pro-enkephalin	ENK-Arg Gly-Leu	8.0	31 kDa	Lys-Arg Lys-Lys	Serine protease	LINDBERG and THOMAS (1990)
Human kidney	Pro-renin	Renin	6.0	30 kDa	Lys-Arg	Cysteine (thiol) protease (cathepsin B)	WANG et al. (1991)
Bovine corpus luteum, neural lobe	Pro-oxytocin synthetic fragments	Oxytocin	7.2	58 kDa	Lys-Arg	?	CLAMAGIRAND et al. (1987) PLEVRAKIS et al. (1989)

[a] In general, most of the cleavages occur on the carboxyl side of the paired basic residues although cleavages on the amino side and in between the basic residues have been found and are indicated in the table. For many of the enzymes studied, the exact bonds within the paired basic residues where processing occurs have not been identified.

residues of POMC (Benjannet et al. 1991) and furin to cleave those in β-NGF (Breshnahan et al. 1991). The enzymes which have been purified to homogeneity include a POMC-converting enzyme from bovine pituitary which is an aspartyl protease (Loh et al. 1985; Loh 1986), a prosomatostatin convertase from anglerfish pancreas (Mackin and Noe 1987), a thiol protease from chromaffin granules which cleaves fragments of pro-ENK and pro-substance P (Maret and Fauchere 1988) and a prooxytocin/neurophysin convertase (Plevrakis et al. 1989). The N-terminal region of the anglerfish protease shows 65% homology to a portion of mouse PC2 (Mackin et al. 1991a). Another thiol protease, identified as cathepsin B, has been shown to process pro-renin at paired basic residues (Wang et al. 1991).

II. Monobasic Residue-Specific Endoproteases

Biosynthesis of hormones such as SS-28, CCK, gastrin, dynorphins, ANP, vasopressin-associated neurophysin and *Aplysia* ELH involve cleavage of the precursors at a single Arg (Devi and Goldstein 1984; Daugherty and Yamada 1989; Gluschankof et al. 1984; Schwartz 1986). Cleavage at a single lysine occurs rarely, a few examples being *Aplysia* pro-FMRF amide and guinea pig pro-pancreatic polypeptide (Schwartz 1986). A number of candidate arginine cleavage enzymes have been identified in tissues which process prohormones and pro-neuropeptides. They include enzymes partially purified from atrium, adrenal medulla, frog skin, brain synaptosomes, rat brain membranes, intestines and anglerfish islets. The properties of these enzymes are listed in Table 3. Most of these proteases are specific for single arginines and there are examples from various classes including aspartic, thiol, serine and metalloproteases. With the exception of those enzymes from frog skin and the anglerfish aspartic protease, all of the monobasic specific endoproteases appear to be neutral enzymes.

III. Tetrabasic Residue-Specific Endoproteases

Tetrabasic cleavage of ACTH at Lys^{15}-Lys^{16}-Arg^{17}-Arg^{18} yields $ACTH_{1-17}$, which is subsequently processed to α-MSH (acetyl-$ACTH_{1-13}$-NH_2) in the pituitary intermediate lobe. In addition, precursors to the enzymes PC1 and PC2 are processed at tetrabasic sites (Christie et al. 1991). Recently, a calcium-activated serine protease which has a pH optimum of 5–6 and is highly specific for tetrabasic residues was isolated from bovine intermediate pituitary secretory vesicle membranes (Estivariz et al. 1992). This enzyme, named acidic ACTH converting enzyme (AACE), was sensitive to the serine protease inhibitors phenylmethanesulfonyl fluoride (PMSF) and diisopropylfluorophosphate (DFP). AACE differ from PC2, a subtilisin-like serine protease which can cleave the tetrabasic site of ACTH as well as dibasic sites in POMC (Smeekens et al. 1991; Thomas et al. 1991).

Table 3. Monobasic residue – specific endoproteases

Source	Substrate	Product	pH optimum	Size	Specificity[a]	Class	Reference
Bovine atria granules	Pro-ANF	ANF	7.5–8.5	70 kDa	Arg ↓ Ser	Serine protease	Wypij and Harris (1988)
Ovine adrenal medulla chromaffin granules	BAM-12P	Adrenorphin	7.8	35–45 kDa	Gly ↓ Arg	Metallo-protease	Tezapsidis and Parish (1989)
Frog skin	Pro-caerulein, pro-xenopsin fragments	Caerulein xenopsin	6.5	60 kDa	Arg ↓ Gly	Metallo protease	Kuks et al. (1989)
Frog skin	Pro-caerulein, pro-xenopsin fragments	Caerulein xenopsin	5.5	25–35 kDa	Arg ↓ Gly	Thiol protease	Darby et al. (1991)
Rat brain synaptosomes	CCK-33	CCK-8	8.0	45 kDa dimer	Arg ↓ Asp	Serine protease	Viereck and Beinfeld (1990)
Bovine pituitary secretory vesicles	Dynorphin B-29	Dyn B-29 (1–13)	7.0–8.0	200 kDa	Thr ↓ Arg	Thiol protease	Devi and Goldstein (1984) Devi et al. (1991)
Bovine adrenal chromaffin granules	Dynorphin A	Dynorphin 1–11	8.0	31 kDa	Lys ↓ Leu	Serine (trysin-like) protease	Shen et al. (1989)
Anglerfish islet secretory granules	Pro-anglerfish somatostatin II	Somatostatin-28	3.5–5.0	39 kDa	Arg ↓ Ser	Aspartic protease	Mackin et al. (1991b)
Rat intestine mucosal secretory granules	Pro-somatostatin synthetic fragment	Somatostatin-28	7.5	67 kDa	Arg ↓ Ser	Serine protease	Beinfeld et al. (1989a) Bourdais et al. (1991)
Rat brain cytosol	Pro-somatostatin synthetic fragment	Somatostatin-28	8.0	80 kDa	Arg ↓ Ser	Metallo-protease	Beinfeld et al. (1989b)

[a] See footnote table 2.

IV. Exopeptidases

Following cleavages by endoproteases at basic residues, the remaining C- or N-terminal basic amino acids are removed by carboxypeptidase B-like and aminopeptidase B-like exopeptidases. A carboxypeptidase B-like enzyme has been purified from adrenal medulla, pituitary and brain, where it is found in secretory granules (FRICKER 1988). The amino acid sequence of enzyme has been derived from its cDNA cloned from rat hypothalamus and bovine pituitary (FRICKER et al. 1986, 1989). This enzyme, named carboxypeptidase H, is also referred to as carboxypeptidase E and enkephalin convertase. The purified soluble form of 50 kDa has an N-terminal signal peptide and a short pro-peptide segment with five arginines immediately preceding the N terminus of the mature enzyme. The rat and bovine enzymes contain four dibasic pairs, although it is not known if they are processed. A 53-kDa membrane-associated form of the enzyme has also been purified. Carboxypeptidase H shows a 17%–20% homology with pancreatic carboxypeptidases but it is distinguished by its low pH optimum (5.5) and its extreme selectivity for basic residues (Arg>Lys>His) (FRICKER 1991). The enzyme is activated by Co^{2+} and can be localized using guanidinoethylmercaptosuccinic acid (HOOK and LOH 1984; FRICKER 1988). The mRNA level of the enzyme is coordinately regulated with prohormone mRNA levels (BONDY et al. 1989; KANAMATSU et al. 1986; RODRIGUEZ et al. 1989). With regards to enkephalin processing, treatment of chromaffin cells with reserpine elevated the level of Co^{2+}-stimulated carboxypeptidase activity with a concomitant increase in the conversion of ENK precursors to products (HOOK et al. 1985).

An aminopeptidase B-like enzyme from bovine pituitary secretory vesicles has been described (GAINER et al. 1984). There appear to be both soluble and membrane-associated forms of the enzyme although their specificity and pH optima are quite similar (CASTRO et al. 1989). The enzyme has a pH optimum of 6.0 and is stimulated by Co^{2+} and Zn^{2+}. There is a sixfold preference for Arg over Lys at the N terminus. Release studies show that the enzyme activity is coordinately secreted with α-MSH, indicating its presence in secretory vesicles. Anglerfish islets contain an aminopeptidase-B which is Co^{2+}-insensitive (MACKIN et al. 1985) and rat brain has an enzyme complex which is inhibited by chelators and heavy metals (GLUSCHANKOF et al. 1987).

V. Enzymes Involved in N- and C-Terminal Modification

1. Acetyltransferase

N-Acetylation can greatly affect the bioactivity and stability of peptides. For example, N-acetylated α-MSH is more potent than the deacetyl form (RUDMAN et al. 1983), while the N-acetylated form of β-endorphin$_{1-31}$ is devoid of opiate activity (AKIL et al. 1985). Several acetyltransferases have been detected in secretory granules. A soluble activity from bovine intermediate lobe is a neutral enzyme which selectively acetylates $ACTH_{1-13}NH_2$ and related peptides but not histones, serotonin or choline (GLEMBOTSKI 1982). Two activities, one with a pH optimum of 7.4 and another at pH 6.3 have been found in rat intermediate lobe. The pH 6.3 activity acetylates both $ACTH_{1-13}NH_2$ and β-endorphin and is specifically localized in secretory granules, whereas the pH 7.4 activity acetylates $ACTH_{1-13}NH_2$ and is also found in other nonendocrine tissues (CHAPELL et al. 1983).

2. Glutaminyl Cyclase

Pyroglutamic acid occurs at the N terminus of thyrotropic hormone (TRH), gonadotropin-releasing hormone (GnRH), and neurotensin. The precursors of these peptides have a glutamine in the position of the pyroglutamic acid. Recently, a soluble enzyme which converts glutamine to pyroglutamate has been found in a secretory granule-associated fraction in pituitary and brain (BUSBY et al. 1987; FISHER and SPIESS 1987). In all tissues examined glutaminyl cyclase exhibited a neutral to alkaline pH optimum and there were multiple forms of the activity ranging in size from 50 kDa upward.

3. Peptidylglycine α-Amidating Monooxygenase

Peptides which have C-terminal amide groups are always followed by glycine in their precursor sequences (EIPPER and MAINS 1988). The glycine (Gly) is converted to an amide by a two-step reaction as shown below:

$$\text{X-Gly} \xrightarrow{\text{E1}} \text{X-}\alpha\text{-hydroxyglycine} \xrightarrow{\text{E2}} \text{X-amide} + \text{glyoxylic acid}$$

where enzyme E1 is peptidylglycine α-hydroxylating monooxygenase (PHM) and enzyme E2 is peptidyl-α-hydroxyglycine α-amidating lyase (PAL). PHM requires molecular oxygen, copper and ascorbic acid as cofactors and has a pH optimum of 5–8 depending on the substrate (EIPPER et al. 1991). PAL is a 50-kDa enzyme with a pH optimum of 5. Its activity is resistant to thiol reagents and urea but is inhibited by EGTA and restored by Cd^{2+}, Cu^{2+}, Zn^{2+} and Ca^{2+} (BRADBURY and SMYTH 1987). Purification of a soluble 38-kDa peptidylglycine α-amidating activity from bovine neurointermediate lobe led to the cloning of cDNA encoding a 108-kDa protein consisting of both PHM and PAL. This precursor is referred to as peptidylglycine α-amidating monooxygenase (PAM). PAM has been cloned for several species and its structure is highly conserved (EIPPER et al. 1987; MIZUNO et al. 1987). There are two forms of the precursor encoded by 2 cDNAs, bPAM-1 and bPAM2. Each of the precursors contains an N-terminal signal peptide and a short pro-peptide. The PHM enzyme is encoded in the N-terminal third of the molecule followed by an intragranular region. This is followed by the PAL catalytic domain, transmembrane domain and a cytoplasmic tail. Several phosphorylation and glycosylation sites are present and paired basic residues flank PHM and PAL.

Two bovine cDNAs (bPAM-1, bPAM-2) and three rat cDNAs (of 4.2- and 3.8-kb lengths) have been reported. These PAM mRNAs are expressed in heart atrium and ventricle and in the central nervous system. PAM activity has been found in a number of endocrine and neuronal tissues such as pituitary, hypothalamus and various neuroendocrine ectopic tumors as well as in frog skin, which secretes large amounts of amidated peptides (EIPPER and MAINS 1988).

VI. Enzymes Involved in Sulfation and Phosphorylation

1. Tyrosine Sulfation Enzyme

Tyrosine sulfation occurs in various secretory proteins and bioactive peptides present in endocrine and neural tissue, e.g., cholecystokinin (CCK), gastrin, Leu-ENK and the secretogranins. Although no consensus sequence has been shown, an acidic amino acid is usually adjacent (N-terminal) to the sulfated tyrosine and several other acidic residues are in the vicinity (HUTTNER 1988). In addition, the area commonly has a strong potential for β-turn and an absence of steric hinderance, e.g., glycosylation. Sulfation occurs before endoproteolytic cleavage of the precursor, prior to exit of the protein from the *trans*-Golgi complex. A tyrosylprotein sulfotransferase, which catalyzes the transfer of sulfate from 3′-phosphoadenosine-5′-phosphate to specific tyrosyl residues in proteins, has been localized to the Golgi apparatus and shown to be present in adrenal medulla, pituitary, brain and various peptide-secreting cell lines (LEE and HUTTNER 1985). Its activity is optimum at pH 6–6.5, is inhibited by the Ca^{2+} chelator EDTA and is stimulated by Mg^{+2} and Mn^{+2} (RENS-DOMIANO and ROTH 1989).

2. Serine Phosphorylating Enzymes

Regulatory peptides known to occur in phosphorylated forms include ACTH (EIPPER and MAINS 1982), ANP, (RITTENHOUSE et al. 1988), a progastrin-derived peptide (DOCKRAY et al. 1987) and a region of pro-ENK A (WATKINSON et al. 1989). For ANP, the modified site resembles substrates for cyclic nucleotide-dependent or calcium-dependent protein kinases which phosphorylate serine residues near basic amino acids (KREBS and BEAVO 1979). For the other peptides, each site has the general sequence, Ser-X-acidic, which is recognized by physiological casein kinase or Ser-X-X-acidic, the consensus site for casein kinase II (KENNELLY and KREBS 1991). Phosphorylation occurs adjacent to a dibasic cleavage site in progastrin, where it is thought to regulate processing. In ANP, phosphorylation alters bioactivity.

F. Processing of Specific Prohormones

I. Yeast Pheromones

The yeast system has provided a relatively simple model which offers the advantage of genetic manipulation. Much work has focused on the 13-residue α-mating factor, multiple copies of which occur in precursors, separated by spacers consisting of Lys-Arg followed by two to three dipeptides of the form X-Ala (FULLER et al. 1988). Proteolysis involves the sequential action of several gene products: a Lys-Arg specific endoprotease (KEX2; FULLER et al. 1989), a basic carboxypeptidase (KEX1; DMOCHOWSKA et al. 1987) and a dipeptidyl aminopeptidase (STE13; JULIUS et al. 1983). Analysis of yeast mutants with defects in this pathway led to the identification of the regulatory genes.

The KEX2 gene codes for an enzyme that shares homology to the active site domain of the bacterial serine protease, subtilisin. However, the purified enzyme was not sensitive to the typical inhibitors DFP and PMSF, but was inhibited by EDTA, heavy metals and thiol-directed reagents. It was Ca^{2+}-dependent but differed from mammalian calpains. A membrane-bound enzyme localized to the Golgi, KEX2 displays a neutral pH optimum. It cleaves after Lys-Arg in the precursors as well as synthetic substrates, and when expressed in mammalian cells it cleaves POMC to γ-lipotropin and β-endorphin. The KEX2 gene codes for a 90-kDa protein with both a signal sequence and a transmembrane domain. Glycosylation leads to an active form of 135 kDa. The KEX1 gene codes for a 729-residue protein with a C-terminal transmembrane domain and two regions with striking homology to carboxypeptidase Y. Its role is to trim the C-terminal Lys-Arg from the intermediates produced by KEX2 action. Mutations in the STE13 gene resulted in the secretion of N-terminally extended forms of α-mating factor containing multiple X-Ala sequences. This gene codes for a 931-residue transmembrane protein with type IV dipeptidyl aminopeptidase activity, i.e., ability to remove N-terminal X-Ala.

In a yeast mutant lacking the KEX2 gene, an aspartic protease encoded by the YAP3 gene could be induced by transfecting pro-α-mating factor cDNA to overexpress this protein. Yeast with the YAP3 gene product cleaved the precursor at paired basic residues to yield α-mating factor. One wonders whether such "back up" enzymes can be induced in mammalian systems as well.

Like α-mating factor, the sea anemone peptide, pGlu-Gly-Arg-Phe-amide, is bordered C-terminally by Lys-Arg or Arg residues; however, three acidic residues flank the N terminus (DARMER et al. 1991). It could be that an aminopeptidase specific for Asp and Glu is used for trimming in this case.

II. Proopiomelanocortin

Precursor processing for proopiomelanocortin (POMC) can be quite complex (Fig. 4). The enzyme first shown to cleave POMC was POMC-converting enzyme (PCE), a 70-kDa glycoprotein purified from bovine pituitary intermediate lobe and shown to cleave at paired basic residues. It was characterized as a Ca^{2+}-activated aspartic protease with an acidic pH optimum (LOH et al. 1985; BIRCH et al. 1991a). Subsequently, two other

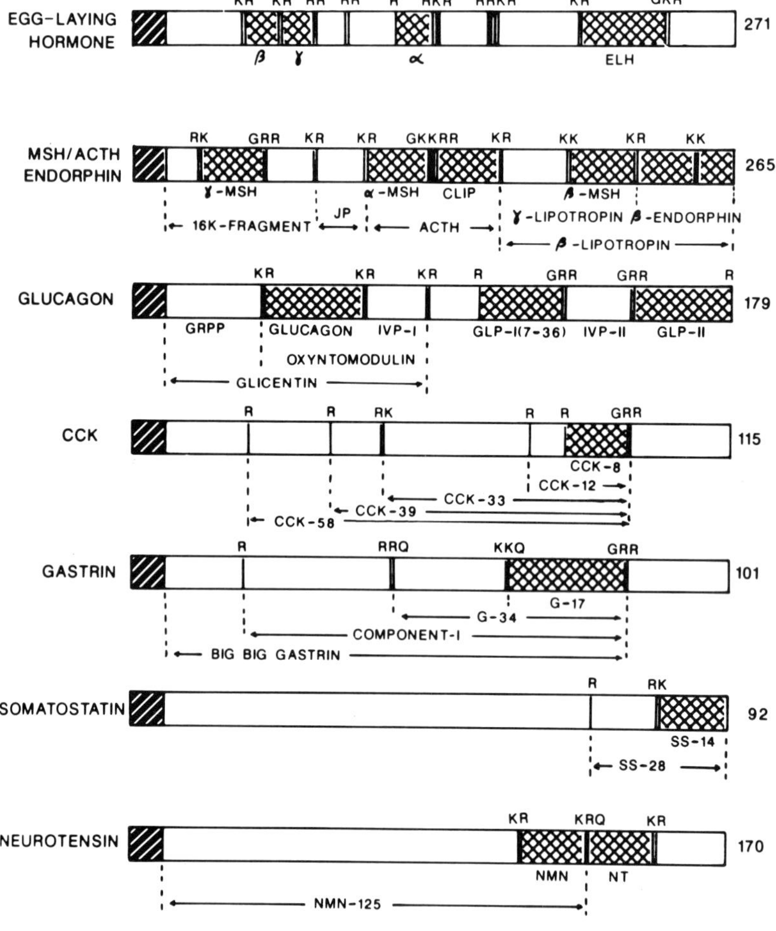

Fig. 4. Diagrammatic representation of the primary structures of selected precursors and the products resulting from processing events. Total amino acids are *indicated to the right*. Signal peptides are *hatched*. Bioactive peptides are *cross-hatched and named below*. Amino acids at known cleavage sites are indicated above using single-letter code (see legend to Fig. 2). Note that dibasic sites bordered by G usually become C-terminal amides and those with Q often convert to N-terminal pyroglutamate

enzymes PC1 and PC2 were cloned (SMEEKENS and STEINER 1990; SEIDAH et al. 1990) and shown by cotransfection with POMC into BSC 40 cells, AtT-20 cells and chromaffin cells to cleave dibasic and tetrabasic sites (see Sects. E.II, E.III). PC1 and PC2 enzymes have been purified from chromaffin granules and anglerfish islet granules and identified by N-terminal sequencing. Characterization of the cDNA encoding PC1 and PC2 indicates that they are Ca^{2+}-activated subtilisin-like serine proteases similar in structure to yeast KEX2 but lacking the transmembrane domain. Indeed, these enzymes were first cloned by using cDNA probes from KEX-2. Recently, a PC2-like enzyme activity which cleaved ACTH at the tetrabasic site to give $ACTH_{1-17}$ has been characterized in bovine pituitary intermediate lobe. This enzyme was similar to KEX2, i.e., it was stimulated by Ca^{2+}, inhibited by the trypsin inhibitor *N*-tosyl-L-lysine chloromethylketone (TLCK), EGTA and Zn^{2+} and it displayed a neutral pH optimum.

An acidic pH optimum, ACTH-converting enzyme (AACE) different from PC2 was also detected in intermediate lobe secretory granules. This AACE was a Ca^{2+}-activated serine protease which cleaved only at the tetrabasic site in ACTH and not at the dibasic sites in POMC (see Sect. E.III). Clearly, there are several enzymes found in this tissue which can process POMC similarly. Since POMC processing may be initiated in the Golgi apparatus (at neutral pH) and may continue in the secretory granules (at acidic pH), the PC1/PC2 enzymes may function at the level of the Golgi, while PCE and AACE act on POMC in the granules.

Several mechanisms for the differential processing of POMC have been proposed. Regional analysis has shown that *N*-POMC in anterior pituitary is 'O'-glycosylated at Thr^{45} whereas the cleaved product *N*-$POMC_{1-49}$ isolated from intermediate lobe was not 'O'-glycosylated (SEGER and BENNETT 1986). This led to the idea that 'O'-glycosylation at Thr^{45} prevents the processing of *N*-POMC at residues 49–50, perhaps by hindering the enzyme. Evidence for this derives from experiments examining the effect of PCE on glycosylated and nonglycosylated forms of *N*-$POMC_{1-77}$ (BIRCH et al. 1991b). Differential processing of ACTH in the anterior/intermediate lobes may also be regulated by variable expression of the enzyme AACE. Its higher level in the intermediate lobe may explain the more extensive processing to α-MSH. Differences in intragranular Ca^{2+} may explain the additional processing of β-LPH in the intermediate vs. the anterior lobes since the effect of the enzyme PCE on β-LPH was shown to be Ca^{2+} dependent (BIRCH et al. 1991a).

III. Gastrointestinal Hormones

1. Glucagon and Related Peptides

As shown in Fig. 4, processing for the 179-amino acid pre-proglucagon can be quite complex, giving rise to a diversity of products in a tissue-specific manner (Holst 1983). In the pancreatic A cell, cleavages at the Lys-Arg sites give rise to glucagon-related polypeptide (GRPP), glucagon, IVP-1 and the remaining C-terminal portion. In contrast, the L cell of the gut produces primarily glicentin, oxyntomodulin, GLP-1, IVP-2 and GLP-2 via a different set of dibasic cleavages. The major form of GLP-1 is the truncated 7–37 sequence, the result of monobasic processing at Arg^6 (Holst et al. 1987). In addition, the terminal glycines of GLP-1 and IVP-2 are converted to amides. Tissue-specific processing is likely to be the result of specific expression of different processing enzymes; however, variable glycosylation and phosphorylation may also contribute (Conlon 1988). Since the precursor appears to be clipped first in the middle, it is tempting to suggest that the two halves may be separately sorted as occurs for ELH. The physiologic functions of these various products include roles in the regulation of acid secretion (GRPP, oxyntomodulin and IVP-1), the stimulation of glycogenolysis and gluconeogenesis (glucagon) and in the control of insulin secretion (GLP-1_{7-36}). Unprocessed GLP-1 and GLP-2, which can potently stimulate adenylate cyclase in brain, might also have roles as neuromodulators.

2. Cholecystokinin

This 115-amino acid precursor contains a single copy of CCK bordered by a monobasic and a dibasic site which are processed in a tissue-specific manner (Fig. 4). In brain, the octapeptide CCK-8 is the primary product, the result of processing at both of these sites. In intestine, the monobasic site is not cleaved and the result is the N-terminally extended CCK-58, although some CCK-39, CCK-33, CCK-22 and CCK-12 are also present (Rehfeld and Hansen 1986). All bioactive forms of CCK are amidated at the C terminus and usually the tyrosine 5 residues from the end is sulfated. Processing of pre-pro CCK is complex and the order of cleavage to yield the various forms has not been elucidated.

Viereck and Beinfeld (1990) have recently reported the isolation from rat brain synaptosomes of an enzyme which can cleave the Arg-Asp bond in CCK-33 to liberate CCK-8. This protease is similar to an activity reported previously (Strauss et al. 1980) and shown to consist of a dimer with the monomeric form being 45 kDa. It has a pH optimum of 8.0, is inhibited by *p*-amino-benzamidine and is classified as a serine protease. This enzyme is monobasic residue-specific and does not cleave at paired basic residues. It does not cleave small fluorescent or chromogenic kallikrein or trypsin substrates or other monobasic residue cleavage sites such as that present in pro-somatostatin to yield SS-28.

3. Gastrin

Gastrin (G) is synthesized from pre-progastrin of 101 amino acids containing a single copy of G-17 flanked on both sides by dibasic residues (Fig. 4). The C-terminal dibasic is preceded by Gly as is typical of amidation sites. As with CCK, alternate processing sites on the N-terminal side can give rise to a number of extended forms, including G-34 (DAUGHERTY and YAMADA 1989). In the antral mucosa of the stomach and in the neural/intermediate lobes of the pituitary, G-17 is the primary product, while in anterior pituitary and in fasting plasma, G-34 predominates. Labeling experiments with [^{3}H] proline in antral slices from rat stomach could not demonstrate interconversion of molecular forms. Using specific radioimmunoassays (RIAs), pro-G was ~0.2% and G-Gly was ~5% of total G-related activity in antral mucosa. Pulse-chase studies with gastrin-containing cells in culture suggest that G-Gly may be the immediate precursor to bioactive forms. Attenuated processing was found in pituitary and pancreas, such that precursor was as much as 90% of total (BARDRAM et al. 1990). Sulfation of G-peptides is variable (~50% in antrum) and it is not required for bioactivity but sulfation may protect G-peptides against degradation in blood. The 1–11 and 1–13 fragments of G-17 have also been identified in antral mucosa, the result of clips at Ala^{11}-Tyr^{12} and Gly^{13}-Trp^{14}. It has been suggested that endopeptidase 24.11 may be the enzyme responsible.

4. Somatostatin

Processing of the 92-amino acid somatostatin (SS) precursor is tissue specific (Fig. 4), yielding primarily the tetradecapeptide SS-14 in brain and pancreas via a dibasic cleavage and SS-28 in gut via processing at a monobasic site (BENOIT et al. 1987). A putative processing enzyme complex, consisting of an endoprotease which cleaves N terminal to the Arg-Lys sequence and a basic aminopeptidase which removes the terminal basic residues, has been demonstrated in rat brain secretory granules (GOMEZ et al. 1988). A monobasic cleavage enzyme that generates SS-28 has been found in rat brain and intestine (BEINFELD et al. 1989a,b).

5. Neurotensin and Neuromedin-N

Two bioactive peptides, neurotensin (NT) and neuromedin-N (NMN), are located adjacent to one another surrounded by Lys-Arg dibasic residues in this 170-residue precursor (Fig. 4). A fourth dibasic (canine, Arg-Arg; rat, Lys-Arg) is apparently not used. In keeping with this, the predicted conformation shows a strong probability for β-turns only in the vicinity of the cleaved sites. Processing is tissue-specific, giving the six-residue NMN in brain and an N-terminally extended NMN-125 in intestine (CARRAWAY and MITRA 1990). Thus, different enzymes may process the Lys-Arg dibasics surrounding NT and the Lys-Arg at the N terminus of NMN. These sites differ in that the former ones are preceded by Leu and followed by neutral residues, while the latter is preceded by Ile and followed by Lys. Based on the ability of pepsin to generate NT— and NMN—immunoreactivity from tissue precursor via cleavages between the C-terminal leucines and the Lys-Arg pairs, a role for a pepsin-related enzyme (e.g., renin) has been postulated (Carraway and Mitra 1987). According to this scheme, an aminopeptidase B-like enzyme would trim the Lys-Arg pairs from the N terminals of the intermediates.

6. Pancreatic Polypeptide

The precursor to pancreatic polypeptide (PP) is similar in structure to pro-peptide YY and pro-NPY (Fig. 4). It has one dibasic (Lys-Arg) and one monobasic (Arg) site, which are cleaved in a sequential fashion (Schwartz 1987). After removal of the signal peptide, PP-Gly is liberated and amidated. Monobasic processing occurs later, giving rise to an eicosapeptide, the function of which is not known. Work with pancreatic cells in culture suggests that the two cleavages are performed by different proteases.

G. Conclusion

We have attempted to present the reader with a novel perspective on peptide hormone processing, one which favors the idea that this could be a much more ubiquitous occurrence than generally recognized. Thus, in addition to describing various mechanisms known to occur within "endocrine" cells, we have also highlighted a few examples of processing in the extracellular space and at the target. We conclude that future work concerning the maturation and effects of bioactive peptides should consider all cells of the organism as potential participants, not simply those known to synthesize and store the active product(s). Although some of these ideas are speculative, we hope that they have stimulated an interest in this area and opened some minds to new concepts regarding the operation of peptide-producing systems.

References

Akil H, Shiomi H, Matthews J (1985) Induction of the intermediate pituitary by stress: synthesis and release of a form of β-endorphin. Science 227:424–426

Andrews PC, Brayton KA, Dixon JE (1989) Post-translational proteolytic processing of precursors to regulatory peptides. Experientia Suppl 56:192–209

Bardram L, Hilsted L, Rehfeld JF (1990) Progastrin expression in mammalian pancreas. Proc Natl Acad Sci USA 87:298–302

Beinfeld MC, Bourdais J, Kuks P, Morel A, Cohen P (1989a) Characterization of an endoprotease from rat small intestinal mucosal secretory granules which generates somatostatin-28 from prosomatostatin by cleavage after a single arginine residue. J Biol Chem 264:4460–4465

Beinfeld MC, Bourdais J, Morel A, Kuks PFM, Cohen P (1989b) Characterization of a somatostatin-28-generating metallo-endoprotease from rat brain cytosol. Biochem Biophys Res Commun 165:968–976

Benjannet S, Rondeau N, Day R, Chretien M, Seidah NG (1991) PC1 and PC2 are proprotein convertases capable of cleaving pro-opiomelanocortin at distinct pairs of basic residues. Proc Natl Acad Sci USA 88:3564–3568

Benoit R, Ling N, Esch F (1987) A new prosomatostatin-derived peptide reveals a pattern for prohormone cleavage at monobasic sites. Science 238:1126–1129

Bienkowski RS (1983) Intracellular degradation of newly synthesized secretory proteins. Biochemistry 214:1–10
Birch NP, Bennett HPJ, Estivariz FE, Loh YP (1991a) Effect of calcium ions on the processing of pro-opiomelanocortin by bovine intermediate lobe pro-opiomelanocortin converting enzyme. Eur J Biochem 201:85–89
Birch NP, Estivariz FE, Bennett HPJ, Loh YP (1991b) Differential glycosylation of *N*-POMC$_{1-77}$ regulates the production of γ3-MSH by purified pro-opiomelanocortin converting enzyme: a possible mechanism for tissue-specific processing. FEBS Lett 290:191–194
Blalock JE (1989) A molecular basis for bidirectional communication between the immune and neuroendocrine systems. Physiol Rev 69:1–32
Bondy CA, Whitnall MH, Brady LS (1989) Regulation of carboxypeptidase H gene expression in magnocellular neurons: response to osmotic stimulation. Mol Endocrinol 3:2086–2092
Bourdais J, Pierotti AR, Boussetta H, Barre N, Devilliers G, Cohen P (1991) Isolation and functional properties of an arginine selective endoprotease from rat intestinal mucosa. J Biol Chem 266:23386–23391
Bradbury AF, Smyth DG (1987) Biosynthesis of the C-terminal amide in peptide hormones. Biosci Rep 7:907–916
Brakch N, Boussetta H, Rholam M, Cohen P (1989) Processing endoprotease recognizes a structural feature at the cleavage site of peptide prohormones. J Biol Chem 264:15912–15916
Breshnahan PA, Leduc R, Thomas L, Thorner J, Gibson H, Brake AJ, Barr PJ, Thomas G (1991) Human furin gene encodes a yeast KEX2-like endoprotease that cleaves pro-β-NGF in vivo. J Cell Biol 111:2851–2859
Brooks VL, Brownfield MS, Reid IA (1982) Measurement and localization of angiotensin-like immunoreactivity in juxtaglomerular cells of the rat. Regul Pept 4:317–324
Busby WH Jr, Quackenbush GE, Humm J, Youngblood WW, Kizer JS (1987) An enzyme(s) that converts glutaminyl-peptides into pyroglutamyl-peptides. J Biol Chem 262:8532–8533
Cambell DJ (1987) Circulating and tissue angiotensin system. J Clin Invest 79:1–6
Carraway RE, Mitra SP (1987) Precursor forms of neurotensin (NT) in cat: processing with pepsin yields NT (3–13) and NT (4–13). Regul Pept 18:139–154
Carraway RE, Mitra SP (1990) Differential processing of neurotensin/neuromedin N precusor(s) in canine brain and intestine. J Biol Chem 265:8627–8631
Carraway RE, Reinecke M (1984) Neurotensin-like peptides and a novel model of the evolution of signaling systems. In: Falkmer S, Håkanson R, Sundler F (eds) Evolution and tumour pathology of the neuroendocrine system. Elsevier, Amsterdam, pp 245–283
Castro MG, Birch NP, Loh YP (1989) Regulated secretion of pro-opiomelanocortin converting enzyme and an aminopeptidase B-like enzyme from dispersed bovine intermediate pituitary cells. J Neurochem 52:1619–1628
Chapell MC, Loh YP, O'Donohue TL (1983) Evidence for an opiomelanotropin acetyltransferase in the rat pituitary neurointermediate lobe. Peptides 3:405–410
Christie DL, Batchelor DC, Palmer DJ (1991) Identification of KEX2 related proteases in chromaffin granules by partial amino acid sequence analysis. J Biol Chem 266:15679–15683
Clamagirand C, Camier M, Fahy C, Clavreul C, Creminon C, Cohen P (1987) C-Terminally extended ocytocin and pro-ocytocin: neurophysin peptide converting enzyme in bovine corpus luteum. Biochem Biophys Res Commun 143:789–796
Clements JA (1989) The glandular kallikrein family of enzymes: tissue-specific expression and hormonal regulation. Endocr Rev 10:393–419
Conlon JM (1988) Proglucagon-derived peptides: nomenclature, biosynthetic relationships and physiological roles. Diabetologia 31:563–566
Darby NJ, Smyth DG (1990) Endopeptidases and prohormone processing. Biosci Rep 10:1–13

Darby NJ, Lackey DB, Smyth DG (1991) Purification of a cysteine endopeptidase which is secreted with bioactive peptides from the epidermal glands of *Xenopus laevis*. Eur J Biochem 195:65–70

Darmer D, Schmutzler C, Diekhoff D, Grimmelikhuijzen CJP (1991) Primary structure of the precursor for the sea anemone neuropeptide Antho-RFamide (<Glu-Gly-Arg-Phe-NH_2). Proc Natl Acad Sci USA 88:2555–2559

Daugherty D, Yamada T (1989) Posttranslational processing of gastrin. Physiol Rev 69:483–502

Davidson HW, Rhodes CJ, Hutton JC (1988) Intraorganellar calcium and pH control proinsulin cleavage in the pancreatic β cell via two distinct site-specific endopeptidases. Nature 333:93–96

Devi L, Goldstein A (1984) Dynorphin converting enzyme with unusual specificity from rat brain. Proc Natl Acad Sci USA 81:1892–1896

Devi L, Gupta P, Fricker LD (1991) Subcellular localization, partial purification and characterization for a dynorphin processing endopeptidase from bovine pituitary. J Neurochem 56:320–321

Dickerson IM, Mains RE (1990) Cell-type specific posttranslational processing of peptides by different pituitary cell lines. Endocrinology 127:133–140

Dickerson IM, Dixon JE, Mains RE (1990) Biosynthesis and posttranslational processing of site-directed endoproteolytic cleavage mutants of pro-neuropeptide Y in mouse pituitary cells. J Biol Chem 265:2462–2469

Diment S, Martin KJ, Stahl PD (1989) Cleavage of parathyroid hormone in macrophage endosomes illustrates a novel pathway for intracellular processing of proteins. J Biol Chem 264:13403–13406

Dmochowska A, Dignard D, Henning D, Thomas DY, Bussey H (1987) Yeast KEX1 gene encodes a putative protease with a carboxypeptidase B-like function involved in killer toxin and α-factor precursor processing. Cell 50:573–584

Dockray GJ, Varro A, Desmond H, Young J, Gregory H, Gregory RA (1987) Post-translational processing of the porcine gastric precursor by phosphorylation of the COOH-terminal fragment. J Biol Chem 262:8643–8647

Egel-Mitani M, Flygenring HP, Hansen MT (1990) A novel aspartic protease allowing KEX2-independent MFα prohormone processing in yeast. Yeast 6:127–137

Eipper BA, Mains RE (1980) Structure and biosynthesis of pro-adrenocorticotropin/endorphin and related peptides. Endocr Rev 1:1–27

Eipper BA, Mains RE (1982) Phosphorylation of pro-adrenocorticotropin/endorphin-derived peptides. J Biol Chem 257:4907–4915

Eipper BA, Mains RE (1988) Peptide α-amidation. Annu Rev Physiol 50:333–344

Eipper BA, Park LP, Dickerson IM, Keutmann HT, Thiele EA, Rodriguez H, Schofield PR, Mains RE (1987) Structure of the precursor to an enzyme mediating COOH-terminal amidation in peptide synthesis. Mol Endocrinol 1:777–790

Eipper BA, Perkins SN, Husten JE, Johnson RC, Keutmann HT, Mains RE (1991) Peptidyl-α-hydroxyglycine α-amidating lyase. J Biol Chem 266:7827–7833

Elbein AD (1987) Inhibitors of the biosynthesis and processing of N-linked oligosaccharide chains. Annu Rev Biochem 56:497–534

Estivariz FE, Birch NP, Loh YP (1989) Generation of Lys-γ3-melanotropin from pro-opiomelanocortin$_{1-77}$ by a bovine intermediate lobe secretory vesicle membrane-associated aspartic protease and purified pro-opiomelanocortin converting enzyme. J Biol Chem 264:17796–17801

Estivariz FE, Friedman TC, Chikuma T, Loh YP (1992) Processing of adrenocorticotrpin by two proteases in bovine intermediate lobe secretory vesicle membranes: a distinct acidic, tetrabasic residue-specific calcium-activated serine protease and a PC2-like enzyme. J Biol Chem 267:7456–7463

Evans EA, Gilmore R, Blobel G (1986) Purification of microsomal signal peptidase as a complex. Proc Natl Acad Sci USA 83:581–585

Feinmark SJ (1988) Cooperative synthesis of leukotrienes by leukocytes and vascular cells. Ann NY Acad Sci 524:122–132

Fisher DA, Lakshmanan J (1990) Metabolism and effects of epidermal growth factor and related growth factors in mammals. Endocr Rev 11:418–439

Fisher JM, Scheller RH (1988) Prohormone processing and the secretory pathway. J Biol Chem 263:16515–16518

Fisher JM, Sossin W, Newcomb R, Scheller H (1988) Multiple neuropeptides derived from a common precursor are differentially packaged and transported. Cell 54:813–822

Fisher WH, Spiess J (1987) Identification of a mammalian glutaminyl cyclase converting glutaminyl into pyroglutamyl peptides. Proc Natl Acad Sci USA 84:3628–3632

Fricker LD (1988) Carboxypeptidase E. Annu Rev Physiol 50:309–312

Fricker LD (1991) Peptide processing exopeptidases: amino- and carboxypeptidases involved with peptide biosynthesis. In: Fricker LD (ed) Peptide biosynthesis and processing. CRC, Boca Raton, pp 199–229

Fricker LD, Evans CJ, Esch FS, Herbert E (1986) Cloning and sequence analysis of cDNA for bovine carboxypeptidase E. Nature 323:461–464

Fricker LD, Adelman JP, Douglass J, Thompson RC, von Strandmann RP, Hutton JC (1989) Isolation and sequence analysis of cDNA for rat carboxypeptidase E (EC 3.4.17.10), a neuropeptide processing enzyme. Mol Endocrinol 3:666–673

Fuller RS, Sterne RE, Thorner J (1988) Enzymes required for yeast prohormone processing. Annu Rev Physiol 50:345–362

Fuller RS, Brake A, Thorner J (1989) Yeast prohormone processing enzyme (KEX2 gene product) is a Ca^{2+}-dependent serine protease. Proc Natl Acad Sci USA 86:1434–1438

Furie B, Furie BC (1991) Molecular basis of γ-carboxylation; role of propeptide in the vitamin K-dependent proteins. Ann NY Acad Sci 614:1–10

Gainer H, Russell JT, Loh YP (1984) An aminopeptidase activity in bovine pituitary secretory vesicles that cleaves the N-terminal arginine from β-lipotropin$_{60-65}$. FEBS Lett 175:135–139

Giraud AS, Williams RG, Dockray GJ (1984) Evidence for different patterns of post-translational processing of pro-enkephalin in the bovine adrenal, colon and striatum indicated by radioimmunoassay using region specific antisera to Met-Enk-Arg-Phe and Met-Enk-Arg-Gly-Leu. Neurosci Lett 46:223–228

Glembotski CC (1982) Acetylation of α-MSH and β-endorphin in the rat intermediate pituitary. J Biol Chem 257:424–426

Gluschankof P, Cohen P (1987) Proteolytic enzymes in the post-translational processing of polypeptide hormone precursors. Neurochem Res 12:951–958

Gluschankof P, Morel A, Gomez S, Nicolas P, Fahy C, Cohen P (1984) Enzymes processing somatostatin precursors: an Arg-Lys-esteropeptidase from the rat brain cortex converting somatostatin-28 into somatostatin-14. Proc Natl Acad Sci USA 81:6662–6666

Gluschankof P, Gomez S, Morel A, Cohen P (1987) Enzymes that process somatostatin precursors. J Biol Chem 262:9615–9620

Goldblatt PJ (1986) What is the role of extrarenal renin? Arch Pathol Lab Med 110:1128–1130

Gomez S, Gluschankof P, Lepage A, Cohen P (1988) Relationship between endo- and exopeptidases in a processing enzyme system: activation of an endoprotease by the aminopeptidase B-like activity in somatostatin-28 convertase. Proc Natl Acad Sci USA 85:5468–5472

Greenbaum LM, Cho C, Barlas A (1989) The role of T-kininogen and H and L kininogens in health and disease. Adv Exp Med Biol 247A:113–116

Harris RB (1989) Processing of pro-hormone precursor proteins. Arch Biochem Biophys 275:315–333

Holst JJ (1983) Gut glucagon, enteroglucagon, gut glucagonlike immunoreactivity, glicentin-current status. Gastroenterology 84:1602–1613

Holst JJ, Orskov C, Nielsen OV, Schwartz TW (1987) Truncated glucagon-like peptide I, an insulin-releasing hormone from the distal gut. FEBS Lett 211: 169–174

Hook VYH, Loh YP (1984) Carboxypeptidase B-like converting enzyme activity in secretory granules of rat pituitary. Proc Natl Acad Sci USA 81:2776–2780

Hook VYH, Eiden LE, Pruss RM (1985) Selective regulation of carboxypeptidase peptide hormone-processing enzyme during enkephalin biosynthesis in cultured bovine adrenomedullary chromaffin cells. J Biol Chem 260:5991–5997

Huttner WB (1988) Tyrosine sulfation and the secretory pathway. Annu Rev Physiol 50:363–376

Julius D, Blair L, Brake A, Sprague G, Thorner J (1983) Yeast α-factor is processed from a larger precursor polypeptide: the essential role of a membrane-bound dipeptidyl aminopeptidase. Cell 32:839–852

Kanamatsu T, Unswerth CD, Diliberto EJ, Viveros OH, Hong JS (1986) Reflex splanchnic nerve stimulation increases levels of pro-enkephalin A mRNA and pro-enkephalin A-related peptides in the rat adrenal medulla. Proc Natl Acad Sci USA 83:9245–9249

Kennelly PJ, Krebs EG (1991) Consensus sequences as substrate specificity determinants for protein kinases and protein phosphatases. J Biol Chem 266:15555–15558

Kizer JS, Tropsha A (1991) A motif found in propeptides and prohormones that may target them to secretory vesicles. Biochem Biophys Res Commun 174:586–592

Krebs EG, Beavo JA (1979) Phosphorylation-dephosphorylation of enzymes. Annu Rev Biochem 48:923–959

Krieger TJ, Hook VYH (1991) Purification and characterization of a novel thiol protease involved in processing the enkephalin precursor. J Biol Chem 266: 8376–8383

Kriegler M, Perez C, DeFay K, Albert I, Lu SD (1988) A novel form of TNF/cachectin is a cell surface cytotoxic transmembrane protein: ramifications for the complex physiology of TNF. Cell 53:45–53

Kuks PF, Creminon C, Leseney A-M, Bourdais J, Morel A, Cohen P (1989) *Xenopus laevis* skin Arg-Xaa-Val-Arg-Gly-endoprotease. J Biol Chem 264: 14609–14612

Kurjan J, Hershowitz I (1984) Structure of a yeast pheronome gene (MFα): a putative α-factor precursor contains tandem copies of mature α-factor. Cell 30:933–943

Lee RWH, Huttner WB (1985) (Glu^{62}, Ala^{30}, Tyr^{8})n serves as a high affinity substrate for tyrosylprotein sulfotransferase. Proc Natl Acad Sci USA 82: 6143–6147

Lindberg I, Thomas G (1990) Cleavage of pro-enkephalin by a chromaffin granule processing enzyme. Endocrinology 126:480–487

Loh YP (1986) Kinetic studies on the processing of human β-lipotropin by bovine pituitary intermediate lobe pro-opiomelanocortin-converting enzyme. J Biol Chem 261:11949–11955

Loh YP, Brownstein MJ, Gainer H (1984) Proteolysis in neuropeptide processing and other neural functions. Annu Rev Neurosci 7:189–222

Loh YP, Parish DC, Tuteja R (1985) Purification and characterization of a paired basic residue-specific pro-opiomelanocortin converting enzyme from bovine pituitary intermediate lobe secretory vesicles. J Biol Chem 260:7194–7205

MacGregor RR, Jilka RL, Hamilton JW (1986) Formation and secretion of fragments of parathormone. J Biol Chem 261:1929–1934

Mackin RB, Noe BD (1987) Direct evidence for two distinct prosomatostatin converting enzymes. J Biol Chem 262:6453–6456

Mackin RB, Mackin JA, Noe BD (1985) Detection and partial characterization of an aminopeptidase B-like enzyme in anglerfish islets. Endocrinology 116:69a

Mackin RB, Noe BD, Spiess J (1991a) Identification of a somatostatin-14-generating pro-peptide converting enzyme as a member of the KEX2/furin/PC family. Endocrinology 129:2263–2265

Mackin RB, Noe BD, Spiess J (1991b) The anglerfish somatostatin-28-generating pro-peptide converting enzyme is an aspartyl protease. Endocrinology 129: 1951–1957

Mains RE, Bloomquist BT, Eipper BA (1991) Manipulation of neuropeptide biosynthesis through the expression of antisense RNA for peptidylglycine α-amidating monooxygenase. Mol Endocrinol 5:187–193

Maret GE, Fauchere JL (1988) Purification of an endopeptidase from bovine adrenal medulla granules which cleaves in vitro at paired but not single basic residues. Anal Biochem 172:248–258

Marti U, Burwen SJ, Jones Al (1989) Biological effects of epidermal growth factor, with emphasis on the gastrointestinal tract and liver: an update. Hepatology 9:126–138

Massague J (1990) Transforming growth factor-α, a model for membrane anchored growth factors. J Biol Chem 265:21393–21396

Matsumura L, Ikegawa R, Takaoka M, Morimoto S (1990) Conversion of porcine big endothelin to endothelin by an extract from the porcine aortic endothelial cells. Biochem Biophys Res Commun 167:203–210

Miyazono K, Hellman U, Wernstedt C, Heldin C-H (1988) Latent high molecular weight complex of transforming growth factor β1. J Biol Chem 263:6407–6415

Mizuno K, Ohsuye K, Wada Y, Fuchimura K, Tanaka S, Matsuo H (1987) Cloning and sequence of cDNA encoding a peptide C-terminal α-amidating enzyme from *Xenopus laevis*. Biochem Biophys Res Commun 148:546–552

Mizuno K, Nakamura T, Ohshima T, Tanaka S, Matsua H (1989) Characterization of KEX2-encoded endopeptidase from yeast *Saccharomyces cerevisiae*. Biochem Biophys Res Commun 159:305–311

Molinari AM, Abbondanza C, Armetta I, Medici N, Minucci S, Moncharmont B, Nigro V (1991) Proteolytic activity of the purified hormone-binding subunit in the estrogen receptor. Proc Natl Acad Sci USA 88:4463–4467

Moore H-P (1987) Factors controlling packaging of peptide hormones into secretory granules. Ann NY Acad Sci 493:50–59

Moore H-P, Gumbiner B, Kelly RB (1983) Chloroquine diverts ACTH from a regulated to a constitutive pathway in AtT-20 cells. Nature 302:434–436

Mueller SG, Paterson AJ, Kudlow JE (1990) Transforming growth factor α in arterioles: cell surface processing of its precursor by elastasis. Mol Cell Biol 10:4596–4602

Nyberg F, Nylander I, Terenius L (1986) Enkephalin-containing polypeptides in human cerebrospinal fluid. Brain Res 371:278–286

Pandey KN, Inagami T (1986) Regulation of renin angiotensins by gonadotropic hormones in cultured murine Leydig tumor cells. J Biol Chem 261:3934–3938

Pandiella A, Massague J (1991) Cleavage of the membrane precursor for transforming growth factor α is a regulated process. Proc Natl Acad Sci USA 88:1726–1730

Parish DC, Tuteja R, Altstein M, Gainer H, Loh YP (1986) Purification and characterization of a paired basic residue-specific prohormone-converting enzyme from bovine pituitary neural lobe secretory vesicles. J Biol Chem 261: 14392–14397

Peach MJ (1977) Renin-angiotensin system: biochemistry and mechanisms of action. Physiol Rev 57:313–370

Pfeffer SR, Rothman JE (1987) Biosynthetic protein transport and sorting by the endoplasmic reticulum and golgi. Annu Rev Biochem 56:829–852

Plevrakis I, Clamagirand C, Creminon C, Brakch N, Rholam M, Cohen P (1989) Prooxytocin/neurophysin convertase from bovine neurohypophysis and corpus luteum secretory granules: complete purification, structure-function relationships and competitive inhibitor. Biochemistry 28:2705–2710

Puca GA, Abbondanza C, Nigro V, Armetta I, Medici N, Molinari AM (1986) Estradiol receptor has proteolytic activity that is responsible for its own transformation. Proc Natl Acad Sci USA 83:5367–5371

Rangaraju NS, Harris RB (1991) Processing enzyme specificity is a consequence of prohormone precursor protein conformation. Arch Biochem Biophys 290: 418–426

Regoli D, Barabe J (1980) Pharmacology of bradykinin and related kinins. Pharmacol Rev 32:1–46

Rehfeld JF, Hansen HF (1986) Characterization of preprachoIecystokinin products in the porcine cerebral cortex. J Biol Chem 261:5832–5840

Renfrew CA, Hubbard AL (1991) Sequential processing of epidermal growth factor in early and late endosomes of rat liver. J Biol Chem 266:4348–4356

Rens-Domiano S, Roth JA (1989) Characterization of tyrosylprotein sulfotransferase from rat liver and other tissues. J Biol Chem 264:899–905

Rettenmier CW, Roussel MF, Ashmun RA, Ralph P, Price K, Sherr CJ (1987) Synthesis of membrane-bound CSF-1 and down-modulation of CSF-1 receptors in NIH 3T3 cells transformed by cotransfection of the human CSF-1 and c-*fms* (CSF-1 receptor) genes. Mol Cell Biol 7:2378–2387

Rholam M, Nicolas P, Cohen P (1986) Precursors for peptide hormones share common secondary structures forming features at the proteolytic processing sites. FEBS Lett 207:1–6

Richter WO, Jacob BG, Schwardt P (1990) Processing of the lipid-mobilizing peptide β-lipotropin in rabbit adipose tissue. Mol Cell Endocrinol 71:229–238

Rittenhouse J, Moberly L, Ahmed H, Marcus F (1988) Phosphorylation in situ of atrial natriuretic peptide prohormone at the cyclic-AMP-dependent site. J Biol Chem 263:3778–3783

Rodriguez C, Brayton KA, Brownstein M, Dixon JE (1989) Rat pre-carboxypeptidase H: cloning, characterization and sequence of the cDNA and regulation of the mRNA by corticotropin-releasing factor. J Biol Chem 264:5988–5995.

Rudman D, Hollins BM, Kutner MH, Moffitt SD, Lynn MJ (1983) Three types of α-melanocyte stimulating hormone: bioactivities and half lives. Am J Physiol 244:E47-54

Sato SM, Mains RE (1988) Plasticity in the ACTH-related peptides produced by primary cultures of neonatal rat pituitary. Endocrinology 122:68–77

Sato Y, Tsuboi R, Lyons R, Moses H, Rifkin DB (1990) Characterization of the activation of latent TGF-β by co-cultures of endothelial cells and pericytes or smooth muscle cells: a self-regulating system. J Cell Biol 111:757–763

Schaudies RP, Gorman RM, Savage CR, Poretz RD (1987) Proteolytic processing of epidermal growth factor within endosomes. Biochem Biophys Res Commun 143:710–715

Schnabel E, Mains RE, Farquhar MG (1989) Proteolytic processing of pro-ACTH/endorphin begins in the golgi complex. Mol Endocrinol 3:1223–1235

Schwartz TW (1986) The processing of peptide precursors: proline-directed arginyl cleavage and other monobasic processing mechanisms. FEBS Lett 200:1–10

Schwartz TW (1987) Cellular peptide processing after a single arginyl residue – studies on the common precursor for pancreatic polypeptide and pancreatic icosapeptide. J Biol Chem 262:5093–5099

Schwartz TW (1990) The processing of peptide precursors. In: Okamoto H (ed) Molecular biology of the islets of Langerhans. Cambridge University Press, Cambridge, pp 153–205

Sealey JE, Atlas SA, Laragh JH (1983) Prorenin in plasma and kidney. Fed Proc 42:2681–2689

Seger MA, Bennett HPJ (1986) Structure and bioactivity of the amino-terminal fragment of pro-opiomelanocortin. J Steroid Biochem 25:703–710

Seidah NG, Hendy GN, Hamelin J, Paquin J, Lazure C, Metters KM, Rossier J, Chretien M (1987) Chromogranin A can act as a reversible processing enzyme inhibitor. FEBS Lett 211:144–150

Seidah NG, Gaspar L, Mion P, Marcinkiewicz M, Mbikay M, Chretien M (1990) cDNA sequence of two distinct pituitary proteins homologous to KEX2 and furin gene products: tissue-specific mRNAs encoding candidates for prohormone processing proteinases. DNA 9:415–424

Seidah NG, Marcinkiewicz S, Benjannet S, Gaspar L, Beaubien G, Mattei MG, Lazure C, Mbikay M, Chretien M (1991) Cloning and primary sequence of a mouse candidate prohormone convertase PC1 homologous to PC2, furin, and kex2: distinct chromosomal localization and messenger RNA distribution in brain and pituitary compared to PC2. Mol Endocrinol 5:111–122

Sessa WC, Kaw S, Hecker M, Vane JR (1991) The biosynthesis of endothelin-1 by human polymorphonuclear leukocytes. Biochem Biophys Res Commun 174: 613–618

Sevarino KA, Stock P, Ventimiglia R, Mandel G, Goodman R (1989) Aminoteminal sequences of prosomatostatin direct intracellular targeting but not processing specificity. Cell 57:11–19

Shen FS, Roberts SF, Lindberg I (1989) A putative processing enzyme for proenkephalin in bovine adrenal chromaffin granule membranes. J Biol Chem 264: 15600–15605

Smeekens SP, Steiner DF (1990) Identification of a human insuloma cDNA encoding a novel mammalian protein structurally related to the yeast dibasic processing protease KEX2. J Biol Chem 265:2997–3000

Smeekens SP, Auruch AS, LaMendola J, Chan SJ, Steiner D (1991) Identification of a cDNA encoding a second putative prohormone convertase related to PC2 in AtT-20 cells and islets of Langerhans. Proc Natl Acad Sci USA 88:340–344

Smyth DG, Darby NJ, Maruthainar K (1988) Sequential formation of β-endorphin-related peptides in porcine pituitary. Neuroendocrinology 47:317–322

Steiner DF, Quinn PS, Chan SJ, Marsh J, Tager HS (1980) Processing mechanisms in the biosynthesis of proteins. Ann NY Acad Sci 343:1–53

Strauss E, Malesci A, Yalow RS (1980) Further characterization of brain cholecystokinin-converting enzymes. Proc Natl Acad Sci USA 77:3669–3671

Stoller TJ, Shields D (1989) The role of paired basic amino acids in mediating proteolytic cleavage of prosomatostatin. J Biol Chem 264:6922–6928

Terenius L, Nyberg F (1988) Neuropeptide-processing, -converting and -inactivating enzymes in human cerebrospinal fluid. Int Rev Neurobiol 30:101–121

Tezapsidis N, Parish DC (1989) Characterization and partial purification of a novel prohormone processing enzyme from ovine adrenal medulla. FEBS Lett 246: 44–48

Thibault S, Garcia R, Gutkowska J, Bilodeau J, Lazure C, Seidah NG, Chretien M, Genest J, Cantin M (1987) The propeptide Asn^{1}-Tyr^{126} is the storage form of rat atrial natriuretic factor. Biochem J 241:265–272

Thim L, Hansen MT, Norris K, Hoegh I, Boel E, Forstrom J, Ammerer G, Fiil NP (1986) Secretion and processing of insulin precursors in yeast. Proc Natl Acad Sci USA 83:6766–6770

Thomas L, Leduc R, Thorne BA, Smeekens SP, Steiner DF, Thomas G (1991) Kex2-like endoproteases PC2 and PC3 accurately cleave a model prohormone in mammalian cells: evidence for a common core of neuroendocrine processing enzymes. Proc Natl Acad Sci USA 88:5297–5301

Thorne BA, Thomas G (1990) An in vivo characterization of the cleavage site specificity of the insulin cell prohormone processing enzymes. J Biol Chem 265:8436–8443

Van den Ouweland AMW, van Juijnhoven HLP, Keizer GD, Dorssers LCJ, van de Ven WJM (1990) Structural homology between the human *fur* gene product and the subtilisin-like protease encoded by yeast KEX2. Nucleic Acids Res 18:664–666

Viereck JC, Beinfeld MC (1990) Partial purification of a CCK cleaving endoprotease from rat brain synaptosomes. Soc Neurosci Abstr 16:392

Wang PH, Do YS, Macaulay L, Shinagowa T, Anderson PW, Baxter JD, Hsueh WA (191) Identification of renal cathepsin B as a human pro-renin-processing enzyme. J Biol Chem 266:12633–12638

Watkinson A, Young J, Vano A, Dockray GJ (1989) The isolation and chemical characterization of phosphorylated enkephalin-containing peptides from bovine adrenal medulla. J Biol Chem 264:3061–3065

Whitacre CC (1990) Immunology, a state of the art lecture. Ann NY Acad Sci 594:1–16

Worthy K, Figueroa CD, Dieppe PA, Bhoola KD (1990) Kallikreins and kinins: mediators in inflammatory joint disease. Int J Exp Pathol 71:587–601

Wypij DM, Harris RB (1988) Atrial granules contain an amino-terminal processing enzyme of atrial natriuretic factor. J Biol Chem 263:7079–7086

Zamir N, Weber E, Palkovits M, Brownstein M (1984) Differential processing of prodynorophins and proenkephalins in specific regions of the rat brain. Proc Natl Acad Sci USA 81:6886–6889

CHAPTER 4

Regulation of Peptide Secretion from Gastroenteric Endocrine Cells

D.L. Barber

A. Introduction

Peptides released from gastroenteric endocrine cells regulate a myriad of gastrointestinal functions, including gastric mucus and acid secretion, intestinal absorption and secretion, pancreatic exocrine secretion, motility, and the release of peptides from other gastroenteric endocrine cells. Characterizing the mechanisms regulating gastroenteric peptide release is therefore critical for understanding the physiology and pathophysiology of the gastrointestinal system. The field of endocrinology began nearly 100 years ago with the identification by Bayliss and Starling (1902) of the gastrointestinal hormone secretin. In the intervening years, the expression, localization, and biologic actions of at least 15 gastroenteric peptides have been characterized. Elucidating the mechanisms regulating gastroenteric peptide release, however, has been difficult. The diffuse tissue distribution of gastroenteric peptides, the complexity of neural, paracrine, and endocrine pathways regulating peptide secretion, and the lack of available model systems to accurately identify secretory events at the cell level have limited our study of gastroenteric peptide release.

B. Gastroenteric Endocrine Cells

I. Localization

Gastroenteric endocrine cells are localized primarily in the mucosa, where they are dispersed within a complex epithelium. Peptide-containing endocrine cells are diffusely distributed along the length of the gastrointestinal tract, from the gastric fundus through the colon, where they occupy the deep and intermediate regions of gastric glands and are distributed along the crypt-villus zone within the intestine. The general morphology of gastroenteric endocrine cells is columnar, with an apical membrane containing prominent microvilli in contact with the lumen. An abundance of dense-core secretory granules are localized in the expanded basal region of the cell. The size (150–400 nm) and morphological shape of these granules is generally peptide specific (Polak 1989). Many gastroenteric peptides such as gastrin, somatostatin (SS), cholecystokinin (CCK), and neurotensin are expressed not only in mucosal epithelial cells but also in neurons within the central, peripheral and enteric nervous systems. This dual localization of peptides in endocrine cells and neurons has made it difficult to interpret release studies using intact tissue, leading to questions regarding not only the site of released peptide, but also the source of peptides regulating gastroenteric endocrine cells.

II. Methods for Studying Peptide Release

Measurements of peptide release, originally made using bioassay systems, are now almost exclusively performed by radioimmunoassay. The predominant means for characterizing the regulation of gastroenteric peptide release include in vivo studies and in vitro models using intact tissue. In vivo determinations of circulating peptide levels have provided a valuable means of assessing the integrated functional response following nerve stimulation and the ingestion of nutrients. In vitro models for studying peptide release, including vascularly perfused organ preparations (Saffouri et al. 1980; Schubert et al. 1985; Holst et al. 1987), perifused tissue segments (Schubert et al. 1988b; Schubert and Hightower 1990) and organ culture preparations (Wolfe et al. 1984; Harty et al. 1985), have identified the localization of peptide release while maintaining the integrity of local neural and paracrine regulation. The use of intact tissue, however, cannot determine the direct action of secretagogues at the level of the endocrine cell and is therefore limited in resolving the functional complexity of the regulatory mechanisms underlying gastroenteric peptide release. Additionally, the transmembrane signaling mechanisms coupling stimulus to secretion cannot be characterized using these methods.

The limitations imposed by studying peptide release from intact tissue have recently been circumvented by the use of primary culture systems containing isolated mucosal cells. Soll et al. (1984b) and Yamada et al. (1984) developed the first viable culture system of isolated gastric endocrine cells, enriched for fundic somatostatin-containing cells. Subsequent to this study, the isolation and culture of antral gastrin cells (Giraud et al. 1987; Sugano et al. 1987) and enteric neurotensin (Barber et al. 1986a,b), enteroglucagon (Buchan et al. 1987), peptide YY (Aponte et al. 1988), and cholecystokinin (Koop and Buchan 1992) cells was successfully developed. These systems have provided a previously unavailable means to study not only the direct regulation of endocrine cells, but also the biochemical and molecular events regulating gastroenteric peptide secretion.

The characteristics of gastroenteric primary culture systems indicate that the morphological and functional integrity of endocrine cells in situ is maintained. Following adherence, generally on a collagen substrate, isolated endocrine cells demonstrate a morphological polarity, with a reformation of an apical microvillus structure and a basolateral accumulation of secretory

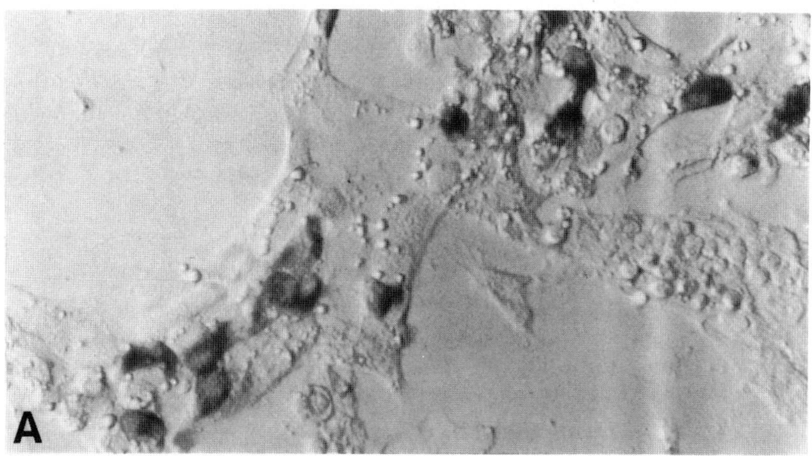

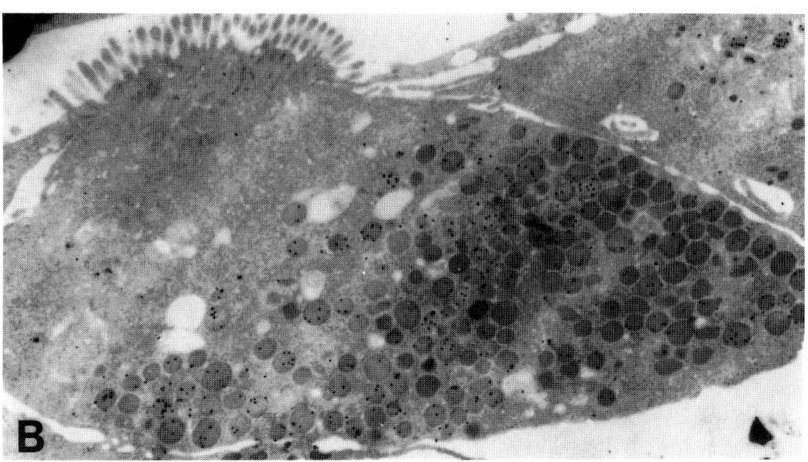

Fig. 1A,B. Neurotensin-immunoreactive cells in culture. Neurotensin immunoreactivity was identified **A** at the light microscopic level with peroxidase-antiperoxidase (×200) and **B** at the electron microscopic level with colloidal gold (×10000). (Electron micrograph, courtesy of A.M.J. Buchan)

vesicles (Fig. 1). Although all of these cell systems are heterogeneous, the percentage of peptide-specific endocrine cells, ranging from 5% for cholecystokinin-containing cultures (Koop and Buchan 1992) to 50% for enteroglucagon-containing cultures (Buchan et al. 1987), is highly enriched. Basal peptide release from primary cultures is generally less than 1% of the total peptide cell content and cell responsiveness is validated by the time- and dose-dependent action of specific secretagogues.

Peptide release from gastroenteric endocrine cells is regulated by receptors for hormones and neurotransmitters, and by ingested nutrients, mechanical distension, and luminal pH. A synopsis on the regulated release of individual gastroenteric peptides is beyond the scope of this review. The focus of this chapter is rather on general pathways and mechanisms that have been characterized to regulate peptide release from multiple types of gastroenteric endocrine cells (Table 1). When indicated, species-specific and tissue-specific differences among these general pathways will be discussed.

C. Neurotransmitter and Peptide Regulation of Release

I. Acetylcholine

Studies in vivo and using isolated perfused organ preparations have demonstrated that activation of vagal pathways either stimulates and/or inhibits the release of gastrin (Debas et al. 1984; Holst et al. 1987; Olesen et al. 1987), SS (McIntosh et al. 1981; Holst et al. 1987; Olesen et al. 1987; Schubert et al. 1982), CCK (Cantor et al. 1986a; Schafmayer et al. 1988), and neurotensin (Feurle et al. 1982; Al-Saffir et al. 1984). These conflicting findings most likely reflect the presence of both cholinergic and noncholinergic transmitters in parasympathetic nerves and they emphasize the difficulty in interpreting release studies using intact tissue. The precise role of cholinergic receptors in regulating gastroenteric peptide release has been more clearly defined by the recent use of isolated cell systems. Findings with cell preparations have revealed a direct cholinergic inhibition of gastric SS (Yamada et al. 1984) and intestinal CCK (Barber et al. 1986c), neurotensin (Barber et al. 1986b, 1987) and enteroglucagon (Buchan et al. 1987; Barber et al. 1987) release. Cholinergic agonists, however, have no effect on basal peptide release from isolated cells, inhibiting only stimulated release. Whether this reflects an inability to experimentally detect an inhibition of a low basal release (usually in the 10- to 100-fmol range), or indicates a physiologic role of cholinergic receptors in vivo, is uncertain. In vivo, cholinergic agonists have been shown to have no effect on basal release of CCK (Lewis and Williams 1990) or SS (Del Tacca et al. 1987), inhibiting only the stimulated release of these peptides. Two exceptions to the general cholinergic inhibition of gastroenteric peptide release are: (1) the recent findings of a dual stimulatory and inhibitory cholinergic regulation of gastric SS release and (2) an indirect cholinergic stimulation of gastrin secretion.

A dual stimulatory and inhibitory cholinergic regulation of SS release was first observed by Schubert et al. (1982), who found that atropine blockade of the vagal-mediated inhibition of gastric SS release revealed a stimulatory vagal mechanism on peptide release. The presence of stimulatory and inhibitory cholinergic receptors on fundic SS cells was later confirmed by studying selective receptor agonists and antagonists in fundic mucosal segments (Schubert and Hightower 1990) and fundic SS cells maintained in

Table 1. Summary of the better-characterized mechanisms regulating gastroenteric peptide release. As described in the text, species differences in these actions may exist

	Gastric peptides			Enteric peptides			
	Gastrin	Somatostatin Fundic	Somatostatin Antral	Cholecystokinin	Neurotensin	Enteroglucagon	Peptide YY
Muscarinic cholinergic	+/–	+/–	+	–	–	–	○
β-Adrenergic	+	+	+	○	+	+	+
Somatostatin	–			–	–	–	–
Gastrin-releasing peptide/bombesin	+	+[a]	+[a]	+	+	+	+
Proteins/amino acids	+	+	+	+	○	○	○
Fatty acids	○	+	+	+	+	+	+

+, stimulates; –, inhibits; ○, no effect or undetermined.
[a] Most likely indirectly mediated.

culture (YAMADA et al. 1984; CHIBA et al. 1987a). Cholinergic inhibition of SS release in the rat (SUE et al. 1985) and dog (YAMADA et al. 1984; CHIBA et al. 1987a; DEL TACCA et al. 1987) is mediated by a pirenzipine-sensitive muscarinic receptor. Muscarinic inhibition of fundic SS cells, most probably mediated by an M_2 receptor subtype, reduces intracellular adenosine-3′,5′-cyclic monophosphate (cAMP) (CHIBA et al. 1987a) and calcium ($[Ca^{2+}]_i$) levels (CHIBA et al. 1989). Pertussis toxin, a bacterial toxin that ADP ribosylates and thereby inactivates inhibitory guanine nucleotide binding proteins (G_i) (BOKOCH et al. 1983), completely reverses cholinergic inhibition of SS release (CHIBA et al. 1987a). Pertussis blockade, however, unmasks a pirenzipine-insensitive, cholinergic stimulation of SS release (CHIBA et al. 1987a; SCHUBERT and HIGHTOWER 1990). This stimulatory cholinergic input is coupled to increases in $[Ca^{2+}]_i$ (CHIBA et al. 1989; SCHUBERT and HIGHTOWER 1990), which are independent of phosphatidylinositol hydrolysis (CHIBA and YAMADA 1987). Hence the fundic SS-containing cell may express both stimulatory and inhibitory muscarinic cholinergic receptors.

In contrast to fundic SS release, cholinergic regulation of antral SS release in some species may be mediated exclusively through a stimulatory muscarinic receptor. In the isolated perfused pig stomach, field stimulation of the vagus decreases fundic SS output (OLSEN et al. 1987) and increases antral SS release (HOLST et al. 1987; OLSEN et al. 1987). BUCHAN et al. (1992) has found that in human antral cell cultures carbachol exerts only a stimulatory action on SS secretion and this effect is mediated by a pirenzipine-insensitive M_3 receptor subtype. Differences in cholinergic regulation of fundic and antral SS release, however, may be species dependent. In rat fundic and antral mucosal segments, the regulation of SS release is identical (SCHUBERT et al. 1988b), including a predominant cholinergic inhibition as well as a concurrent cholinergic stimulation of SS release from both tissues (SCHUBERT and HIGHTOWER 1990).

Cholinergic regulation of gastrin release is also complex. Studies on gastrin release in vivo and using isolated perfused stomach preparations indicate that there are both excitatory and inhibitory vagal pathways that involve both cholinergic and noncholinergic mechanisms. Vagal stimulation by sham feeding, insulin hypoglycemia or electric field stimulation increases gastrin release (KONTUREK et al. 1980; DEBAS et al. 1984; HOLST et al. 1987; SCHUBERT and MAKHLOUF 1987). However, atropine has either no effect on gastrin release following vagal stimulation (NISHI et al. 1985) or enhances this response (FAROUQ and WALSH 1975; FELDMAN and WALSH 1980). In contrast, gastrin release in response to muscarinic agonists is inhibited by atropine, a nonselective cholinergic antagonist (MARTINDALE et al. 1982; SAFFOURI et al. 1980), and by the M_1/M_2 antagonist, pirenzipine (SUE et al. 1985; DEL TACCA et al. 1987).

The most likely pathway for the atropine-insensitive vagal stimulation of gastrin secretion is a direct noncholinergic input from nerve fibers expressing gastrin-releasing peptide (GRP), the mammalian homolog of the amphibian skin peptide bombesin. Vagal stimulation of the isolated perfused pig antrum induces an atropine-insensitive release of GRP (HOLST et al. 1987), and the perfusion of antibodies to GRP (HOLST et al. 1987) and bombesin (SCHUBERT et al. 1985) blocks gastrin release in response to vagal stimulation. Additionally, a direct stimulatory effect of bombesin on gastrin release is observed in antral gastrin cells in culture (GIRAUD et al. 1987; SUGANO et al. 1987; CAMPOS et al. 1990).

In contrast, the atropine-sensitive, cholinergic stimulation of gastrin release is suggested to be mediated indirectly by SS (Fig. 2; DUVAL et al. 1981; SCHUBERT et al. 1982). Somatostatin has a well-characterized inhibitory paracrine action on gastrin release (see Sect. C.III) and a direct cholinergic inhibition of SS release is observed in isolated fundic cells (YAMADA et al. 1984). Additionally, perfusion of SS antibodies in the isolated rat stomach increases gastrin secretion (SAFFOURI et al. 1979; SHORT et al. 1985; MCINTOSH et al. 1991). Hence, the atropine-sensitive, cholinergic inhibition of SS release may result in a stimulation of gastrin secretion through the removal of an inhibitory SS input on the G cell. Further support for an indirect cholinergic regulation of gastrin release is the inability to observe a gastrin response to cholinergic agonists in isolated human gastrin cell preparations maintained in culture (BUCHAN, personal communication).

II. Norepinephrine

In almost all instances, norepinephrine acts on gastroenteric endocrine cells to stimulate peptide release through a β-adrenoreceptor. Findings from in vivo studies and isolated perfused organ preparations demonstrate a β-

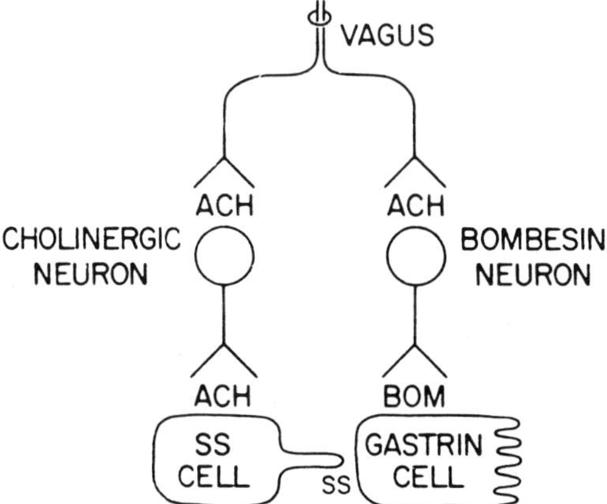

Fig. 2. Model describing the potential indirect cholinergic regulation of gastrin release. The atropine-sensitive, cholinergic regulation of gastrin release is suggested to be indirectly mediated by somatostatin. The atropine-insensitive vagal stimulation of gastrin release is most probably mediated by nerve fibers expressing bombesin-like peptide. (From DUVAL et al. 1981, published by The American Physiological Society)

adrenergic stimulation of gastrin (STADIL and REHFELD 1973; CHRISTENSEN and STADIL 1976; PETERS et al. 1982; KOOP et al. 1983), SS (GOTO et al. 1981; KOOP et al. 1983), and enteroglucagon (LICKLEY et al. 1981) release. Studies using isolated endocrine cell preparations have confirmed and expanded these findings and indicate a direct β-adrenergic stimulatory action on the release of human antral gastrin (BUCHAN 1991), canine fundic SS (SOLL et al. 1984b; YAMADA et al. 1984), canine intestinal neurotensin (BARBER et al. 1986b, 1987), enteroglucagon (BUCHAN et al. 1987; BARBER et al. 1987), and peptide YY (APONTE et al. 1988).

β-Adrenergic stimulation of human gastrin- (BUCHAN 1991) and canine neurotensin-containing cells (BARBER et al. 1989) is mediated through a terbutaline-sensitive β_2-adrenoceptor subtype. β-Adrenergic receptors are coupled to multiple effector pathways including the activation of adenylyl cyclase (LEFKOWITZ and CARON 1988) and voltage-dependent Ca^{2+} channels (YATANI et al. 1987) as well as the inhibition of Na^+ channels (SCHUBERT et al. 1989) and Mg^{2+} efflux (ERDOS and MAGUIRE 1983). In primary cultures of neurotensin- and enteroglucagon-containing canine ileal cells, β-adrenergic agonists have recently been shown to also stimulate Na-H exchange, inducing an intracellular alkalinization in a HEPES buffer (Fig. 3; BARBER et al. 1989). This previously unidentified action of β-adrenergic agonists on Na-H exchange appears to be a ubiquitous property of the receptor as it is observed in multiple cell types that endogenously express the β-adrenergic receptor (BARBER et al. 1989; GANZ et al. 1990) as well as in cells that have been transfected with either β_1- or β_2-receptor subtypes (GANZ et al. 1990; BARBER 1991, 1992). Receptor-regulated Na-H exchange is a well-established proliferative signal (GRINSTEIN and ROTHSTEIN 1986); however, its functional importance in the postmitotic gastroenteric endocrine cell is undetermined. Through its role in regulating intracellular pH, the exchanger may indirectly modulate pH-sensitive processes that regulate secretory events, such as protein phosphorylation (SIFFERT and AKKERMAN 1988), assembly of cytoskeletal proteins (TOOZE and BURKE 1987; PARTON et al. 1991) and membrane vesicle movement (COSSON et al. 1989; TOOZE and BURKE 1987). DIAL et al. (1991) suggest that alkalinization of the secretory granule may be a stimulatory signal for gastrin release. In addition to the β-adrenergic receptor, activity of the Na-H exchanger is regulated by prostaglandin E_1 (PGE_1), somatostatin, and D_2-dopamine receptors on enteric endocrine cells (BARBER et al. 1989, 1991) and by gastrin and CCK receptors on pancreatic acinar cells (CARTER et al. 1987; BASTIE and WILLIAMS 1990), further suggesting that this signaling mechanism may have a functional significance in regulating cells of the gastrointestinal tract.

III. Somatostatin

Somatostatin, originally isolated from sheep hypothalamus as a tetradecapeptide (BRAZEAU et al. 1973), is expressed throughout the gastrointestinal

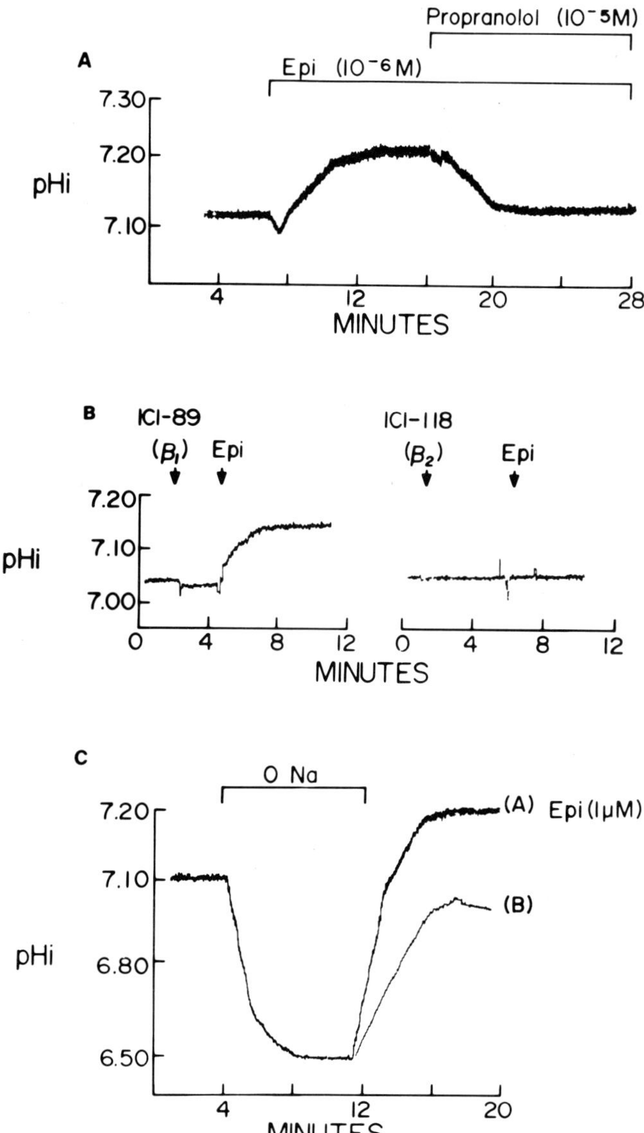

Fig. 3A–C. Effect of epinephrine (*Epi*) on intracellular pH (pH_i) in enteric endocrine cells. **A** Time course of epinephrine-induced alkalinization and the reversal of this effect by propranolol. **B** Time course for effects of epinephrine in the presence of the β_1-adrenergic antagonist ICI 89 406 and the β_2-antagonist ICI 118 551. **C** Rate of pH_i recovery from an acid load induced by replacing extracellular Na^+ with *N*-methyl-D-glucamine in the absence (**B**) and presence (**A**) of epinephrine. (From BARBER et al. 1989)

tract as both SS-14 and SS-28. The general inhibitory action of SS on gastroenteric peptide secretion is well established; however, its localization in endocrine cells of the gastric (Larsson et al. 1979) and intestinal (Polak et al. 1977) mucosa, and well as in neurons of the submucous and myenteric plexus (Costa et al. 1977), has complicated determinations on the physiological importance of SS as an endocrine, paracrine or neural regulator of peptide release.

Probably the best-characterized action of SS is that of a paracrine regulator of antral gastrin secretion. Morphologically, antral SS-containing mucosal cells demonstrate cytoplasmic extensions that terminate towards the basolateral domain of gastrin cells (Larsson et al. 1979). Functionally, SS inhibits gastrin release from intact tissue (Bloom et al. 1974; Saffouri et al. 1980; Harty et al. 1985) and from isolated gastrin cells in culture (Giraud et al. 1987; Sugano et al. 1987; Campos et al. 1990; DelValle and Yamada 1990). An important functional linkage between gastrin and SS, as suggested by Saffouri et al. (1980), is indicated by the ability of anti-SS antibodies to increase gastrin secretion in the isolated perfused rat stomach (Saffouri et al. 1979; Short et al. 1985; McIntosh et al. 1991) and in rat antral mucosal segments (Wolfe et al. 1984). Hence, both the tonic and regulated release of gastrin is dependent on the status of SS secretion. As discussed above, this SS-dependent regulation of gastrin is best exemplified by the role of SS in the cholinergic regulation of gastrin release.

In the jejunum, SS is thought to act as a paracrine regulator of CCK release (Abucam and Reichlin, 1990; Lewis and Williams 1990), although the majority of intestinal SS-like immunoreactivity is localized in enteric plexuses (Costa et al. 1977). In humans (DeJong et al. 1987), dogs (Misumi et al. 1988), and rats (Lewis and Williams 1990), SS acts in vivo to inhibit stimulated CCK release. In rats, SS also inhibits the expression of mRNA encoding CCK (Kanayma and Liddle 1990). A direct inhibition of CCK release by SS is suggested from the work of Koop and Buchan (1992), using isolated canine jejunal cells in culture. Similar to its effect on gastrin, SS may tonically inhibit the CCK-containing cell in vivo, as resting levels of circulating CCK are diminished following the administration of anti-SS antibodies (Abucam and Reichlin 1990).

In the ileum, SS inhibits bombesin- and lipid-stimulated neurotensin release in vivo (Rokaeus 1982; Ferris et al. 1985). Studies with isolated canine ileal cells in culture indicate that this effect is directly mediated through SS receptors on the neurotensin-containing cell (Barber et al. 1986b, 1987, 1989, 1991). Canine ileal cell cultures have also been used to demonstrate a direct action of SS in inhibiting enteroglucagon release (Buchan et al. 1987; Barber et al. 1987, 1991).

In the gastrointestinal tract, SS inhibits peptide release that has been stimulated by nutrients (Ferris et al. 1985; Misumi et al. 1988; DelValle and Yamada 1990; Lewis and Williams 1990; Barber et al. 1991), by receptor-mediated increases in intracellular cAMP (Barber et al. 1986b,

1987; BUCHAN et al. 1987) and $[Ca^{2+}]_i$ (GIRAUD et al. 1987; SUGANO et al. 1987; CAMPOS et al. 1990), and by postreceptor-induced increases in cAMP accumulation (HARTY et al. 1985; BARBER et al. 1987; KOOP and BUCHAN 1992), and $[Ca^{2+}]_i$ (BARBER et al. 1987). The ability of SS to inhibit peptide release following the activation of multiple intracellular messenger systems reflects the divergent signaling pathways regulated by the SS receptor. The transmembrane signaling mechanisms coupled to the SS receptor, characterized predominantly in pituitary endocrine cells, include an attenuation of intracellular cAMP (JAKOBS and SCHULTZ 1983; REISINE et al. 1985) and $[Ca^{2+}]_i$ (LEWIS ct al. 1985; KOCH et al. 1985) and an increase in K^+ conductance, which induces membrane hyperpolarization (YAMASHITA et al. 1986). All of these signaling actions of the SS receptor are mediated by an inhibitory GTP-binding protein that is sensitive to pertussis toxin. In a number of tissue-specific cells, including enteric and pituitary endocrine cells, SS also inhibits Na-H exchange, inducing an intracellular acidification (BARBER et al. 1989; BARBER 1991). In contrast to the action of SS on other effector pathways, SS inhibits Na-H exchange and peptide release independently of changes in cAMP accumulation and independently of a pertussis toxin-sensitive G_i protein (Fig. 4). The mechanisms whereby SS receptors regulate multiple effectors have not been determined. However, the identification of distinct receptor subtypes in the brain (TRAN et al. 1985; RAYNOR et al. 1991) and the recent cloning of subtype-specific SS receptor isoforms (YAMADA et al. 1992) may help to identify whether there is a structural basis for this divergence in function.

IV. Gastrin-Releasing Peptide and Bombesin

In the gastrointestinal tract, gastrin-releasing peptide (GRP), the mammalian homolog of the amphibian skin peptide bombesin, is localized exclusively in neurons. GRP exerts a general stimulatory action on the release of gastroenteric peptides extending from the antrum (GIRAUD et al. 1987; SUGANO et al. 1987) to the colon (APONTE et al. 1988). Probably the best-characterized action of this peptide, reflected in its name, is to stimulate antral gastrin release. As previously discussed, the atropine-insensitive release of gastrin following vagal stimulation is most likely mediated by GRP. GRP or bombesin stimulates gastrin release in vivo (TAYLOR et al. 1979; BUNNETT et al. 1985) and in the isolated perfused stomach (DUVAL et al. 1981; MARTINDALE et al. 1982; HOLST et al. 1987). A direct stimulatory effect of bombesin on gastrin release has been confirmed using primary cultures of antral gastrin-containing cells (GIRAUD et al. 1987; SUGANO et al. 1987; CAMPOS et al. 1990). In antral cell preparations, bombesin-stimulated gastrin release is completely inhibited by selective analogs (BUCHAN et al. 1990b), antagonists (CAMPOS et al. 1989), and bombesin-specific antibodies (Fig. 5; GIRAUD et al. 1987). Findings from in vivo studies (BUNNETT et al. 1985) and from isolated cell preparations

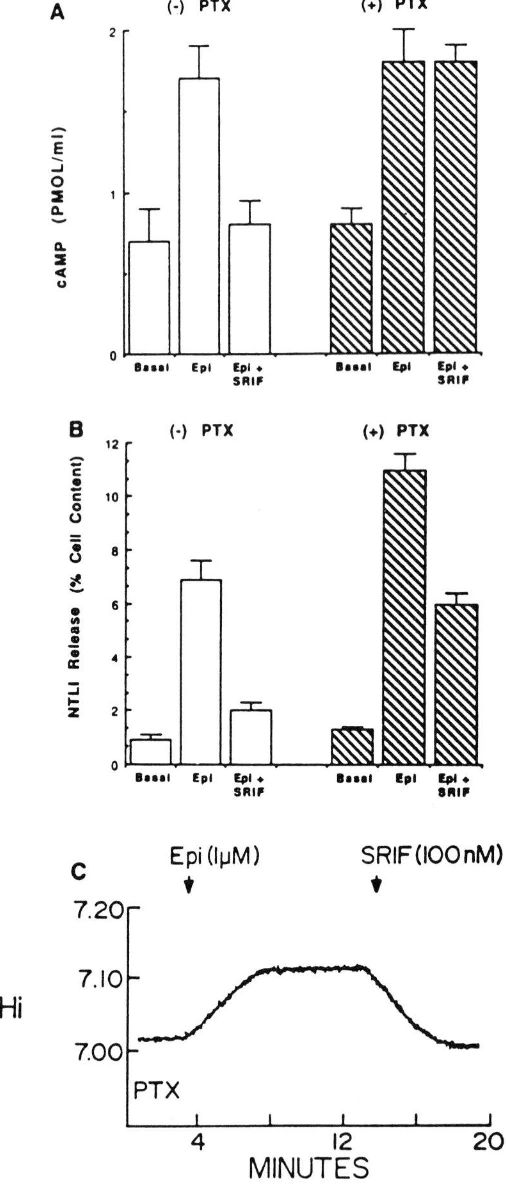

Fig. 4A–C. Effect of pertussis toxin (PTX) on somatostatin (SRIF) actions in enteric endocrine cells. **A** Somatostatin regulation of cAMP in the absence (−) and presence (+) of pertussis toxin in basal state, with epinephrine (Epi), and epinephrine plus somatostatin. **B** Somatostatin regulation of neurotensin release in the absence (−) and presence (+) of pertussis toxin in basal state, with epinephrine, and epinephrine plus somatostatin. **C** Somatostatin-induced acidification of intracellular pH (pH_i) in the presence of pertussis toxin. Somatostatin was added once a steady-state level of epinephrine-induced alkalinization was achieved. (From BARBER et al. 1989)

(GIRAUD et al. 1987; CAMPOS et al. 1990) suggest an extremely potent secretory action, as femtomolar concentrations of bombesin effectively stimulate gastrin release (Fig. 5). The recent identification of high-affinity, high-capacity binding sites for [^{125}I]Tyr^{4}-bombesin in fractions of isolated antral cells enriched for gastrin cells (VIGNA et al. 1988) confirms the presence of high-affinity bombesin receptors on the gastrin cell.

Like gastrin, the atropine-resistant increase in gastric SS secretion following stimulation of the vagus is thought to be mediated by vagal fibers expressing GRP. This increase in SS secretion may be the predominant mechanism mediating bombesin-induced inhibition of acid secretion. Somatostatin release in response to vagal activation or electric field stimulation is inhibited by bombesin and GRP antagonists (HOLST et al. 1987; SCHUBERT et al. 1991) and by specific GRP antibodies (HOLST et al. 1987). In isolated perfused stomach preparations, GRP or bombesin stimulates SS release (CHIBA et al. 1980; DUVAL et al. 1981; MCINTOSH et al. 1981; MARTINDALE et al. 1982; HOLST et al. 1987) and this action may involve both fundic and antral SS (SCHUBERT et al. 1988b). It is uncertain, however, if these peptides have a direct effect on SS release. In the isolated perfused rat stomach, bombesin-stimulated SS release is inhibited by atropine (MARTINDALE et al. 1982; GUO et al. 1990) and by gastrin antiserum (GUO et al. 1990). Additionally, in cell cultures enriched for either fundic (CHIBA et al. 1988) or antral (BUCHAN et al. 1990) somatostatin-containing cells, bombesin has no effect on peptide release. GUO et al. (1990) suggests that the stimulatory action of GRP or bombesin on SS release in intact tissue may be indirectly mediated through both neural and paracrine pathways.

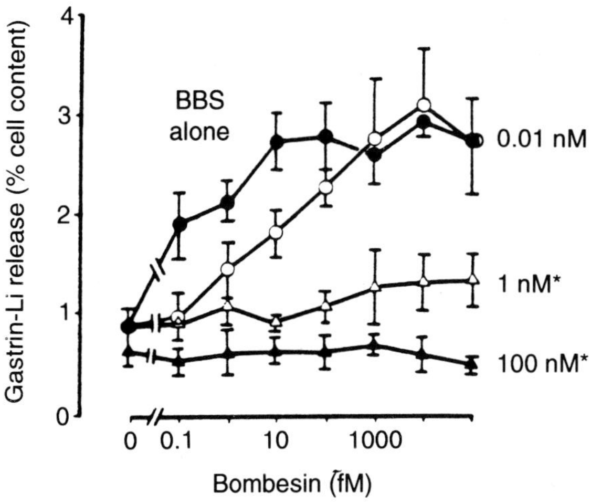

Fig. 5. Effect of bombesin and bombesin-specific antibody on gastrin release. Dose response to bombesin alone (BBS alone; ●) and in the presence of bombesin antibody at 0.01 nM (○), 1.0 nM (△), and 100 nM (▲). (From GIRAUD et al. 1987, published by The American Physiological Society)

Gastrin-releasing peptide stimulates the release of several enteric peptides. In intact jejunum, CCK secretion is increased following administration of GRP or bombesin in man (Ghatei et al. 1982), pig (Cantor et al. 1986b), and rat (Lewis and Williams 1990). In the ileum, bombesin-stimulated neurotensin release is observed in vivo in man (Ghatei et al. 1982) and rat (Rokaeus 1982), and in primary cultures of canine neurotensin-containing cells (Barber et al. 1986a). Bombesin also stimulates the release of peptide YY in primary cultures of isolated colon cells (Aponte et al. 1988).

V. Gastrin and Cholecystokinin

Gastrin and cholecystokinin (CCK) are structurally related peptides that share a biologically active pentapeptide amide sequence. Although specific receptors for each of these peptides have been identified, pancreatic acinar cells express CCK receptors that can bind gastrin with low affinity (Gardner and Jensen 1989) and the gastrin-responsive receptor on parietal cells binds gastrin and CCK with equal affinities (Soll et al. 1984a). Hence, differentiating the effect of these two peptides in regulating gastroenteric cells is complicated by the presence of receptors having affinities for both peptides.

The cross-reactivity of receptors for gastrin and CCK is exemplified by the ability of both peptides to stimulate fundic somatostatin release. Gastrin-17 stimulates SS release in the isolated perfused stomach (Chiba et al. 1980; Guo et al. 1990) and in primary cultures of fundic somatostatin-containing cells (Soll et al. 1984b; Yamada et al. 1984). In fundic mucosal cell cultures, cholecystokinin-8 is more effective than gastrin in stimulating SS secretion; however, these peptides are equipotent in displacing binding of ^{125}I-[Leu^{15}]gastrin-17 (Fig. 6; Soll et al. 1985). Chiba et al. (1987b) have determined that the potency of CCK over gastrin in stimulating SS release is paralleled by the greater efficacy of CCK compared with gastrin in stimulating phosphatidylinositol hydrolysis in fundic cell cultures. Hence, fundic SS-containing cells may express a common gastrin/CCK receptor that can be occupied with similar affinities by both peptides but is activated more effectively by CCK.

In contrast, antral somatostatin cells may express CCK-specific receptors. Gastrin has no effect on SS release in the isolated perfused pig antrum (Holst et al. 1987) or in human antral somatostatin-containing cells in culture (Buchan et al. 1990). Cholecystokinin, however, effectively stimulates SS release from antral cell cultures (Buchan et al. 1990a). These findings, together with the previously described differences in the cholinergic regulation of fundic and antral SS release, clearly indicate distinct regulatory mechanisms for SS-containing cells in the gastric fundus and antrum.

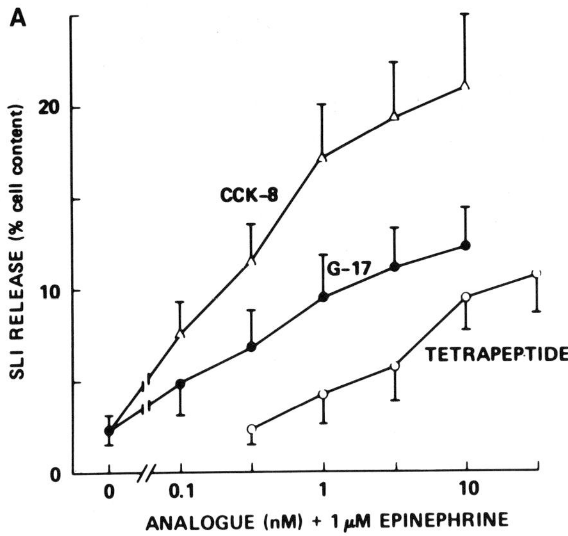

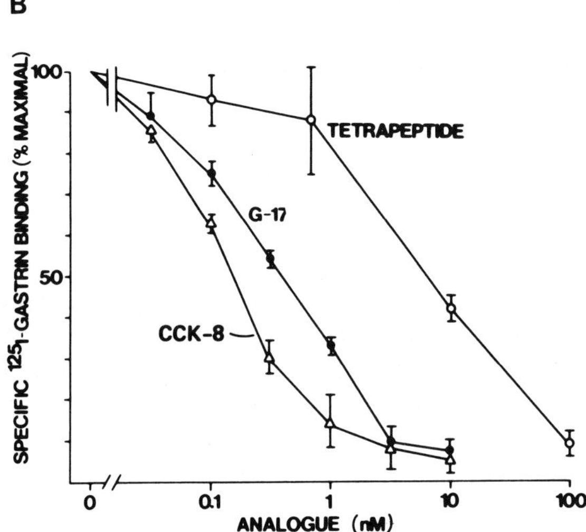

Fig. 6A,B. Effect of gastrin and cholecystokinin on canine fundic cells. Dose-dependent action of tetrapeptide, gastrin-17 (*G-17*), and cholecystokinin-8 (*CCK-8*) on **A** somatostatin (*SLI*) release in fundic mucosal cell cultures and **B** inhibition of ^{125}I-[Leu^{15}]gastrin-17 binding to isolated fundic cells enriched for somatostatin-containing cells. (From SOLL et al. 1985, published by The American Physiological Society)

D. Nutrient Regulation of Peptide Release

Endocrine cells of the gastrointestinal tract are regulated not only by receptors for neurotransmitters and hormones, but also by ingested nutrients. Studies in vivo or using perfused organ preparations indicate that ingestion or perfusion of proteins, amino acids, lipids and fatty acids, but not carbohydrates, stimulates the release of several gastroenteric peptides. It has been suggested that nutrient-regulated peptide release is mediated by local neural or hormonal pathways, and studies using intact tissues have been unable to address this possibility. The recent use of isolated cell preparations, however, has demonstrated that nutrient-stimulated peptide release probably reflects a direct action on the gastroenteric endocrine cell.

I. Proteins and Amino Acids

Gastric acid secretion in response to a meal may in part be mediated through the regulation of gastrin and somatostatin release by nutrients. Although circulating levels of SS increase following ingestion of a mixed meal, this response is mediated preddominantly by lipids and to a lesser extent by proteins and amino acids (Penman et al. 1981; Lucey et al. 1984; Ensinck et al. 1990). In contrast, individual amino acids, but not intact protein, are a major stimulant of gastrin release (Blair et al. 1975; Taylor et al. 1982; Lichtenberger 1982), most probably through a direct action on antral G cells (Lichtenberger et al. 1982; DelValle and Yamada 1990). Lichtenberger et al. (1982) determined that amines are more effective than individual amino acids in stimulating gastrin release and suggested that amino acids must be decarboxylated to be effective stimulants. Using isolated G cell cultures, DelValle and Yamada (1990) have confirmed that amines have a greater efficacy over amino acids in stimulating gastrin release; however, these compounds may act through different pathways as SS inhibits gastrin release in response to amino acids but not amines. The action of amino acids is suggested to be indicative of a receptor-mediated mechanism while amines, being hydrophobic, may freely permeate the plasma membrane to act at an intracellular site. The recent finding that amines, but not most amino acids, directly stimulate gastrin release from isolated secretory granules (Dial et al. 1991) further suggests the existence of dual pathways in protein-stimulated gastrin release.

Proteins and amino acids also stimulate jejunal CCK release and this response is thought to be a major stimulatory signal in postprandial pancreatic exocrine secretion. In humans (Himeno et al. 1983; Owyang et al. 1986) and dog (Meyer and Grossman 1974; Konturek et al. 1986) amino acids, particularly the aromatic compounds phenylalanine and tryptophan, are effective stimulants of CCK release. In dogs, amino acid-stimulated release appears to be a direct action on CCK cells (Barber et al. 1986c; Koop and Buchan 1992) and may be stereospecific for L- but not D-isomers (Meyer and Grossman 1974; Barber et al. 1986c). In contrast, intact protein, but not individual amino acids, stimulates CCK release in rats (GREEN and MIYASAKA, 1983; LIDDLE et al. 1986; DOUGLAS et al. 1988; LEWIS and WILLIAMS 1990). The ineffectiveness of individual amino acids to stimulate CCK release in rats is suggested to be due to a negative feedback regulation of pancreatic exocrine secretion, whereby intraluminal proteases attenuate CCK release (GREEN and MIYASAKA 1983; LOUIE et al. 1986), possibly through a trypsin-sensitive, CCK-releasing peptide (MIYASAKA et al. 1989). This regulatory loop in the rat is suggested by findings that CCK release is stimulated by the inactivation or diversion of pancreatic proteases (GREEN and MIYASAKA 1982), as well as by trypsin inhibitors (BRAND and MORGAN 1981; LIDDLE et al. 1984). The species-specific significance of this negative feedback mechanism, however, has been questioned by OWYANG et al. (1986), who find that trypsin also inhibits phenylalanine-stimulated CCK release in man, indicating that this negative feedback loop may not be unique to the rat.

II. Fatty Acids

A well-characterized role of nutrients in gastrointestinal endocrine cell function is the ability of lipids to stimulate the in vivo release of several peptides, including SS (PENMAN et al. 1981; LUCEY et al. 1984; ENSINCK et al. 1990), CCK (KONTUREK et al. 1986; LEWIS and WILLIAMS 1990), secretin (WATANABE et al. 1986; LI et al. 1990; RHEE et al. 1991), neurotensin (FERRIS et al. 1981; ROSELL and ROKAEUS 1979), peptide YY (TAYLOR 1985; APONTE et al. 1985) and enteroglucagon (HOLST 1978; OHNEDA et al. 1975). The stimulatory signal in regulating peptide release is either emulsified fats or fatty acids, but not intact lipids (LIDDLE et al. 1985; WATANBE et al. 1986; LEWIS and WILLIAMS 1990). Fatty acid-induced release of gastroenteric peptides indirectly mediates the action of lipids on multiple gastrointestinal functions including gall-bladder contraction (LIDDLE et al. 1985), pancreatic exocrine secretion (LI et al. 1990; GREEN et al. 1989; KONTUREK et al. 1986), gastric acid secretion (SEAL et al. 1987; RHEE et al. 1991) and the intestinal translocation of luminal fats (ARMSTRONG et al. 1985).

The functional importance of luminal vs. circulating fatty acids in stimulating peptide release has been difficult to determine. Peptides localized in the proximal intestine, including secretin and CCK, are released by luminal perfusion of fatty acids (KONTUREK et al. 1986; LEWIS and WILLIAMS 1990; RHEE et al. 1991). However, 90% of fat absorption is completed in the proximal intestine, which makes a luminal action on peptide release from the distal ileum questionable. Observations that fat infused into the proximal, but not the distal, small intestine stimulates the release of peptides from the ileum (READ et al. 1984; FUJIMURA et al. 1989) has led to the suggestion that ileal peptide release in response to fatty acids may be mediated by intramural neural pathways. The recent use of primary culture systems, however, has identified that fatty acids directly stimulate the release of neurotensin and enteroglucagon in cells isolated from the distal ileum (BARBER et al. 1991), as well as the release of peptide YY in colonic enterocytes (APONTE et al. 1988). These findings, together with infusion studies using intact tissue, suggest that circulating, rather than luminal, fatty acids regulate peptide release from the distal gut.

The effect of fatty acids on neurotensin and enteroglucagon release, in addition to being direct, is selective for long-chain unsaturated fatty acids and stereospecific for *cis* rather than *trans* isomers (BARBER et al. 1991). The cellular mechanisms mediating fatty acid activation, identified in enteric endocrine (BARBER et al. 1991) and other tissue-specific cells (CHOW and JONDAL 1990; NISHIKAWA et al. 1988; TOUNY et al. 1990), include an increase in $[Ca^{2+}]_i$ and an activation of protein kinase C. In enteric endocrine cells, fatty acid-induced increases in $[Ca^{2+}]_i$ are stereospecific for *cis* rather than *trans* isomers (Fig. 7) and are mediated by the mobilization of intracellular Ca^{2+} rather than an influx of extracellular Ca^{2+} (BARBER et al. 1991). The precise signaling pathway whereby exogenous fatty acids regulate $[Ca^{2+}]_i$ and protein kinase C, however, is uncertain. Fatty acids selectively and stereospecifically stimulate protein kinase C in purified enzyme preparations (SEKIGUCHI et al. 1987; ALLEN and KATZ 1991), which has led to questions regarding the role of phospholipids in generating fatty acid effects on $[Ca^{2+}]_i$ and protein kinase C (MURIKAMI et al. 1986; NISHIKAWA et al. 1988; CHOW and JONDAL 1990). Additionally, although fatty acid-stimulated peptide release appears indicative of receptor-mediated response, a selective and stereospecific fatty acid binding site has not been identified. A tissue-specific plasma membrane protein that transports both saturated and unsaturated exogenous fatty acids has been isolated from enterocytes (STREMMEL 1988); however, the expression of this protein in endocrine cells and its putative role in mediating fatty acid-induced peptide release remain undetermined.

E. Regulation of Release by Luminal pH

The release of gastroenteric peptides by intraluminal pH may be an important signal in the feedback regulation of gastric acid and pancreatic bicarbonate secretion. Intraduodenal acidification stimulates secretin (FAHRENKRUG et al. 1978; CHEY and KONTUREK 1982; CHEN et al. 1985) and CCK (CHEN et al. 1985) release and acid-induced increases in duodenal peptide secretion correlate with acid-induced increases in pancreatic bicarbonate output (CHEY and KONTUREK 1982; CHEN et al. 1985).

Acidification of the gastric lumen inhibits gastric acid secretion and this effect may be mediated in part by an inhibition of gastrin release and/or a stimulation of SS release. Intraluminal acid inhibits antral gastrin release (WALSH et al. 1975; LUCEY et al. 1989) and pharmacologically blocking the parietal cell H^+,K^+-ATPase increases circulating gastrin levels (LARSSON et al. 1986; ALLEN et al. 1986; LUCEY et al. 1989). Whether acid-induced inhibition of release reflects a direct effect on the gastrin cell is uncertain.

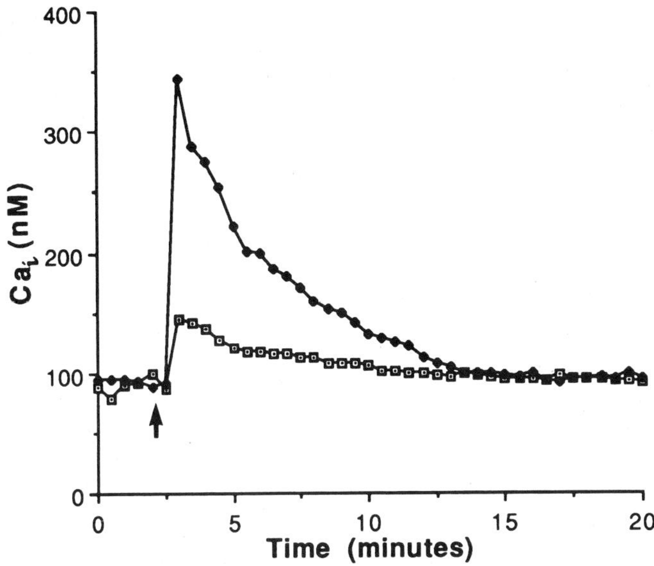

Fig. 7. Stereospecific effect of fatty acids on $[Ca^{2+}]_i$ in enteric endocrine cells. $[Ca^{2+}]_i$ was determined using the fluorescent Ca^{2+}-sensitive dye fura-2. Changes in $[Ca^{2+}]_i$ over time were monitored after the addition of oleic acid (*cis*-11 ■) or elaidic acid (*trans*-11; □). (From BARBER et al. 1991, published by The American Physiological Society)

Hydrochloric acid stimulates gastric SS release (HOLST et al. 1983; SCHUBERT et al. 1988a; LUCEY et al. 1989) and cimetidine reduces postprandial increases in SS secretion (LUCEY et al. 1989). Through its inhibitory action, SS could mediate the acid-inhibited release of gastrin. The action of luminal acidity on peptide release, however, is complex and may be only a modulatory component in the integrated postprandial response to nutrients and local neural and paracrine pathways (SCHUBERT et al. 1988a; LUCEY et al. 1989).

F. Summary

Elucidating the mechansims directly reulating gastroenteric endocrine cells has been limited by the complex and diffuse tissue distribution of peptides in the gastrointestinal tract. The recent use of primary cultures enriched for peptide-specific endocrine cells has provided a previously unavailable means for characterizing the direct, tissue-specific regulation of peptide release and the transmembrane signaling mechanisms coupling stimulus to secretion. Determinations with isolated cell preparations, however, must be correlated with findings using intact tissue to assess the integrated action of nutrients and neural, endocrine and paracrine pathways in regulating the release of gastroenteric peptides in vivo.

References

Abucham J, Reichlin S (1990) Cysteamine induces cholecystokinin release from the duodenum. Gastroenterology 99:1633–1640

Allen BG, Katz S (1991) Isolation and characterization of the calcium- and phospholipid-dependent protein kinase (protein kinase C) subtypes from bovine heart. Biochemistry 30:4334–4343

Allen JM, Bishop AE, Daly MJ, Larsson H, Carlsson E, Polak JM, Bloom SR (1986) Effect of inhibition of acid secretion on the regulatory peptides in the rat stomach. Gastroenterology 90:970–977

Al-Saffir A, Theodorsson-Norheim E, Rosell S (1984) Nervous control of the release of neurotensin-like immunoreactivity from the small intestine of the rat. Acta Physiol Scand 122:1–6

Aponte GW, Fink AS, Meyer JH, Tatemoto K, Taylor IL (1985) Regional distribution and release of peptide YY with fatty acids of different chain lengths. Am J Physiol 249:G745–G750

Aponte GW, Taylor IL, Soll AH (1988) Primary culture of PYY cells from canine colon. Am J Physiol 254:G829–G836

Armstrong MJ, Ferris CF, Leeman SE (1985) Neurotensin increases the translocation of [^{3}H]oleic acid from intestinal lumen into the lymph in rats. In: Bonfils S (ed) Regulatory peptides: mode of action on digestive, nervous and endocrine systems. Elsevier, Amsterdam (INSERM symposium, vol 25)

Barber DL (1991) Mechanisms of receptor-mediated regulation of Na-H exchange. Cell Signal 3:387–397

Barber DL, Buchan AMJ, Walsh JH, Soll AH (1986a) Isolated canine ileal mucosal cells in short-term culture: a model for study of neurotensin release. Am J Physiol 250:G374–G384

Barber DL, Buchan AMJ, Walsh JH, Soll AH (1986b) Regulation of neurotensin release from canine enteric primary cell cultures. Am J Physiol 250:G385–G390

Barber DL, Walsh JH, Soll AH (1986c) Release and characterization of cholecystokinin from isolated canine jejunal cells. Gastroenterology 91:627–636

Barber DL, Gregor M, Soll AH (1987) Somatostatin and muscarinic inhibition of canine enteric endocrine cells: cellular mechanisms. Am J Physiol 253:G684–G689

Barber DL, McGuire ME, Ganz MB (1989) β-Adrenergic and somatostatin receptors regulate Na-H exchange independent of cAMP. J Biol Chem 264: 21038–21042

Barber DL, Cacace AM, Raucci DT, Ganz MB (1991) Fatty acids stereospecifically stimulate neurotensin release and increase $[Ca^{2+}]_i$ in enteric endocrine cells. Am J Physiol 261:G497–G503

Barber DL, Ganz MB, Bongiorno PB, Strader CD (1992) Mutant constructs of the β-adrenergic receptor that are uncoupled from adenylyl cyclase retain functional coupling to Na-H exchange. Mol Pharmacol 41:1056–1060

Bastie M-J, Williams JA (1990) Gastrointestinal peptides activate Na^{+}-H^{+} exchanger in AR42J cells by increasing its affinity for intracellular H^{+}. Am J Physiol 258:G958–G966

Bayliss W, Starling EH (1902) Mechanism of pancreatic secretion. J Physiol (Lond) 28:325–353

Blair EL, Greenwell JR, Grunk ER (1975) Gastrin response to meals of different composition in normal subjects. Gut 16:773–776

Bloom SR, Mortimer RH, Thorner MO, Besser GM, Hall R, Gomez-Pan A, Roy VM, Russel RCG, Coy DH, Kastin AJ, Schally AV (1974) Inhibition of gastrin and gastric acid secretion by growth-hormone release-inhibiting hormone. Lancet 2:1106–1109

Bokoch GM, Katada T, Northrup JK, Hewlett EL, Gilman AG (1983) Identification of the predominant substrate for ADP-ribosylation by islet activating protein. J Biol Chem 258:2072–2075

Brand SJ, Morgan RGH (1981) The release of rat intestinal cholecystokinin after oral trypsin inhibitor measured by bio-assay. J Physiol (Lond) 319:325–343

Brazeau P, Vale W, Burgus R, Ling N, Butcher M, Rivier J, Guillemin R (1973) Hypothalamic polypeptide that inhibits the secretion of immunoreactive pituitary growth hormone. Science 179:77–79

Buchan AMJ (1991) Effect of sympathomimetics on gastrin secretion from antral G cells in culture. J Clin Invest 87:1382–1386

Buchan AMJ, Barber DL, Gregor M, Soll AH (1987) Morphologic and physiologic studies of canine ileal enteroglucagon-containing cells in short-term culture. Gastroenterology 93:791–800

Buchan AMJ, Curtis SB, Meloche TM (1990a) Release of somatostatin immunoreactivity from human antral D cells in culture. Gastroenterology 99:690–696

Buchan AMJ, Meloche M, Coy DH (1990b) Inhibition of bombesin-stimulated gastrin release from isolated human G cells by bombesin analogs. Pharmacology 41:237–245

Buchan AMJ, MacLeod MD, Meloche RM, Kwok YN (1992) Muscarinic regulation of somatostatin release from primary cultures of human antral epithelial cells. Pharmacology 44:33–40

Bunnett NW, Clark B, Debas HT, del Milton RC, Kovacs TOG, Orloff MS, Pappas TN, Reeve JR, Rivier JE, Walsh JH (1985) Canine bombesin-like gastrin releasing peptides stimulate gastrin release and acid secretion in the dog. J Physiol (Lond) 365:121–130

Campos RV, Buchan AMJ, Pederson RA, McIntosh CHS (1989) Inhibition of bombesin-stimulated gastrin release from isolated canine G cells by bombesin antagonists. Can J Physiol Pharmacol 67:1520–1524

Campos RV, Buchan AMJ, Meloche RM, Pederson RA, Kwok YN, Coy DH (1990) Gastrin secretion from human antral G cells in culture. Gastroenterology 99: 36–44

Cantor P, Holst JJ, Knuhtsen S, Rehfeld JF (1986a) The effect of vagal stimulation on the release of cholecystokinin in anaesthetized pigs. Scand J Gastroenterol 21:1069–1072

Cantor P, Holst JJ, Knuhtsen S, Rehfeld JF (1986b) Effect of neuroactive agents on cholecystokinin release from the isolated, perfused porcine duodenum. Acta Physiol Scand 130:627–632

Carter KJ, Rutledge PL, Steer ML, Silen W (1987) Secretogogue-induced changes in intracellular pH and amylase release in mouse pancreatic acini. Am J Physiol 253:G690–G696

Chen YF, Chey WY, Chang T-M, Lee KY (1985) Duodenal acidification releases cholecystokinin. Am J Physiol 249:G29–G33

Chey WY, Konturek S (1982) Plasma secretin and pancreatic secretion in response to liver extract meal with varied pH and exogenous secretin in dog. J Physiol (Lond) 324:263–272

Chiba T, Yamada T (1987) Mechanisms for the stimulatory and inhibitory effects of carbamylcholine on canine gastric D-cells. Biochem Biophys Res Comm 147: 140–144

Chiba T, Taminato T, Kadowaki S, Inoue Y, Mori K, Seino Y, Abe H, Chihara K, Matsukura S, Fujita T, Goto Y (1980) Effects of various gastrointestinal peptides on gastric somatostatin release. Endocrinology 106:145

Chiba T, Raffoul K, Yamada T (1987a) Divergent stimulatory and inhibitory actions of carbamylcholine on gastric D-cells. J Biol Chem 252:8467–8469

Chiba T, Sugano K, Park J, Yamada T (1987b) Potential mediation of somatostatin secretion from canine fundic D-cells by protein kinase C. Am J Physiol 253: G62–G67

Chiba T, Park J, Yamada T (1988) Regulation of somatostatin release from dispersed canine fundic D-cells. In: Reichlin S (ed) Somatostatin, basic and clinical status. Plenum, New York, p 229

Chiba T, Fujita T, Yamada T (1989) Carbachol inhibits stimulant-induced increases in fundic D-cell cytosolic Ca^{2+} concentration. Am J Physiol 257:G308–G312

Chow SC, Jondal M (1990) Polyunsaturated free fatty acids stimulate an increase in cytosolic Ca^{2+} by mobilizing the inositol 1,4,5-trisphosphate-sensitive Ca^{2+} pool in T cells through a mechanism independent of phosphoinositide turnover. J Biol Chem 265:902–907

Cosson P, de Curtis I, Pouyssegur J, Griffiths G, Davoust J (1989) Low cytoplasmic pH inhibits endocytosis and transport from the *trans*-Golgi network to the cell surface. J Cell Biol 108:377–387

Costa M, Patel Y, Furness JB, Arimura A (1977) Evidence that some intrinsic neurons of the intestine contain somatostatin. Neurosci Lett 6:215–222

Christensen KC, Stadil F (1976) On the beta adrenergic contribution to gastric acid and gastrin responses to hypoglycaemia in man. Scand J Gastreenterol 37:81–86

Debas HT, Hollinshead J, Seal A, Soon-Shiong P, Walsh JH (1984) Vagal control of gastrin release in the dog: pathways for stimulation and inhibition. Surgery 95:34–37

DeJong AJL, Klamer M, Jansen JBMJ, Lamers CBHW (1987) Effect of atropine and somatostatin on bombesin-stimulated plasma gastrin, cholecystokinin and pancreatic polypeptide in man. Regul Pept 17:285–293

DelTacca M, Soldani G, Polloni A, Bernardini C, Costa F, Billini M (1987) The effects of the antimuscarinic drugs pirenzepine and atropine on plasma portal levels of somatostatin and gastrin in the dog. J Endocrinol Invest 10:507

DelValle J, Yamada T (1990) Amino acids and amines stimulate gastrin release from canine antral G-cells via different pathways. J Clin Invest 85:139–143

Dial EJ, Cooper LC, Lichtenberger LM (1991) Amino acid- and amine-induced gastrin release from isolated rat endocrine granules. Am J Physiol 260:G175–G181

Douglas BR, Woutersen RA, Jansen JBMJ, DeJong AJL, Lamers CBHW (1988) The influence of different nutrients on plasma cholecystokinin levels in the rat. Experientia 44:21–23

DuVal JW, Saffouri B, Weir GC, Walsh JH, Arimura A, Makhlouf GM (1981) Stimulation of gastrin and somatostatin secretion from the isolated rat stomach by bombesin. Am J Physiol 241:G242–G247

Ensinck JW, Vogel RE, Laschansky EC, Francis BH (1990) Effect of ingested carbohydrate, fat, and protein on the release of somatostatin-28 in humans. Gastroenterology 98:633–638

Erdos JJ, Maguire ME (1983) Hormone-sensitive magnesium transport in murine S49 lymphoma cells: characterization and specificity for magnesium. J Physiol (Lond) 337:351–371

Fahrenkrug J, Schaffalitzky De Muckadell OB, Rune SJ (1978) pH threshold for release of secretin in normal subjects and in patients with duodenal ulcer and patients with chronic pancreatitis. Scand J Gastroenterol 13:177–186

Farooq O, Walsh JH (1975) Atropine enhances serum gastrin response to insulin in man. Gastroenterology 68:662–666

Feldman M, Walsh JH (1980) Acid inhibition of sham feeding stimulated gastrin release and gastric acid secretion: effect of atropine. Gastroenterology 78: 772

Ferris CF, Hammer RA, Leeman SE (1981) Elevation of plasma neurotensin during lipid perfusion of rat small intestine. Peptides 2:263–266

Ferris CF, Parker MC, Armstrong MJ, Leeman SE (1985) Inhibition of neurotensin release by cyclic hexapeptide analog of somatostatin. Peptides 6:5–8

Feurle GE, Baca I, Knauf W (1982) Atropine depresses release of neurotensin and its effect on the exocrine pancreas. Regul Pept 4:75–82

Fujimura M, Khalil T, Sakamoto T, Greeley GH Jr, Salter MG, Townsend CM Jr, Thompson JC (1989) Release of neurotensin by selective perfusion of the jejunum with oleic acid in dogs. Gastroenterology 96:1502–1505

Ganz MB, Pachter JA, Barber DL (1990) Multiple receptors coupled to adenylate cyclase regulate Na-H exchange independent of cAMP. J Biol Chem 265:8989–8992

Gardner JD, Jensen RT (1989) Recetors for gut peptides and other secretagogues on pancreatic acinar cells. In: Schultz SG (ed) The gastrointestinal system. American Physiological Society, Bethesda, p 171 (Handbook of physiology)

Ghatei MA, Jung RT, Stevenson JC, Hillyard CJ, Adrian TE, Lee YC, Christofides ND, Sarson DL, Mashiter K, MacIntyre I, Bloom SR (1982) Bombesin: action on gut hormones and calcium in man. J Clin Endocrinol Metab 54:980–985

Giraud AS, Soll AH, Cuttitta F, Walsh JH (1987) Bombesin stimulation of gastrin release from canine gastrin cells in primary culture. Am J Physiol 252:G413–G420

Goto Y, Berelowitz M, Frohoman LA (1981) Effect of catecholamines on somatostatin secretion by isolated perfused rat stomach. Am J Physiol 240:E274–E278

Green GM, Miyasaka K (1983) Rat pancreatic response to intestinal infusion of intact and hydrolyzed protein. Am J Physiol 245:G394–G398

Green GM, Taguchi S, Friestman J, Chey WY, Liddle RA (1989) Plasma secretin, CCK, and pancreatic secretion in response to dietary fat in the rat. Am J Physiol 256:G1016–G1021

Grinstein S, Rothstein A (1986) Mechanisms of regulation of the Na^+/H^+ exchanger. J Membr Biol 90:1–12

Guo Y-S, Thompson JC, Singh P (1990) Role of gastrin in bombesin-stimulated somatostatin release. Gastroenterology 99:1297–1302

Harty RF, Maico DG, McGuigan JE (1985) Postreceptor inhibition of antral gastrin release by somatostatin. Gastroenterology 88:675–680

Himeno S, Tarui S, Kanayama S, Kuroshima T, Shinomura Y, Hayashi C, Tateishi K, Imagawa K, Hashimura E, Hamaoka T (1983) Plasma cholecystokinin responses after ingestion of liquid meal and intraduodenal infusion of fat, amino acids, or hydrochloric acid in man: analysis with region specific radioimmunoassay. Am J Gastroenterol 78:703–707

Holst JJ (1978) Extrapancreatic glucagons. Digestion 17:168–190

Holst JJ, Jensen SL, Knuhtsen S, Nielson OV, Rehfeld J (1983) Effect of vagus, gastric inhibitory peptide and HCl on gastrin and somatostatin release from perfused pig antrum. Am J Physiol 244:G515–G522

Holst JJ, Knuhtsen S, Orskov C, Skak-Nielsen T, Poulsen SS, Nielsen OV (1987) GRP-producing nerves control antral somatostatin and gastrin secretion in pigs. Am J Physiol 253:G767–G774

Jakobs K, Schultz G (1983) Occurrence of a hormone-sensitive inhibitory coupling component of the adenylate cyclase in S49 lymphoma *cyc* variant. Proc Natl Acad Sci USA 80:3899–3903

Kanayma S, Liddle RA (1990) Somatostatin regulates duodenal cholecystokinin and somatostatin messenger RNA. Am J Physiol 258:G358–G364

Koch B, Dorflinger L, Schonbrunn A (1985) Pertussis toxin blocks both cAMP-mediated and cAMP-independent actions of somatostatin. J Biol Chem 260: 13138–13145

Konturek SJ, Obtulowicz W, Kwiecien N, Dobrzanska M, Swierczek J, Oleksy J (1980) Effect of pirenzepine and atropine on gastric secretory and plasma hormonal responses to sham feeding in duodenal ulcer patients before and after resection of antrum and duodenal bulb. Scand J Gastroenterol 9:351

Konturek SJ, Tasler J, Bilski J, DeJong AJL, Jansen JBMJ, Lamers CB (1986) Physiological role and localization of cholecystokinin release in dogs. Am J Physiol 250:G391–397

Koop H, Behrens I, Bothe E, Koschwitz H, McIntosh CHS, Pederson RA, Arnold R, Creutzfeldt A, Creutzfeldt W (1983) Adrenergic control of rat gastric somatostatin and gastrin release. Scand J Gastroenterol 18:65–71

Koop I, Buchan AMJ (1992) CCK release from isolated canine epithelial cells in short term culture. Gastroenterology 102:28–34

Larsson H, Carlsson E, Mattsson H, Lundell L, Sundler F, Sundell G, Wallmark B, Watanabe T, Håkanson T (1986) Plasma gastrin and gastric enterochromaffinlike cell activation and proliferation: studies with omeprazole and ranitidine in intact and antrectomized rats. Gastroenterology 90:970–977

Larsson LI, Goltermann N, Demagistris L, Rehfeld JF, Schwartz TW (1979) Somatostatin cell processes as pathways for paracrine secretion. Science 205: 1393–1395

Lefkowitz R, Caron M (1988) Adrenergic receptors. J Biol Chem 263:4993–4996

Lewis LD, Williams JA (1990) Regulation of cholecystokinin secretion by food, hormones, and neural pathways in the rat. Am J Physiol 258:G512–G518

Lewis D, Weight F, Luini A (1985) A guanine nucleotide binding protein mediating the inhibition of voltage-dependent calcium current by somatostatin in a pituitary cell line. Proc Natl Acad Sci USA 83:9035–9039

Li P, Lee KY, Chang TM, Chey WY (1990) Hormonal mechanism of sodium oleate-stimulated pancreatic secretion in rats. Am J Physiol 259:G960–G965

Lichtenberger LM (1982) Importance of food in the regulation of gastrin release and formation. Am J Physiol 243:G429–G441

Lichtenberger LM, Delansorne R, Graziani LA (1982) Importance of amino acid uptake and decarboxylation in gastrin release from isolated G-cells. Nature 295:698–700

Lickley HLA, Kemmer FW, Gray DE et al. (1981) Chromatographic pattern of extrapancreatic glucagon and glucagon-like immunoreactivity before and after stimulation by epinephrine and participation of glucagon in epinephrine-induced hepatic glucose overproduction. Surgery 90:186–194

Liddle RA, Goldfine ID, Williams JA (1984) Bioassay of plasma cholecystokinin in rats: effects of food, trypsin inhibitor, and alcohol. Gastroenterology 87: 542–549

Liddle RA, Goldfine I, Rosen M, Taplitz R, Williams J (1985) Cholecystokinin bioactivity in human plasma: molecular forms, responses to feeding, and relationship to gallbladder contraction. J Clin Invest 75:1144–1152

Liddle RA, Green GM, Conrad CK, Williams JA (1986) Proteins but not amino acids, carbohydrates, or fats stimulate cholecystokinin secretion in the rat. Am J Physiol 251:G243–G248

Louie DS, May D, Miller P, Owyang C (1986) Cholecystokinin mediates feedback regulation of pancreatic enzyme secretion in rats. Am J Physiol 250:G252–G259

Lucey MR, Fairclough PD, Wass JAH, Swasowski P, Medbak S, Webb J, Rees LH (1984) Response of circulating somatostatin, insulin, gastrin and GIP to intraduodenal infusion of nutrients in normal man. Clin Endocrinol (Oxf) 21: 209–217

Lucey MR, Wass JAH, Rees LH, Dawson AM, Fairclough PD (1989) Relationship between gastric acid and elevated plasma somatostatinlike immunoreactivity after a mixed meal. Gastroenterology 97:867–872

Martindale R, Kauffman GL, Levin S, Walsh JH, Yamada T (1982) Differential regulation of gastrin and somatostatin secretion from isolated perfused rat stomachs. Gastroenterology 83:240–244

McIntosh CHS, Pederson RA, Koop H, Brown JC (1981) Gastric inhibitory polypeptide stimulated secretion of somatostatinlike immunoreactivity from the stomach: inhibition by acetylcholine or vagal stimulation. Can J Physiol Pharmacol 59:468–472

McIntosh CHS, Tang CL, Malcolm AJ, Ho M, Kwok YN, Brown JC (1991) Effect of a purified somatostatin monoclonal antibody and its fab fragments on gastrin release. Am J Physiol 260:G489–G498

Meyer JH, Grossman MI (1974) Comparison of D- and L-phenylalanine as pancreatic stimulants. Am J Physiol 195:1058–1061

Misumi A, Shiratori K, Lee KY, Barkin JS, Chey WY (1988) Effect of SMA 201–995, a somatostatin analogue, on the exocrine pancreatic secretion and gut hormone release in dogs. Surgery 103:450–455

Miyasaka K, Guan D, Liddle RA, Green GM (1989) Feedback regulation by trypsin: evidence for intraluminal CCK-releasing peptide. Am J Physiol 257: G175–G181

Murakami K, Chan SY, Routtenberg A (1986) Protein kinase C activation by *cis*-fatty acid in the absence of Ca^{2+} and phospholipids. J Biol Chem 261:15424–15429

Nishi S, Seino Y, Takemura J, Ishida H, Seno M, Chiba T, Yanaihara C, Yanaihara N, Imura H (1985) Vagal regulation of GRP, gastric somatostatin and gastrin secretion in vitro. Am J Physiol 248:E425–429

Nishikawa M, Hidaka H, Shirakawa S (1988) Possible involvement of direct stimulation of protein kinase C by unsaturated fatty acids in platelet activation. Biochem Pharmacol 37:3079–3089

Ohneda A, Yanbe A, Maruhama Y, Ishii S, Kai Y, Abe R, Yamagata S (1975) Characterization of circulating immunoreactive glucagon in response to intraduodenal administration of fat in dogs. Gastroenterology 68:715–721

Olsen M, Holst JJ, Sottimano C, Nielsen OV (1987) Autonomic nervous control of fundic secretion of somatostatin and antral secretion of gastrin and somatostatin in pigs. Digestion 36:24–35

Owyang C, May D, Louie DS (1986) Trypsin suppression of pancreatic enzyme secretion. Gastroenterology 91:637–643

Parton RG, Dotti CG, Bacallao R, Kurtz I, Simons K, Prydz S (1991) pH-induced microtubule-dependent redistribution of late endosomes in neuronal and epithelial cells. J Cell Biol 113:261–274

Penman E, Was JAH, Medbak S, Morgan L, Lewis JM, Besser JM, Rees LH (1981) Response of circulating immunoreactive somatostatin to nutritional stimuli in normal subjects. Gastroenterology 81:692–699

Peters MN, Walsh JH, Ferrari J, Feldmen M (1982) Adrenergic regulation of distention-induced gastrin release in humans. Gastroenterology 82:659–663

Polak JM (1989) Endocrine cells of the gut. In: Schultz SG (ed) The gastrointestinal system. American Physiological Society, Bethesda, p 79 (Handbook of physiology)

Polak JM, Pearse AGE, Grimelius L, Bloom SR, Arimure A (1977) Growth-hormone release-inhibiting hormone in gastrointestinal and pancreatic D cells. Lancet 2:1220–1224

Raynor K, Wang H-L, Dichter M, Reisine T (1991) Subtypes of brain somatostatin receptors couple to multiple cellular effector systems. Mol Pharmacol 40: 248–253

Read NW, McFarlane A, Kinsman RI, Bates TE, Blackhall NW et al. (1984) Effect of infusion of nutrient solutions into the ileum on gastrointestinal transit and plasma levels of neurotensin and enteroglucagon. Gastroenterology 86:274–280

Reisine T, Zhang Y, Sekura R (1985) Pertussis toxin treatment blocks the inhibition of somatostatin and increases the forskolin stimulation of cAMP accumulation and ACTH secretion from mouse anterior pituitary tumor cells. J Pharmacol Exp Ther 232:275–282

Rhee JC, Chang TM, Lee KY, Jo YH, Chey WY (1991) Mechanism of oleic acid-induced inhibition on gastric acid secretion in rats. Am J Physiol 260:G564–G570

Rökaeus A (1982) Inhibition of bombesin- and fat-stimulated release of neurotensin-like immunoreactivity by somatostatin in the rat. Br J Pharmacol 77:439P

Rosell S, Rokaeus A (1979) The effect of ingestion of amino acids, glucose and fat on circulating neurotensin-like immunoreactivity (NTLI) in man. Acta Physiol Scand 107:263–267

Saffouri B, Weir GC, Bitar KN, Makhlouf GM (1979) Stimulation of gastrin secretion from the vascularly perfused rat stomach by somatostatin antiserum. Life Sci 20:1749–1754

Saffouri B, Weir GC, Bitar KN, Makhlouf GM (1980) Gastrin and somatostatin secretion by perfused rat stomach: functional linkage of antral peptides. Am J Physiol 238:G495–G501

Schafmayer A, Nustede R, Pompino A, Kohler H (1988) Vagal influence on cholecystokinin and neurotensin release in conscious dogs. Scand J Gastroenterol 23:315–320

Schubert B, VanDongen AMJ, Kirsch GE, Brown AM (1989) β-adrenergic inhibition of cardiac sodium channels by dual G-protein pathways. Science 245:516–519

Schubert ML, Hightower J (1990) Functionally distinct muscarinic receptors on gastric somatostatin cells. Am J Physiol 258:G982–G987

Schubert ML, Makhlouf GM (1987) Neural regulation of gastrin and somatostatin secretion in rat gastric antral mucosa. Am J Physiol 253:G721–G725

Schubert ML, Bitar KN, Makhlouf GM (1982) Regulation of gastrin and somatostatin secretion by cholinergic and noncholinergic intramural neurons. Am J Physiol 243:G442–G447

Schubert ML, Saffouri B, Walsh JH, Makhlouf GM (1985) Inhibition of neurally mediated gastrin secretion by bombesin antiserum. Am J Physiol 248:G456–G462

Schubert ML, Edwards NF, Makhlouf GM (1988a) Regulation of gastric somatostatin secretion in mouse by luminal acidity: a local feedback mechanism. Gastroenterology 94:317–322

Schubert ML, Saffouri B, Makhlouf GM (1988b) Identical patterns of somatostatin secretion from isolated antrum and fundus of rat stomach. Am J Physiol 254:G20–G24

Schubert ML, Hightower J, Coy DH, Makhlouf GM (1991) Regulation of acid secretion by bombesin/GRP neurons of the gastric fundus. Am J Physiol 260:G156–G160

Seal AM, Meloche RM, Liu YQE, Buchan AMJ, Brown JC (1987) Effects of monoclonal antibodies to somatostatin on somatostatin-induced and intestinal fat-induced inhibition of gastric acid secretion. Gastroenterology 93:1187–1192

Sekiguchi K, Tsukuda M, Ojita K, Kikkawa U, Nishizuka Y (1987) Three distinct forms of rat brain protein kinase C: differential response to unsaturated fatty acids. Biochem Biophys Res Commun 145:797–802

Short GM, Doyle JW, Wolfe MM (1985) Effect of antibodies to somatostatin on acid secretion and gastrin release by the isolated perfused rat stomach. Gastroenterology 88:984–987

Siffert W, Akkerman JW (1988) Na^+/H^+ exchange as a modulator of platelet activation. Trends Biochem Sci 13:245–249

Soll AH, Amirian DA, Thomas LP, Reedy TJ, Elashoff JD (1984a) Gastrin receptors on isolated canine parietal cells. J Clin Invest 73:1434–1447

Soll AH, Yamada T, Park J, Thomas LP (1984b) Release of somatostatinlike immunoreactivity from canine fundic mucosal cells in primary culture. Am J Physiol 247:G558–G566

Soll AH, Amirian DA, Park J, Elashoff JD, Yamada T (1985) Cholecystokinin potently releases somatostatin from canine fundic mucosal cells in short-term culture. Am J Physiol 248:G569–G573

Stadil F, Rehfeld JR (1973) Release of gastrin by epinephrine in man. Gastroenterology 65:210–215

Stremmel W (1988) Uptake of fatty acids by jejunal mucosal cells is mediated by a fatty acid binding membrane protein. J Clin Invest 82:2001–2010

Sue R, Toomey ML, Todisco A, Soll AH, Yamada T (1985) Pirenzepine-sensitive muscarinic receptors regulate gastric somatostatin and gastrin. Am J Physiol 248:G184–G187

Sugano K, Park J, Soll AH, Yamada T (1987) Stimulation of gastrin release by bombesin and canine gastrin-releasing peptides. J Clin Invest 79:935–942

Taylor IL (1985) Distribution and release of peptide YY in dog measured by specific radioimmunoassay. Gastroenterology 88:731–737
Taylor IL, Walsh JH, Carter D, Wood J, Grossman MI (1979) Effect of atropine and bethanechol on bombesin-stimulated release of pancreatic polypeptide and gastrin in dog. Gastroenterology 77:714–718
Taylor IL, Byrne WJ, Christie DL, Ament ME, Walsh JH (1982) Effect of individual L-amino acids on gastric acid secretion and serum gastrin and pancreatic polypeptide release in humans. Gastroenterology 83:273–278
Tooze J, Burke B (1987) Accumulation of adrenocorticotropin secretory granules in the midbody of telophase AtT20 cells: evidence that secretory granules move anterogradely along microtubules. J Cell Biol 104:1047–1057
Touny SE, Khan W, Hannun Y (1990) Regulation of platelet protein kinase C by oleic acid. J Biol Chem 265:16437–16443
Tran V, Beal M, Martin J (1985) Two types of somatostatin receptors differentiated by cyclic somatostatin analogs. Science 228:492–495
Vigna SR, Giraud AS, Soll AH, Walsh JH, Mantyh PW (1988) Bombesin receptors on gastrin cells. Ann NY Acad Sci 547:131–137
Walsh J, Richardson C, Frodtran J (1975) pH dependence of acid secretion and gastrin release in normal and ulcer patients. J Clin Invest 55:462–469
Watanabe S, Chey WY, Lee KY, Chang TM (1986) Secretin is released by digestive products of fat in dogs. Gastroenterology 90:1008–1017
Wolfe MM, Jain DK, Reel M, McGuigan JE (1984) Effects of carbachol on gastrin and somatostatin release in rat antral tissue culture. Gastroenterology 87:86–93
Yamada T, Soll AH, Park J, Elashoff J (1984) Autonomic regulation of somatostatin release: studies with primary cultures of canine fundic mucosal cells. Am J Physiol 247:G567–G573
Yamada Y, Post SR, Wang K, Taggert SH, Bell GI, Sieno S (1992) Cloning and functional characterization of a family of human and mouse somatostatin receptors expressed in brain, gastrointestinal tract and kidney. Proc Natl Acad Sci USA 89:251–255
Yamashita N, Shibuya N, Ogata E (1986) Hyperpolarization of the membrane potential caused by somatostatin in dissociated human pituitary adenoma cells that secrete growth hormone. Proc Natl Acad Sci USA 83:6198–6202
Yatani A, Codina J, Imoto Y, Reeves JP, Birnbaumer L, Brown AM (1987) A G protein directly regulates mammalian cardiac calcium channels. Science 238: 1288–1291

CHAPTER 5

Peptide Receptors and Signal Transduction in the Digestive Tract

M. LABURTHE, P. KITABGI, A. COUVINEAU, and B. AMIRANOFF

A. Introduction

The digestive tract is equipped with a diffuse endocrine tissue which synthesizes and releases a variety of peptide hormones. It is also richly innervated by extrinsic and intrinsic neurons which store many neuropeptides and classical neurotransmitters (see Chap. 1). Since the discovery of the first hormone in intestine, i.e., secretin (BAYLISS and STARLING 1902), the digestive tract has been a major source for the discovery and isolation of new regulatory peptides (MCDONALD 1991a,b). They have been identified thereafter in many other tissues, in particular the central nervous system, giving rise to the so-called brain-gut axis concept (MCDONALD 1991a,b). The reverse is also true since peptides initially discovered in brain have been thereafter characterized in the gut.

This chapter deals with the description of receptors for regulatory peptides which are produced by and display target cells within the digestive tract. These regulatory peptides cannot be considered strictly as gastrointestinal hormones since they behave as hormones, neurotransmitters or paracrine agents and are also present in many other parts of the organism. Receptors relevant to digestive tract physiology will be mainly addressed, but characterization in other organs will be mentioned when it highlights some of their basic properties.

B. General Considerations and Methodology

The general properties of receptors and the methodologies used for their functional and molecular characterization will be briefly outlined.

I. Receptor Concept and General Properties

The concept of the receptor emerged early in the last century in the minds of pharmacologists who wished to explain the selectivity of drug actions. This concept was thereafter extended to the action of regulatory agents such as hormones, neurotransmitters or growth factors. Purification and cloning of many receptors (see below) now provide a molecular reality to the operational definition of receptors as part of the cell with which a regulatory agent interacts to produce a biological response.

The expression of a limited set of neurohormonal receptors by a cell with a given phenotype plays a pivotal role in the ability of this cell to sample its neuroendocrine environment. A receptor for a regulatory peptide is usually a glycoprotein located at the plasma membrane that fulfills three main purposes: (1) it recognizes an extremely low concentration of peptide in its immediate vicinity, e.g., *affinity*. The higher the affinity, the lower the concentration of peptide necessary to give half-maximal occupation (K_d) of the receptor population. Dissociation constants (K_d) are usually in the range of 0.01–1 n*M*. (2) It discriminates its natural ligand(s) in a vast excess of other peptides in the extracellular fluids, e.g., *specificity*. (3) It translates the peptide-receptor interaction into a biological response (see Sect. C).

Direct binding studies of radioactive peptides to various preparations of their target tissues (see below) identify receptors by their affinity and specificity. The term receptor is defined in this chapter as the peptide-recognizing component that can be characterized functionally and identified at the molecular level by a radioactive probe of the peptide. This definition excludes receptors that have been described only by pharmacological techniques.

II. Methods for Study of Receptors

Direct characterization of receptors requires suitable receptor preparations, radiolabeled ligands and finally a method for separation of bound and unbound ligands (KAHN 1976; GARDNER 1979; SPIEGEL 1991).

1. Tissue Preparations

With the exception of receptor autoradiography studies, most binding assays require a suitable receptor preparation: isolated intact cells, cells in primary culture, cultured cell lines, cell or tissue homogenates, crude or highly purified plasma membranes. It is not within the scope of this chapter to describe these preparations inasmuch as they depend on tissues. Reviews pertaining to the advantages and limitations of each preparation are available (GARDNER 1979; LABURTHE and AMIRANOFF 1989). Peptide receptors can also be studied in solution after solubilization from the plasma membrane by nondenaturing detergents (see below).

2. Radiolabeled Ligand

Most studies use ^{125}I for labeling peptides because the specific activity of this radioisotope is high (2,125 Ci/mmol) and its half-life is sufficiently long that the tracer can be used for several weeks. Different methods make it possible to incorporate ^{125}I into tyrosine (Tyr) residues of peptides (BOLTON 1977). Under some conditions, ^{125}I can also be incorporated in histidine (His) residues for peptides which do not contain Tyr such as porcine secretin (MILUTINOVIC et al. 1976). The major problem with ^{125}I-labeled peptides is the biological activity of the tracers. When Tyr residues are not in the active site of the peptide, there is usually no problem. If there are two or more Tyr residues but only one can be labeled without altering the biological activity of the peptide, the problem can be solved by purifying the pertinent monoiodinated form of the peptide by high-performance liquid chromatography (HPLC), e.g., for NPY (SHEIKH et al. 1989) or ion-exchange chromatography, e.g., for neurotensin (SADOUL et al. 1984). If there is no Tyr or His residue in the peptide, the problem can be solved by synthesizing a peptide analog in which a Tyr residue has been added without altering the biological activity of the peptide (WYNN et al. 1983). Alternatively, but usually less preferably, the peptide can be labeled by the ^{125}I-labeled Bolton-Hunter reagent (BOLTON and HUNTER 1977). Finally, the iodination procedure can also oxidize methionine residues, which may be important for the biological activity of the peptide (LAMBERT et al. 1982).

In any case, each peptide has its own problem of iodination and the reader should refer to the corresponding literature.

3. Separation of Bound and Unbound Ligand

For all particulate receptor preparations, two methods are commonly used, centrifugation and filtration (GARDNER 1979). The choice of method depends on practical issues for which a particular method is more suitable, e.g., rapidity (filtration), handling of many samples (centrifugation), or use of sticky ligands (centrifugation). For solubilized receptor preparations, the separation methods depend mainly on the chemical properties of the ligand and include polyethylene glycol precipitation of ligand-receptor complex, adsorption of free ligand to charcoal and gel filtration (HOLLENBERG 1990). More recently, adsorption of free ligand to Whatman filters pretreated with polyethylenimine has proven a very useful method (BRUNS et al. 1983).

III. Molecular Characterization and Purification

The various approaches and methodologies that have been developed to characterize peptide receptors at the molecular level have been extensively reviewed (RUOHO et al. 1984; PILCH and CZECH 1984; HOLLENBERG 1990; VINCENT and KITABGI 1991).

With the exception of radiation-inactivation, which allows the determination of the molecular weight of functional receptors in their membrane environment (JUNG 1984), other biochemical methods require the solubilization of receptors. Since detergents used for extracting receptors from plasma membranes may alter the functionality of receptors in solution, a preliminary approach to molecular characterization often consists of covalently linking a radioligand to the receptor protein. The covalent complex can be thereafter analyzed by sodium dodecyl sulfate polyacrylamide gel electrophoresis (SDS-PAGE) for determination of its molecular weight. Two covalent labeling procedures have been developed: (1) Affinity cross-linking, in which a cross-linker is added to a preexisting peptide-receptor complex to effect covalent linkage (PILCH and CZECH 1984). Various cross-linkers reacting with primary amines, sulfhydryls or carboxylic acids are available. (2) Photoaffinity labeling, in which a photoreactive derivative of the peptide is prepared and covalent linkage is obtained by photolysis of the ligand-receptor complex (RUOHO et al. 1984).

Covalent labeling techniques are fruitful preliminary approaches but have several limitations for characterizing and purifying functional receptors. Cross-linking of peptides to receptors usually has a low yield, i.e., <10% (PILCH and CZECH 1984), which is poorly compatible with further purification of the covalent complex. Moreover, such irreversible complexes are not physiologically relevant and are analyzed in denaturing detergent solutions by SDS-PAGE. Therefore, in order to document the nature and molecular form(s) of receptors in a functional state and with the goal of purifying the receptor protein by affinity chromatography (see below), the solubilization of functional receptors in nondenaturing detergent solutions has been developed (HOLLENBERG 1990; VINCENT and KITABGI 1991). The analysis of solubilized receptors by molecular sieving and sucrose density gradient ultracentrifugation provides the hydrodynamic parameters of solubilized receptors including Stokes radius, sedimentation coefficient, frictional ratio and molecular weight (COUVINEAU et al. 1986; CALVO et al. 1989; VOISIN et al. 1991). Moreover, such analysis has made it possible to characterize oligomeric forms of receptors which contain guanine nucleotide-binding proteins (G proteins) physically associated with the receptor protein in a functional complex, and also to identify the corresponding G proteins (COUVINEAU et al. 1986, 1990b; CALVO et al. 1989).

Purification of peptide receptors is not compatible with classical biochemical procedures and must involve a highly specific step such as ligand affinity chromatography (VINCENT and KITABGI 1991; HOLLENBERG 1990) and/or immunoaffinity using an antibody raised against the receptor (REYL-DESMARS et al. 1989). Ligand affinity gels used to purify receptors are usually prepared by coupling peptides to various supports with spacer arms (VINCENT and KITABGI 1991). This procedure yields gels containing 1–100 nmol ligand/ml gel and in many cases does not allow one to control the site of covalent linking of the peptide to the gel, resulting in the presence of biologically inactive peptide on the matrix. Recently, a newly designed affinity chromatography procedure has been developed by direct solid-phase synthesis of the peptide on polyacrylamide resin (COUVINEAU et al. 1990a). This procedure yields approximately 15 μmol peptide/ml gel and provides a specific orientation of the peptide which is linked to the resin by its C terminus. It has permitted the purification of vasoactive intestinal peptide (VIP) receptors from porcine liver by one-step affinity chromatography (COUVINEAU et al. 1990a).

Extensive reviews on the solubilization and purification of peptide receptors have been published (HOLLENBERG 1990; VINCENT and KITABGI 1991).

IV. Receptor Cloning Strategies

An essential step for understanding the molecular basis of receptor function is the determination of the primary structure of the receptor protein. Because receptors are large proteins, their complete amino acid sequence cannot be obtained from conventional sequencing techniques. For all the receptors whose amino acid sequence is known, this has been achieved by cloning and sequencing the corresponding complementary DNA (cDNA) and deducing the amino acid sequence from the cDNA nucleotide sequence. Several strategies can be applied to the cloning of a receptor cDNA (Fig. 1). They may be divided into two categories: those which require receptor purification and those which do not. They will be briefly outlined in this section. These methods have led to the cloning of a number of hormone, growth factor, neuropeptide and neurotransmitter receptors. It is beyond the scope of this paper to review all the relevant literature. Only those studies pertaining to receptors for peptides found in the gastrointestinal tract or illustrating a particular point will be cited.

In the first type of strategy, receptor purification in sufficient amounts is required to generate partial amino acid sequence data or to raise polyclonal or monoclonal antireceptor antibodies. The partial peptide sequences serve to synthesize oligonucleotides which may be used as primers for the polymerase chain reaction (PCR) with an appropriate template (cDNA from a receptor-rich tissue). The synthetic oligonucleotides or the PCR-generated DNA are then used as probes for screening a cDNA library constructed from a tissue that abundantly expresses the receptor. Alternatively, the cDNA library is constructed in an expression vector and the library is screened by reacting the expressed proteins with the antireceptor antibodies. These two methods were independently and simultaneously used to clone the lutropin-choriogonadotropin receptor (LOOSFELT et al. 1989; MCFARLAND et al. 1989). Recently, the purification-partial sequence strategy was successfuly applied to the cloning of the gastrin-releasing peptide (GRP)-preferring bombesin receptor (BATTEY et al. 1991).

Receptor purification is often a difficult task and cloning strategies that do not require receptor purification have been devised. One such strategy is based on the ability of *Xenopus laevis* oocytes to functionally express foreign microinjected mRNA coding for a variety of proteins including membrane receptors. A class of receptors that can be readily expressed and detected at the level of a single oocyte are the G protein-coupled receptors that activate the phospholipase C pathway in their tissue of origin. Expres-

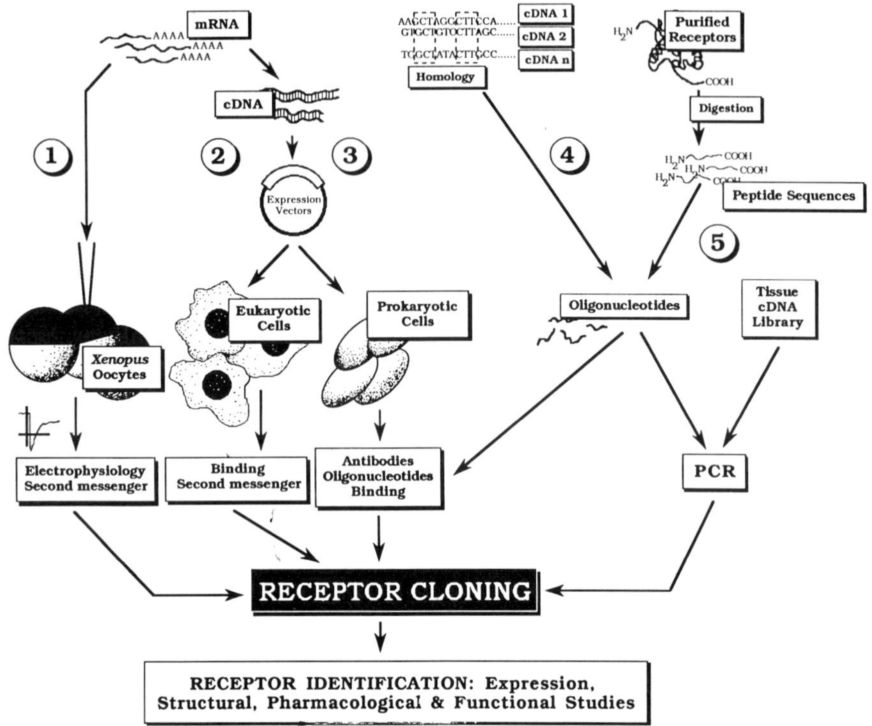

Fig. 1. Strategies for receptor cloning. Different strategies have been developed to clone various plasma membrane receptors. The classical approach from protein to DNA (⑤) has been used initially to clone several receptors but requires receptor purification and information about the amino acid sequence of some receptor fragments. Since receptor purification is a very difficult task, other strategies which take advantage of the tremendous development of molecular biology techniques and avoid the purification step have been increasingly used. They include expression cloning (① to ③) and homology cloning (④). The five most utilized strategies used for screening positive clones during receptor cloning are schematized here (the black box "receptor cloning" contains all the methods involved in the cloning itself; SAMBROOK et al. 1989):

① Expression by injection of mRNA into *Xenopus* oocytes. Measurement of electrophysiological response or second messenger level in injected oocytes upon agonist application reflects the expression of receptors. This strategy has been used for cloning the neurokinin-2 receptor (MASU et al. 1987).

② Expression of mRNA in eukaryotic cells. To identify the cDNA clone encoding the receptor, a cDNA library constructed with a mammalian expression vector is expressed in eukaryotic cells. The transfected cells are then assayed for their ability to bind radioactive ligands or to produce second messengers upon agonist application. This strategy has been used for cloning the secretin receptor (ISHIHARA et al. 1991).

③ Expression of mRNA in bacteria. To identify the cDNA clone encoding the receptor, a cDNA library constructed with a bacteriophage expression vector is expressed in prokaryotic cells. The transfected bacteria are then screened by reacting the expressed proteins with an antireceptor antibody. This strategy has been used for cloning the LH/hCG receptor (LOOSFELT et al. 1989; MCFARLAND et al. 1989).

sion in the oocyte of such a receptor will lead to stimulation by the receptor agonist of inositol trisphosphate production, calcium mobilization and activation of oocyte calcium-dependent chloride channels. The resulting depolarization can be measured in a single oocyte by electrophysiological means. Functional receptor expression is possible because the oocyte provides the necessary environment and all the machinery required for receptor functioning, including the correct trafficking and insertion of the foreign receptor into the oocyte plasma membrane, as well as a native phospholipase C system composed of the enzyme and the G protein(s) that couple the foreign receptor to the enzyme. Thus, the oocyte provides a functional assay for receptor cDNA or mRNA expression and allows one to select and clone the active mRNA species. This strategy has recently been successfully employed by NAKANISHI and his group for the cloning of the tachykinin NK2 (neurokinin A/substance K-preferring), NK1 (substance P-preferring) receptors (MASU et al. 1987; YOKOTA et al. 1989), and the neurotensin receptor (TANAKA et al. 1990).

Most of the cloned hormone and regulatory peptide receptors including those mentioned above belong to the superfamily of G protein-coupled receptors. These receptors are all composed of a single polypeptide chain with 7 transmembrane segments of 20–25 amino acids in length, an N-terminal extracellular domain, a C-terminal cytoplasmic domain, and 3 extracytoplasmic and 3 cytoplasmic loops (Fig. 2). They all share sequence homology, particularly in the transmembrane domains. Based on these homologies, it has been possible to use DNA probes derived from a cloned receptor to screen at low stringency a library presumably containing the cDNA of related receptor(s). This strategy has quite recently led to the cloning of the tachykinin NK3 (neurokinin B/neuromedin K-preferring) receptor (SHIGEMOTO et al. 1990) and the neuromedin B-preferring bombesin receptor (WADA et al. 1991). It may also be worth mentioning here that the histamine H_2 gastric receptor has been cloned using this "homology" approach (GANTZ et al. 1991). It should be noted that unexpected receptors

---

Alternatively, synthetic oligonucleotides can be used for screening (see strategies ④ and ⑤). Finally, the recent demonstration that the β_3-adrenergic receptor cDNA can be expressed in a functional form in bacteria (STROSBERG 1991) suggests the possible direct screening by binding assay in bacteria.

④ Homology cloning. Based on sequence homologies between previously cloned receptors, selected oligonucleotides are synthesized and used (a) to directly screen tissue cDNA libraries; (b) as primers for PCR in order to amplify related receptor sequences from cDNA libraries. This strategy has been used for cloning the neurokinin-3 receptor (SHIGEMOTO et al. 1990).

⑤ The protein to DNA approach. After receptor purification and sequencing of receptor fragments, oligonucleotides are synthesized and used to screen tissue cDNA libraries expressed in bacteria (see ③). A variation on this theme is to use selected oligonucleotides as primers for the PCR to amplify the receptor sequence. This strategy has been used for cloning the GRP/bombesin receptor (BATTEY et al. 1991)

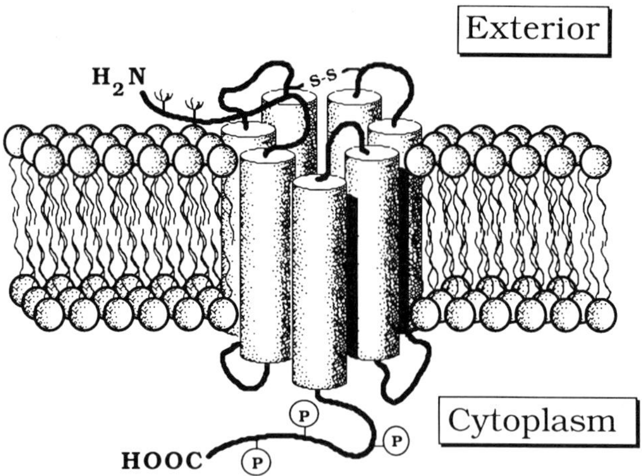

Fig. 2. The seven-transmembrane model for G protein-coupled receptors. The cylinders spanning the lipid bilayer represent the seven transmembrane helical domains (displaying maxima in hydropathy pattern) that are structurally conserved in G protein-coupled receptors. On the extracellular loops, S-S represents the proposed disulfide bridge between two extracellular loops which seems to be highly conserved in G protein-coupled receptors. Consensus N-glycosylation sites are marked on the N-terminal extracellular domain (ψ) and have been deduced from amino acid sequences of G protein-coupled receptors. On the cytoplasmic side, loops are involved in coupling to transducing and desensitization systems. On the C-terminal cytoplasmic domain are shown the various sites of phosphorylation (***P***) by cAMP-dependent kinase (see LEFKOWITZ 1991; STROSBERG 1991 for details)

may be cloned by this approach. Thus, while looking for NK2 (neurokinin A/substance K)-related receptors, MATSUDA et al. (1990) recently cloned a protein of the G protein-coupled receptor family that subsequently proved to be a cannabinoid (marijuana constituent) receptor. Finally, this strategy may also result in the cloning of putative receptors for which the endogenous ligand is not yet discovered or remains to be determined (LIBERT et al. 1989).

C. Signal Transduction by Receptors

Evolution has selected several strategies for signal transduction (SPIEGEL 1991). The receptor may be itself the effector such as an ion channel (for example, the nicotinic receptor for acetylcholine is a sodium channel) or an enzyme (for example, the insulin receptor is a tyrosine kinase). Alternatively, the receptor may be coupled to an effector by a G protein. In this case, the effector may be an enzyme such as adenylyl cyclase or phospholipase C which catalyze the synthesis of intracellular second messengers (see below). It may also be an ion channel. Extensive reviews on the general properties of receptors (GERSHENGORN 1991; SPIEGEL 1991) and G proteins (BIRNBAUMER et al. 1990) have been published recently. We will briefly outline here the adenylyl cyclase – adenosine-3′,5′-cyclic monophosphate (cAMP) pathway and the phospholipase C – phosphatidyl inositol pathway.

I. Adenylyl Cyclase: Cyclic AMP Pathway

A major transduction mechanism by which occupancy of a receptor by a peptide transmits the information within the target cell is the adenylyl cyclase-catalyzed synthesis of cAMP. Recent reviews have described this mechanism in detail (GILMAN 1989; BIRNBAUMER et al. 1990; GERSHENGORN 1991).

Adenylyl cyclase is a plasma membrane-bound enzyme consisting of 1134 amino acid residues (KRUPINSKI et al. 1989), which catalyzes the formation of cAMP from adenosine triphosphate (ATP). The intracellular second messenger cAMP acts subsequently by binding to the regulatory subunits of cAMP-dependent protein kinase which in turn phosphorylates a number of substrate proteins involved in the biological response (GERSHENGORN 1991). G_s and G_i proteins are responsible for coupling stimulatory and inhibitory receptors to adenylyl cyclase, respectively (GILMAN 1989; BIRNBAUMER et al. 1990; SIMON et al. 1991). G proteins comprise a family whose members play a key role in transmembrane signaling pathways (BIRNBAUMER et al. 1990). They are hetero-trimers composed of α-, β- and γ-subunits. Individual G proteins are distinguished by their α subunits which bind and hydrolyze guanosine triphosphate (GTP) (see below). The mechanism whereby occupancy of a stimulatory receptor by an agonist promotes adenylyl cyclase activation is the following. In the basal state, G_s proteins exist as αβγ-trimers when the nucleotide binding site is occupied by guanosine diphosphate (GDP). Interaction of the G_s protein with the activated receptor promotes the exchange of GDP for GTP and the subsequent dissociation of the α-GTP complex from the βγ-dimer. As another consequence of this interaction, the affinity of the receptor for the agonist is lowered dramatically. The α_s-GTP complex then activates adenylyl cyclase but detailed insights into the nature of this interaction are still lacking. Due to the intrinsic GTPase activity of α_s, α_s-GTP is rapidly converted to α_s-GDP, which reassociates with the βγ-dimer to form the inactive heterotrimeric form of G_s. As a result adenylyl cyclase activation ceases unless agonists continue to be present to initiate another cycle of activation. This cycle can be blocked by cholera toxin which catalyzes the incorporation of ADP-ribose into α_s. This results in persistent activation of adenylyl cyclase because of inhibition of the GTPase activity of α_s.

The mechanism of inhibition of adenylyl cyclase is less clear. The current hypothesis implies that activation of α_i by inhibitory receptors occupied by agonists frees βγ-dimers, which can interact with endogenous

α_s, thus inactivating α_s (SIMON et al. 1991). This scheme has been criticized and other hypotheses have been proposed (SIMON et al. 1991). Three types of α_i-subunits named α_{i1}, α_{i2} and α_{i3} have been characterized and cloned (JONES and REED 1987; BIRNBAUMER et al. 1990).

II. Phosphatidyl Inositol Signaling

Another major transduction mechanism by which agonist-receptor complexes transmit their information to the target cell is the phospholipase C-catalyzed hydrolysis of membrane phophatidylinositols. A large number of reviews have covered this topic in detail (BERRIDGE 1987; NISHIZUKA 1986, 1988; MAJERUS et al. 1990). The essential features of this signaling pathway will be briefly outlined here. The binding of an agonist to its receptor triggers, via G protein(s), the translocation of phospholipase C (PLC) enzymes from a cytoplasmic pool to the plasma membrane. The mechanisms of the translocation are not yet understood. Once associated with the membrane, these enzymes hydrolyze several phosphatidylinositols such as phosphatidylinositol 4,5-bisphosphate to yield inositol phosphates and 1,2-diacylglycerol. It is now well established that inositol (1,4,5)trisphosphate [Ins(1,4,5)P_3] is one of the inositol phosphates formed, and together with diacylglycerol may act as a second messenger in a bifurcating pathway of cellular events. Ins(1,4,5)P_3 binds to a receptor located on the endoplasmic reticulum, from which it triggers waves of calcium release. These transient rises in intracellular calcium activate in turn a variety of cellular events (e.g., ion channel opening, activation of a variety of calmodulin-dependent and -independent enzymes). Diacylglycerol activates protein kinase C enzymes which phosphorylate a number of intracellular or membrane proteins. Depending on the cell system, these two signaling pathways may contribute independently or synergistically to the final cellular response which can be as diverse as muscle contraction, endocrine and exocrine secretion, protein synthesis or cell proliferation.

Several G proteins appear to be involved in coupling receptors to PLC activation. In some systems they are sensitive to pertussis toxin, which blocks the agonist effect, while in other systems they are not. The pertussis toxin sensitive G proteins might belong to the G_i or G_o families (BIRNBAUMER et al. 1990). The pertussis toxin-insensitive G proteins are yet to be identified. Recently, a pertussis toxin-insensitive G protein designated G_q has been purified from bovine brain and shown to stimulate PLC activity in vitro (SMRCKA et al. 1991). It remains to demonstrate that G_q is involved in the hormonal regulation of PLC.

D. Peptide Receptors in the Digestive Tract

We are detailing here the properties of three peptide receptors which play an important role in digestive tract physiology and illustrate various signal transduction pathways, i.e., VIP, galanin and neurotensin receptors. Table 1 gives a short survey of all peptide receptors that are relevant to digestive tract physiology.

Table 1. Peptide receptors in the digestive tract: a brief survey

Cholecystokinin		
Type of receptor	CCK_A	CCK_B
Mol. wt. (kDa)	80 [2]–100 [3]	51 [4]
Main target tissues	Pancreas [5], gallbladder [6], Brain [1]	Brain [1]
Signal transduction	IP3 ↗ DAG ↗ [7]	?
Associated G protein	Yes, but not characterized [8]	?
Purified	From rat pancreas [9]	Not yet
Cloned	From rat pancreas; 444 aa; 7TM [10]	From rat brain; 452 aa; 7TM [11]
Natural agonist	CCK8 ≫ gastrin, CCK4	CCK8 > Gastrin, CCK4
Antagonist	Yes [7]	Yes [12]

References: [1] Moran et al. (1986); [2] Pearson et al. (1987); [3] Fourmy et al. (1989); [4] Sakamoto et al. (1984); [5] Innis and Snyder (1980); [6] Shaw et al. (1987); [7] Jensen et al. (1989); [8] Williams and McChesnay (1987); [9] Duong et al. (1989); [10] Wank et al. (1992a); [11] Wank et al. (1992b); [12] Freidinger (1989).

Galanin	
Type of receptor	Galanin
Mol. wt. (kDa)	54 [1]
Main target tissues	Endocrine pancreas [1–3], brain [4]
Signal transduction	cAMP ↘ [3], ATP-sensitive K^+ channel ↗ [5]
Associated G protein	Gi [1–3], Gk [5]
Purified	From porcine brain [6]
Cloned	Not yet
Natural agonist	Galanin
Antagonist	Yes [7]

References: [1] Amiranoff et al. (1989a); [2] Lagny-Pourmir et al. (1989a); [3] Amiranoff et al. (1988); [4] Servin et al. (1987); [5] Dunne et al. (1989); [6] Chen et al. (1992); [7] Bartfai et al. (1992).

Gastric inhibitory polypeptide	
Type of receptor	GIP
Mol. wt. (kDa)	59 [1,2]
Main target tissues	Pancreatic β cell [1–3]
Signal transduction	cAMP ↗ [3]
Associated G protein	Gs (?) [4]
Purified	Not yet
Cloned	Not yet
Natural agonist	GIP
Antagonist	None

References: [1] Couvineau et al. (1984b); [2] Amiranoff et al. (1986); [3] Amiranoff et al. (1984); [4] Amiranoff et al. (1985).

Table 1. (continued)

Gastrin	
Type of receptor	Gastrin
Mol. wt. (kDa)	78 [1]
Main target tissues	Gastrointestinal smooth muscles, mucosal gastric cells [2,3], pancreas [4]
Signal transduction	IP3 ↗ DAG ↗ [5]
Associated G protein	?
Purified	Not yet
Cloned	From canine parietal cell; 453 aa; 7TM [6]
Natural agonist	Gastrin > CCK8, cerulein [3]
Antagonist	Yes [7]

References: [1] RAMANI and PREISSMAN (1989); [2] SOLL et al. (1984); [3] TAKEUSHI et al. (1979); [4] YU et al. (1987); [5] ROCHE et al. (1991); [6] KOPIN et al. (1992); [7] MARTINEZ et al. (1984).

Gastrin-releasing peptide/bombesin	
Type of receptor	GRP/bombesin
Mol. wt. (kDa)	75–85 [1,2]
Main target tissues	Exocrine pancreas, gastrointestinal tract, brain
Signal transduction	IP3 ↗ DAG ↗ [3]
Associated G protein	Pertussis toxin-insensitive G [4]
Purified	From mouse fibroblast Swiss 3T3 cells [1]
Cloned	From mouse fibroblast Swiss 3T3 cells; 384 aa; 7 TM [5]
Natural agonist	GRP/bombesin > litorin, neuromedin B [6]
Antagonist	Yes [6]

References: [1] FELDMAN et al. (1990); [2] HUANG et al. (1990); [3] TAKUWA et al. (1987); [4] FISHER and SCHONBRUNN (1988); [5] BATTEY et al. (1991); [6] JENSEN and COY (1991).

Glucagon	
Type of receptor	Glucagon
Mol. wt. (kDa)	53 [1]–63 [2]
Main target tissues	Liver, fat cells [1–5]
Signal transduction	cAMP ↗ [3]
Associated G protein	Gs [4]
Purified	Partially [5]
Cloned	Not yet
Natural agonist	Glucagon > oxyntomodulin[a] [3]
Antagonist	Yes [6]

References: [1] JOHNSON et al. (1981); [2] HERBERG et al. (1984); [3] BATAILLE et al. (1988); [4] PADRELL et al. (1987); [5] IWANIJ and VINCENT (1990); [6] EPAND et al. (1981).

[a] An oxyntomodulin-preferring receptor which recognizes glucagon with low affinity has been identified in isolated rat oxyntic gastric glands [3].

Table 1. (continued)

Growth hormone releasing peptide	
Type of receptor	GRF
Mol. wt. (kDa)	70 [1]
Main target tissues	Hypophysis [2]
Signal transduction	cAMP ↗ [3]
Associated G protein	Gs (?)
Purified	Not yet
Cloned	Not yet
Natural agonist	GRF > VIP [3]
Antagonist	Yes [4]

References: [1] VELICELEBI et al. (1986); [2] SEIFERT et al. (1985); [3] REYL-DESMARS et al. (1985); [4] ROBBERECHT et al. (1985).

Neuropeptide Y	
Type of receptor	Y1, Y2 [1] and Y3 [2]
Mol. wt. (kDa)	70 (Y1) [1]; 50 (Y2) [1]
Main target tissues	Brain [1]; kidney [2]; heart [3]; intestine (low affinity) [4]
Signal transduction	cAMP ↘ [5]; Ca^{2+}? [6]
Associated G protein	Gi [5]; Go [6]
Purified	From rabbit kidney (42 type) [7]
Cloned	From Drosophila melanogaster; 449 aa; 7TM [7]; From human brain; 384 aa; 7TM [8,9]; From rat brain; 349 aa; 7TM [10]; From bovine brain; 353 aa; 7TM [11]
Natural agonist	NPY > PYY, PP [4]
Antagonist	Yes [3]

References: [1] SHEIKH and WILLIAMS (1990); [3] GIMPL et al. (1990); [2] BALASUBRAMANIAM and SHERIFF (1990); [4] NGUYEN et al. (1990); [5] KASSIS et al. (1987); [6] EWALD et al. (1988); [7] SHEIKH et al. (1991).

Neurotensin	
Type of receptor	NT
Mol. wt. (kDa)	49–110 [1,2]
Main target tissues	Brain, stomach, intestine [3]
Signal transduction	cAMP ↘, IP3 ↗ DAG ↗ [3–5]
Associated G protein	Gi and pertussis toxin-insensitive G [4,5]
Purified	From mouse brain [6]
Cloned	From rat brain; 424 aa; 7 TM [7]
Natural agonist	NT > xenopsin, neuromedin N [3]
Antagonist	None

References: [1] MAZELLA et al. (1985a); [2] MAZELLA et al. (1985b); [3] KITABGI et al. (1985); [4] AMAR et al. (1987); [5] BOZOU et al. (1989); [6] MAZELLA et al. (1989); [7] TANAKA et al. (1990).

Table 1. (continued)

Opiates			
Type of receptor	Mu [1]	Delta [1]	Kappa [1]
Mol. wt. (kDa)	65 [2]	53 [2]	63 [6]
Main target tissues	Brain, digestive tract	Brain, digestive tract	Brain, digestive tract
Signal transduction	cAMP ↘, K^+ channel ↗ [3]	cAMP ↘, K^+ channel ↗ [3]	cAMP ↘, K^+ channel ↗ [3]
Associated G protein	Gi [3]	Gi [3]	Gi [3]
Purified	From bovine striatum [4]	From mouse neuroblastoma × rat glioma [5]	From human placenta [6]
Cloned	Not yet	Not yet	Not yet
Natural agonist	β-Endorphin> Dynorphin (1–8) Dynorphin A and B Met- and Leu-enkephalin	Met- and Leu-enkephalin> β-Endorphin Dynorphin (1–8) Dynorphin A and B	Dynorphin A and B> Dynorphin (1–8)
Antagonist	Yes [1]	Yes [1]	Yes [1]

References: [1] LESLIE (1987); [2] HOWARD et al. (1986); [3] CHILDERS (1988); [4] GIOANNINI et al. (1985); [5] SIMONDS et al. (1985); [6] AHMED et al. (1989).

Pancreatic polypeptide	
Type of receptor	PP [1]
Mol. wt. (kDa)	67 [2]
Main target tissues	Small intestine [1], nervous tissue [2,3]
Signal transduction	?
Associated G protein	?
Purified	Not yet
Cloned	Not yet
Natural agonist	PP > PYY, NPY [1]
Antagonist	None

References: [1] GILBERT et al. (1988); [2] INUI et al. (1990); [3] SCHWARTZ et al. (1987).

Peptide YY	
Type of receptor	PYY [1,2]
Mol. wt. (kDa)	44 [3]–50 [4]
Main target tissues	Small intestine [1,2], brain [4]
Signal transduction	cAMP ↘ [5]
Associated G protein	Gi (?)
Purified	Only solubilized [3]
Cloned	Yes [6]
Natural agonist	PYY > NPY, PP [1,2]
Antagonist	None

References: [1] LABURTHE et al. (1986b); [2] LABURTHE (1990); [3] VOISIN et al. (1991); [4] INUI et al. (1988); [5] SERVIN et al. (1989); [6] LIARHAMMAR et al. (1992).

Table 1. (continued)

Pituitary adenylate cyclase activating polypeptide	
Type of receptor	PACAP
Mol. wt. (kDa)	57 [1]–65 [2]
Main target tissues	Brain, liver, pancreatic cell line [1–4]
Signal transduction	cAMP ↗ [3]
Associated G protein	Gs [?]
Purified	Only solubilized [4]
Cloned	Not yet
Natural agonist	PACAP > VIP, secretin, helodermin [1–4]
Antagonist	None

References: [1] OHTAKI et al. (1990); [2] ROBBERECHT et al. (1991); [3] MIYATA et al. (1989); [4] MASUDA et al. (1990).

Secretin	
Type of receptor	Secretin
Mol. wt. (kDa)	51 [1]–62 [2]
Main target tissues	Exocrine pancreas [3], stomach [4], brain [5]
Signal transduction	cAMP ↗ [6]
Associated G protein	Gs [4,5]
Purified	Not yet
Cloned	From mouse neuroblastoma × rat glioma; 449 aa; 7 TM [7]
Natural agonist	Secretin > helodermin, helospectins, PHI, VIP, GRF [3]
Antagonist	Yes [6]

References: [1] GOSSEN et al. (1989); [2] BAWAB et al. (1988); [3] ZHOU et al. (1989); [4] GESPACH et al. (1980); [5] ISHIHARA et al. (1991); [6] HAFFAR et al. (1991).

Somatostatin	
Type of receptor	Somatostatin
Mol. wt. (kDa)	56 [1]–100 [2]
Main target tissues	Ubiquitous [3]
Signal transduction	CAMP ↘ [4], K^+ channel ↗ [5]
Associated G protein	Gi [3], Gk (?) [5]
Purified	From the human gastric cancer cell line HGT-1 [6]
Cloned	From human islets: two forms 369 and 391 aa; 7TM [7]; From rat brain; 369 aa; 7TM [8]
Natural agonist	Somatostatin 14 and 28 [3]

References: [1] KNUHTSEN et al. (1990); [2] MURTY et al. (1989); [3] PATEL et al. (1990); [4] REYL-DESMARS et al. (1986); [5] YATANI et al. (1990); [6] REYL-DESMARS et al. (1989); [7] YAMADA et al. (1992); [8] KLUXEN et al. (1992).

Table 1. (continued)

Tachykinins (substance P, substance K, neuromedin K)			
Type of receptor	NK1 [1] (Substance P)	NK2 [1] (Substance K)	NK3 (Neuromedin K)
Mol. wt. (kDa)	53 [2]	43[a] [3]	51[a] [4]
Main target tissues	Brain, ileum, carotid artery [5]	Brain, duodenum, pulmonary artery [5]	Brain, portal vein [5]
Signal transduction	IP3 ↗ DAG ↗ [1]	IP3 ↗ DAG ↗ [1]	IP3 ↗ DAG ↗ [1]
Associated G protein	Yes [6]	Yes [3]	Yes [4]
Purified	Not yet	Not yet	Not yet
Cloned	From rat brain; 407 aa; 7 TM [6]	From bovine stomach; 384 aa; 7 TM [3]	From rat brain; 452 aa; 7 TM [4]
Natural agonist	Substance P > Substance K, Neuromedin K	Substance K > Substance P, Neuromedin K	Neuromedin K > Substance K, Substance P
Antagonist	Yes [5]	Yes [5]	Yes [5]

References: [1] Quirion and Dam (1988); [2] Boyd et al. (1991); [3] Masu et al. (1987); [4] Shigemoto et al. (1990); [5] Regoli et al. (1990); [6] Yokota et al. (1989).
[a] Molecular weight of the peptide core calculated from the amino acid sequence of cloned receptor.

Vasoactive intestinal peptide	
Type of receptor	VIP
Mol. wt. (kDa)	46–73 [1]
Main target tissues	Ubiquitous [1]
Signal transduction	cAMP ↗ [1]
Associated G protein	Gs [2]
Purified	From porcine liver [3]
Cloned	From rat lung; 459 aa; 7 TM [4]
Natural agonist	VIP > PACAP, helodermin; helospectins, PHI, GRF, secretin [5–7]
Antagonist	Yes [5–8]

References: [1] Laburthe and Couvineau (1988); [2] Couvineau et al. (1986); [3] Couvineau et al. (1990a); [4] Ishihara et al. (1992); [5] Laburthe et al. (1986a); [6] Pandol et al. (1986); [7] Waelbroeck et al. (1985); [8] Turner et al. (1986).

I. Vasoactive Intestinal Peptide

Vasoactive intestinal peptide (VIP) was originally isolated from hog small intestine (Said and Mutt 1970) and later found to be widely distributed in the central and peripheral nervous system (Said and Rosenberg 1976). VIP belongs to a growing family of structurally related peptides (see Mutt 1988) that now comprises secretin, glucagon, gastric inhibitory polypeptide (GIP), peptide histidine isoleucineamide (PHI), growth hormone-releasing factor (GRF), pituitary adenylate cyclase activating peptide (PACAP) (Miyata et al. 1989) and peptides isolated from the venom of the gila monster, including helodermin and helospectins.

The largest amounts of VIP in the organism are found in the digestive tract, where VIP is strictly a neuropeptide. The VIP-containing neurons innervate epithelial cells, exocrine glands, smooth muscles and other neurons. All these points are developed in other chapters of this handbook. The watery diarrhea syndrome or Verner-Morrison syndrome, where secreting tumors release massive amounts of VIP into the blood (BLOOM et al. 1973), highlights one of its most important biological actions, i.e., the stimulation of intestinal secretion of water and electrolytes (see LABURTHE and DUPONT 1982; LABURTHE and AMIRANOFF 1989). In consonance with the widespread distribution of VIP, multiple biological effects of the neuropeptide in the digestive tract have been documented (see SAID 1986). They include stimulation of water and electrolytes secretion (pancreas, intestine, gallbladder), of enzyme secretion (pancreas) and of mucus secretion (intestine), inhibition of absorption (intestine and colon), increased bile flow, and inhibition of acid and pepsin secretion (stomach). VIP also exerts dual effects (relaxation and/or contraction) on smooth muscles in esophagus, stomach and small and large intestines, with predominant relaxant actions (see KHALIL et al. 1987). Finally, VIP has a small glycogenolytic effect in liver (KERINS and SAID 1973) and regulates enterocytic metabolism by inhibiting glucose oxidation (VIDAL et al. 1988) and glycolysis (ROSSI et al. 1989) and by stimulating long-chain fatty acid oxidation (VIDAL et al. 1989).

Vasoactive intestinal peptide interacts with specific receptors located on the surface of target cells. They have been characterized throughout the organism using $[^{125}I]$VIP as a radioligand. In the digestive tract the following organs or tissues are equipped with VIP receptors: intestinal epithelium (LABURTHE et al. 1978a; PRIETO et al. 1979), colonic epithelium (BROYART et al. 1981), gastric epithelium (GESPACH et al. 1984), exocrine pancreas (CHRISTOPHE et al. 1976), gallbladder epithelium (DUPONT et al. 1981), parotid cells (INOUE et al. 1985) and liver (DESBUQUOIS 1974; BATAILLE et al. 1974). Autoradiographic studies also visualized the presence of VIP-binding sites in almost all layers of the gastrointestinal segments, i.e., mucosa, muscularis mucosa, smooth muscles of arterioles, circular and longitudinal smooth muscles of the muscularis externa, myenteric plexus and lymph nodules (ZIMMERMAN et al. 1989). We will focus here on studies dealing with VIP receptor characterization in tissues of gastrointestinal origin, with special reference to intestinal epithelium.

Depending on target cells, species and/or receptor preparations, one or two classes of VIP-binding sites have been identified (references herein). The high-affinity receptor usually displays a K_D of about 0.1–1 nM. Activation of VIP receptors is linked to stimulation of adenylyl cyclase activity (AMIRANOFF et al. 1978) through G_s proteins (LABURTHE and COUVINEAU 1988; CALVO et al. 1989; see also Sect. C.I for the mechanism of adenylyl cyclase activation). The receptor-G_s protein coupling is further evidenced by the ability of GTP to enhance the dissociation of VIP from receptor (AMIRANOFF et al. 1980). The VIP-induced increase of intracellular cyclic

AMP production in target cells (LABURTHE et al. 1978a, 1979; ROBBERECHT et al. 1976; LABURTHE and COUVINEAU 1988) probably underlies most of the biological actions of VIP. A few exceptions exist since the abundant VIP receptors found in liver stimulate adenylyl cyclase with low efficacy (ROUYER-FESSARD et al. 1989a) and fail to alter significantly cAMP levels in isolated hepatocytes (LABURTHE et al. 1978b). VIP, at micromolar concentrations, is also able to mobilize intracellular calcium through breakdown of phosphoinositides (MALHOTRA et al. 1988). The physiological significance of this latter effect observed in a few nervous tissue preparations (MALHOTRA et al. 1988; AUDIGIER et al. 1988), but not yet in the digestive tract, remains unclear.

The study of structure-activity relationship of peptides for interacting with VIP receptors has taken advantage of the existence of the large family of VIP-related peptides. PHI, PHM, secretin, GRF (rat or human), helodermin and PACAP compete with $[^{125}I]$VIP for binding to receptors in various target cells (BATAILLE et al. 1980; RAUFMAN et al. 1982; LABURTHE et al. 1983, 1985, 1986a; CHRISTOPHE et al. 1988; ROBBERECHT et al. 1991). It must be stressed that some of these peptides, e.g., secretin, GRF and PACAP, as well as other members of the VIP family of related peptides, have their own specific receptors (see Fig. 3, Table 1). Marked species specificity in the interaction of these peptides with VIP receptors has been evidenced by studies of receptors in rat and human intestine (LABURTHE et al. 1983, 1985, 1986a). The human VIP receptor is more discriminating since the cross-reactivity of VIP-related peptides is generally lower in human than in rat tissues. VIP-related peptides that interact with VIP receptors behave as full agonists with the exception of human GRF (hGRF), which is a partial agonist in rat intestine (LABURTHE et al. 1983, 1986a). The use of VIP fragments (COUVINEAU et al. 1984a) and analogs (ROBBERECHT et al. 1986) has given further insight into the structural requirement for interaction with VIP receptors. The entire sequence of VIP is needed for binding to VIP receptors with high affinity. Deleting the N-terminal histidine results in a peptide having a 100-fold lower affinity for VIP receptors. Shorter fragments either have a very low affinity or do not interact at all with VIP receptors (COUVINEAU et al. 1984a). VIP analogs modified in the N-terminal domain also display a sharp decrease in affinity for receptors (ROBBERECHT et al. 1986).

Based on initial observations indicating that hGRF behaves as a mixed VIP agonist/antagonist in the rat intestine (LABURTHE et al. 1983), VIP antagonists have been designed by modifying the hGRF molecule, e.g., Ac-Tyr^1 hGRF (LABURTHE et al. 1986a) or (N-Ac-Tyr^1, D-Phe^2)-GRF_{1-29}-NH_2 (WAELBROECK et al. 1985). A VIP analog, i.e., (4 Cl-D-Phe^6, Leu^{17})VIP, has also been shown to behave as a VIP antagonist (PANDOL et al. 1986). Finally, VIP_{10-28} is also claimed to be an antagonist in the colon carcinoma cell line HT-29 (TURNER et al. 1986). It is worth noting that they are weak VIP antagonists which remain rather uninteresting for pharmacological

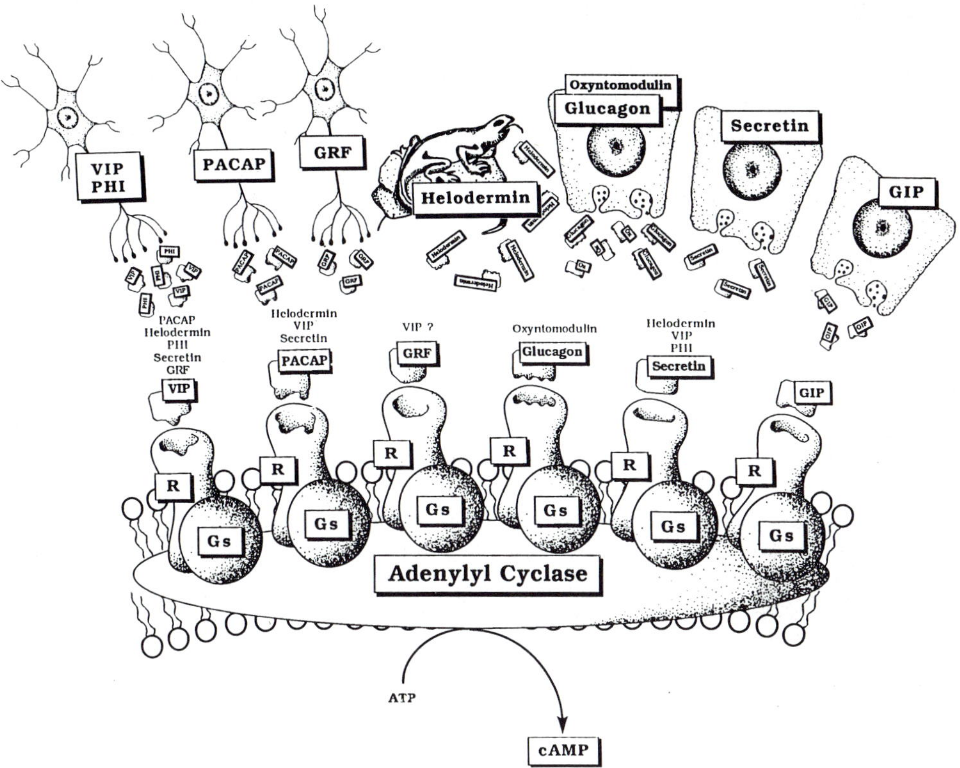

Fig. 3. Schematic representation of receptors for VIP and related peptides. The amino acid sequences of VIP and related peptides and their cellular localization in neurons (VIP, PHI, PACAP, GRF) or endocrine cells (glucagon, oxyntomodulin, secretin, GIP) are described elsewhere (McDONALD 1991b). Helodermin isolated from gila monster (*Heloderma suspectum*) venom appears also to be present in mammalian tissues (McDONALD 1991b).

Specific receptors for VIP, PACAP, GRF, glucagon, secretin and GIP have been identified and characterized (see Sect. D.I; Table 1). Occupation of these receptors by their natural ligands triggers, via G_s proteins, the activation of adenylyl cyclase. cAMP production in target cells probably underlies most of the biological actions of these peptides (Table 1). With regard to the receptor-binding sites, some receptors for VIP-related peptides display common pharmacological properties. For example, VIP and secretin receptors recognize the same array of peptides. On a pharmacological basis, they are distinguishable from each other only by their ability to bind either VIP with higher affinity than other peptides in the case of VIP receptors or secretin with higher affinity than other peptides in the case of secretin receptors. These cross-reactivities also apply, though to a lesser extent, to other receptors of this family. The exception is the gastric inhibitory peptide (GIP) receptor, which only recognizes GIP (see Sect. D.I, Table 1 and LABURTHE and COUVINEAU 1988 for details) studies since they have a K_i in the micromolar range and the GRF derivative antagonists have GRF agonist properties. Moreover, the VIP antagonist properties of these peptides vary according to species (LABURTHE et al. 1986a) and tissues (ROUYER-FESSARD et al. 1989a). In spite of the fact that VIP receptor heterogeneity is likely, there is still modest pharmacological evidence for the existence of VIP receptor subtypes in different tissues (CHRISTOPHE et al. 1988; ROUYER-FESSARD et al. 1989a). Recent structural characterization of VIP receptors gives further evidence for molecular heterogeneity of VIP receptors (see below), but the definitive answer to this issue has to await the cloning of VIP receptor genes from different tissues.

The molecular properties of VIP receptors have been investigated by means of cross-linking (LABURTHE et al. 1984; LABURTHE and COUVINEAU 1988) and photoaffinity labeling (ROBICHON et al. 1987; MARTIN et al. 1988) techniques (see Sect. B.III). Various cross-linking patterns have been reported with regard to the molecular weight of the major VIP-binding protein, e.g., from 46 to 73 kDa (LABURTHE and COUVINEAU 1988). Molecular weight heterogeneity among VIP receptors was observed at an early stage when considering different tissues within a given species (COUVINEAU and LABURTHE 1985a) or different species within a given tissue (COUVINEAU and LABURTHE 1985b). As far as the digestive tract is concerned, the following molecular weights have been determined: 73 kDa in rat small intestine (LABURTHE et al. 1984); 63–64 kDa in human colon (COUVINEAU and LABURTHE 1985a) or human colonic cancer cell lines (COUVINEAU et al. 1985; MULLER et al. 1985; LABURTHE et al. 1987); 48–53 kDa in rat liver (COUVINEAU and LABURTHE 1985b; NGUYEN et al. 1986); 53 kDa in porcine liver (COUVINEAU et al. 1990b); 63 kDa in rat pancreatic membranes (CHRISTOPHE et al. 1988); and 66 kDa in a tumoral rat pancreatic cell line (SVOBODA et al. 1988). This molecular heterogeneity extends to target cells outside the digestive tract (see LABURTHE and COUVINEAU 1988). VIP receptors are glycoproteins (NGUYEN et al. 1986; COUVINEAU et al. 1990b; LABURTHE and COUVINEAU 1988; EL BATTARI et al. 1987; DICKINSON et al. 1986) and a possible cause of the above-mentioned molecular heterogeneity is the existence of VIP receptor proteins with different carbohydrate chains (DICKINSON et al. 1986). Alterations of the molecular weight of VIP receptors have been observed under pathophysiological circumstances such as ontogenic development of rat intestine (CHASTRE et al. 1987) and rat liver regeneration following partial hepatectomy (GUIJARRO et al. 1991).

In the digestive tract, VIP receptors have been solubilized in an active form from rat liver (COUVINEAU et al. 1986, 1990a; GUIJARRO et al. 1989), porcine liver (COUVINEAU et al. 1990b), rat intestine (CALVO et al. 1989) and human colonic adenocarcinoma cells (EL BATTARI et al. 1988). Some of these studies provide evidence for the existence in solution of high molecular weight oligomeric forms of VIP receptors which arise from the physical association of receptors and G_s proteins (COUVINEAU et al. 1986, 1990a; CALVO et al. 1989). Up to now, VIP receptors have been purified only from porcine liver (Couvineau et al. 1990a). A cDNA encoding a VIP receptor protein has been cloned from rat lung (Ishihara et al. 1992; see also Chap. 13). A cDNA clone encoding a human VIP receptor was claimed to be identified in a library from the Nalm 6 line of leukemic pre-β lymphoblasts (Sreedharan et al. 1991). However, data from several groups (Nagata et al. 1992; Cook et al. 1992) provided strong evidence suggesting that this clone does *not* encode a VIP receptor and that it should be considered an orphan G protein – coupled receptor.

A remarkable property of VIP receptors is their widespread expression in a variety of tissues within the digestive tract (see above). As shown in rat small intestine crypt-villus axis, VIP receptors do emerge early during differentiation of epithelial cells from stem cells (Laburthe and Amiranoff 1989; Voisin et al. 1990). Moreover, the epithelial VIP receptor in intestine is included in the program of gene expression leading to different phenotypes, i.e., enterocyte (Laburthe and Amiranoff 1989) and mucus-secreting goblet cell (Laburthe et al. 1989). This is consistent with the dual role of VIP in controlling hydroelectrolytic secretion and mucus output as evidenced in differentiated clonal cell lines with different phenotypes (Laburthe and Amiranoff 1989; Laburthe et al. 1989; Rouyer-Fessard et al. 1989b). The expression of VIP receptors also extends to cancer cells in human colon (Laburthe et al. 1978a,b, 1980, 1987) and pancreas (Estival et al. 1983). The study of human colon adenocarcinoma cell lines in culture has revealed that the expression of VIP receptors may be dependent on the differentiation of cancer cells (Laburthe et al. 1987). VIP may even control a new function of colonic cancer cells, i.e., stimulation of glycogenolysis (Rousset et al. 1981), which is not held by normal epithelial cells from human adults (Zweibaum et al. 1991). Although antiproliferative and differentiation-promoting effects of VIP in colon cancer cells have been claimed (Hoosein et al. 1989), the significance and relevance of VIP receptors in human colon cancer remain largely unknown (Laburthe and Amiranoff 1989; Zweibaum et al. 1991).

II. Galanin

Galanin is a 29 amino acid peptide isolated from porcine intestine extracts by a chemical strategy that detects the C-terminal amidated structure of peptides (Tatemoto et al. 1983). It is widely distributed in central and peripheral nervous systems of numerous species. Analysis of cDNAs encoding porcine and rat galanin reveals that galanin is synthesized initially as part of a large precursor peptide that includes a signal peptide, galanin itself and a 59–60 amino acid galanin mRNA-associated peptide (Rökaeus and Brownstein 1986; Kaplan et al. 1988). The deduced amino acid sequence of rat galanin is 90% similar to porcine galanin, with all three amino acid differences in the C-terminal heptapeptide (Kaplan et al. 1988).

In consonance with its widespread localization, galanin exerts numerous biological functions including regulation of gastrointestinal motility (BAUER et al. 1989), modulation of the pituitary-hypothalamus axis (OTTLECZ et al. 1986) and inhibition of pancreatic endocrine secretions (DUNNING et al. 1986) (for a review of the different actions of galanin, see RÖKAEUS 1987). At the present time, the best-recognized biological effect of galanin is the inhibition of insulin release leading to a sustained hyperglycemia in vivo (McDONALD et al. 1985). Therefore, galanin receptors have been first identified and characterized in a pancreatic β-cell tumor (AMIRANOFF et al. 1987) and later in an insulin-secreting pancreatic β-cell line in culture Rin m5F (AMIRANOFF et al. 1988; LAGNY-POURMIR et al. 1989). More recently, a pancreatic δ-cell line has been shown to express specific galanin receptors (AMIRANOFF et al. 1991). Meanwhile, biochemical and/or autoradiographic studies have revealed the existence of specific binding sites for galanin in the central nervous system (SERVIN et al. 1987; FISONE et al. 1989), the intestinal and gastric smooth muscles (ROSSOWSKI et al. 1990; KING et al. 1989) and the visual cortex (ROSIER et al. 1991), but the role of galanin and its mechanism of action in those tissues have not been investigated in detail.

We will essentially describe here the pancreatic galanin receptor in relation to its role on the inhibition of insulin release. The properties of galanin receptors in other tissues will be discussed when they are available. The inhibitory mechanism of galanin on the pancreatic β-cell is believed to involve the binding of galanin to specific membrane receptors, the decrease in the intracellular second messenger cAMP and/or an activation of an ATP-sensitive K^+ channel, and/or a more direct action of galanin at the level of the exocytotic sites of insulin secretory granules.

The radioiodination of galanin preserves its biological activity (LAGNY-POURMIR et al. 1989a) and direct radioligand binding assays have made it possible to identify target cells for galanin in endocrine pancreas and brain. Remarkable similarities between these receptors may be pointed out. In these models, one class of high-affinity galanin-binding sites has been found with a K_D in the 1-nM range and a maximal capacity ranging from 30 to 100 fmol/mg protein.

Synthetic fragments of galanin and chemically modified sequences of galanin have been used to probe the structural requirement of the galanin-binding site in the pancreatic β-cell line Rin m5F (LAGNY-POURMIR et al. 1989b; GALLWITZ et al. 1990). The whole galanin amino acid sequence is required for full binding to receptor and consequently for full biological activity. Removal of the first amino acid glycine (Gly) residue from the NH_2 terminus (Gal_{2-29}) reduced the affinity fourfold, while further removal of the second residue tryptophan totally inactivates the peptide. Moreover, its replacement by Tyr, phenylalanine or alanine or by a nonaromatic residue dramatically alters the peptide-binding activity. In addition, exchange of the L- to D-stereoisomer of this amino acid in position 2 leads to a dramatically impaired binding activity, such impairment being much lower when alanine in position 7 is concerned. In contrast, the carboxy terminus of galanin appears to be less crucial since removal of 14 residues at the carboxy terminus (Gal_{1-15}) reduced the affinity only by tenfold while the efficacy of the peptide is not changed. Moreover, based on the observed similar activities of Gal_{1-29} and $Gal_{1-22}{}^{23}Cys$ on binding to receptor (GALLWITZ et al. 1990), it was suggested that the seven C-terminal amino acids of galanin do not play an important role in the interaction of galanin with its receptor. The structural requirement of galanin receptor for its ligand in the pancreatic β cell compares well with that observed for the brain galanin receptor (LAGNY-POURMIR et al. 1989b). It contrasts, however, markedly with that described in rat smooth muscle, which appears to exhibit more affinity for the C-terminal sequence of the peptide (ROSSOWSKI et al. 1990), suggesting the existence of galanin receptor subtypes.

The molecular characterization of pancreatic galanin receptor has been investigated using cross-linking techniques (see Sect. B.III). In membranes from Rin m5F cells (AMIRANOFF et al. 1989a), hamster pancreatic β-cell tumor (AMIRANOFF et al. 1987) and the pancreatic δ-cell line Rin 14B (AMIRANOFF et al. 1991), the covalently labeled receptor behaves as a monomeric protein of M_r 54000. An M_r 54000 protein has been also reported for the cross-linked galanin receptor in brain (SERVIN et al. 1987).

As with many peptide receptors negatively coupled to adenylate cyclase (see below), galanin receptors in pancreatic β- or δ-cells are coupled to a G_i protein. Indeed, guanine nucleotides inhibited the binding of [^{125}I]galanin to its receptor because of a marked increase of the dissociation constant (LAGNY-POURMIR et al. 1989a; AMIRANOFF et al. 1991). In addition, pertussis toxin, known to inactivate the G_i protein, converts galanin receptors from a high-affinity state to a low-affinity state. This series of data clearly indicate that galanin receptors are coupled to a pertussis toxin-sensitive G_i protein. Moreover, the successful solubilization by 3-[(3-cholaminopropyl)dimethylammonio)-1-propanesulfonate (CHAPS) of the brain galanin receptor tightly associated with the α_i-subunit of a G protein further emphasizes the close association of the galanin receptor with a guanine nucleotide regulatory protein (CHEN et al. 1992).

There is abundant evidence that galanin controls cAMP levels in the pancreatic β cell (AMIRANOFF et al. 1988). In Rin m5F cells, galanin inhibits forskolin and GIP-stimulated cAMP levels in dose-dependent manner from $10^{-10}\,M$ to $10^{-7}\,M$, which is compatible with the affinity of receptors for galanin. A maximal 50%–60% inhibition is observed. Galanin directly inhibits membrane adenylyl cyclase with an IC_{50} (0.3 nM) closely related to that observed for its effect on cAMP production. Pertussis toxin treatment of cells totally abolishes the pattern of galanin-inhibited cAMP levels in the Rin m5F cells, indicating the involvement of a pertussis toxin-sensitive G protein in the functional coupling of galanin receptors with the cyclic AMP production system. As described for the galanin receptor, the entire sequence of the peptide is required for its full activity in inhibiting the cAMP production system in Rin m5F cells (AMIRANOFF et al. 1989b).

Pancreatic β-cell ATP-sensitive K^+ channels which are target proteins for sulfonylureas are implicated in the regulatory process of insulin release.

As postulated by different groups, galanin may inhibit insulin release through activation of the ATP-sensitive K^+ channels (DE WEILLE et al. 1988; DUNNE et al. 1989). Such an event triggers a cascade of events which lead from hyperpolarization of membranes to the closing of a voltage-dependent Ca^{2+} channel with a subsequent decrease in cytosolic Ca^{2+} (AHREN et al. 1986; NILSSON et al. 1989) and a decrease in insulin release. Finally, a cAMP- and ATP-sensitive K^+channel-independent mechanism responsible for the action of galanin on the pancreatic β cell has also been suggested. This mechanism involves a direct interference with insulin exocytosis (ULLRICH and WOLLHEIM 1989) likely mediated by protein kinase C (SHARP et al. 1989). Although not demonstrated, it is tempting to conclude that these mechanisms are operating in concert in action of galanin on the pancreatic β cell (Fig. 4). However, whether they are triggered by single or multiple types of galanin receptor remains to be determined.

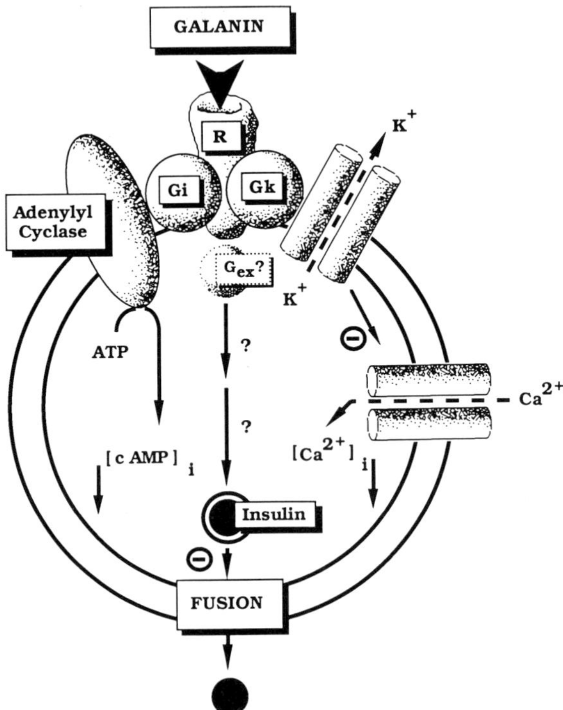

Fig. 4. Postulated mechanisms of action of galanin in pancreatic β cells. Occupation of galanin receptors by galanin leads to:
(1) Inhibition of adenylyl cyclase through a pertussis toxin-sensitive inhibitory G_i protein. (2) Activation of ATP-sensitive K^+ channels through a pertussis toxin-sensitive G_k protein. This effect leads, via membrane hyperpolarization, to closing of Ca^{2+} channels with a subsequent decrease of intracellular Ca^{2+}. (3) Inhibition of exocytosis of insulin through a pertussis toxin-sensitive substrate, here schematically proposed as a distinct G protein (G_{ex}) which is coupled to exocytosis in an as yet undefined manner

III. Neurotensin

Neurotensin (NT) is a tridecapeptide that was originally isolated from bovine hypothalamus (Carraway and Leeman 1973) and later from bovine and human intestine (Kitabgi et al. 1976; Carraway et al. 1978; Hammer et al. 1980). NT belongs to a family of biologically active peptides that comprises xenopsin isolated from the skin of *Xenopus laevis* (Araki et al. 1973), neuromedin N purified from porcine spinal cord (Minamino et al. 1984) and mammalian xenopsin recently isolated from rat brain, liver and gastrointestinal tissues (Carraway et al. 1990). Variants of NT, neuromedin N and xenopsin have also been characterized in avian species (Carraway et al. 1988). In the mammalian gut, NT is essentially found in endocrine N cells dispersed throughout the jejunoileal mucosa (see Rosell and Rökaeus 1981 for review). Immunoreactive NT, comprising intact NT and N-terminal NT fragments, is released in the circulation upon food ingestion, and is thought to act as a paracrine and endocrine modulator of digestive functions (gastrointestinal motility, gastric acid secretion), and in particular fat digestion (reviewed in Rosell and Rökaeus 1981; Ferris 1989). All these points have been developed in more detail throughout this volume. Like all hormones and neuropeptides, NT interacts with specific receptors located on the surface of its target cells. NT receptors have been characterized using radioligand binding techniques in a variety of cell lines and tissues (see Kitabgi et al. 1985 for review). We will focus here on those studies dealing with NT receptor characterization in tissues of the digestive tract.

Neurotensin has been shown to affect the contractility of a variety of gastrointestinal preparations in vitro (Kitabgi 1982; Kitabgi et al. 1985). Consistent with this, radioligand binding studies have identified NT receptors on smooth muscle and nerve cells in the digestive tract of several mammalian species. Specific [^{3}H]NT binding has been reported in membranes from guinea pig ileum (Kitabgi and Freychet 1979; Goedert et al. 1984) and iodinated NT binding has been characterized in membranes from rat gastric fundus and canine small intestine (Ahmad et al. 1987a). Tritiated NT bound to a single class of binding sites with K_D in the nanomolar range, whereas iodinated ligands, due to their higher specific activity, reveal the additional existence of a higher-affinity class of sites with K_D in the 100-pM range. Autoradiographic studies with [^{3}H]NT have shown the presence of NT receptors in both the circular and longitudinal smooth muscle layers of the guinea pig ileum (Goedert et al. 1984). Iodinated NT binding has also been observed, though not characterized, in neuronal membranes of deep muscular and submucous plexuses from canine and porcine small intestine (Ahmad et al. 1987a; Seybold et al. 1990). Although NT receptors appear to be absent from normal enterocytes and colonocytes in adults (Laburthe, unpublished observations), high-affinity NT binding has been demonstrated in the HT 29 cell line derived from a human colon carcinoma (Kitabgi et al. 1980). A single high-affinity class of sites was detected in HT 29 cell membranes with a K_D value (0.27 nM) similar to that found for the high-affinity sites in smooth muscle membranes (BOZOU et al. 1989). As shall be seen, HT 29 cells have provided an extremely useful model to study NT-receptor interactions.

A general feature of central and peripheral high-affinity NT receptors is their sensitivity to Na^+ and GTP. Thus, in rat gastric fundus, canine intestinal circular smooth muscle and HT 29 cell membranes, physiological concentrations of Na^+ decreased by about tenfold the affinity of NT for its high-affinity binding sites (KITABGI et al. 1984; AHMAD et al. 1987a; BOZOU et al. 1989). A similar effect is observed with micromolar concentrations of GTP. Therefore, it may be expected that in physiological situations the affinity of NT for its receptors will be in the nanomolar range. In agreement with this, [^{3}H]NT or [^{125}I-Tyr^3]NT binding to intact HT 29 cells in physiological media occurs to a single population of receptors with a K_D of 1.2 nM (KITABGI et al. 1980; AMAR et al. 1986), i.e., a fivefold higher value than the K_D of NT for binding to HT 29 cell membranes (BOZOU et al. 1989). This may also explain why EC_{50} values of NT for contracting or relaxing a variety of isolated gastrointestinal preparations are in the nanomolar range (KITABGI 1982; KITABGI et al. 1985). The properties of the modulation of NT binding by monovalent cations and guanyl nucleotides in gastrointestinal tissues are quite similar to those observed with NT receptors of neural origin (BOZOU et al. 1986), and have been taken as an indication that NT receptors are coupled to G proteins (see Sect. C). This will be discussed later.

The structural requirements of gastrointestinal NT receptors toward a variety of NT fragments and analogs are known mainly from in vitro bioassay studies with isolated smooth muscle preparations of the guinea pig ileum (KITABGI et al. 1980) and of the rat gastric fundus (QUIRION et al. 1980a,b), and from binding assays with rat fundus (KITABGI et al. 1984) and HT 29 cell membranes (KITABGI et al. 1980). In general, the results of these studies are in good agreement, i.e., the relative potencies of NT analogs correlate well in these various systems. The data can be summarized as follows: (1) the C-terminal hexapeptide sequence of NT contains all the structural requirements for full binding and biological potency; (2) the amino acids in the C-terminal pentapeptide sequence of NT cannot be replaced with corresponding D-amino acids without a considerable loss of binding and biological activity; (3) the positive charges on the side chains of Arg^8 and Arg^9, the aromaticity of residue 11, and the side chain methyl groups of Ile^{12} and Leu^{13} are important for the interaction of NT with its receptors. These structural features are similar to those found for central NT receptors (KITABGI et al. 1985). Given the homology of natural NT-related peptides with the C-terminal hexapeptide sequence of NT, it is not surprising that neuromedin N and xenopsin have been found to interact with central and peripheral NT receptors (CHECLER et al. 1982, 1986). Most biologically active peptides occur as families of structurally related molecules and it seems to be the rule rather than the exception that peptides of the same family recognize distinct subtypes of functional receptors. However, there is no clear pharmacological evidence as yet for the existence of NT receptor subtypes that would preferentially interact with members of the NT family. Finally, it should be pointed out that there exists at present no useful blocker for the NT receptor. Clearly, the synthesis and screening of new peptide and nonpeptide analogs of NT will be necessary to search for compounds that could disclose putative NT receptor subtypes or that would display antagonist properties.

The molecular properties of rat gastric fundus NT receptors have been investigated by means of photoaffinity labeling and cross-linking techniques (see Sect. B.III), and target size analysis by radiation inactivation. Photoaffinity labeling and cross-linking led to specific covalent labeling of a single protein with an apparent molecular weight of 110000 (MAZELLA et al. 1985b). A similar value of 103000 was obtained from radiation inactivation studies (AHMAD et al. 1987b). These results are consistent with the rat gastric NT receptor being composed of a single polypeptide chain of about 100 kDa. A similar size is reported for solubilized, active NT receptors purified by affinity chromatography from newborn mouse brain (MAZELLA et al. 1989). In contrast, rat brain NT receptors are composed of two subunits of M_r 49000 and 51000 as evidenced by both photoaffinity labeling and cross-linking techniques (MAZELLA et al. 1985a). Recently, a functional rat brain NT receptor has been cloned using the *Xenopus laevis* oocyte expression system (see Sect. B.IV). This receptor is a 424 amino acid protein with a theoretical molecular weight of 47000 (TANAKA et al. 1990). It possesses three potential N-glycosylation sites in its N-terminal region. Not unexpectedly, the NT receptor belongs to the family of G protein-coupled receptors (Fig. 2) and shares 19%–24% sequence homology with other G protein-coupled receptors such as the rat NK1-substance P, the rat m_5-muscarinic, the hamster α_1-adrenergic and the rat $5HT_2$-serotonin receptors (TANAKA et al. 1990). The NT receptor mRNA has been shown to be present, in addition to brain tissues, in rat heart, duodenum, small intestine and liver tissues (TANAKA et al. 1990). The molecular relationship of this receptor with the 49–50 KDa receptor covalently labeled in rat brain and the 100-KDa receptor observed in rat gastric and mouse brain tissues remains unclear at present. Further molecular studies and cloning will be needed to understand the apparent molecular heterogeneity of NT receptors.

The main second messenger pathway activated by NT appears to be the stimulation of phospholipase C with subsequent production of cellular inositol phosphates and Ins(1,4,5)P_3-induced increase in cytosolic Ca^{2+} (see Sect. C). It is this general property of NT to stimulate phospholipase C which has led to the cloning of the rat brain NT receptor using the *Xenopus oocyte* mRNA expression assay (TANAKA et al. 1990). NT stimulation of inositol phosphate and Ca^{2+} levels have been extensively studied in HT 29 cells (AMAR et al. 1986; BOZOU et al. 1989; TURNER et al. 1990) and in the neuroblastoma N1E115 cell line (AMAR et al. 1987; SNIDER et al. 1986). NT has also been shown to stimulate guanosine-3′,5′-cyclic monophosphate (cGMP) and to inhibit cAMP production in N1E115 cells (AMAR et al. 1985; BOZOU et al. 1986; GILBERT and RICHELSON 1984). However, the peptide has no effect on cyclic nucleotide levels in guinea pig ileal smooth muscle tissues (KITABGI and FREYCHET 1979) and in HT 29 cells (AMAR et al. 1986). The stimulatory effect of NT on cGMP levels in N1E115 cells most likely results from the following cascade of events: NT-stimulated phospholipase C activity, increased Ins(1,4,5)P_3 production, rise in cytosolic Ca^{2+}, Ca-calmodulin activation of the nitric oxide synthesizing pathway, and nitric oxide stimulation of soluble guanylate cyclase (MARSAULT and FRELIN 1991). The reason why NT does not stimulate cGMP formation in HT 29 cells (although it activates phospholipase C in this system) is that the cells lack nitric oxide synthase and soluble guanylate cyclase (MARSAULT and FRELIN 1991).

We have seen that the cloned NT receptor belongs to the superfamily of G protein-coupled receptors, implying that it functions through G proteins. Consistent with this, we have mentioned earlier that a general property of the NT receptor is the negative modulation of its affinity by GTP, a common feature to G protein-coupled receptors. Apparently, the NT receptor can interact with several G proteins. Thus, NT stimulation of phospholipase C is partially reversed by pertussis toxin in N1E115 cells (AMAR et al. 1987) and is totally insensitive to the toxin in HT 29 cells (BOZOU et al. 1989). On the other hand, NT inhibition of cAMP production in N1E115 cells is reversed by pertussis toxin (BOZOU et al. 1986). This indicates that in these systems NT receptors interact with both pertussis toxin-sensitive G proteins (such as G_i) and pertussis toxin-insensitive G proteins.

From what precedes, intracellular Ca^{2+} appears to be the major mediator of NT action. This is consistent with the known cellular effects of the peptide. For instance, NT can elicit either contraction or relaxation of gastrointestinal smooth muscle preparations depending on the tissue (KITABGI 1982; KITABGI et al. 1985). There is strong evidence that both types of response are mediated by a rise in intracellular [Ca^{2+}]. Electrophysiological studies and studies with Ca^{2+} channel blockers have shown that Ca^{2+} channels are activated in those tissues which contract in response to NT (KITABGI 1982; KITABGI et al. 1985; DONOSO et al. 1986; SNAPE et al. 1987). Whether Ca^{2+} channel activation is a direct consequence of NT-receptor interaction or is secondary to an initial mobilization of intracellular Ca^{2+} stores is not clear at present. On the other hand, NT has been shown to activate Ca^{2+}-dependent K channels in muscle preparations that relax in response to the peptide (KITABGI et al. 1985). In this case, the activation of hyperpolarizing K channels is probably the consequence of internal Ca^{2+} mobilization as tissue relaxation still occurs in Ca^{2+}-free medium (KULLAK et al. 1987). Finally, it is interesting to note here that, in a subclone (CL.16E) of HT 29 cells that differentiate into mucus-secreting cells, NT and neuromedin N have recently been shown to stimulate mucus secretion through Ca^{2+} mobilization (AUGERON et al. 1991).

E. Conclusion and Perspectives

During the 1980s there has been a tremendous acceleration in the pace at which new peptides have been discovered owing to technological advances in the peptide purification and sequence analysis but also to the development of new routes to discovery of peptides (see McDONALD 1991a for review). Following on from peptide isolation, new receptors have been promptly identified by classical binding assays using radioiodinated probes (see Table 1). In some cases, the discovery of a receptor has even preceded the characterization of the biological role of the corresponding peptide as in the case of peptide YY, for example (LABURTHE et al. 1986b; LABURTHE 1990). On the other hand, some peptides are still orphans of their own specific receptor and biological function. This is the case for PHI, which is a VIP receptor agonist but does not appear yet to play a specific role aside from its VIP-like effects (see Sect. D.I).

While classical routes of investigation have made it possible to further discover new peptide receptors during the last decade, molecular biology now emerges as the preferred way to the structural characterization and discovery of receptors (Fig. 1). Because receptors are large proteins expressed in very low amounts by target cells, their amino acid sequence can be obtained only by cloning and sequencing the corresponding cDNA. Receptor cloning has clearly represented a breakthrough in understanding receptor structure and function (Table 1). The power of molecular biology now expands amply beyond this technical achievement since putative receptors for which the ligand is unknown or is even not yet isolated have already been cloned (LIBERT et al. 1989; WATSON and JAMES 1989). From the preceding discussion, it may be hypothesized that new peptides will be isolated on the basis of their ability to interact with these putative receptors whose cDNA sequences will be used for transfection into cell lines, i.e., the inverse of the classical approach. Another major task of molecular biololgy, in particular the polymerase chain reaction (PCR) method, will be the identification of receptor subtype genes which will make it possible to investigate the precise pharmacology of each receptor subtype after transfection of the genes into immortalized cell lines which previously lacked any member of the receptor family (WATSON and JAMES 1989). The recent availability of bacterial clones expressing functional receptors after transfection will further allow the development of this strategy (STROSBERG 1991).

In conclusion, it is likely that most of the presently known peptide receptors as well as still unidentified receptors will be cloned in the next few years since “hardly a week seems to pass without news of the cloning and sequencing of cDNA for one or another G protein-coupled receptor" (LEFKOWITZ 1991), e.g., the family to which belong most peptide receptors (see Table 1). These powerful new means of investigation permit the use of novel techniques to uncover the physiological and pathophysiological roles of these peptides in the digestive tract.

Acknowledgements. We gratefully acknowledge the important contributions of collaborators and colleagues to the work reviewed here. The space contraints of this chapter and the many publications on peptide receptors prevent a comprehensive survey of the literature. We apologize to any of our colleagues whose studies have not been noted specifically.

Research in our laboratories was supported by the Institut National de la Santé et de la Recherche Médicale, the Centre National de la Recherche Scientifique, the Ministère de la Recherche et de la Technologie, the Fondation pour la Recherche Médicale, the Conseil Scientifique de la Faculté de Médecine Xavier Bichat, the Association pour la Recherche sur le Cancer and the Association Française de Lutte contre la Mucoviscidose.

References

Ahmad S, Berezin I, Vincent JP, Daniel EE (1987a) Neurotensin receptors in canine intestinal smooth muscle: preparation of plasma membranes and characterization of (Tyr^{3}-^{125}I)-labelled neurotensin binding. Biochim Biophys Acta 896: 224–238

Ahmad S, Kwan CY, Vincent JP, Grover AK, Jung CY, Daniel EE (1987b) Target size analysis of neurotensin receptors. Peptides 8:195–197

Ahmed MS, Zhou D, Cavinato AG, Maulik D (1989) Opioid binding properties of the purified *kappa* receptor from human placenta. Life Sci 44:861–871

Ahren B, Arkhammar P, Berggren PO, Nilsson T (1986) Galanin inhibits glucose-stimulated insulin release by a mechanism involving hyperpolarization and lowering of cytoplasmic free Ca^{++} concentration. Biochem Biophys Res Commun 140:1059–1063

Amar S, Mazella J, Checler F, Kitabgi P, Vincent JP (1985) Regulation of cyclic GMP levels by neurotensin in neuroblastoma clone N1E115. Biochem Biophys Res Commun 129:117–125

Amar S, Kitabgi P, Vincent JP (1986) Activation of phosphatidylinositol turnover by neurotensin in the human colonic adenocarcinoma cell line HT 29. FEBS Lett 201:31–36

Amar S, Kitabgi P, Vincent JP (1987) Stimulation of inositol phosphate production by neurotensin in neuroblastoma N1E115 cells: implication of GTP-binding proteins and relationship with the cyclic GMP response. J Neurochem 49:999–1006

Amiranoff B, Laburthe M, Dupont C, Rosselin G (1978) Characterization of a vasoactive intestinal peptide (VIP)-sensitive adenylate cyclase in rat intestinal epithelial cell membranes. Biochim Biophys Acta 544:474–481

Amiranoff B, Laburthe M, Rosselin G (1980) Characterization of specific binding sites for vasoactive intestinal peptide (VIP) in rat intestinal epithelial cell membranes. Biochim Biophys Acta 267:215–224

Amiranoff B, Vauclin-Jacques N, Laburthe M (1984) Functional GIP receptors in a hamster pancreatic beta cell line, In 111: specific binding and biological effects. Biochem Biophys Res Commun 123:671–676

Amiranoff B, Vauclin-Jacques N, Laburthe M (1985) Interaction of gastric inhibitory polypeptide (GIP) with the insulin-secreting pancreatic cell line In 111: characteristics of GIP binding sites. Life Sci 36:807–813

Amiranoff B, Couvineau A, Vauclin-Jacques N, Laburthe M (1986) Gastric inhibitory polypeptide receptor in hamster pancreatic beta cells. Direct cross-linking, solubilization and characterization as a glycoprotein. Eur J Biochem 159:353–358

Amiranoff B, Servin AL, Rouyer-Fessard C, Couvineau A, Tatemoto K, Laburthe M (1987) Galanin receptors in a hamster pancreatic beta cell tumor: identification and molecular characterization. Endocrinology 121:284–289

Amiranoff B, Lorinet AM, Lagny-Pourmir I, Laburthe M (1988) Mechanism of galanin-inhibited insulin release: occurrence of a pertussis toxin-sensitive inhibition of adenylate cyclase. Eur J Biochem 177:147–152

Amiranoff B, Lorinet AM, Laburthe M (1989a) Galanin receptors in the rat pancreatic beta cell line Rin m5F: molecular characterization by chemical cross-linking. J Biol Chem 264:20714–20717

Amiranoff B, Lorinet AM, Yanaihara N, Laburthe M (1989b) Structural requirement for galanin action in the pancreatic beta cell line Rin m5F. Eur J Pharmacol 163:205–207

Amiranoff B, Lorinet AM, Laburthe M (1991) A clonal rat pancreatic delta cell line (Rin 14B) expresses a high number of galanin receptors negatively coupled to a pertussis toxin-sensitive cAMP production pathway. Eur J Biochem 195:459–463

Araki K, Tachibana S, Uchiyama M, Nakajima T, Yasuhara T (1973) Isolation and structure of a new active peptide "Xenopsin" on the smooth muscle, especially on a strip of fundus from rat stomach, from the skin of *Xenopus laevis*. Chem Pharm Bull (Tokyo) 21:2801–2804

Audigier S, Barberis C, Jard S (1988) Vasoactive intestinal polypeptide increases inositol phospholipid breakdown in the rat superior cervical ganglion. Ann NY Acad Sci 527:579–582

Augeron C, Voisin T, Maoret JJ, Berthon B, Laburthe M, Laboisse CL (1992) Neurotensin and neuromedin N stimulate mucin output in human goblet cells (CL.16E) via a neurotensin receptor. Am. J. Physiol. 262:G470–G476

Balasubramaniam A, Sheriff S (1990) Neuropeptide Y (18–36) is a competitive antagonist of neuropeptide Y in rat cardiac ventricular membranes. J Biol Chem 265:14724–14727

Bartfai T, Fisone G, Langel U (1992) Galanin and galanin antagonists: molecular and biochemical perspectives. Trends Pharmacol Sci 13:312–317

Bataille D, Freychet P, Rosselin G (1974) Interaction of glucagon, gut glucagon, vasoactive intestinal polypeptide and secretin with liver and fat cell plasma membranes: binding to specific sites and stimulation of adenylate cyclase. Endocrinology 95:713–721

Bataille D, Gespach C, Laburthe M, Amiranoff B, Tatemoto K, Vauclin N, Rosselin G (1980) Porcine peptide having N-terminal histidine and C-terminal isoleucine amide: VIP-like and secretin-like effects in different tissues from rat. FEBS Lett 114:240–242

Bataille D, Blache P, Mercier F, Jarrousse C, Kervran A, Dufour M, Mangeat P (1988) Glucagon and related peptides. Molecular structure and biological specificity. Ann NY Acad Sci 527:168–185

Battey JF, Way JM, Corjay MH, Shapira H, Kusano K, Harkins R, Wu JM, Slattery T, Mann E, Feldman RI (1991) Molecular cloning of the bombesin/gastrin-releasing peptide receptor from Swiss 3T3 cells. Proc Natl Acad Sci USA 88:395–399

Bauer FE, Zintel A, Kenny MJ, Calder D, Ghatei MA, Bloom SR (1989) Inhibitory effect of galanin on postprandial gastrointestinal motility and gut hormone release in humans. Gastroenterology 97:260–264

Bawab W, Gespach C, Marie JC, Chastre E, Rosselin G (1988) Pharmacology and molecular identification of secretin receptors in rat gastric glands. Life Sci 42:791–798

Bayliss WM, Starling EH (1902) The mechanism of pancreatic secretion. J Physiol (Lond) 28:325–353

Berridge MJ (1987) Inositol triphosphate and diacylglycerol: two interacting second messengers. Annu Rev Biochem 56:159–193

Birnbaumer L, Abramowitz J, Brown AM (1990) Receptor-effector coupling by G proteins. Biochim Biophys Acta 1031:163–224

Bloom SR, Polak JM, Pearse AGE (1973) Vasoactive intestinal peptide and watery diarrhea syndrome. Lancet 2:14–16

Bolton AE (1977) Radioiodination techniques. In: The Radiochemical Center. Review 18. Printanium, Amersham, pp 5–71

Bolton AE, Hunter WM (1973) The labelling of proteins to high specific radioactivity by conjugation to a ^{125}I-containing acylating agent. Biochem J 133:529–538

Boyd ND, White CF, Cerpa R, Kaiser ET, Leeman SE (1991) Photoaffinity labeling of the substance P receptor using a derivative of substance P containing *p*-benzoylphenylalanine. Biochemistry 30:336–342

Bozou JC, Amar S, Vincent JP, Kitabgi P (1986) Neurotensin-mediated inhibition of cyclic AMP formation in neuroblastoma N1E115 cells: involvement of the inhibitory GTP-binding component of adenylate cyclase. Mol Pharmacol 29: 489–496

Bozou JC, Rochet N, Magnaldo I, Vincent JP, Kitabgi P (1989) Neurotensin stimulates inositol trisphosphate-mediated calcium mobilization but not protein kinase C activation in HT 29 cells. Involvement of a G protein. Biochem J 264:871–878

Broyart JP, Dupont C, Laburthe M, Rosselin G (1981) Characterization of vasoactive intestinal peptide (VIP) receptors in human colonic epithelial cells. J Clin Endocrinol Metab 52:715–721

Bruns RF, Lawson-Wendling K, Pugsley TA (1983) A rapid filtration assay for soluble receptor using polyethylenimine-treated filters. Anal Biochem 132: 74–81

Calvo JR, Couvineau A, Guijarro L, Laburthe M (1989) Solubilization and hydrodynamic characterization of guanine nucleotide-sensitive vasoactive intestinal peptide-receptor complexes from rat intestine. Biochemistry 28:1667–1672

Carraway RE, Leeman SE (1973) The isolation of a new hypotensive peptide, neurotensin, from bovine hypothalami. J Biol Chem 248:6854–6861

Carraway RE, Kitabgi P, Leeman SE (1978) The amino acid sequence of radioimmunoassayable neurotensin from bovine intestine. J Biol Chem 253:7996–7998

Carraway RE, Cochrane DE, Mitra SP (1988) Xenopsin-related peptide generated in avian gastric extracts. Regul Pept 22:303–314

Carraway RE, Mitra SP, Muraki K (1990) Isolation and structures of xenopsin-related peptides from rat stomach, liver and brain. Regul Pept 29:229–239

Chastre E, Emami S, Rosselin G, Gespach C (1987) Ontogenic development of vasoactive intestinal peptide receptors in rat intestinal cells and liver. Endocrinology 121:2211–2221

Checler F, Labbé C, Granier C, van Rietschoten J, Kitabgi P, Vincent JP (1982) [Trp^{11}]-neurotensin and xenopsin discriminate between rat and guinea-pig neurotensin receptors. Life Sci 31:1145–1150

Checler F, Vincent JP, Kitabgi P (1986) Neuromedin N: high affinity interaction with brain neurotensin receptors and rapid inactivation by brain synaptic peptidases. Eur J Pharmacol 126:239–244

Chen YH, Couvineau A, Laburthe M, Amiranoff B (1992) Solubilization and molecular characterization of active galanin receptors from rat brain. Biochemistry 31:2415–2422

Chen Y, Fournier A, Couvineau A, Laburthe M, Amiranoff B (1993) Purification of a galanin receptor from pig brain. Submitted

Childers SR (1988) Opioid-coupled second messenger systems. In: Pasternak GW (ed) The opiate receptors. Humana, Clifton, pp 231–271

Christophe J, Conlon TP, Gardner JD (1976) Interaction of porcine vasoactive intestinal peptide with dispersed pancreatic acinar cells from guinea pig. Binding of the radioiodinated peptide. J Biol Chem 251:4629–4634

Christophe J, Svoboda M, Waelbroeck M, Winand J, Robberecht P (1988) Vasoactive intestinal peptide receptors in pancreas and liver. Ann NY Acad Sci 527:238–256

Cook JS, Wolsing DH, Lameh J, Olson CA, Correa PE, Sadee W, Blumenthal EM, Rosenbaum JS (1992) Characterization of a RDC1 gene which encodes the canine homolog of a proposed human VIP receptor. Expression does not correlate with an increase in VIP binding sites. FEBS Lett. 300:149–152

Couvineau A, Laburthe M (1985a) The human vasoactive intestinal peptide receptor: molecular identification by covalent cross-linking in colonic epithelium. J Clin Endocrinol Metab 61:50–55

Couvineau A, Laburthe M (1985b) The rat liver vasoactive intestinal peptide (VIP) binding site: molecular characterization by covalent cross-linking and evidence for differences from the intestinal receptor. Biochem J 225:473–479

Couvineau A, Rouyer-Fessard C, Fournier A, St Pierre S, Pipkorn R, Laburthe M (1984a) Structural requirements for VIP interaction with specific receptors in human and rat intestinal membranes: effect of nine partial sequences. Biochem Biophys Res Commun 121:493–498

Couvineau A, Amiranoff B, Vauclin-Jacques N, Laburthe M (1984b) The GIP receptor on pancreatic beta cell tumor: covalent cross-linking and molecular characterization. Biochem Biophys Res Commun 122:283–288

Couvineau A, Rousset M, Laburthe M (1985) Molecular identification and structural requirement of vasoactive intestinal peptide (VIP) receptors in the human colon adenocarcinoma cell line HT-29. Biochem J 231:139–143

Couvineau A, Amiranoff B, Laburthe M (1986) Solubilization of the liver vasoactive intestinal peptide receptor. Hydrodynamic characterization and evidence for an association with a functional GTP regulatory protein. J Biol Chem 261:14482–14489

Couvineau A, Voisin T, Guijarro L, Laburthe M (1990a) Purification of VIP receptor from porcine liver by a newly designed one-step affinity chromatography. J Biol Chem 265:13386–13390

Couvineau A, Rouyer-Fessard C, Voisin T, Laburthe M (1990b) Functional and immunological evidence for stable association of solubilized VIP receptor and Gs protein from rat liver. Eur J Biochem 187:605–609

Desbuquois B (1974) The interaction of vasoactive intestinal polypeptide and secretin with liver cell membranes. Eur J Biochem 46:439–450

De Weille J, Schmid-Antomarchi H, Fosset M, Lazdunski M (1988) ATP-sensitive K^+ channels that are blocked by hypoglycemia-inducing sulfonylureas in insulin-secreting cells are activated by galanin, a hyperglycemia-inducing hormone. Proc Natl Acad Sci USA 85:1312–1316

Dickinson KEJ, Schachter M, Miles CMM, Coy DH, Sever PS (1986) Characterization of vasoactive intestinal peptide (VIP) receptors in mammalian lung. Peptides 7:791–800

Donoso MV, Huidobro-Toro JP, Kullak A (1986) Involvement of calcium channels in the contractile activity of neurotensin but not acetylcholine: studies with calcium channel blockers and Bay K 8644 on the rat fundus. Br J Pharmacol 88:837–844

Dunne MJ, Bullet MJ, Li GD, Wollheim CB, Petersen OH (1989) Galanin activates nucleotide-dependent K^+ channels in insulin secreting cells via a pertussis toxin-sensitive G protein. EMBO J 8:413–420

Dunning BE, Ahren BB, Veith RC, Böttcher G, Sundler F, Taborsky GJ (1986) Galanin: a novel pancreatic neuropeptide. Am J Physiol 251:127–133

Duong LT, Hadac EH, Miller LJ, Vlasuk GP (1989) Purification and characterization of the rat pancreatic cholecystokinin receptor. J Biol Chem 264:17990–17996

Dupont C, Broyart JP, Broer Y, Chenut B, Laburthe M, Rosselin G (1981) Importance of the vasoactive intestinal peptide receptor in the stimulation of cyclic adenosine 3′,5′: monophosphate in gallbladder epithelial cells of man. Comparison with the guinea pig. J Clin Invest 67:742–752

El Battari A, Luis J, Martin JM, Fantini J, Muller JM, Marvaldi J, Pichon J (1987) The vasoactive intestinal peptide receptor on intact colonic adenocarcinoma cells (HT29-D4). Evidence for its glycoprotein nature. Biochem J 242:185–191

El Battari A, Martin JM, Luis J, Pouzol O, Secchi J, Marvaldi J, Pichon J (1988) Solubilization of the active vasoactive intestinal peptide receptor from human colonic adenocarcinoma cells. J Biol Chem 263:17685–17689

Epand RM, Rosselin G, Hui Bon Hoa D, Cote TE, Laburthe M (1981) Structural requirements for glucagon receptor binding and activation of adenylate cyclase in liver: study of chemically modified forms of the hormone, including *N*-alpha-trinitrophenyl glucagon, an antagonist. J Biol Chem 256:1128–1132

Estival A, Mounielou P, Trocheris V, Scemama JL, Clemente F, Hollande E, Ribet A (1983) Presence of VIP receptors in a human pancreatic adenocarcinoma cell line, modulation of the cAMP response during cell proliferation. Biochem Biophys Res Commun 111:958–963

Ewald DA, Sternweiss PC, Miller RJ (1988) Guanine nucleotide-binding protein Go-induced coupling of neuropeptide Y receptors to Ca^{2+} channels in sensory neurons. Proc Natl Acad Sci USA 85:3633–3637

Feldman RI, Wu JM, Jensen JC, Mann E (1990) Purification and characterization of the bombesin/gastrin-releasing peptide receptor from Swiss 3T3 cells. J Biol Chem 265:17364–17372

Ferris CF (1989) Neurotensin. In: Makhlouf GM (ed) The gastrointestinal system, vol 2: Neural and endocrine biology. American Physiological Society, Bethesda, pp 559–586 (Handbook of physiology, sect 6)

Fisher JB, Schonbrunn A (1988) The bombesin receptor is coupled to a guanine nucleotide-binding protein which is insensitive to pertussis and cholera toxin. J Biol Chem 263:2808–2816

Fisone G, Langel U, Carlquist M, Bergman T, Consolo S, Hökfelt T, Unden A, Andell S, Bartfai T (1989) Galanin receptor and its ligands in the rat hippocampus. Eur J Biochem 181:269–276

Fourmy D, Lopez P, Poirot S, Jimenez J, Dufresne M, Moroder L, Powers SP, Vaysse N (1989) A new probe for affinity labelling pancreatic cholecystokinin receptor with minor modification of its structure. Eur J Biochem 185:397–403

Freidinger RM (1989) Non-peptide ligands for peptide receptors. Trends Pharmacol Sci 10:270–274

Gallwitz B, Schmidt WE, Schwarzhoff R, Creutzfeldt W (1990) Galanin: structural requirement for binding and signal transduction in Rin m5F insulinoma cells. Biochem Biophys Res Commun 172:268–275

Gantz I, Schäffer M, del Valle J, Logsdon C, Campbell V, Uhler M, Yamada T (1991) Molecular cloning of a gene encoding the histamine H_2 receptor. Proc Natl Acad Sci USA 88:429–433

Gardner JD (1979) Receptors for gastrointestinal hormones. Gastroenterology 76:202–214

Gershengorn MC (1991) Signal transduction by cell surface receptors. In: Introduction to endocrine investigation 1991. Techniques and concepts. Endocrine Society Press, Bethesda, MD, pp 81–84

Gespach C, Bataille D, Dupont C, Rosselin G, Wunsch E, Jaeger R (1980) Evidence for a cyclic AMP system highly sensitive to secretin in gastric glands isolated from rat fundus and antrum. Biochim Biophys Acta 630:433–441

Gespach C, Emami S, Rosselin G (1984) Gastric inhibitory peptide (GIP), pancreatic glucagon and vasoactive intestinal peptide (VIP) are cAMP-inducing hormones in the human gastric cancer cell line HGT-1. Homologous desensitization of VIP receptor activity. Biochem Biophys Res Commun 120:641–649

Gilbert JA, Richelson E (1984) Neurotensin stimulates formation of cyclic GMP in murine neuroblastoma clone N1E115. Eur J Pharmacol 99:245–246

Gilbert WR, Frank BH, Gavin JR, Gingerich RL (1988) Characterization of specific pancreatic polypeptide receptors on basolateral membranes of the canine small intestine. Proc Natl Acad Sci USA 85:4745–4749

Gilman AG (1989) G proteins and regulation of adenylyl cyclase. JAMA 262:1819–1825

Gimpl G, Gerstberger R, Mauss U, Klotz KN, Lang RE (1990) Solubilization and characterization of active neuropeptide Y receptors from rabbit kidney. J Biol Chem 265:18142–18147

Gioannini TL, Howard AD, Hiller JH, Simon EJ (1985) Purification of an active opioid-binding protein from bovine striatum. J Biol Chem 260:15117–15121

Goedert M, Hunter J, Ninkovic M (1984) Evidence for neurotensin as non-adrenergic, non-cholinergic neurotransmitter in guinea-pig ileum. Nature 311:59–62

Gossen D, Poloczek P, Svoboda M, Christophe J (1989) Molecular architecture of secretin receptors: the specific covalent labelling of a 51 kDa peptide after cross-linking of (^{125}I)iodo secretin to intact pancreatic acini. FEBS Lett 243:205–208

Guijarro L, Couvineau A, Calvo JR, Laburthe M (1989) Solubilization of active and stable receptors for vasoactive intestinal peptide from rat liver. Regul Pept 25:37–50

Guijarro L, Couvineau A, Rodriguez-Pena MS, Arilla E, Laburthe M, Prieto JC (1992) Vasoactive intestinal peptide receptor in rat liver after partial hepatectomy. Biochem J 285:515–520

Haffar BM, Hocart SJ, Coy DH, Mantey S, Chiang HCV, Jensen RT (1991) Reduced peptide bond pseudopeptide analogues of secretin. A new class of secretin receptor antagonists. J Biol Chem 266:316–322

Hammer RA, Leeman SE, Carraway SE, Williams RH (1980) Isolation of human intestinal neurotensin. J Biol Chem 255:2476–2480

Herberg JT, Codina J, Rich KA, Rojas FJ, Iyengar R (1984) The hepatic glucagon receptor. J Biol Chem 259:9285–9294

Herzog H, Hort YJ, Ball HJ, Hayes G, Shine J, Selbie LA (1992) Cloned human neuropeptide Y receptor coupled to two different second messenger systems. Proc. Natl. Acad. Sci. USA 89:5794–5798

Hollenberg M (1990) Receptor solubilization, characterization and isolation. In: Yamamura H (ed) Methods in neurotransmitter receptor analysis. Raven, New York, pp 111–145

Hoosein NM, Black BE, Brattain DE, Brattain MG (1989) Promotion of differentiation in human colon carcinoma cells by vasoactive intestinal polypeptide. Regul Pept 24:15–26

Howard AD, Sarne Y, Gioannini TL, Hiller JM, Simon EJ (1986) Identification of distinct binding site subunits of mu and delta opioid receptors. Biochemistry 25:357–360

Huang SC, Yu DH, Wank SA, Gardner JD, Jensen RT (1990) Characterization of the bombesin receptor on mouse pancreatic acini by chemical cross-linking. Peptides 11:1143–1150

Innis RB, Snyder S (1980) Distinct cholecystokinin receptors in brain and pancreas. Proc Natl Acad Sci USA 76:521–535

Inoue Y, Kaku K, Kaneko T, Yanaihara N, Kanno T (1985) Vasoactive intestinal peptide binding to specific receptors on rat parotid acinar cells induces amylase secretion accompanied by intracellular accumulation of cyclic adenosine 3′-5′-monophosphate. Endocrinology 116:686–692

Inui A, Okita M, Inoue T, Sakatani N, Oya M, Morioka H, Shii K, Yokono K, Baba S (1988) Characterization of peptide YY receptors in the brain. Endocrinology 124:402–409

Inui A, Okita M, Miura M, Hirosue Y, Nakajima M, Inoue T, Oya M, Baba S (1990) Characterization of the receptor for peptide YY and avian polypeptide in chicken and pig brain. Endocrinology 127:934–941

Ishihara T, Shigemoto R, Mori K, Takahashi K, Nagata S (1992) Functional expression and tissue distribution of a novel receptor for vasoactive intestinal peptide. Neuron 8:811–819

Ishihara T, Nakamura S, Kaziro Y, Takahashi T, Takahashi K, Nagata S (1991) Molecular cloning and expression of a cDNA encoding the secretin receptor. EMBO J 10:1635–1641

Iwanij V, Vincent AC (1990) Characterization of the glucagon receptor and its functional domains using monoclonal antibodies. J Biol Chem 265:21302–21308

Jensen RT, Coy DH (1991) Progress in the development of potent bombesin receptor antagonists. Trends Pharmacol Sci 12:13–19

Jensen RT, Wank SA, Rowley WH, Sato S, Gardner JD (1989) Interaction of CCK with pancreatic acinar cells. Trends Pharmacol Sci 10:418–422

Johnson GL, MacAndrew VI, Pilch PF (1981) Identification of the glucagon receptor in rat liver membranes by photoaffinity crosslinking. Proc Natl Acad Sci USA 78:875–878

Jones DT, Reed RR (1987) Molecular cloning of five GTP-binding protein cDNA species from rat olfactory neuroepithelium. J Biol Chem 262:14241–14249

Jung CY (1984) Molecular weight determination by radiation inactivation. In: Venter JC, Harrison LC (eds) Molecular and chemical characterization of membrane receptors. Liss, New York, pp 193–208 (Receptor biochemistry and methodology, vol 3)

Kahn CR (1976) Membrane receptors for hormones and neurotransmitters. J Cell Biol 70:261–286

Kaplan LM, Spindel ER, Isselbacher KJ, Chin WW (1988) Tissue-specific expression of the rat galanin gene. Proc Natl Acad Sci USA 85:1065–1069

Kassis S, Olasmaa M, Terenius L, Fishman PH (1987) Neuropeptide Y inhibits cardiac adenylate cyclase through a pertussis toxin-sensitive G protein. J Biol Chem 262:3429–3431

Kerins C, Said SI (1973) Hyperglycemic and glycogenolytic effects of vasoactive intestinal polypeptide. Proc Soc Exp Biol Med 142:1014–1017

Khalil T, Alinder GA, Rayford PL (1987) Vasoactive intestinal peptide. In: Thompson JC, Greeley GH, Rayford PL, Townsend CM (eds) Gastrointestinal endocrinology. McGraw-Hill, New York, pp 260–272

King SC, Slater P, Turnberg LA (1989) Autoradiographic localization of binding sites for galanin and VIP in small intestine. Peptides 10:313–317

Kitabgi P (1982) Effects of neurotensin on intestinal smooth muscle: application to the study of structure-activity relationships. Ann NY Acad Sci 400:37–55

Kitabgi P, Freychet P (1979) Neurotensin: contractile activity, specific binding, and lack of effect on cyclic nucleotides in intestinal smooth muscle. Eur J Pharmacol 55:35–42

Kitabgi P, Carraway RE, Leeman SE (1976) Isolation of a tridecapeptide from bovine intestinal tissue and its partial characterization as neurotensin. J Biol Chem 251:7053–7058

Kitabgi P, Poustis C, Granier C, van Rietschoten J, Rivier J, Morgat JL, Freychet P (1980) Neurotensin binding to extraneural and neural receptors: comparison with biological activity and structure-activity relationships. Mol Pharmacol 18:11–19

Kitabgi P, Kwan CY, Fox JET, Vincent JP (1984) Characterization of neurotensin binding to rat gastric smooth muscle receptor sites. Peptides 5:917–923

Kitabgi P, Checler F, Mazella J, Vincent JP (1985) Pharmacology and biochemistry of neurotensin receptors. Rev Clin Basic Pharmacol 5:397–486

Kluxen FW, Bruns C, Lübbert H (1992) Expression cloning of a rat brain somatostatin receptor cDNA. Proc. Natl. Acad. Sci. USA 89:4618–4622

Knuhtsen S, Esteve JP, Cambillau C, Colas B, Susini C, Vaysse N (1990) Solubilization and characterization of active somatostatin receptors from rat pancreas. J Biol Chem 265:1129–1133

Kopin AS, Lee YM, McBride EW, Miller LJ, Lu M, Lin HY, Kolakowski LF, Beinborn M (1992) Expression cloning and characterization of the canine parietal cell gastrin receptor. Proc. Natl. Acad. Sci. USA 89:3605–3609

Krause J, Eva C, Seeburg PH, Sprengel R (1992) Neuropeptide Y1 subtype pharmacology of a recombinantly expressed neuropeptide receptor. Mol. Pharmacol. 41:817–821

Krupinski J, Coussen F, Bakalyar HA, Tag W-J, Feinstein PG, Orth K, Slaughter C, Reed RR, Gilman AG (1989) Adenylyl cyclase amino acid sequence: possible channel- or transporter-like structure. Science 244:1558–1564

Kullak A, Donoso MV, Huidobro-Toro JP (1987) Extracellular calcium dependence of the neurotensin-induced relaxation of intestinal smooth muscles: studies with calcium channel blockers and BAY K-8644. Eur J Pharmacol 135:297–305

Laburthe M (1990) Peptide YY and neuropeptide Y in the gut: availability, biological actions and receptors. Trends Endocrinol Metab 1:168–174

Laburthe M, Amiranoff B (1989) Peptide receptors in intestinal epithelium. In: Makhlouf GM (ed) The gastrointestinal system, vol 2: Neural and endocrine biology. American Physiological Society, Bethesda, pp 215–243 (Handbook of physiology, sect 6)

Laburthe M, Couvineau A (1988) Molecular analysis of vasoactive intestinal peptide receptors. A comparison with receptors for VIP-related peptides. Ann NY Acad Sci 527:296–313

Laburthe M, Dupont C (1982) VIPergic control of intestinal epithelium in health and disease. In: Said SI (ed) Vasoactive intestinal polypeptide, Raven, New York, pp 407–423 (Advances in peptide hormone research series)

Laburthe M, Rousset M, Boissard C, Chevalier G, Zweibaum A, Rosselin G (1978a) Vasoactive intestinal peptide (VIP): a potent stimulator of adenosine 3′5′: cyclic monophosphate accumulation in gut carcinoma cell lines in culture. Proc Natl Acad Sci USA 75:2772–2775

Laburthe M, Bataille D, Rousset M, Besson J, Broer Y, Zweibaum A, Rosselin G (1978b) The expression of cell surface receptors for VIP, secretin and glucagon in normal and malignant cells of the digestive tract. In: Nicholls P (ed) Membrane proteins. Pergamon, Oxford, pp 271–290

Laburthe M, Prieto JC, Amiranoff B, Dupont C, Hui Bon Hoa D, Rosselin G (1979) Interaction of vasoactive intestinal peptide with isolated intestinal epithelial cells from rat. Eur J Biochem 96:239–248

Laburthe M, Rousset M, Chevalier G, Boissard C, Dupont C, Zweibaum A, Rosselin G (1980) Vasoactive intestinal peptide (VIP) control of cyclic AMP levels in seven human colorectal adenocarcinoma cell lines in culture. Cancer Res 40:2529–2533

Laburthe M, Amiranoff B, Boige N, Rouyer-Fessard C, Tatemoto K, Moroder L (1983) Interaction of GRF with VIP receptors and stimulation of adenylate cyclase in rat and human intestinal epithelial membranes. Comparison with PHI and secretin. FEBS Lett 159:89–92

Laburthe M, Breant B, Rouyer-Fessard C (1984) Molecular identification of VIP receptors in rat intestinal epithelium by covalent cross-linking: evidence for two classes of binding sites with different structural and functional properties. Eur J Biochem 139:181–187

Laburthe M, Couvineau A, Rouyer-Fessard C, Moroder M (1985) Interaction of PHM, PHI and 24-glutamine PHI with human VIP receptors from colonic epithelium: comparison with rat intestinal receptors. Life Sci 36:991–995

Laburthe M, Couvineau A, Rouyer-Fessard C (1986a) Study of species specificity in growth hormone releasing factor (GRF) interaction with vasoactive intestinal peptide (VIP) receptors using GRF and VIP receptors from rat and human:

evidence that Ac-Tyr[1] hGRF is a competitive VIP antagonist in the rat. Mol Pharmacol 29:23–27

Laburthe M, Chenut B, Rouyer-Fessard C, Tatemoto K, Couvineau A, Servin A, Amiranoff B (1986b) Interaction of peptide YY with rat intestinal epithelial plasma membranes: binding of the radioiodinated peptide. Endocrinology 118:1910–1917

Laburthe M, Rousset M, Rouyer-Fessard C, Couvineau A, Chantret I, Chevalier G, Zweibaum A (1987) Development of vasoactive intestinal peptide-responsive adenylate cyclase during enterocytic differentiation of Caco-2 cells in culture. Evidence for an increased receptor level. J Biol Chem 262:10180–10184

Laburthe M, Augeron C, Rouyer-Fessard C, Roumagnac I, Maoret JJ, Grasset E, Laboisse C (1989) Functional VIP receptors in the human mucus-secreting colonic epithelial cell line CI.16E. Am J Physiol 256:G443–G450

Lagny-Pourmir I, Amiranoff B, Lorinet AM, Tatemoto K, Laburthe M (1989a) Characterization of galanin receptors in the insulin secreting cell line Rin m5F: evidence for coupling with an inhibitory GTP-regulatory protein (G_i). Endocrinology 124:2635–2641

Lagny-Pourmir I, Lorinet AM, Yanaihara N, Laburthe M (1989b) Structural requirement for galanin interaction with receptors from pancreatic beta cells and from brain tissue of the rat. Peptides 10:757–761

Lambert DT, Stachelek C, Varga JM, Lerner AB (1982) Iodination of β-melanotropin. J Biol Chem 257:8211–8215

Larhammar D, Blomqvist AG, Yee F, Jazin E, Yoo H, Wahlestedt C (1992) Cloning and functional expression of a human Neuropeptide Y/Peptide YY receptor of the Y1 type. J. Biol. Chem. 267:10935–10938

Lefkowitz RJ (1991) Thrombin receptor. Variation on a theme. Nature 351:353–354

Leslie FM (1987) Methods used for the study of opioid receptors. Pharmacol Rev 39:197–249

Li XJ, Wu YA, North RA, Forte M (1992) Cloning, functional expression and developmental regulation of a Neuropeptide Y receptor from Drosophila melanogaster. J. Biol. Chem. 267:9–12

Libert F, Parmentier M, Lefort A, Dinsart C, van Sande J, Maenhaut C, Simons MJ, Dumont JE, Vassart G (1989) Selective amplification and cloning of four new members of the G protein-coupled receptor family. Science 244:569–572

Loosfelt H, Misrahi M, Atger M, Salesse R, Vu Hai-Luu MT, Jolivet A, Guiochon-Mantel A, Sar S, Jallal B, Garnier J, Milgrom E (1989) Cloning and sequencing of porcine LH-hCG receptor cDNA: variants lacking transmembrane domain. Science 245:525–528

Majerus PW, Ross TS, Cunningham TW, Caldwell KK, Jefferson AB, Bansal VS (1990) Recent insights in phosphatidylinositol signaling. Cell 63:459–465

Malhotra RK, Wakade TD, Wakade AR (1988) Vasoactive intestinal polypeptide and muscarine mobilize Ca^{2+} through breakdown of phosphoinositides to induce catecholamine secretion. Role of IP3 in exocytosis. J Biol Chem 263:2123–2126

Marsault R, Frelin C (1992) Activation by nitric oxid of guanylate cyclase in endothelial cells from capillaries. J. Neurochem 59:942–949

Martin JM, Darbon H, Luis J, El Battari A, Marvaldi J, Pichon J (1988) Photoaffinity labelling of the vasoactive intestinal peptide binding site on intact human colonic adenocarcinoma cell line (HT29-D4). Synthesis and use of photosensitive vasoactive intestinal peptide derivatives. Biochem J 250:679–685

Martinez J, Magous M, Lignon MF, Laur J, Castro B, Bali JP (1984) Synthesis and biological activity of new peptide segments of gastrin exhibiting gastrin antagonist property. J Med Chem 27:1597–1601

Masu Y, Nakayama K, Tamaki H, Harada Y, Kuno M, Nakanishi S (1987) cDNA cloning of bovine substance-K receptor through oocyte expression system. Nature 329:836–838

Masuda Y, Ohtaki T, Kitada C, Tsuda M, Arimura A, Fujino M (1990) Solubilization of receptor for pituitary adenylate cyclase activating polypeptide from bovine brain. Biochem Biophys Res Commun 172:709–714

Matsuda LA, Lolait SJ, Brownstein MJ, Young AC, Bonner TI (1990) Structure of a cannabinoid receptor and functional expression of the cloned cDNA. Nature 346:561–564

Mazella J, Kitabgi P, Vincent JP (1985a) Molecular properties of neurotensin receptors in rat brain. Identification of subunits by covalent labeling. J Biol Chem 260:508–514

Mazella J, Kwan CY, Kitabgi P, Vincent JP (1985b) Covalent labeling of neurotensin receptors in rat gastric fundus plasma membranes. Peptides 6:1137–1141

Mazella J, Chabry J, Zsurger N, Vincent JP (1989) Purification of the neurotensin receptor from mouse brain by affinity chromatography. J Biol Chem 264:5559–5563

McDonald TJ (1991a) A historical overview of the gastroenteropancreatic regulatory peptides. In: Daniel EE (ed) Neuropeptide function in the gastrointestinal tract. CRC, Boca Raton, pp 2–18

McDonald TJ (1991b) Gastroenteropancreatic regulatory peptide structures: an overview. In: Daniel EE (ed) Neuropeptide function in the gastrointestinal tract. CRC, Boca Raton, pp 19–86

McDonald TJ, Dupre J, Tatemoto K, Greenberg GR, Radzuik J, Mutt V (1985) Galanin inhibits insulin secretion and induces hyperglycaemia in dogs. Diabetes 34:192–196

McFarland KC, Sprengel R, Phillips HS, Köhler M, Rosemblit N, Nikolics K, Segaloff DL, Seeburg PH (1989) Lutropin-choriogonadotropin receptor: an unusual member of the G protein-coupled receptor family. Science 245:494–499

Meyerhof W, Wulfsen I, Schönrock C, Fehr S, Richter D (1992) Molecular cloning of a somatostatin-28 receptor and comparison of its expression pattern with that of a somatostatin-14 receptor in rat brain. Proc. Natl. Acad. Sci USA 89: 10267–10271.

Milutinovic S, Schulz I, Rosselin G (1976) The interaction of secretin with pancreatic membranes. Biochim Biophys Acta 432:113–127

Minamino N, Kangawa K, Matsuo H (1984) Neuromedin N: a novel neurotensin-like peptide identified in porcine spinal cord. Biochem Biophys Res Commun 122:542–549

Miyata A, Arimura A, Dahl RR, Minamino N, Uehara A, Jiang L, Culler M, Coy DH (1989) Isolation of a novel 38 residue-hypothalamic polypeptide which stimulates adenylate cyclase in pituitary cells. Biochem Biophys Res Commun 164:567–574

Moran TH, Robinson PH, Goldrich MS, McHugh PR (1986) Two brain cholecystokinin receptors: implications for behavioral actions. Brain Res 362:175–179

Muller JM, Luis J, Fantini J, Abadie F, Giannellini J, Marvaldi J, Pichon J (1985) Covalent cross-linking of vasoactive intestinal peptide (VIP) to its receptor in intact colonic adenocarcinoma cell in culture (HT-29). Eur J Biochem 151: 411–417

Murty KK, Srikant CB, Patel YC (1989) Evidence for multiple protein constituents of the somatostatin receptor in pituitary tumor cells: affinity cross-linking and molecular characterization. Endocrinology 125:948–955

Mutt V (1988) Vasoactive intestinal polypeptide and related peptides. Isolation and chemistry. Ann NY Acad Sci 527:1–19

Nagata S, Ishihara T, Robberecht P, Libert F, Parmentier M, Christophe J, Vassart G (1992) Doubt expressed about identity of remaining orphan clone. RDC1 may not be VIP receptor. Trends Pharmacol. Sci 13:102–103

Nguyen TD, Williams JA, Gray GM (1986) Vasoactive intestinal peptide receptor on liver plasma membranes: characterization as a glycoprotein. Biochemistry 25:361–368

Nguyen TD, Heintz GG, Kaiser LM, Staley CA, Taylor IL (1990) Neuropeptide Y: differential binding to rat intestinal laterobasal membranes. J Biol Chem 265:6416–6422

Nilsson T, Arkhammar P, Rorsman P, Berggren PO (1989) Suppression of insulin release by galanin and somatostatin is mediated by a G protein: an effect

involving repolarization and reduction in cytoplasmic free Ca^{++} concentration. J Biol Chem 264:7302–7309
Nishizuka Y (1986) Studies and perspectives of protein kinase C. Science 233: 305–312
Nishizuka Y (1988) The molecular heterogeneity of protein kinase C and its implications for cellular regulation. Nature 334:661–665
Ohtaki T, Watanabe T, Ishibashi Y, Kitada C, Tsuda M, Gottschall PE, Arimura A, Fujino M (1990) Molecular identification of receptor for pituitary adenylate cyclase activating polypeptide. Biochem Biophys Res Commun 171:838–844
Ottlecz A, Samson WK, McCann SM (1986) Galanin: evidence for a hypothalamic site of action to release growth hormone. Peptides 7:51–55
Padrell E, Herberg JT, Monstirsky B, Floyd G, Premont RT, Iyengar R (1987) The hepatic glucagon receptor: a comparison study of regulatory and structural properties. Endocrinology 120:2316–2325
Pandol SJ, Dharmsathaphorn K, Schoeffield MS, Vale W, Rivier J (1986) Vasoactive intestinal peptide receptor antagonist [4CI-D-Phe^6, Leu^{17}]VIP. Am J Physiol 250:G553–G557
Patel YC, Murthy KK, Escher EE, Banville D, Spiess J, Srikant CB (1990) Mechanism of action of somatostatin: an overview of receptor function and studies of the molecular characterization and purification of somatostatin receptor proteins. Metabolism 39 Suppl 2:63–69
Pearson RK, Miller LJ, Powers SP, Hadac EM (1987) Biochemical characterization of the pancreatic cholecystokinin receptor using monofunctional photoactivable probes. Pancreas 2:79–84
Pilch PF, Czech MP (1984) Affinity cross-linking of peptide hormones and their receptors. In: Venter JC, Harrison LC (eds) Receptor biochemistry and methodology, vol 1. Liss, New York, pp 161–175
Prieto JC, Laburthe M, Rosselin G (1979) Interaction of vasoactive intestinal peptide with isolated intestinal epithelial cells from rat. Eur J Biochem 96: 229–237
Quirion R, Dam TV (1988) Multiple neurokinin receptors: recent developments. Regul Pept 22:18–25
Quirion R, Regoli D, Rioux F, St Pierre S (1980a) The stimulatory effects of neurotensin and related peptides in rat stomach strips and guinea-pig atria. Br J Pharmacol 68:83–91
Quirion R, Regoli D, Rioux F, St Pierre S (1980b) Structure activity studies with neurotensin: analysis of positions 9, 10 and 11. Br J Pharmacol 69:689–692
Ramani N, Preissman M (1989) Molecular identification and characterization of the gastrin receptor in guinea pig gastric glands. Endocrinology 124:1881–1887
Raufman JP, Jensen RT, Sutlif VE, Pisano JJ, Gardner JD (1982) Actions of Gila monster venom on dispersed acini from guinea pig pancreas. Am J Physiol 242:6470–6474
Regoli D, Dion S, Drapeau G (1990) Receptor for neuropeptides (neurokinins): functional studies. In: Daniel EE (ed) Neuropeptide function in the gastrointestinal tract. CRC, Boca Raton, pp 193–207
Reyl-Desmars F, Baird A, Zyitin FN (1985) GRF is a highly potent activator of adenylate cyclase in the normal human, bovine and rat pituitary: interaction with somatostatin. Biochim Biophys Res Commun 127:977–985
Reyl-Desmars F, Laboisse C, Lewin MJM (1986) A somatostatin receptor negatively coupled to adenylate cyclase in the human gastric cell line HGT-1. Regul Pept 16:207–215
Reyl-Desmars F, Le Roux S, Linard C, Ben Kouka F, Lewin MJM (1989) Solubilization and immunopurification of a somatostatin receptor from the human gastric tumoral cell line HGT-1. J Biol Chem 264:18789–18795
Rimland J, Xin W, Sweetnam P, Saijoh K, Nestler EJ, Duman RS (1992) Sequence and expression of a neuropeptide Y receptor cDNA. Mol. Pharmacol. 40:869–876

Robberecht P, Conlon TP, Gardner JD (1976) Interaction of porcine vasoactive intestinal peptide with dispersed pancreatic acinar cells from the guinea pig: structural requirement for effects of vasoactive intestinal peptide and secretin on cellular adenosine 3′:5′-monophosphate. J Biol Chem 251:4635–4639

Robberecht P, Coy DH, Waelbroeck M, Heiman ML, Camus JC, Christophe J (1985) Structural requirements for the activation of rat anterior pituitary adenylate cyclase by growth hormone-releasing factor (GRF): discovery of [*N*-acetyl-Tyr^1,D-Arg^2(1-29)-NH_2 as a GRF antagonist on membranes. Endocrinology 117:1759–1764

Robberecht P, Coy DH, de Neef P, Camus JC, Cauvin A, Waelbroeck M, Christophe J (1986) Interaction of vasoactive intestinal peptide (VIP) and N-terminally modified VIP analogs with rat pancreatic, hepatic and pituitary membranes. Eur J Biochem 159:45–49

Robberecht P, Gourlet P, Cauvin A, Buscail L, de Neef P, Arimura A, Christophe J (1991) PACAP and VIP receptors in rat liver membranes. Am J Physiol 260:G97–G102

Robichon A, Kuks PFM, Besson J (1987) Characterization of vasoactive intestinal peptide receptors by a photoaffinity label. J Biol Chem 262:11539–11545

Roche S, Bali JP, Galleyrand JC, Magous R (1991) Characterization of a gastrin-type receptor on rabbit gastric parietal cells using L 365,260 and L 364,718. Am J Physiol 260:G182–G188

Rökaeus A (1987) Galanin: a newly isolated biologically active neuropeptide. Trends Neurosci 10:158–164

Rökaeus A, Brownstein MJ (1986) Construction of a porcine adrenal medullary cDNA library and nucleotide sequence analysis of two clones encoding a galanin precursor. Proc Natl Acad Sci USA 83:6287–6291

Rosell S, Rökaeus Å (1981) Actions and possible hormonal functions of circulating neurotensin. Clin Physiol 1:3–21

Rosier AM, Vandesande F, Orban GA (1991) Laminar and regional distribution of galanin binding sites in cat and monkey visual cortex determined by in vitro receptor autoradiography. J Comp Neurol 305:264–272

Rossi I, Monge L, Feliu JE (1989) Short-term regulation of glycolysis by vasoactive intestinal peptide in epithelial cells isolated from rat small intestine. Biochem J 262:397–402

Rossowski WJ, Rossowski TM, Zacharia S, Ertan A, Coy DH (1990) Galanin binding sites in rat gastric and jejunal smooth muscle membranes preparations. Peptides 11:333–338

Rousset M, Laburthe M, Chevalier G, Boissard C, Rosselin G, Zweibaum A (1981) Vasoactive intestinal peptide (VIP) control of glycogenolysis in the human colon carcinoma cell line HT-29 in culture. FEBS Lett 126:38–40

Rouyer-Fessard C, Couvineau A, Voisin T, Laburthe M (1989a) Ac-Tyr^1 hGRF discriminates between VIP receptors from rat liver and intestinal epithelium. Life Sci 45:829–833

Rouyer-Fessard C, Augeron C, Grasset E, Maoret JJ, Laboisse CL, Laburthe M (1989b) VIP receptors and control of short circuit current in the human intestinal clonal cell line Cl.19A. Experientia 45:1102–1105

Ruoho AE, Rashidbaigi A, Roeder PE (1984) Approaches to the identification of receptors utilizing photoaffinity labeling. In: Venter JC, Harrison LC (eds) Receptor biochemistry and methodology. vol 1. Liss, New York, pp 119–160

Sadoul JL, Mazella J, Amar S, Kitabgi P, Vincent JP (1984) Preparation of neurotensin selectively iodinated on the tyrosine 3 residue. Biological activity and binding properties on mammalian neurotensin receptors. Biochem Biophys Res Commun 120:812–819

Sakamoto C, Williams JA, Goldfine ID (1984) Brain CCK receptors are structurally distinct from pancreas CCK receptors. Biochem Biophys Res Commun 124: 497–502

Said SI (1986) Vasoactive intestinal peptide. J Endocrinol Invest 9:191–200

Said SI, Mutt V (1970) Polypeptide with broad biological activity: isolation from small intestine. Science 169:1217–1218

Said SI, Rosenberg RN (1976) Vasoactive intestinal polypeptide: abundant immunoreactivity in neural cell lines and normal nervous tissues. Science 192:907–908

Sambrook J, Fritsch EF, Maniatis T (1989) Molecular cloning. A laboratory manual. In: Nolan C (ed) vols 1–3. Cold Spring Harbor Laboratory Press, Cold Spring Harbor

Schwartz TW, Sheikh SP, O'Hare MMT (1987) Receptors on phaeochromocytoma cells for two members of the PP-fold family – NPY and PP. FEBS Lett 225: 209–214

Seifert H, Perrin M, Rivier J, Vale W (1985) Binding sites for growth hormone releasing factor on rat anterior pituitary cells. Nature 313:487–489

Servin AL, Amiranoff B, Rouyer-Fessard C, Tatemoto K, Laburthe M (1987) Identification and molecular characterization of galanin receptor sites in rat brain. Biochem Biophys Res Commun 144:298–306

Servin AL, Rouyer-Fessard C, Balasubramaniam A, St Pierre S, Laburthe M (1989) Peptide YY and neuropeptide Y inhibit vasoactive intestinal peptide-stimulated cyclic AMP production in rat small intestine: structural requirement of peptides for interacting with peptide YY-preferring receptors. Endocrinology 124: 692–700

Seybold VS, Treder BG, Aanonsen LM, Parsons AM, Brown DR (1990) Neurotensin binding sites in porcine jejunum: biochemical characterization and intramural localization. Synapse 6:81–90

Sharp GW, LeMarchand-Brustel Y, Yada T, Russo LL, Bliss CR, Cormont M, Monge L, van Obberghen E (1989) Galanin can inhibit insulin release by a mechanism other than membrane hyperpolarization or inhibition of adenylate cyclase. J Biol Chem 264:7302–7309

Shaw MJ, Hadac EH, Miller JL (1987) Preparation of enriched plasma membranes from bovine gallbladder muscularis for characterization of cholecystokinin receptors. J Biol Chem 262:14313–14318

Sheikh SP, Williams JA (1990) Structural characterization of Y1 and Y2 receptors for neuropeptide Y and peptide YY by affinity cross-linking. J Biol Chem 265:8304–8310

Sheikh SP, Hansen AP, Williams JA (1991) Solubilization and affinity purification of the Y2 receptor for neuropeptide Y and peptide YY from rabbit kidney. J Biol. Chem 266:23959–23966

Sheikh SP, O'Hare MMT, Tortora O, Schwartz TW (1989) Binding of monoiodinated neuropeptide Y to hippocampal membranes and human neuroblastoma cell lines. J Biol Chem 264:6648–6654

Shigemoto R, Yokota Y, Tsuchida K, Nakanishi S (1990) Cloning and expression of a rat neuromedin K receptor cDNA. J Biol Chem 265:623–628

Simon MI, Strathmann MP, Gautam N (1991) Diversity of G proteins in signal transduction. Science 252:802–808

Simonds WF, Burke TR, Rice KC, Jacobson AE, Klee WA (1985) Purification of the opiate receptor of NG108-15 neuroblastoma-glioma hybrid cells. Proc Natl Acad Sci USA 82:4974–4978

Smrcka AV, Hepler JR, Brown KO, Sternweis PC (1991) Regulation of polyphosphoinositide-specific phospholipase C activity by purified G_q. Science 251:804–807

Snape WJ Jr, Tan ST, Kao HW, Hyman PE (1987) Mechanism of neurotensin depolarization of rabbit colonic smooth muscle. Regul Pept 18:287–297

Snider RM, Forray C, Pfenning M, Richelson E (1986) Neurotensin stimulates inositol phospholipid metabolism and calcium mobilization in murine neuroblastoma clone N1E-115. J Neurochem 47:1214–1218

Soll AH, Amirian DA, Thomas LP, Reedy TJ, Elashoff JD (1984) Gastrin receptors on isolated parietal cells. J Clin Invest 73:1434–1447

Spiegel AM (1991) Introduction to receptors. In: Introduction to endocrine investigation 1991. Techniques and concepts. The Endocrine Society Press, Bethesda, MD, pp 105–109

Sreedharan SP, Robichon A, Peterson KE, Goetzl EJ (1991) Cloning and expression of the human vasoactive intestinal peptide receptor. Proc Natl Acad Sci USA 88:4986–4990

Strosberg AD (1991) Structure/function relationship of proteins belonging to the family of receptors coupled to GTP-binding proteins. Eur J Biochem 196:1–10

Svoboda M, de Neef P, Tastenoy M, Christophe J (1988) Molecular characteristics and evidence for internalization of vasoactive intestinal peptide (VIP) receptors in the tumoral rat pancreatic acinar cell line AR 4-2 J. Eur J Biochem 176: 707–713

Takeuchi K, Speir GR, Johnson LR (1979) Mucosal gastrin receptor. IV. Binding specificity. Am J Physiol 239:G395–G399

Takuwa N, Takuwa Y, Bollag WE, Rasmussen H (1987) The effects of bombesin on polyphosphoinositide and calcium metabolism in Swiss 3T3 cells. J Biol Chem 262:182–188

Tanaka K, Masu M, Nakanishi S (1990) Structure and functional expression of the cloned rat neurotensin receptor. Neuron 4:847–854

Tatemoto K, Rökaeus A, Jörnvall H, McDonald TJ, Mutt V (1983) Galanin, a novel biologically active peptide from porcine intestine. FEBS Lett 164:124–128

Turner JT, Jones SB, Bylund DB (1986) A fragment of vasoactive intestinal peptide, VIP(10–28), is an antagonist of VIP in the colon carcinoma cell line, HT29. Peptides 7:849–854

Turner JT, James-Kracke MR, Camden JM (1990) Regulation of the neurotensin receptor and intracellular calcium mobilization in HT29 cells. J Pharmacol Exp Ther 253:1049–1056

Ullrich S, Wollheim CB (1989) Galanin inhibits insulin secretion by direct interference with exocytosis. FEBS Lett 247:401–404

Velicelebi G, Patthi S, Provow S, Akong M (1986) Covalent cross-linking of growth hormone-releasing factor to pituitary receptors. Endocrinology 118:1278–1283

Vidal H, Comte B, Beylot M, Riou JP (1988) Inhibition of glucose oxidation by vasoactive intestinal peptide in isolated rat enterocytes. J Biol Chem 263: 9206–9211

Vidal H, Beylot M, Comte B, Vega F, Riou JP (1989) Vasoactive intestinal peptide stimulates long-chain fatty acid oxidation and inhibits acetyl-coenzyme A carboxylase activity in isolated rat enterocytes. J Biol Chem 264:4901–4906

Vincent JP, Kitabgi P (1991) Receptors for neuropeptides: receptor isolation studies and molecular biology. In: Daniel EE (ed) Neuropeptide function in the gastrointestinal tract. CRC, Boca Raton, pp 231–247

Voisin T, Rouyer-Fessard C, Laburthe M (1990) Distribution of the common peptide YY/ neuropeptide Y receptor along rat intestinal villus-crypt axis. Am J Physiol 258:G753–G759

Voisin T, Couvineau A, Rouyer-Fessard C, Laburthe M (1991) Solubilization and hydrodynamic properties of active peptide YY receptor from rat jejunal crypts. Characterization as a Mr 44 000 glycoprotein. J Biol Chem 266:10762–10767

Wada E, Way J, Shapira H, Kusano K, Lebacq-Verheyden AM, Coy D, Jensen R, Battey J (1991) cDNA cloning, characterization, and brain region-specific expression of a neuromedin-B-preferring bombesin receptor. Neuron 6:421–431

Waelbroeck M, Robberecht P, Coy DH, Camus JC, de Neef P, Christophe J (1985) Interaction of growth hormone-releasing factor (GRF) and 14 GRF analogs with rat pancreatic VIP receptors. Discovery of (*N*-Ac-Try^1, D-Phe^2)-GRF(1-29)-NH2 as a vasoactive intestinal peptide (VIP) antagonist. Endocrinology 116: 2643–2649

Wank SA, Harkins R, Jensen RT, Shapira H, De Weerth A, Slattery T (1992a) Purification, molecular cloning and functional expression of the cholecystokinin receptor from rat pancreas. Proc. Natl. Acad. Sci. USA 89:3125–3129

Wank SA, Pisegna JR, de Weerth A (1992b) Brain and gastrointestinal cholecystokinin receptor family: structure and functional expression. Proc. Natl. Acad. Sci. USA 89:8691–8695

Watson SP, James W (1989) PCR and the cloning of receptor subtype genes. Trends Pharmacol Sci 10:346–348

Williams JA, McChesnay DJ (1987) Cholecystokinin induces the interaction of its receptor with a guanine nucleotide binding protein. Regul Pept 18:109–117

Wynn PC, Aguilera G, Morell J, Catt KJ (1983) Properties and regulation of high affinity pituitary receptors for corticotropin-releasing factor. Biochem Biophys Res Commun 110:602–608

Yamada Y, Post SR, Wang K, Tager HS, Bell GI, Seino S (1992) Cloning and functional characterization of a family of human and mouse somatostatin receptors expressed in brain, gastrointestinal tract and kidney. Proc. Natl. Acad. Sci. USA 89:251–255

Yatani A, Birnbaumer L, Brown AM (1990) Direct coupling of the somatostatin receptor to potassium channels by a G protein. Metabolism 39 Suppl 2:91–95

Yokota Y, Sasai Y, Tanaka K, Fujiwara T, Tsuchida K, Shigemoto R, Kakizuka A Ohkubo H, Nakanishi S (1989) Molecular characterization of a functional cDNA for rat substance P receptor. J Biol Chem 264:17649–17652

Yu DH, Noguchi M, Zhou ZC, Villanueva ML, Gardner JD, Jensen RT (1987) Characterization of gastrin receptors on guinea pig pancreatic acini. Am J Physiol 253:G793–G801

Zhou ZC, Gardner JD, Jensen RT (1989) Interaction of peptides related to VIP and secretin with guinea pig pancreatic acini. Am J Physiol 256:G283–G290

Zimmerman RP, Gates TS, Mantyh CR, Vigna SR, Welton ML, Passaro EP, Mantyh PW (1989) Vasoactive intestinal polypeptide receptor binding sites in the human gastrointestinal tract: localization by autoradiography. Neuroscience 3:771–783

Zweibaum A, Laburthe M, Grasset E, Louvard D (1991) The use of cultured cell lines in studies of intestinal cell differentiation and function. In: Field M, Frizzell RA (eds) Intestinal absorption and secretion. American Physiological Society, Bethesda, pp 223–255 (Handbook of physiology, vol 4)

CHAPTER 6

Proteolytic Inactivation of Neurohormonal Peptides in the Gastrointestinal Tract

J.M. CONLON

A. Introduction

The gastrointestinal tract is a major site of production and inactivation of neurohormonal peptides. While biochemical and morphological aspects of the synthesis of gastrointestinal peptides have been studied in detail, relatively little work has been done to investigate the mechanisms by which these peptides are degraded following release. The neurohormonal peptides discussed in this article have different cellular distributions both along and within the different layers of the stomach and gut. Consequently, an analysis of the mechanisms by which these peptides are inactivated must take into account the microenvironment into which they are released.

Gastrin and cholecystokinin (CCK) are found in mucosal endocrine-like cells with highest concentration in the upper GI tract but neuronal CCK has also been identified in the rat ileum and colon. Neurotensin is localized primarily to endocrine cells in the distal small intestine but the peptide is also found in nerve fibers in the myenteric plexus of the upper GI tract and in the circular muscle layer of the cecum. In the stomach, somatostatin is found primarily in mucosal D cells but somatostatin-containing neurons are found through the intestine particularly in the myenteric and submucous plexuses. The enkephalins, tachykinins, and vasoactive intestinal polypeptide (VIP) are found exclusively in neurons distributed throughout the GI tract with a particular abundance of fibers in the myenteric and submucous plexi and in the circular smooth layer (SCHULTZBERG et al. 1980). As well as release into blood and interstitial fluid, the endocrine cells of the stomach and gut release biologically active peptides into the gastrointestinal lumen. The presence of gastrin and somatostatin in gastric juice and the release of gastrin, somatostatin, neurotensin and CCK into the lumen of the intestine has been described (reviewed by RAO 1991). Substance P- and VIP-containing fibers innervate blood vessels in the mucous and submucous layers and project into the mucosa and these peptides are also released into the gastrointestinal lumen in response to nerve stimulation.

In this light, the review will analyze the proteolytic degradation of the primarily endocrine peptides of the GI tract, such as neurotensin and gastrin, in terms of the contribution of peptidases localized in the basolateral and brush-border membranes of epithelial cells in the mucosal layer and in the membranes on the luminal side of the capillaries of the vascular beds of the stomach and gut. An analysis of the mechanisms of inactivation of gastrointestinal neuropeptides, such as the tachykinins and enkephalins, will consider the role of peptidases located in the synaptic cleft and in the plasma membrane and interstitial fluid surrounding neighboring cells in the muscle layers of the stomach and gut.

B. Proteolytic Enzymes Implicated in the Inactivation of Neurohormonal Peptides in the Gastrointestinal Tract

I. Endopeptidases

1. Endopeptidase 24.11

Endopeptidase 24.11 (EC 3.4.24.11), also referred to as "enkephalinase" and neutral endopeptidase, is very widely distributed in mammalian tissues and a general role for the enzyme in the hydrolysis and inactivation of biologically active peptides at cell surfaces has been proposed (MATSAS et al. 1983; TURNER et al. 1985). The term "enkephalinase" was considered inappropriate as the enzyme shows no particular specificity for the enkephalins. Evidence for the involvement of endopeptidase 24.11 in the catabolism of gastrointestinal peptides is very strong. Within the GI tract, the enzyme is found in highest concentration in the brush border membranes of intestinal mucosal cells and also, in much lower concentration, in the basolateral membrane (DANIELSEN et al. 1980; GEE et al. 1983). Using fluorogenic substrates in conjugation with specific inhibitors, CHECLER et al. (1987a) have shown that endopeptidase 24.11 is present in plasma membrane-enriched preparations of circular and longitudinal smooth muscle from dog ileum and BARELLI et al. (1989) have shown activity in synaptosomal fractions prepared from the myenteric, deep muscular and submucous plexi of dog ileum. Well-characterized closed membrane vesicles from pig (SCHAFER et al. 1986a) and guinea pig (NAU et al. 1986) ileal smooth muscle were also associated with a high activity of the enzyme. A monoclonal antibody raised against endopeptidase 24.11 from pig kidney has been used to purify the enzyme by immunoaffinity chromatography from pig intestinal microvilli (FULCHER et al. 1983) and pig fundic muscle (BUNNETT et al. 1988b). Endopeptidase 24.11 from the pig GI tract is a single polypeptide chain glycoprotein of apparent molecular mass of 90000. The gastrointestinal enzyme differs from its counterparts isolated from pig kidney and brain in the nature and extent of their glycosylation.

There does not appear to be a consensus amino acid sequence in a polypeptide substrate for recognition by endopeptidase 24.11. The enzyme preferentially hydrolyzes peptide bonds at the N-terminal side of hydrophobic residues such as phenylalanine, leucine, isoleucine, valine, tryptophan

Table 1. Sites of hydrolysis of some gastrointestinal peptides by purified endopepeptidase 24.11

Peptide	Cleavage site(s)
Met-enkephalin	Tyr-Gly-Gly↓Phe-Met
Leu-enkephalin	Tyr-Gly-Gly↓Phe-Met
β-Endorphin	Tyr-Gly-Gly↓Phe-Met-Thr-Ser-Glu-Lys-Ser- Gln-Thr-Pro↓Leu-Val-Thr↓Leu↓Phe-Lys-Asn- Ala-Ile↓Ile-Lys-Asn-Ala-Tyr-Lys-Lys-Gly-Glu
Substance P	Arg-Pro-Lys-Pro-Gln-Gln↓Phe↓Phe-Gly↓Leu-Met.NH_2
Neurokinin A	His-Lys-Thr-Asp-Ser↓Phe-Val-Gly↓Leu-Met.NH_2
Neurotensin	pGlu-Leu-Tyr-Asn-Lys-Pro-Arg-Arg-Pro↓Tyr↓Ile-Leu
Somatostatin	Ala-Gly-Cys-Lys-Asn↓Phe↓Phe-Trp-Lys-Thr↓Phe-Thr-Ser-Cys
Neuromedin C	Gly-Asn-His↓Trp-Ala↓Val-Gly-His↓Leu-Met.NH_2
Gastrin-17	pGlu-Gly-Pro-Trp↓Leu-Glu-Glu-Glu-Glu-Glu- Ala↓Tyr-Gly↓Trp↓Met-Asp↓Phe.NH_2
CCK-8	Asp-Tyr(SO_3H)Met-Gly↓Trp-Met-Asp↓Phe.NH_2
VIP	His-Ser-Asp↓Ala↓Val-Phe-Thr-Asp-Asn-Tyr- Thr-Arg↓Leu-Arg-Lys-Gln↓Met-Ala-Val-Lys- Lys↓Tyr-Leu-Asn-Ser-Ile↓Leu-Asn.NH_2

and tyrosine. Studies with purified endopeptidase 24.11 from pig kidney, however, have shown that there is considerable variation in k_{cat}/K_m values for various peptide substrates tested (Matsas et al. 1984a). Substance P, for example, is hydrolyzed very efficiently (K_m = 31.9 μM; k_{cat} = 5062 min^{-1}), whereas gonadotropin-releasing hormone (GnRH) is a relatively poor substrate (K_m = 755 μM; k_{cat} = 840 min^{-1}). It appears, therefore, that amino acid residues in a polypeptide remote from the scissile bond do influence the rate of hydrolysis. The sites of cleavage of some gastrointestinal peptides by purified brain and/or gastric endopeptidase 24.11 are shown in Table 1. The enzyme resembles thermolysin (EC 3.4.24.4) in its sensitivity to inhibition by naturally occurring phosphoramidon [*N*-(α-rhamnopyranosyloxyhydroxy-phosphinyl)-L-Leu-L-Trp] (Fulcher et al. 1982). Evidence has been provided, however, that both the epithelial layer

(NAU et al. 1985) and the longitudinal muscle layer (NAU et al. 1986) of the small intestine contains a phosphoramidon-insensitive form of endopeptidase 24.11 in addition to the phosphoramidon-sensitive component. Thiorphan (D,L,3-mercapto-2-benzylpropanoyl-glycine) was the first synthetic inhibitor of the enzyme to be described and numerous analogs based upon this structure have since been made (reviewed in ROQUES and BEAUMONT 1990). Orally active inhibitors such as SCH 34826 ((*S*)-*N*-[*N*-[1-[[2,2-dimethyl-1,3-dioxolan-4-yl)methoxy]carbonyl]-2-phenylethyl-L-phenylalanine]-ß-alanine) clearly have great therapeutic potential (CHIPKIN et al. 1988).

The gene directing the synthesis of endopeptidase 24.11 in the human (MALFROY et al. 1988), rat (MALFROY et al. 1987) and rabbit (DEVAULT et al. 1987) has now been cloned and sequenced. The primary structure of endopeptidase 24.11 has been very strongly conserved between species but shows little similarity to thermolysin. Nevertheless, amino acid residues involved in the active site of thermolysin have their counterparts in endopeptidase 24.11 (ERDOS and SKIDGEL 1989). Thus the zinc-coordinating His^{142} and His^{146} residues in thermolysin correspond with His^{583} and His^{587} in endopeptidase 24.11 (DEVAULT et al. 1988b) and in both enzymes a glutamic acid residue catalyzes the nucleophilic attack of a water molecule on the carbonyl group of the scissile bond (DEVAULT et al. 1988a).

2. Endopeptidase 24.15

Endopeptidase 24.15 is a zinc-containing metalloendopeptidase that was first isolated from the cytosolic fraction of rat brain (ORLOWSKI et al. 1983) but is also found in high concentration in testes and pituitary. In the brain, 20%–25% of the total activity is associated with membrane fractions, particularly synaptosomes (ACKER et al. 1987). The enzyme from rat brain synaptosomes generates Leu-enkephalin from dynorphin-(1–8) and from α- and β-neo-endorphin and Met-enkephalin from Met-enkephalin-Arg^6-Gly^7-Leu^8. A physiologically important role for endopeptidase 24.15 in the inactivation of luteinizing hormone releasing hormone (LHRH) in the hypothalamus has been proposed (MOLINEAUX et al. 1988). Specificity studies have indicated that the enzyme preferentially cleaves substrates in which the P_1 residue is aromatic and the highest k_{cat}/K_m ratios were observed with substrates with a hydrophobic residue in the P_3' position (ORLOWSKI et al. 1988). As discussed in Sect. C.I, endopeptidase 24.15 activity has been detected in plasma membrane-enriched fractions of circular muscle from dog ileum (CHECLER et al. 1987a) and synaptosomal fractions prepared from myenteric, deep muscular and submucous (highest activity) plexuses (BARELLI et al. 1989). Neurotensin-metabolizing peptidase activity in purified plasma membranes from rat fundus, which resulted in cleavage of the Arg^8-Arg^9 bond, has also been ascribed to endopeptidase 24.15 (CHECLER et al. 1987a). The concentration of endopeptidase 24.15 in the gastrointestinal tract is much lower than the concentration of endopeptidase 24.11

and a major role for the former enzyme in the inactivation of gut neurohormonal peptides is improbable. However, the availability of selective and nontoxic inhibitors of endopeptidase 24.15, e.g., *N*-[1(R,S)-carboxy-2-phenylethyl]-alanylalanyl-phenylalanine-*p*-aminobenzoate (CHU and ORLOWSKI 1984; ORLOWSKI et al. 1988) will permit a fuller assessment of its importance in the GI tract.

3. Endopeptidase 24.16

Endopeptidase 24.16 is a membrane-associated peptidase purified from detergent-solubilized rat brain synaptic membranes (CHECLER et al. 1986) and an homogenate of rat ileum (BARELLI et al. 1988). The peptidase was identified by its ability to convert neurotensin into neurotensin$_{1-10}$ by cleavage of the Pro^{10}-Tyr^{11} bond. The enzyme is a metalloendopeptidase of molecular mass of 70–75 kDa that displays a high affinity for neurotensin (K_m = 2.6 μM) and the structurally related peptides neuromedin N and xenopsin but will also cleave other small peptides, e.g., dynorphin$_{1-13}$, substance P, bradykinin. The importance of this enzyme in the inactivation of peptides in the gastrointestinal tract remains to be assessed.

II. Exopeptidases

1. Peptidyl Dipeptidase A

Peptidyl dipeptidase A (EC 3.4.15.1) is best known for its ability to generate angiotensin II from angiotensin I ("angiotensin-converting enzyme") and to inactivate bradykinin ("kininase II") in the pulmonary circulation. The enzyme has, however, a widespread distribution in mammalian tissues. The mucosal brush border of human and pig intestine is a rich source of the enzyme with a relatively uniform distribution along the gut (WARD et al. 1980). Peptidyl dipeptidase A is not confined to the mucosal layer of the intestine and membrane vesicles prepared from the longitudinal muscle layer of guinea pig ileum showed an 11-fold enrichment in the activity of the enzyme (NAU et al. 1986). Similarly, membrane vesicles derived from the longitudinal and circular smooth muscle of pig small intestine were enriched sixfold relative to the tissue homogenate (SCHAFER et al. 1986a). The primary specificity of peptidyl dipeptidase A is the removal of dipeptides from the C terminus of oligopeptides where the penultimate residue is not proline and the terminal residue is not aspartic acid or glutamic acid (BUNNING et al. 1983; ERDOS 1987). Maximum rates of reaction are seen with substrates containing an aromatic residue in the P_1 position. Examples illustrating the sites of cleavage of some gastrointestinal peptides by the purified enzyme are shown in Table 2. Cleavages involving peptidyl dipeptidase A are usually characterized by dependence of the reaction rate on the concentration of chloride ions (BUNNING and RIORDAN 1983).

Table 2. Sites of hydrolysis of some gastrointestinal peptides by purified peptidyl dipeptidase A (angiotensin-converting enzyme)

Peptide	Cleavage site(s)
Met-enkephalin	Tyr-Gly-Gly↓Phe-Met
Leu-enkephalin	Tyr-Gly-Gly↓Phe-Leu
Substance P	Arg-Pro-Lys-Pro-Gln-Gln-Phe-Phe↓Gly↓Leu-Met.NH_2
Neurokinin A	Not hydrolyzed
Neurokinin B	Not hydrolyzed
Neurotensin	pGlu-Leu-Tyr-Glu-Asn-Lys-Pro-Arg-Arg-Pro-Tyr↓Ile-Leu

Although peptides with unsubstituted C termini are preferred substrates, the enzyme will display endopeptidase activity towards certain C-terminally α-amidated peptides (YOKOSAWA et al. 1983). Peptidyl dipeptidase A from human kidney hydrolyzed substance P at the Phe^8-Gly^9 and Gly^9-Leu^{10} bonds to generate the (1–8) and (1–9) fragments in the ratio 4:1 (SKIDGEL et al. 1984). Only the C-terminal dipeptide, however, was released from the free acid form of substance P. Purified isoenzymes from rat lung and brain corpus striatum displayed different specificities towards α-amidated peptide substrates (STRITTMATTER et al. 1985). The lung enzyme utilized a pathway involving initial release of the C-terminal tripeptide followed by sequential degradation of the (1–8) fragment by removal of dipeptides whereas the brain enzyme also utilized a pathway involving sequential removal of dipeptides only from substance P to give the (1–5) fragment. Peptidyl dipeptidase A from pig kidney did not hydrolyze neurokinin A (HOOPER et al. 1985) or neurokinin B (HOOPER and TURNER 1985). Similarly, neither of two molecular forms of peptidyl dipeptidase A from pig striatum, differing only in the extent of their glycosylation, hydrolyzed neurokinin A or neurokinin B (HOOPER and TURNER 1987). Neurokinin A was, however, reported to be a substrate for the enzyme from rat striatum but not for the enzyme from rat lung (STRITTMATTER et al. 1985).

The primary structure of human peptidyl dipeptidase A has been deduced from the nucleotide sequence of DNA complementary to mRNA from vascular endothelial cells (SOUBRIER et al. 1988) and the recombinant enzyme has been expressed in transfected Chinese hamster ovary cells (WEI et al. 1991). The amino acid sequence of the enzyme displayed no clear similarity with other proteins but peptidyl dipeptidase A shares with thermolysin and endopeptidase 24.11 conserved histidine and glutamic acid residues at the active site. Several highly effective inhibitors of peptidyl dipeptidase A have been developed which have proved valuable in assessing the involvement of the enzyme in the in vivo inactivation of neurohormonal peptides. These include captopril (D-3-mercapto-2-methylpropanoyl-L-proline) and enalapril (*N*-(*S*)-1-(ethoxycarbonyl)-3-phenylpropyl)-L-alanyl-L-proline maleate) (WYVRATT et al. 1983). Several naturally occurring peptides, e.g., the family of bradykinin-potentiating peptides from the venom of the snake *Agkistrodon halys blomhoffi* (KATO and SUZUKI 1971) and an octapeptide from the muscle of the tuna, *Neothunnus macropterus*, act as inhibitors of the enzyme. As discussed in Sect. D.I, the contribution of peptidyl dipeptidase A to inactivation pathways in the GI tract may be relatively minor as unfavorable k_{cat}/K_m values for many GI peptides means that the enzyme is unable to compete successfully for substrate with endopeptidase 24.11.

2. Aminopeptidases

Aminopeptidases, acting synergistically with endopeptidases such as endopeptidase 24.11, play an important role in the inactivation of neurohormonal peptides in the GI tract. The involvement of aminopeptidases is particularly important in the case of peptides with unsubstituted N termini for which removal of the first residue results in loss of biological activity, e.g., enkephalins (Sect. D.II). Much of the aminopeptidase activity associated with the gut is cytosolic but several membrane-bound ectoenzymes have been identified and characterized. These include aminopeptidase N (formerly aminopeptidase M) (EC 3.4.11.2), aminopeptidase P (EC 3.4.11.9), aminopeptidase A (EC 3.4.11.7) and aminopeptidase W.

Aminopeptidase N is a zinc-containing glycoprotein that is localized predominantly to the brush-border membrane of the intestine (GRAY and SANTIAGO 1977). The enzyme has a broad specificity but hydrolysis of peptides with N-terminal aspartate, glutamate and proline residues proceeds slowly. Aminopeptidase N is inhibited by the bacterial peptides amastatin ([(2S,3R)-3-amino-3-hydroxy-5-methylhexanoyl]-L-valyl-L-valyl-L-aspartic acid) and bestatin ([(2S,3R)-3-amino-3-hydroxy-4-phenylbutanoyl]-L-leucine) (UMEZAWA et al. 1976) and less efficiently by puromycin (BARCLAY and PHILLIPPS 1980). Bestatin-sensitive aminopeptidase activity was detected in membranes prepared from the longitudinal muscle layer of bovine (HAZATO et al. 1985) and guinea pig (NAU et al. 1986) small intestine and the demonstration that aminopeptidase N is a major component of synaptic membranes from pig brain striatum (MATSAS et al. 1985) suggests that the enzyme may be important in terminating the action of peptides released from the neurons of the gut myenteric plexus. Recently, aminopeptidase N has been purified from a Triton X-100 solubilized preparation of pig intestinal muscle and shown to be similar to the enzyme from mucosa (TERASHIMA et al. 1991). Aminopeptidase P is an integral membrane protein of the intestinal brush border (LASCH et al. 1988) and specifically cleaves the N-terminal amino acid from peptide substrates with an adjacent proline (Xaa-Pro. . . .). Aminopeptidase P from rat brain liberated the N-terminal arginine residue from substance P and the N-terminal lysine residue from substance P_{3-11} (HARBECK and MENTLEIN 1991). The enzyme is strongly inhibited by metal chelating agents, e.g., 1,10-phenanthroline but common inhibitors of aminopeptidases such as amastatin, bestatin or puromycin were ineffective (Harbeck and Mentlein 1991). Aminopeptidase A shows a strong specificity for substrates with an N-terminal aspartate or glutamate residue (Benajiba and Maroux 1980) and aminopeptidase W will preferentially cleave substrates with an N-terminal aromatic residue (Gee and Kenny 1987). Prolyl aminopeptidase (EC 3.4.11.5) removes an N-terminal proline residue from short peptides but a recent report (Turzynski and Mentlein 1990) concluded that, in the rat, it is identical to the cytosolic enzyme leucyl aminopeptidase (EC 3.4.11.1).

3. Dipeptidyl Aminopeptidase IV

The cryptic epithelium of the mucosal layer of the small intestine contains a high concentration of dipeptidyl aminopeptidase IV (EC 3.4.14.5) (Kenny et al. 1976; Grossrau 1979). This enzyme contains a serine residue at its active site and will remove dipeptides from the N terminus of substrates with a proline (or less effectively alanine) residue adjacent to the N-terminal amino acid (Kato et al. 1978). Several studies have reported the sequential removal of Arg-Pro and Lys-Pro from substance P in reactions catalyzed by this enzyme (Kato et al. 1978; Conlon and Sheehan 1983) but the high K_m value for this reaction (2.0 mM) casts doubt upon the physiological relevance of the observation. Dipeptidyl aminopeptidase IV, as well as aminopeptidase N, is associated with the plasma membrane of the vascular endothelium (Palmieri and Ward 1983) and so an involvement of these enzymes in the inactivation of peptides released by endocrine cells and enteric neurons into blood is a possibility.

4. Carboxypeptidases

The presence of an α-amidated C-terminal amino acid residue in many neurohormonal peptides, e.g., substance P, CCK, and gastrin, affords some protection against the action of carboxypeptidases. Nevertheless, carboxypeptidase P (EC 3.4.17.–), which removes C-terminal amino acids linked to proline, and an exopeptidase which removes C-terminal basic amino acids and resembles carboxypeptidase N (kininase I), are present in intestinal as well as kidney brush-border membrane preparations (Hooper and Turner 1988) and may be involved in the inactivation of gastrointestinal peptides.

C. Pathways of Proteolytic Inactivation of Some Gastrointestinal Hormones

I. Neurotensin

Several studies have implicated the small intestine as a site of both synthesis and degradation of neurotensin (NT). NT-like immunoreactivity is released into the human circulation in response to ingestion of nutrients but biologically inactive metabolites (principally NT_{1-8} and NT_{1-11} are the major molecular forms in both portal and peripheral venous blood and intact neurotensin (a tridecapeptide) represents only a minor component (HAMMER et al. 1982). Similarly, plasma collected from the superior mesenteric vein of rats during perfusion of the small intestine with oleic acid contained an appreciable concentration of the (1–8) and (1–11) fragments and these radioactive metabolites were detected in the superior mesenteric vein during infusion of [^{3}H]neurotensin into the superior mesenteric artery (FERRIS et al. 1985). Synthetic NT was also efficiently cleared by passage through the vascular bed of the intestine (and kidney) of the sheep but not by the lung and liver (SHULKES et al. 1983). NT_{1-13} is, however, the predominant molecular form in acidic extracts of the intestines of several mammalian species and the intact peptide was released from isolated canine ileal mucosal cells in short-term culture (BARBER et al. 1986). NT is relatively stable in whole blood ($t_{1/2} > 30$ min at 37°C) and so degradation probably occurs concomitant with, or shortly after, release in the interstitial fluid bathing the secretory and target cells or in the capillary beds of the intestine, for example.

In a system that involved introduction of radiolabeled NT directly into the gastric submucosa of conscious rats and collection of metabolites in surgically implanted dialysis fibers, catabolism of NT by proteases in interstitial fluid of the stomach was demonstrated (BUNNETT et al. 1984). The half-life of the intact peptide was between 9 and 15 min and the (1–8), (9–13), (1–11) metabolites and free tyrosine were identified in a dialysate of the submucosal interstitial fluid. In a related study using the system (ORLOFF et al. 1986), NT catabolism was partially but significantly inhibited by enalapril, suggesting that peptidyl dipeptidase A may be involved in the initial stages of inactivation. A study by CHECLER et al. (1988) has identified a major role for endopeptidase 24.11 in the in vivo inactivation of NT in the dog ileum. Intraarterial perfusion of segments of ileum with [^{3}H]neurotensin resulted in rapid ($t_{1/2}$ between 2 and 6 min) conversion of the peptide into metabolites that were identified as NT_{1-7}, NT_{1-8}, NT_{1-10}, NT_{1-11}, and NT_{11-13} and free tyrosine. Pretreatment of the tissue with the inhibitor of endopeptidase 24.11, thiorphan, resulted in a significant protection of intact NT and a major decrease in the formation of the (1–11) metabolite. In contrast, captopril provided no protection for intact NT but reduced formation of the (1–8) metabolite. It was suggested, therefore, that peptidyl dipeptidase A participates only in the secondary conversion of NT_{1-10} into NT_{1-8}.

Incubation of NT with dispersed epithelial cells from pig jejunoileum resulted in cleavage of the peptide at several sites by cell surface proteolytic enzymes (SHAW et al. 1987). The apparent K_m for degradation was $23 \pm 3\,\mu M$ and V_{max} was 584 ± 16 pmol/10^6 cells per minute. The principal sites of cleavage were at the Tyr^{11}-Ile^{12} bond, generating NT_{1-11}, and at the

Pro^{10}-Tyr^{11} bond generating NT_{1-10}. Formation of these metabolites was completely inhibited by phosphoramidon ($K_i = 6\,nM$) but not by captopril or *p*-chloromercuriphenylsulfonic acid, an inhibitor of thiol peptidases. The C-terminal fragments, NT_{11-13} and NT_{12-13}, were metabolized to free amino acids by exopeptidases. Incubation of NT with purified endopeptidase 24.11 from pig stomach also resulted in formation of the (1–10) and (1–11) fragments. A minor pathway of degradation by the enterocytes involved a phosphoramidon-insensitive cleavage of the Tyr^3-Glu^4 bond, generating NT_{1-3} and NT_{4-13}. NT_{1-8}, the major circulating metabolite, was not formed when intact NT was incubated with enterocytes but was produced when NT_{1-11} was used as substrate. The results of this in vitro study are consistent with the findings in vivo of CHECLER et al. (1988).

Neurotensin inactivation by peptidases within the wall of the intestine has been investigated by CHECLER et al. (1987a) using purified plasma membrane preparations from the circular and longitudinal smooth muscle of dog ileum. Endopeptidase 24.11 activity in both membrane preparations was responsible for formation of NT_{1-11} and NT_{1-10}, as in the epithelial layer. It was suggested, however, that a second neutral endopeptidase (probably endopeptidase 24.16) contributed to the formation of the (1–10) fragment. Neurotensin was also a substrate for an uncharacterized carboxypeptidase that hydrolyzed the Ile^{12}-Leu^{13} bond. The activities of proline endopeptidase and endopeptidase 24.15 were detected in membranes from circular muscle only and it was suggested that these enzymes were responsible for generating the (1–7) and (1–8) fragments respectively. Secondary processing of inactive degradation products by peptidyl dipeptidase A resulted in the conversion of NT_{1-10} to the (1–8) fragment and free tyrosine was generated from NT_{11-13}. An enzyme in the circular muscle membranes with the specificity of dipeptidyl aminopeptidase IV converted NT_{9-13} to NT_{11-13}. 5-Oxoprolyl-peptide hydrolase (EC 3.4.19.3) activity was not detected in the muscle membrane preparation and the presence of a pyroglutamyl residue in NT appears to protect the intact peptide from aminopeptidases in both the muscle and epithelial layers of the gut.

II. Gastrin

Carboxyl-terminally α-amidated gastrin-17, present in sulfated and unsulfated forms, represents the main component in antral mucosa of mammals but analysis of the molecular forms of immunoreactive gastrin in plasma taken from the antral vein of dogs (DOCKRAY et al. 1982) and pigs (POWER et al. 1986) showed an appreciably different distribution from that in antral mucosal extracts. A higher proportion of shorter C-terminal fragments of gastrin-17 were found in the venous effluent of the stomach. Gastrin-17 is cleaved relatively slowly by enzymes in plasma ($t_{1/2} = 35\,min$ in dog plasma) and so it was suggested that the peptide was metabolized at or before the time it enters the circulation by enzymes in the basolateral borders of the G cell, in the extracellular fluid or on the luminal side of the endothelial membrane of antral capillaries (DOCKRAY et al. 1982). Infusion of phosphoramidon into the gastroepiploic artery of the pig increased the proportion of intact gastrin-17 in antral venous plasma, suggesting that endopeptidase 24.11 was involved in the in vivo conversion to C-terminal metabolites (POWER et al. 1987). However, BUNNETT et al. (1988a) have suggested that, as gastrin-17 is only a moderately good substrate for the enzyme (V_{max} = 0.06 nmol/min per microgram enzyme compared with 3.4 nmol/min per microgram enzyme for cholecystokinin-8), endopeptidase 24.11 cannot be solely responsible for the postsecretory degradation of gastrin.

Incubation of extracts of pig antrum with purified endopeptidase 24.11 from kidney resulted in disappearance of gastrin-like immunoreactivity (POWER et al. 1987). Unsulfated human gastrin-17 was hydrolyzed at the Trp^{4}-Leu^{5}, Ala^{11}-Tyr^{12}, Gly^{13}-Trp^{14}, Trp^{14}-Met^{15} and Asp^{16}-Phe^{17} bonds by purified endopeptidase 24.11 from pig stomach (Table 1) BUNNETT et al. 1988). Cleavage of the Asp-Phe bond will result in complete inactivation of the peptide. Amongst other metabolites, the (1–11), (1–13) and (1–16) fragments of human gastrin-17 were identified in the digest. These particular metabolites, together with the (5–17) fragment, were identified in plasma taken from healthy human volunteers during infusion of gastrin-17 (DESCHODT-LANCKMAN et al. 1988) and it was suggested that the action of endopeptidase 24.11 contributes to the molecular heterogeneity of circulating gastrin. The Trp^{4}-Met^{5} bond of pig unsulfated gastrin-17 is not cleaved by gastric endopeptidase 24.11 but hydrolysis of the Ala^{11}-Tyr^{12}, Gly^{13}-Trp^{14}, and Asp^{16}-Phe^{17} bonds was observed (BUNNETT et al. 1988). Consistent with these observations, the (1–11) and (1–13) fragments of gastrin-17 have been identified in an extract of pig antrum (POWER et al. 1988). Although unsulfated and sulfated gastrin-17 are present in approximately equal amounts in porcine antrum, the unsulfated form of the (1–13) fragment predominates, suggesting that unsulfated gastrin-17 is a better substrate for the enzyme.

III. Cholecystokinin

Like gastrin, gut cholecystokinin (CCK) is extremely heterogeneous with multiple molecular forms arising from a single gene transcript by different pathways of post-translational processing. This molecular heterogeneity may be in part artifactual, occurring during tissue extraction and peptide characterization. Conversion of larger molecular forms to smaller peptides during gel permeation chromatography has been described and an enzyme in acid extracts of rat intestine that converts CCK-33 to CCK-12 has been identified (TURKELSON et al. 1990).

Cholecystokinin-8, the predominant molecular form of neuronal CCK, is a good substrate for purified endopeptidase 24.11 from pig stomach

(BUNNETT et al. 1988) and cleavages at the Gly^4-Trp^5 and Asp^7-Phe^8 bonds were identified. Hydrolysis at these sites was also reported during incubations of the peptide with endopeptidase 24.11 from kidney (MATSAS et al. 1984; NAJDOVSKI et al. 1985). Splitting of the Asp-Phe bond results in complete inactivation of the peptide and this cleavage proceeded four times more rapidly than at the Gly-Trp bond (MATSAS et al. 1984b). Indirect evidence suggests that aminopeptidase activity in the GI tract may be important in modulating the action of CCK. Incubation of CCK-8 with smooth muscle strips from cat intestine resulted in a time-dependent loss of immunoreactivity (PRAISSMAN et al. 1982). In contrast, CCK that was acetylated on its lone N-terminal amino group was completely resistant to degradation under the same conditions. Deletion of the N-terminal aspartyl residue results in a peptide with greatly reduced biological potency (JENSEN et al. 1980). Hydrolysis of CCK-8 by membrane-bound enzymes in pig brain corpus striatum also involves the action of bestatin-sensitive aminopeptidases acting together with endopeptidase 24.11 (MATSAS et al. 1984).

IV. Somatostatin

Incubation of somatostatin (SS)-14 with plasma membrane vesicles from the basolateral and brush border faces of pig small intestinal epithelial cells resulted in rapid degradation of the peptide (WEBER et al. 1986). Cleavages between Ala^1-Gly^2, Phe^6-Phe^7, Phe^7-Trp^8 and Thr^{10}-Phe^{11} were observed indicative of aminopeptidase and endopeptidase action. An involvement of endopeptidase 24.11 was not conclusively established in this study but phosphoramidon-sensitive cleavages of the Phe^6-Phe^7 and Thr^{10}-Phe^{11} bonds of SS-14 were observed during incubation of the peptide with synaptic membranes from rat hippocampus and with purified endopeptidase 24.11 from rat brain (SAKURADA et al. 1990). The extreme susceptibility of SS-14 to aminopeptidase activity in blood has been documented (MCMARTIN and PUPDON 1978) and it is probable that appreciable conversion to des-[Ala^1]-SS takes place in the vascular beds of the GI tract. Des-[Ala^1]- and des-[Ala^1Gly^2]-SS-14 were also identified in extracts of freshly prepared enterocytes from pig jejunum (SCHAFER et al. 1986b). Somatostatin-14 is very unstable in the lumen of the small intestine and is completely degraded within 5 min by a dilute preparation of rat intestinal juice (PETERS and MCMARTIN 1982). Studies with the more stable analogue cyclo(-Asn-Phe-Phe-D-Trp-Lys-Thr-Phe-γ-aminobutyrate-) have shown that there is a single primary cleavage site at Lys-Thr and the resulting linear peptide is rapidly degraded (ALLEN et al. 1984). The analog containing a D-Lys substitution was even more stable in rat intestinal juice, pointing the way towards the design of potentially therapeutically valuable orally administered SS-based agents.

D. Pathways of Proteolytic Inactivation of Some Gastrointestinal Neuropeptides

I. Tachykinins

The catabolism of substance P (SP) by peptidases in the stomach wall of the conscious rat has been studied by ORLOFF et al. (1986) using the system of implanted dialysis fibers described in Section C.I. [Prolyl^{2,4}-3,4(N)^{3}H] SP, injected directly into the wall of the gastric corpus, was rapidly ($t_{1/2} < 5$ min) converted into major metabolites that were identified as the (1–2) and (3–4) fragments together with minor metabolites identified as the (1–6), (1–7) and (1–8) fragments. Unexpectedly, pretreatment of the tissue with phosphoramidon provided no consistent protection for intact [^{3}H]substance P whereas catabolism was partially inhibited by captopril and enalapril. The authors concluded, therefore, that peptidyl dipeptidase A but not endopeptidase 24.11 was involved in the initial stages of substance P breakdown in this preparation. The formation of the (1–2) and (3–4) metabolites is suggestive of the action of dipeptidyl aminopeptidase IV but the unavailability of nontoxic inhibitors of this enzyme precluded confirmation of its involvement.

In a study by NAU et al. (1985), the mechanism of proteolytic inactivation of substance P in the epithelial layer of the gut was investigated by incubation of the peptide with dispersed enterocytes and with purified brush-border and basolateral membranes from pig jejunum. The pattern of metabolites produced by the three preparations was very similar but the rate of degradation by the brush-border membranes was approximately fivefold greater than the rate produced by an equivalent concentration of basolateral membranes, indicating that enzymes in the brush-border were probably responsible for the greater part of the cell-mediated degradation. Rapid cleavages between the Gln^6-Phe^7, Phe^7-Phe^8 and Gly^9-Leu^{10} bonds were observed. The rate of formation of the (1–9) fragment exceeded that of the other metabolites, indicating the cleavage of the Gly^9-Leu^{10} bond may be the physiologically important mechanism for the inactivation of the substance P released into the mucosal layer of the gut. Degradation was strongly but not completely (90%) inhibited by 1 μM phosphoramidon, demonstrating a major involvement of endopeptidase 24.11. The apparent K_m value for the formation of the (1–9) fragment (55 μM) was comparable to the apparent K_m for degradation of substance P by the purified enzyme from kidney (32 μM). Despite the high concentration of peptidyl dipeptidase A in the epithelial layer of the gut, catabolism of substance P was unaffected by captopril and the concentration of chloride ions and formation of the (1–8) fragment, indicative of the action of the enzyme, was not observed. Although the apparent K_m for the degradation of substance P by purified peptidyl dipeptidase A (25 μM) (ERDOS and SKIDGEL 1989) is similar to the K_m for degradation by endopeptidase 24.11, the k_{cat} value for the latter enzyme (5062 min^{-1}) is considerably greater than the k_{cat} value for peptidyl dipeptidase A (225 min^{-1}). As a consequence, the peptidyl dipeptidase A in the mucosal layer is unable to compete with endopeptidase 24.11 for available substrate. Formation of SP_{3-11} by epithelial cell membranes, indicative of the action of dipeptidyl aminopeptidase IV, was observed only at very high substrate concentrations (>100 µmol/l) consistent with the high K_m value (2 m*M*) observed in the hydrolysis of SP by purified dipeptidyl aminopeptidase IV from human submaxillary gland (KATO et al. 1978).

The proteolytic inactivation of SP in the longitudinal muscle layer of the pig jejunum followed the same pathway as in the epithelial layer but the rate of degradation was approximately 100-fold less (NAU et al. 1986). Incubation of SP with vesicles derived from the longitudinal muscle of pig intestine and showing a 21-fold enrichment in the activity of the plasma membrane marker enzyme 5′-nucleotidase resulted in hydrolysis of the Gln^6-Phe^7, Phe^7-Phe^8 and Gly^9-Leu^{10} bonds. Incubation of neurokinin A (NKA) with the vesicles under the same conditions resulted in cleavage of the Gly^8-Leu^9 bond. For both substrates, hydrolyses at these sites were potently, but not completely, inhibited by phosphoramidon in a dose-dependent manner (ID_{50} for inhibition of degradation of substance P was 23 n*M* and for neurokinin A was 13 n*M*). Substance P was resistant to the action of aminopeptidases in the longitudinal muscle preparation but neurokinin A was rapidly metabolized to the [des-His^1] and des[His^1Lys^2] fragments in a reaction that was insensitive to phosphoramidon but completely inhibited by bestatin. Consistent with the study using dispersed enterocytes (NAU et al. 1986), degradation of substance P and neurokinin A in the muscle layer of the gut was not inhibited by enalapril and was unaffected by chloride concentration, suggesting that dipeptidyl peptidase A was not important in the inactivation process.

A strong indication that endopeptidase 24.11 plays a physiologically important role in the inactivation of tachykinins in the smooth muscle layer of the gut is provided by the study of DJOKIC et al. (1989). The inhibitor leucine-thiorphan (10^{-5} *M*) significantly increased the potency of substance P in contracting isolated segments of smooth muscle from rat and ferret duodenum and ileum. In the rat duodenum, for example, the ED_{50} value for SP decreased from 2.9×10^{-7} *M* to 2.2×10^{-8} *M* in the presence of inhibitor. Similarly, phosphoramidon potentiated, in a concentration-dependent manner, the contraction of the rat duodenum and ileum produced by 10^{-7} *M* SP. In this system, inhibitors of peptidyl dipeptidase A and aminopeptidases did not modulate the action of the peptide.

II. Enkephalins

Methionine and leucine enkephalin represent good substrates for endopeptidase 24.11 (MATSAS et al. 1983; BUNNETT et al. 1990), for peptidyl dipeptidase A (DEFENDINÍ et al. 1982) and for membrane-associated aminopeptidases in brain tissue (De la Baume et al. 1983; Chaillet et al. 1983; Giros et al. 1985). β-Endorphin is a substrate for purified endopeptidase 24.11 and the sites of cleavage are shown in Table 1 (Graf et al. 1985). Hydrolysis of the Gly^3-Phe^4 bond (by endopeptidase 24.11 and peptidyl dipeptidase A) or the Tyr^1-Gly^2 bond (by aminopeptidases) will result in inactivation of the enkephalins. Several studies designed to study the relative importance of the three enzymatic activities for enkephalin degradation in the GI tract have concluded that aminopeptidases probably play the most important role.

The metabolism of Leu-enkephalin in the stomach wall of rats was studied in vivo by delivering the peptide via an infusion catheter directly into the gastric submucosa of conscious animals and in vitro by incubating the peptide with membranes prepared from either gastric muscle or mucosa (Bunnett et al. 1990). In each system, Leu-enkephalin was rapidly inactivated by cleavage of the Tyr^1-Gly^2 bond. Degradation was partially inhibited by amastatin and bestatin but phosphoramidon and captopril were without effect. The analog [D-Ala^2]Leu-enkephalin was resistant to the action of aminopeptidases and it was suggested that a gastric carboxypeptidase was important in the metabolism of this peptide. The importance of exopeptidases in the inactivation of enkephalin in the muscle layer of the intestine was demonstrated by Hazato et al. (1985). Enzymes that degraded Leu-enkephalin were partially purified from an homogenate of longitudinal muscle from bovine small intestine and identified as bestatin-sensitive and bestatin-insensitive aminopeptidases, a dipeptidyl aminopeptidase, a dipeptidyl carboxypeptidase and a carboxypeptidase. The activities of all these enzymes were inhibited by metal chelators and the microbial peptidase inhibitor arphamenine A. The relative unimportance of endopeptidase 24.11 and peptidyl dipeptidase A as regulators of the biological activity of enkephalins in the gut was demonstrated by Geary et al. (1982). Electrically stimulated contractions of the guinea pig ileum are inhibited by enkephalins and proteolysis is believed to be the mechanism by which action of the peptides is terminated. Neither thiorphan ($10^{-7}\,M$–$10^{-4}\,M$) nor captopril ($3 \times 10^{-5}\,M$) modified the abilities of Met- and Leu-enkephalin to inhibit field-stimulated contraction despite the fact that these inhibitors reduced enkephalin degradation in broken cell preparations of guinea pig intestinal tissues. In a similar study by Aoki et al. (1984), bestatin potentiated the ability of the enkephalin to relax the guinea pig ileum, confirming the hypothesis that aminopeptidases play a physiological role in terminating the actions of the enkephalin in the gut wall. Evidence that aminopeptidases play an important role in modulating the action of enkephalins in the central nervous system is provided by the observation that intracerebroventricularly administered bestatin potentiates the antinociceptive action of Met-enkephalin in the mouse hot-plate test (Chaillet et al. 1983). Met- and Leu-enkephalin were efficiently metabolized by membrane-bound aminopeptidase N derived from pig and rabbit cerebral microvessels (Churchill et al. 1987). Other small opioid peptides such as Met-enkephalin-Arg^6-Phe^7 were also hydrolyzed but β-endorphin was resistant to hydrolysis. Blood vessels in the mucous and submucous layers of the rat and guinea pig intestines are innervated by enkephalin-containing fibers (SCHULTZBERG et al. 1980) and so vascular aminopeptidase N may be important in the inactivation of opioid peptides released by enteric neurons into blood.

III. Vasoactive Intestinal Polypeptide

High-affinity receptors for vasoactive intestinal peptide (VIP) are present on the basolateral membrane of intestinal epithelial cells of all mammalian species studied and binding of the ligand results in rapid internalization. Incubation of VIP with dispersed pig enterocytes at 37°C for times as short as 30s resulted in the formation of [des-His^1]VIP by the action of amastatin- and bestatin-sensitive aminopeptidase(s) (NAU et al. 1987). As [des-His^1] VIP has only 1% of the bioactivity of the intact peptide, formation of this metabolite will effectively terminate the action of VIP in the epithelial layer of this intestine. In this system, VIP was also a substrate for cell-surface endopeptidase(s) in reactions that were inhibited by phosphoramidon. VIP was rapidly hydrolyzed by human recombinant endopeptidase 24.11 (GOETZL et al. 1989). As expected, cleavages at sites N-terminal to hydrophobic amino acids were observed but bonds near the N and C termini were hydrolyzed most rapidly. The relative rates of hydrolysis were Ala^4 = Val^5 > Tyr^{22} = Ile^{26} ≫ Leu^{13} = Met^{17} (Table 1). VIP was also a good substrate for purified endopeptidase 24.11 from porcine fundic muscle and the pattern of metabolites produced was similar to that produced by the human recombinant enzyme (NAU and CONLON, unpublished data). A physiological role for endopeptidase 24.11 in modulating the action of VIP in the gut has not yet been shown but phosphoramidon potentiated the VIP-induced relaxation of guinea pig trachea containing an intact epithelium (RHODEN and BARNES 1989).

E. Conclusion

The data summarized in this review support the general hypothesis of TURNER et al. (1985) that neurohormonal peptides of the GI tract are not inactivated by substrate-specific peptidases such as enkephalinase or somatostatinase, but rather by relatively few, well-defined proteolytic enzymes with a widespread distribution in the plasma membrane and synaptic membrane. A crucial role for endopeptidase 24.11 in the degradation of peptides in the epithelial layer of the gut has been established. Evidence suggests that this enzyme is also responsible for terminating the action of tachykinins released from enteric neurons as previously shown in the brain. Endopeptidase 24.11 activity probably contributes to the molecular heterogeneity of gastrin- and neurotensin-related peptides in the venous outflow of the stomach and gut but an involvement of other, less well characterized endopeptidases is probable. The presence of an N-terminal pyroglutamyl residue in gastrin-17 and NT and a penultimate proline residue in SP appears to protect these peptides from aminopeptidase activity. However, aminopeptidase N in both the mucosal and muscle layers of the gut and in the enteric vascular bed is important in the primary inactivation of those peptides requiring an intact N-terminus for biological activity, e.g., enkephalins and VIP. Despite a high concentration in the GI tract, peptidyl dipeptidase A (angiotensin-converting enzyme) may play only a secondary role in the inactivation pathway by hydrolyzing already inactive metabolites. The potent and nontoxic inhibitors of endopeptidase 24.11 and aminopeptidase N that are now available may become increasingly important in the pharmacological regulation of gastrointestinal function.

References

Acker GR, Molineaux CJ, Orlowski M (1987) Synaptosomal membrane-bound form of endopeptidase 24.15 generates Leu-enkephalin from dynorphin$^{1-8}$, α- and β-neoendorphin, and Met-enkephalin from Met-enkephalin-Arg^6-Gly^7-Leu^8. J Neurochem 48:284–292

Allen M, McMartin C, Peters GE, Wade R (1984) The mechanism of degradation of cyclo(-Asn-Phe-Phe-D-Trp-Lys-Thr-Phe-Gaba-) and the relative stabilities of this and other octapeptide somatostatin analogues in rat intestinal juice. Regul Pept 10:29–35

Aoki K, Kajiwara M, Oka T (1984) The role of bestatin-sensitive aminopeptidase, angiotensin-converting enzyme and thiorphan sensitive "enkephalinase" in the potency of enkephalins in the guinea pig ileum. Jpn J Pharmacol 36:59–65

Barber DL, Buchan AMJ, Walsh JH, Soll AH (1986) Isolated canine ileal mucosal in short-term culture: a model for study of neurotensin release. Am J Physiol 250:G374–G384

Barclay RK, Phillipps MA (1980) Inhibition of enkephalin-degrading aminopeptidase activity by certain peptides. Biochem Biophys Res Commun 96:1732–1738

Barelli H, Vincent JP, Checler F (1988) Peripheral inactivation of neurotensin. Isolation and characterization of a metallopeptidase from rat ileum. Eur J Biochem 175:481–489

Barelli H, Ahmad S, Kostka P, Fox JET, Daniel EE, Vincent JP, Checler F (1989) Neuropeptide-hydrolysing activities in synaptosomal fractions from dog ileum myenteric, deep muscular and submucous plexi. Their participation in neurotensin inactivation. Peptides 10:1055–1061

Benajiba A, Maroux S (1980) Purification and characterization of an aminopeptidase A from hog intestinal brush border membranes. Eur J Biochem 107:381–388

Bunnett NW, Mogard M, Orloff MS, Corbet HJ, Reeve JR, Walsh JH (1984) Catabolism of neurotensin in interstitial fluid of the rat stomach. Am J Physiol 246:G675–G682

Bunnett NW, Debas HT, Turner AJ, Kobayashi R, Walsh JH (1988a) Metabolism of gastrin and cholecystokinin by endopeptidase 24.11 from the pig stomach. Am J Physiol 255:G676–G678

Bunnett NW, Turner AJ, Hryszko J, Kobayashi R, Walsh JH (1988b) Isolation of endopeptidase-24.11 (EC 3.4.24.11, "enkephalinase") from the pig stomach. Gastroenterology 95:952–957

Bunnett NW, Walsh JH, Debas HT (1990) Metabolism of enkephalin in the stomach wall of rats. Am J Physiol 258:G143–G151
Bunning P, Riordan JF (1983) Activation of angiotensin converting enzyme by monovalent anions. Biochemistry 22:110–116
Bunning P, Holmquist B, Riordan JF (1983) Substrate specificity and kinetic characteristics of angiotensin converting enzyme. Biochemistry 22:103–110
Chaillet P, Marcais-Collado H, Costentin J, Yi CC, de la Baume S, Schwartz JC (1983) Inhibition of enkephalin metabolism by, and antinociceptive activity of, bestatin, an aminopeptidase inhibitor. Eur J Pharmacol 86:329–336
Checler F, Vincent JP, Kitabgi P (1986) Purification and characterization of a novel neurotensin-degrading peptidase from rat brain synaptic membranes. J Biol Chem 261:11274–11281
Checler F, Ahmad S, Kostka P, Barelli H, Kitabgi P, Fox JET, Kwan CY, Daniel EE, Vincent JP (1987a) Peptidases in dog-ileum circular and longitudinal smooth muscle plasma membranes. Eur J Biochem 166:461–468
Checler F, Barelli H, Kwan CY, Kitabgi P, Vincent JP (1987b) Neurotensin-metabolizing peptidases in rat fundus plasma membranes. J Neurochem 49: 507–512
Checler F, Kostolanska B, Fox JA (1988) In vivo inactivation of neurotensin in dog ileum: major involvement of endopeptidase 24.11. J Pharmacol Exp Ther 244:1040–1043
Chipkin RE, Berger JG, Billard W, Iorio IC, Chapman R, Barnett A (1988) Pharmacology of SCH 34826, an orally active enkephalinase inhibitor analgesic. J Pharmacol Exp Ther 245:829–838
Chu TG, Orlowski M (1984) Active-site directed *N*-carboxymethyl peptide inhibitors of a soluble metalloendopeptidase from rat brain. Biochemistry 23:3598–3603
Churchill L, Bausback HH, Gerritsen ME, Ward PE (1987) Metabolism of opioid peptides by cerebral microvascular aminopeptidase M. Biochim Biophys Acta 923:35–41
Conlon JM, Sheehan L (1983) Conversion of substance P to C-terminal fragments in human plasma. Regul Pept 7:335–345
Danielsen EM, Vyas JP, Kenny AJ (1980) A neutral endopeptidase in the microvillar membrane of pig intestine. Biochem J 191:645–648
Defendini R, Zimmerman EA, Weare JA, Alhenc-Gelas F, Erdos EG (1982) Hydrolysis of enkephalins by human converting enzyme and localization of the enzyme in neuronal components of the brain. In: Costa E, Trabucchi M (eds) Regulatory peptides: from molecular biology to function. Raven, New York, pp 271–280
De la Baume S, Yi CC, Schwartz JC, Chaillet P, Marcais-Collado H, Costentin J (1983) Participation of both "enkephalinase" and aminopeptidase activities in the metabolism of endogenous enkephalins. Neuroscience 8:143–151
Deschodt-Lanckman M, Pauwels S, Najdovski T, Dimaline R, Dockray GJ (1988) In vitro and in vivo degradation of human gastrin by endopeptidase 24.11. Gastroenterology 94:712–21
Devault A, Lazure C, Nault C, Le Moual H, Seidah NG, Chretein M, Kahn P, Powell J, Mallet J, Beaumont A, Roques BP, Crine P, Boileau G (1987) Amino acid sequence of rabbit kidney neutral endopeptidase 24.11 (enkephalinase) deduced from a complementary DNA. EMBO J 6:1317–1322
Devault A, Nault C, Zollinger M, Fournie-Zaluski MC, Roques B, Crine P, Boileau G (1988a) Expression of neutral endopeptidase (enkephalinase) in heterologous COS-1 cells. J Biol Chem 263:4033–4040
Devault A, Sales V, Nault G, Beaumont A, Roques B, Crine P, Boileau G (1988b) Exploration of the catalytic site of endopeptidase 24.11 by site-directed mutagenesis. Histidine residues 583 and 587 are essential for catalysis. FEBS Lett 231:54–58

Djokic TD, Sekizawa K, Borson DB, Nadel JA (1989) Neutral endopeptidase inhibitors potentiate substance P-induced contraction in gut smooth muscle. Am J Physiol 256:G39–G43

Dockray GJ, Gregory RA, Tracy HJ, Zhu WY (1982) Postsecretory processing of heptadecapeptide gastrin: conversion to C-terminal immunoreactive fragments in the circulation of the dog. Gastroenterology 83:224–232

Erdos EG (1987) The angiotensin I-converting enzyme. Lab Invest 56:345–348

Erdos EG, Skidgel RA (1989) Neutral endopeptidase 24.11 (enkephalinase) and related regulators of peptide hormones. FASEB J 3:145–151

Ferris CF, Carraway RE, Hammer RA, Leeman SE (1985) Release and degradation of neurotensin during perfusion of rat small intestine with lipid. Regul Pept 12:101–111

Fulcher IS, Matsas R, Turner AJ, Kenny AJ (1982) Effect of inhibitors of kidney neutral endopeptidase and enkephalin hydrolysis by synaptic membranes. Biochem J 203:519–522

Fulcher IS, Chaplin MF, Kenny AJ (1983) Endopeptidase-24.11 purified from pig intestine is differently glycosylated from that in kidney. Biochem J 215:317–323

Geary LE, Wiley KS, Scott WL, Cohen ML (1982) Degradation of exogenous enkephalin in the guinea-pig ileum: relative importance of aminopeptidase, enkephalinase and angiotensin converting enzyme activity. J Pharmacol Exp Ther 221:104–111

Gee NS, Kenny AJ (1987) Proteins of the kidney microvillar membrane. Enzymic and molecular properties of aminopeptidase W. Biochem J 246:97–102

Gee NS, Matsas R, Kenny AJ (1983) A monoclonal antibody to kidney endopeptidase-24.11. Biochem J 214:377–386

Giros B, Gros C, Solhonne B, Schwartz B (1985) Characterization of aminopeptidases responsible for inactivating endogenous [Met^5] enkephalin in brain slices using peptidase inhibitors and anti-aminopeptidase M antibodies. Mol Pharmacol 29:281–287

Goetzel EJ, Sreedharan SP, Turck CW, Bridenbaugh R, Malfroy B (1989) Preferential cleavage of amino- and carboxyl-terminal oligopeptides from vasoactive intestinal polypeptide by human recombinant enkephalinase (neutral endopeptidase, EC 3.4.24.11). Biochem Biophys Res Commun 158:850–854

Gossrau R (1979) Peptidasen II. Zur Lokalisation der Dipeptidylpeptidase IV (DPP IV). Histochemische und biochemische Untersuchung. Histochemistry 60: 231–248

Graf L, Paldi A, Patthy A (1985) Action of neutral metalloendopeptidase ("enkephalinase") on β-endorphin. Neuropeptides 6:13–19

Gray GM, Santiago NA (1977) Intestinal surface amino-oligopeptidases. I. Isolation of two weight isomers and their subunits from rat brush border. J Biol Chem 252:4922–4928

Hammer RA, Carraway RE, Leeman SE (1982) Elevation of plasma neurotensin-like immunoreactivity after a meal. J Clin Invest 70:74–81

Harbeck HT, Mentlein R (1991) Aminopeptidase P from rat brain. Purification and action on bioactive peptides. Eur J Biochem 198:451–458

Hazato T, Shimamura M, Kase R, Iijima M, Katayama T (1985) Separation of enkephalin-degrading enzymes from longitudinal muscle layer of bovine small intestine. Biochem Pharmacol 34:3179–3183

Hooper NM, Turner AJ (1985) Neurokinin B is hydrolysed by synaptic membranes and by endopeptidase-24.11 ("enkephalinase") but not by angiotensin converting enzyme. FEBS Lett 190:133–136

Hooper NM, Turner AJ (1987) Isolation of two differentially glycosylated forms of peptidyl-dipeptidase A (angiotensin converting enzyme) from pig brain: a re-evaluation of their role in neuropeptide metabolism. Biochem J 241: 625–633

Hooper NM, Turner AJ (1988) Ectoenzymes of the kidney microvillar membrane. Aminopeptidase P is anchored by a glycosyl-phosphatidylinositol moiety. FEBS Lett 229:340–344

Hooper NM, Kenny AJ, Turner AJ (1985) The metabolism of neuropeptides. Neurokinin A (substance K) is a substrate for endopeptidase-24.11 but not for peptidyl dipeptidase A (angiotensin-converting enzyme). Biochem J 231: 357–361

Hooper NM, Hyrszko J, Turner AJ (1990) Purification and characterization of pig kidney aminopeptidase P – a glycosyl-phosphatidylinositol-anchored ectoenzyme. Biochem J 267:509–515

Jensen RT, Lemp GF, Gardiner JD (1980) Interaction of cholecystokinin with specific membrane receptors on pancreatic acinar cells. Proc Natl Acad Sci USA 77:2079–2083

Kato H, Suzuki T (1971) Bradykinin-potentiating peptides from the venom of *Agkistrodon halys blomhoffi*. Isolation of five bradykinin potentiators and the amino acid sequence of two of them, potentiators B and C. Biochemistry 10:972–980

Kato T, Nagatsu T, Fukasawa K, Harada M, Nagatsu I, Sakakibara S (1978) Successive cleavage of N-terminal Arg^1-Pro^2 and Lys^3-Pro^4 from substance P but no release of Arg^1-Pro^2 from bradykinin by X-Pro dipeptidyl aminopeptidase. Biochim Biophys Acta 525:417–422

Kenny AJ, Booth AG, George SG, Ingram J, Kershaw D, Wood EJ, Young AR (1976) Dipeptidyl peptidase IV, a kidney brush-border serine peptidase. Biochem J 155:169–182

Kohama Y, Matsumoto S, Oka H, Teramoto T, Okabe M, Mimura T (1988) Isolation of angiotensin-converting enzyme inhibitor from tuna muscle. Biochem Biophys Res Commun 155:332–337

Lasch J, Koelsch R, Steinmetzer T, Neumann U, Demuth HU (1988) Enzymic properties of intestinal aminopeptidase P: a new continuous assay. FEBS Lett 227:171–174

Malfroy B, Schofeld PR, Kuang WJ, Seeburg PH, Mason AJ, Henzel WJ (1987) Molecular cloning and amino acid sequence of rat enkephalinase. Biochem Biophys Res Commun 144:59–66

Malfroy B, Kuang WJ, Seeburg PH, Mason AJ, Schofield PR (1988) Molecular cloning and amino acid sequence of human enkephalinase (neutral endopeptidase). FEBS Lett 229:206–210

Matsas R, Fulcher IS, Kenny AJ, Turner AJ (1983) Substance P and [Leu^5]-enkephalin are hydrolysed by an enzyme in pig caudate synaptic membranes that is identical with the endopeptidase of kidney microvilli. Proc Natl Acad Sci USA 80:3111–3115

Matsas R, Kenny AJ, Turner AJ (1984a) The metabolism of neuropeptides. The hydrolysis of peptides, including enkephalins, tachykinins and their analogues, by endopeptidase-24.11. Biochem J 223:433–440

Matsas R, Turner AJ, Kenny AJ (1984b) Endopeptidase-24.11 and aminopeptidase activity in brain synaptic membranes are jointly responsible for the hydrolysis of cholecystokinin octapeptide (CCK-8). FEBS Lett 175:124–128

Matsas R, Stephenson SL, Hryszko J, Kenny AJ, Turner AJ (1985) The metabolism of neuropeptides. Phase separation of synaptic membrane preparations with Triton X-114 reveals the presence of aminopeptidase N. Biochem J 231:445–449

McMartin C, Purdon G (1978) Early fate of somatostatin in the circulation of the rat after intravenous injection. J Endocrinol 77:67–74

Molineaux CJ, Lasdun A, Michaud C, Orlowski M (1988) Endopeptidase-24.15 is the primary enzyme that degrades luteinizing hormone releasing hormone both in vitro and in vivo. J Neurochem 51:624–633

Najdovski T, Collette N, Deschodt-Lankman M (1985) Hydrolysis of the C-terminal octapeptide of cholecystokinin by rat kidney membranes: characterization of the cleavage by solubilized endopeptidase-24.11. Life Sci 37:827–834

Nau R, Schafer G, Conlon JM (1985) Proteolytic inactivation of substance P in the epithelial layer of the intestine. Biochem Pharmacol 34:4019–4023

Nau R, Schafer G, Deacon CF, Cole T, Agoston DV, Conlon JM (1986) Proteolytic inactivation of substance P and neurokinin A in the longitudinal muscle layer of guinea pig small intestine. J Neurochem 47:856–864

Nau R, Ballmann M, Conlon JM (1987) Binding of vasoactive intestinal polypeptide to dispersed enterocytes results in rapid removal of the NH_2-terminal histidyl residue. Mol Cell Endocrinol 52:97–103

Orloff MS, Turner AJ, Bunnett NW (1986) Catabolism of substance P and neurotensin in the rat stomach wall is susceptible to inhibitors of angiotensin converting enzyme. Regul Pept 14:21–31

Orlowski M, Michaud C, Chu T (1983) A soluble metallopeptidase from rat brain. Purification of the enzyme and determination of specificity with synthetic and natural peptides. Eur J Biochem 135:81–88

Orlowski M, Michaud C, Molineaux CJ (1988) Substrate-related potent inhibitors of brain metalloendopeptidase. Biochemistry 27:597–602

Palmieri FE, Ward PE (1983) Mesentery vascular metabolism of substance P. Biochim Biophys Acta 755:522–525

Palmieri FE, Petrelli JJ, Ward PE (1985) Vascular, plasma membrane aminopeptidase M. Metabolism of vasoactive peptides. Biochem Pharmacol 34:2309–2317

Peters GE, McMartin C (1982) The breakdown of somatostatin in rat intestinal juice. Scand J Gastroenterol 18 Suppl 82:215–217

Power DM, Bunnett N, Dimaline R (1986) Chromatographic and immunochemical studies on postsecretory processing of gastrin in the pig. Am J Physiol 251: G300–G307

Power DM, Bunnett N, Turner AJ, Dimaline R (1987) Degradation of endogenous heptadecapeptide gastrin by endopeptidase 24.11 in the pig. Am J Physiol 253:G33–G39

Power DM, Dimaline R, Balaspiri L, Dockray GJ (1988) A novel gastrin-processing pathway in mammalian antrum. Biochim Biophys Acta 954:141–147

Praissman M, Fara JW, Praissman LA, Berkowitz JM (1982) Preparation of an *N*-acetyl-octapeptide of cholecystokinin. The role of N-acetylation in protecting the octapeptide from degradation by smooth muscle tissues. Biochim Biophys Acta 716:240–248

Rao RK (1991) Biologically active peptides in the gastrointestinal lumen. Life Sci 48:1685–1704

Rhoden KJ, Barnes PJ (1989) Epithelial modulation of non-adrenergic, non-cholinergic and vasoactive intestinal peptide-induced responses: role of neutral endopeptidase. Eur J Pharmacol 171:247–250

Roques BP, Beaumont A (1990) Neutral endopeptidase-24.11 inhibitors: from analgesics to antihypertensives. Trends Pharmacol Sci 11:245–249

Sakurada C, Yokosawa H, Ishii SI (1990) The degradation of somatostatin by synaptic membrane of rat hippocampus is initiated by endopeptidase-24.11. Peptides 11:287–292

Schafer G, Nau R, Cole T, Conlon JM (1986a) Specific binding and proteolytic inactivation of bradykinin by membrane vesicles from pig intestinal smooth muscle. Biochem Pharmacol 35:3719–3725

Schafer G, Richter G, Conlon JM (1986b) Conversion of somatostatin-28 to somatostatin-14 during maturation of epithelial cells in the porcine jejunum. Biochim Biphys Acta 885:240–247

Shulkes A, Fletcher DR, Hardy KJ (1983) Organ and plasma metabolism of neurotensin in sheep. Am J Physiol 245:E457–E462

Schultzberg M, Hokfelt T, Nilsson G, Terenius L, Rehfeld JF, Brown M, Elde R, Goldstein M, Said S (1980) Distribution of peptide- and catecholamine-containing neurons in the gastro-intestinal tract of rat and guinea-pig: immunohistochemical studies with antisera to substance P, vasoactive intestinal polypeptide, enkephalins, somatostatin, gastrin/cholecystokinin, neurotensin and dopamine β-hydroxylase. Neuroscience 5:689–744

Shaw C, Goke R, Bunnett NW, Conlon JM (1987) Catabolism of neurotensin in the epithelial layer of porcine small intestine. Biochim Biophys Acta 924:167–174

Skidgel RA, Engelbrecht S, Johnson AR, Erdos EG (1984) Hydrolysis of substance P and neurotensin by converting enzyme and neutral endopeptidase. Peptides 5:769–776

Soubrier F, Alhenc-Gelas F, Hubert C, Allegrini J, John M, Tregear G, Corvol P (1988) Two putative active centers in human angiotensin I-converting enzyme revealed by molecular cloning. Proc Natl Acad Sci USA 85:9386–9390

Strittmatter SM, Thiele EA, Kapiloff MS, Snyder SH (1985) A rat brain isozyme of angiotensin-converting enzyme. Unique specificity for amidated peptide substrates. J Biol Chem 260:9825–9832

Terashima H, Rossen AP, Bunnett NW (1991) Purification and characterization of aminopeptidase M from intestinal muscle and mucosa. Gastroenterology 100: A670

Thiele EA, Strittmatter SM, Snyder SH (1985) Substance K and substance P as possible endogenous substrates of angiotensin converting enzyme in the brain. Biochem Biophys Res Commun 128:317–324

Turkelson CM, Solomon TE, Hamilton J (1990) A cholecystokinin-metabolizing enzyme in rat intestine. Peptides 11:213–219

Turner AJ, Matsas R, Kenny AJ (1985) Are there neuropeptide-specific peptidases? Biochem Pharmacol 34:1347–1356

Turzynski, Mentlein R (1990) Prolyl aminopeptidase from rat brain and kidney. Action on peptides and identification as leucyl aminopeptidase. Eur J Biochem 190:509–515

Umezawa H, Aoyagi T, Suda H, Hamada M, Takeuchi T (1976) Bestatin, an inhibitor of aminopeptidase B, produced by actinomycetes. J Antibiot (Tokyo) 29:97–99

Ward PE, Sheridan MA, Hammon KJ, Erdos EG (1980) Angiotensin I converting enzyme (kininase II) of the brush border of the human and swine intestine. Biochem Pharmacol 29:1525–1529

Weber M, Cole T, Conlon JM (1986) Specific binding and degradation of somatostatin by membrane vesicles from pig gut. Am J Physiol 250:G679–G685

Wei L, Alhenc-Gelas F, Soubrier F, Michaud A, Corvol P, Clauser E (1991) Expression and characterization of recombinant human angiotensin I-converting enzyme. J Biol Chem 266:5540–5546

Wyvratt MJ, Tischler MH, Ikeler TJ, Springer JP, Tristam EW, Patchett AA (1983) Bicyclic inhibitors of angiotensin-converting enzyme. In: Hruby VJ, Rich DH (eds) Peptides: structure and function. Pierce Chemical, New York, pp 551–554

Yokosawa H, Ogura Y, Ishii SI (1983) Purification and inhibition by neuropeptides of angiotensin-converting enzyme from rat brain. J Neurochem 41:403–410

CHAPTER 7

Peptidergic Regulation of Gastric Acid Secretion

C.S. Chew

A. Introduction

Hydrochloric acid (HCl) secretion by the gastric parietal cell is controlled by a complex interplay of neural, paracrine and endocrine pathways with peptides in the central nervous system (CNS) and the gastrointestinal (GI) tract playing an important role in all of these processes. The classic approach to defining and characterizing secretory control mechanisms separates events controlling secretion into cephalic, gastric and intestinal phases. Within each phase excitatory and inhibitory pathways are considered. In the intact organism these phases overlap, greatly increasing the complexity of the process. The challenge to investigators past and present has been to devise methods that allow accurate quantitation and characterization of these events. No single experimental model can accomplish these goals and it is only through integration of results obtained from the various in vivo and in vitro models in different species that we can hope to understand all of the mechanisms involved.

A number of different models are presently used to study hierarchies of secretory control and each has its limitations. In vivo models allow assessment of certain hormonal and extrinsic neural influences. Classic in vivo experiments rely on injection or infusion of agonists and antagonists and/or ablation of extrinsic neural or hormonal influences followed by measurement of secretory products. In vivo infusion of specific monoclonal antibodies is also being used to define involvement of putative peptide hormones in control of HCl secretion. Depending on the in vivo model, acid secretory measurements may be complicated, for example, by the buffering effects of food, changes in blood flow, titration of acids other than HCl, gastric emptying and volume changes which may activate intragastric neurons. Moreover, infusion of agents suspected to have a paracrine/neurocrine action are unlikely to reproduce local effects and may produce general effects that are completely unrelated to the acid secretory response.

In vitro studies of the vascularly or luminally perfused intact stomach and chambered gastric mucosa are useful for the study of intrinsic neural and paracrine control mechanisms and avoid complications associated with extrinsic neural influences and unknown circulating factors. These models also allow more precise control of acid secretory measurements; however, such preparations have limited viability, and because they retain much of the neural/paracrine complexity present in the intact organism, it is not feasible to study intracellular activation events associated with a particular cell type. In contrast, isolated cell models, particularly those highly enriched in the cell of interest, are ideal for the study of intracellular control mechanisms as well as for identification and characterization of receptor subtypes. Like the more intact in vitro models, isolated cells also have limited viability (which can be significantly improved, however, by placing cells in short-term culture). Removal of most or all extrinsic influences can be an advantage but may also be a disadvantage because cellular responses may not accurately reflect in vivo activities.

In order to provide an overview in which to consider our present, somewhat limited knowledge of cellular mechanisms associated with peptide regulation of gastric acid secretion, some of the relevant data from both in vivo and in vitro experiments are considered in this chapter. The first section reviews progress toward defining interacting control mechanisms using in vivo and isolated stomach models. The second addresses cellular mechanisms associated with actions of peptides on specific cell types within the gastric mucosa including the parietal cell, which is located within gastric pits in the fundic (oxyntic) mucosa, gastrin-containing G cells in the gastric antrum, somatostatin-containing D cells that are located near G cells in the antrum and parietal cells in the fundus, and histamine-containing endocrine-like (ECL) cells which are found in the fundus near parietal cells. As much as possible, cellular mechanisms are considered within the framework of experimental findings from in vivo and more intact in vitro models. To guide the reader through the complex levels of interaction, Figs. 1 and 2 are provided as summaries of known and postulated interactions of various secretory factors with different cell types within the gastric antrum and fundus. Figure 3 summarizes current knowledge of intracellular activation and inhibitory mechanisms within the parietal cell.

B. Acid Secretory Control Mechanisms In Vivo

I. Extrinsic Innervation and Central Control

The stomach is innervated by both vagal parasympathetic and splanchnic sympathetic pathways (Goyal 1983); however, extrinsic neural control of gastric acid secretion appears to be mediated mainly by the vagus (Hirschowitz 1989). Vagal fibers innervating the stomach originate in the medulla from neurons of the vagal dorsal motor nucleus, the nucleus tractus solitarius and the nucleus ambiguus (Laughton and Powley 1987; Shapiro and Miselis 1985; Taché 1987). Within the stomach vagal fibers terminate on neurons contained within interconnected intrinsic plexi. The majority of vagal fibers innervating the gastrointestinal tract are afferent with only

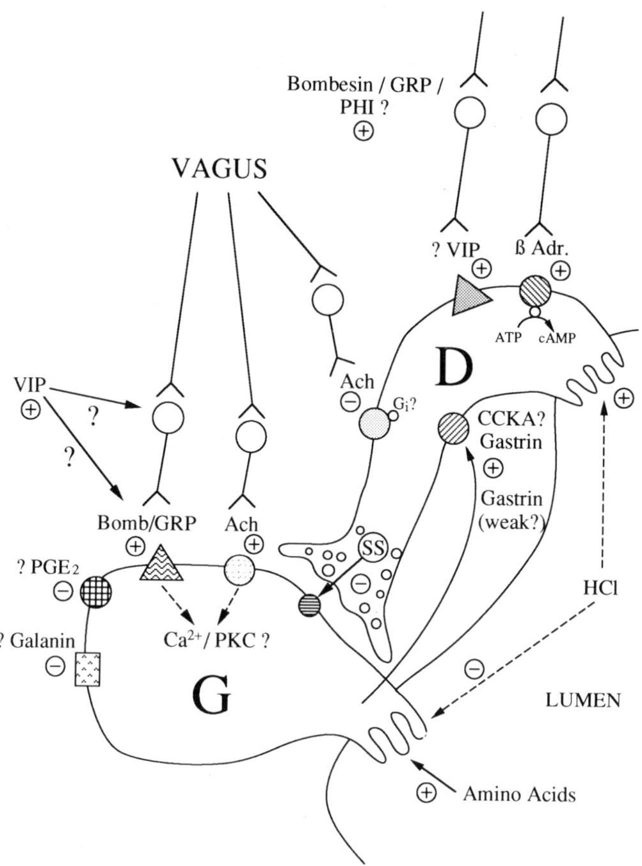

Fig. 1. Model of neural, hormonal and paracrine mechanisms controlling gastrin and SS secretion in the gastric antrum. Vagal stimulation activates stimulatory cholinergic (*Ach*) and noncholinergic pathways (β-adrenergic (*β-ADR*)] resulting in release of gastrin from G cells and cholinergic inhibition of somatostatin (*SS*) release from D cells. SS inhibits gastrin release via a paracrine pathway but may also act as a hormone upon release from D cells in the small intestine (not shown). Gastrin release from G cells by acetylcholine and bombesin (*Bomb*) may be mediated via calcium-dependent pathways. Gastrin may control its own release from G cells via a paracrine pathway involving SS release. CCK released from the small intestine may also act on the D-cell gastrin receptor to cause SS release. Cholinergic inhibition of D cells appears to involve activation of the inhibitory, pertussis toxin-sensitive G protein, G_i. The preganglionic vagal neurotransmitter is acetylcholine which may act on postganglionic M_1-type receptors (not shown). VIP could have an indirect stimulatory effect on G cells and a direct stimulatory effect on D cells. Several interneurons may be interposed between vagal preganglionic fibers and D and G cells. These cells may also be innervated by neuronal circuits arising in the fundus and intestine. Other factors that increase gastrin release include GABA and serotonin both of which also inhibit SS release. Substance P also inhibits SS release. In addition, SS release is stimulated by dopamine, secretin, GIP, GLP-1 and CGRP. The sites of action of these peptides are unknown. Since secretin is in the same family as VIP, it is possible that this peptide acts on VIP receptors (see Chap. 5)

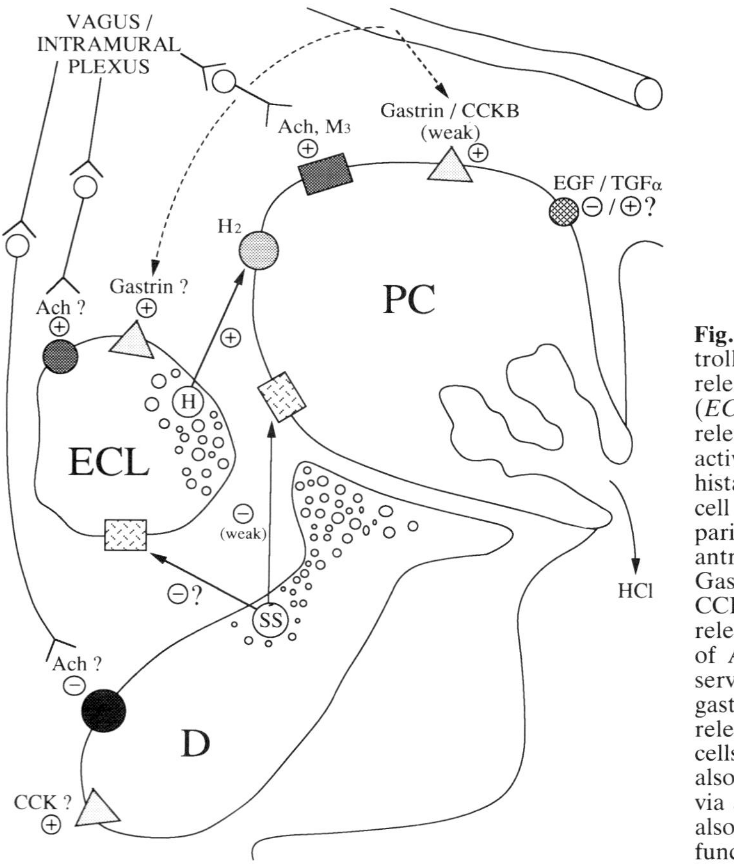

Fig. 2. Model of neural, hormonal and paracrine mechanisms controlling parietal cell HCl secretion. Stimulation of the vagus indirectly releases acetylcholine (*Ach*) near D cells, histamine-containing (*ECL*) cells and parietal cells (*PC*). Ach inhibits somatostatin (*SS*) release from D cells and stimulates parietal cell HCl secretion via activation of muscarinic M_3-type receptors. Ach also stimulates histamine (*H*) release from ECL cells. Histamine stimulates parietal cell HCl secretion via a paracrine pathway involving activation of parietal cell histamine H_2-type receptors. Gastrin released from antral G cells probably also causes histamine release from ECL cells. Gastrin weakly stimulates parietal cells via a direct activation of CCK-B type receptors. CCK (and gastrin?) may also stimulate SS release from D cells. Histamine potentiates the stimulatory effects of Ach and gastrin on parietal cells (not shown). SS release could serve as a negative feedback mechanism to prevent hypersecretion of gastric acid. Such a mechanism may involve inhibition of histamine release from ECL cells and a weak, direct inhibitory effect on parietal cells. TGFα is synthesized within parietal cells. EGF/TGFα may also be synthesized in other unidentified cells, thereby allowing action via autocrine, paracrine or hormonal mechanisms. EGF/TGFα might also serve to maintain certain aspects of differentiated parietal cell function. Additional parietal cell control mechanisms are shown in Fig. 3

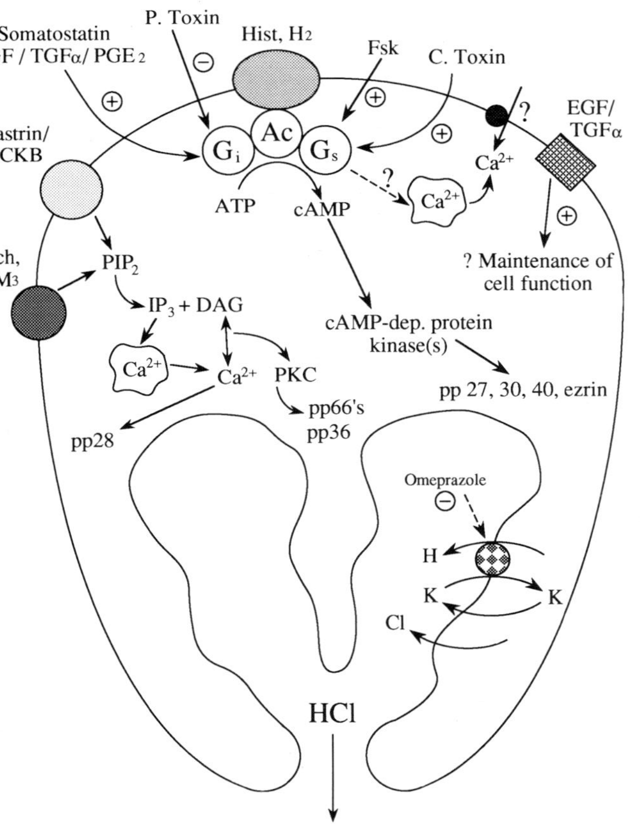

Fig. 3. Intracellular activation mechanisms in the parietal cell. Gastrin/CCK and acetylcholine (*Ach*) stimulate HCl secretion via calcium-dependent pathways. Both agonists cause influx of extracellular calcium by unidentified mechanisms (not shown) and release of calcium from intracellular stores. Elevations in intracellular Ca^{2+} concentration ($[Ca^{2+}]_i$) are mediated by inositol 1,4,5-trisphosphate (IP_3), which is generated by breakdown of phosphatidylinositol 4,5 bisphosphate (*PIP_2*). Diacylglycerol (*DAG*), which is also formed in this reaction, activates protein kinase C (*PKC*). Activation of protein kinase C may be enhanced by Ca^{2+}. Activated PKC phosphorylates proteins of 36 and 66 kDa. A 28-kDa protein is phosphorylated by an unknown kinase in response to increased $[Ca^{2+}]_i$. Histamine stimulates HCl secretion by H_2-histamine receptor-dependent activation of adenylyl cyclase to produce increased cellular cAMP content. Elevation of cAMP activates cAMP-dependent protein kinase(s). Several proteins (27, 30, 40, 80 kDa) which undergo cAMP-dependent increases in phosphorylation have been detected. The 80-kDa phosphoprotein, which is localized within the apical membrane in close proximity to the H^+, K^+-ATPase, has been tentatively identified as the cytoskeletal protein ezrin. Histamine-stimulated secretion is negatively regulated by prostaglandins (*PGE_2*), EGF/TGFα and SS by activation of $G_{i\alpha}$, a negative regulatory subunit of adenylyl cyclase. Pertussis toxin can overcome this inhibition by NAD-dependent ADP ribosylation of the α-subunit of G_i. The stimulatory subunit of adenylyl cyclase, $G_{s\alpha}$, can be activated by forskolin (*F_sk*) and cholera toxin (*C. Toxin*), which act by different mechanisms. Activation of $G_{s\alpha}$ results in elevation of cAMP and stimulation of HCl secretion. Histamine also increases $[Ca^{2+}]_i$ in a subpopulation of parietal cells via H_2-receptor activation. The function of this calcium-release mechanism is unknown but does not appear to be involved in control of HCl secretion. EGF/TGFα weakly inhibit secretion and may also serve to maintain differentiated parietal cell functions. HCl secretion is potently inhibited by H_2-histamine receptor blockers, suggesting a major role for histamine in stimulation of secretion. The H^+, K^+-ATPase or proton pump is inhibited directly by covalent omeprazole binding. Specificity of omeprazole is conferred by a requirement for an acidic environment for conversion of omeprazole into a cationic sulfonamide which reacts with the catalytic α-subunit of the H^+, K^+-ATPase

a small percentage of the abdominal vagal innervation being efferent (Hoffman and Schnitzlein 1961; Gwyn et al. 1985).

Central modulation of gastric secretion has been recognized since the early 1800s when the French physiologist, Pierre Jean Georges Cabanis, noted that emotional conditions modified gastrointestinal function. Approximately 100 years later, Pavlov defined the cephalic or "psychic phase" as the fraction of gastric acid secretion which is activated by the sight and smell of food. Application of electrophysiologic and retrograde-anterograde neuronal tracing techniques as well as central injections of brain peptides suggest that a number of centrally acting peptides are involved in the extrinsic neural modulation of gastric acid secretion. For example, microinjection of members of the gastrin/cholecystokinin (CCK) family or thyrotropin-releasing hormone (TRH)-related peptides into the lateral hypothalamus stimulates gastric acid secretion whereas microinjection of calcitonin, corticotropin-releasing factor (CRF) or calcitonin gene-related peptide (CGRP) inhibits secretion. Opioid peptides acting on central μ-type receptors also suppress gastric acid secretion. Bombesin, which is thus far the most potent central inhibitor of acid secretion, inhibits secretion when injected into medullary and spinal sites and the paraventricular nucleus but stimulates secretion when injected into the preoptic hypothalamus. Neuropeptide Y also has divergent actions depending on the injection site (Taché 1987, 1988; Taché and Yang 1990a,b; Hirschowitz 1989; Debas 1987). Tachykinins active on NK_3-type receptors have recently been added to the expanding list of central inhibitors (Improta and Broccardo 1990). As yet the physiologic relevance of most of these peptides has not been established; however, evidence from studies in rats supports a role for CRF in stress-induced inhibition of gastric acid secretion. Medullary TRH also appears to be emerging as a physiologically relevant activator of gastric vagal pathways (Taché 1987; Taché and Yang 1990a; Yanagisawa et al. 1990).

Current estimates, based on sham feeding protocols (experimental models designed to mimic the cephalic phase in which food is ingested but not allowed to reach the stomach), indicate the maximal acid secretory response to cephalic input is ~50%–60% of that attainable with exogenously administered pentagastrin (Stenquist 1979; Konturek et al. 1978a; Feldman and Richardson 1986). In the context of the normal acid secretory response to ingestion of a meal, the cephalic phase is estimated to contribute approximately one-third to one-half of the total (Richardson et al. 1977). This phase of gastric acid secretion appears to be mediated mainly by vagal release of acetylcholine at preganglionic nerve terminals within the stomach. Cholinergic receptors on postganglionic neurons within the stomach have been proposed to be of the muscarinic M_1 type because in vivo acid secretion is potently inhibited by the M_1-receptor blocker, pirenzepine (Tatsuta et al. 1981; Feldman 1984; Konturek et al. 1980a; Eklund et al. 1987; Yanagisawa et al. 1990). However, it should be emphasized that this hypothesis is based on in vivo studies in which it is not possible to distinguish directly the site(s) of pirenzepine action. Moreover, BLACK (1990) has pointed out that pirenzepine is much more water-soluble than atropine and plasma concentrations of atropine must be ~30× higher than pirenzepine to achieve the same degree of inhibition. Thus, the assumption that pirenzipine is exerting an exclusive effect on M_1-type receptors on postganglionic neurons may or may not be correct.

II. Intrinsic Innervation

The stomach, like most of the gastrointestinal tract, is extensively innervated by a vast collection of interconnected intrinsic autonomic ganglia and associated nerve fibers. The two major intramural plexi include the myenteric plexus, which is located between the outer longitudinal and inner circular muscle layers, and the submucosal plexus, which is located in the submucosa just above the mucosal layer. These plexuses serve an important role in the regulation of both secretion and motility; however, due to their complexity they are as yet only poorly characterized (see Chap. 1). Intramural plexuses receive input from both extrinsic parasympathetic and sympathetic pathways and contain at least 20 potential neurotransmitters, many of which are peptides that are also present in the central nervous system as well as in endocrine/paracrine-like cells within the GI tract.

Known or postulated acid secretory-related functions of relevant peptides are discussed in the following sections.

III. Cephalic and Gastric Phases of Secretion

Within the stomach the cascade of events associated with cephalic and gastric phase activation are complex and incompletely understood. In 1942 UVNÄS postulated that vagal stimulation activates parietal cells directly and further stimulates HCl secretion indirectly by release of antral gastrin. Since then this hypothesis has undergone many modifications. For example, it is now known that parietal cells are not directly innervated by the vagus but rather by neurons within the myenteric plexus which receive input from vagal efferents (RADKE et al. 1980). Moreover, although a number of studies have established that vagal stimulation releases gastrin (BONFILS et al. 1979), the degree of involvement of antral gastrin in the cephalic phase has been controversial since 1968 when the first reliable radiommunoassay for gastrin was established (MCGUIGAN and TRUDEAU 1968). Depending on the species, sham feeding reportedly causes either a small increase (NILSSON et al. 1972; TEPPERMAN et al. 1972; FELDMAN and WALSH 1980; FELDMAN et al. 1985) or no increase (STENQUIST et al. 1979; KONTUREK et al. 1978a) in plasma gastrin levels. In studies where the major sources of gastrin (the antrum and upper duodenum) are removed gastric acid secretion following sham feeding is reduced but not abolished (KNUTSON and OLBE 1974; PRESHAW 1970). Recently, antral vagotomy studies in rats (HIRSCHOWITZ and FONG 1990) and other studies in rats utilizing a neutralizing monoclonal antibody against gastrin (Yang et al. 1989) have provided strong evidence that gastrin does not play a major role in vagal or TRH-stimulated, vagally mediated secretion in this species. Whether this is true in other species remains to be determined.

As food enters the stomach, gastric phase acid secretion is increased not only by the presence of food and certain beverages (Ippoliti et al. 1976; McArthur et al. 1982; Lenz et al. 1983; Feldman and Richardson 1981; Peterson et al. 1986), but also by distention of the stomach wall. Distension activates long (vagovagal) and short (intramural) reflexes producing a moderate increase in gastric acid secretion. With low-grade distension there is little increase in serum gastrin levels (Strunz and Grossman 1978); however, stronger distension does increase gastrin release (Schubert and Shamburek 1990). Nutrients in the stomach, particularly specific amino acids such as tryptophan and phenylalanine in man (Taylor et al. 1982) and cysteine, tryptophan and phenylalanine in dogs (Strunz et al. 1978), produce a higher rate of secretion which may be up to 75% of that achieved with maximal pentagastrin stimulation (Richardson et al. 1977). This nutrient-activated phase of acid secretion is also referred to as the "chemical" phase of secretion (Walsh 1988; Hirschowitz 1989).

In contrast to the cephalic phase, the major secretory stimulus in the gastric phase appears to be gastrin which is released from antral G cells not only in response to vagal stimulation but also in response to food in the stomach (Walsh 1987, 1988). This conclusion has been reinforced by recent immunoneutralization studies in which a monoclonal antibody to the amidated carboxyl terminal of gastrin/CCK was shown to block the acid secretory response to peptone meals in chronic fistula dogs (Kovacs et al. 1989). Secretion initiated by gastric distension, which had been thought to be mediated exclusively via a gastrogastric cholinergic reflex (Debas et al. 1974; Grossman 1981), was also abolished by the gastrin antibody. A possible role for CCK was discounted since CCK has little or no stimulatory effect on in vivo HCl secretion and antagonizes gastrin-stimulated secretion in vivo in this species (Walsh 1987).

1. Gastrin Release Mechanisms

Mechanisms associated with neural and meal-stimulated gastrin release from antral G cells involve several different pathways including, but not limited to, the neurotransmitters, gastrin-releasing peptide (GRP)/bombesin and acetylcholine, the inhibitory neuroparacrine-humoral peptide somatostatin (SS), and stimulatory local factors including protein digestion products and luminal acidity (Schubert and Shamburek 1990; Debas 1987; Hirschowitz 1989; Walsh et al. 1988). The involvement of noncholinergic neurotransmitter(s) in gastrin release has been suspected for some years. For example, electrical stimulation of the vagus leads to atropine-resistant gastrin release

(STENQUIST 1979; FELDMAN et al. 1979b). In addition, local neural reflexes appear to have a noncholinergic component because gastrin release stimulated by luminal protein digestion products in isolated perfused rat stomachs is blocked by the axonal conduction blocker tetrodotoxin (TTX), but is only partially inhibited by atropine (SAFFOURI et al. 1984b).

There is now reasonably good evidence supporting a role for bombesin/GRP as a noncholinergic neurotransmitter in neurally mediated gastrin secretion. In several species nerve fibers containing bombesin-like immunoreactivity terminate near endocrine cells in the antrum (VIGNA et al. 1987) as well as in the acid-secreting oxyntic (fundic) gland mucosa (DOCKRAY et al. 1979; EKBLAD et al. 1985a; JAIN et al. 1985; HOLST et al. 1987a). Electrical stimulation of the vagus releases both GRP-like immunoreactivity and gastrin in isolated perfused pig antrum and neither response is blocked by atropine (HOLST et al. 1987b). In dogs and humans in vivo administration of bombesin increases gastrin release and stimulate HCl secretion (BASSO et al. 1974; VARNER et al. 1981; IMPICCIATORE et al. 1974; BUNNETT et al. 1985). Immunoneutralization with bombesin/GRP antibodies blocks neurally stimulated gastrin release in vascularly perfused rat and pig stomachs (SCHUBERT et al. 1985; HOLST et al. 1987b). Bombesin also stimulates gastrin secretion in the isolated rat stomach (DUVAL et al. 1981; MARTINDALE et al. 1982) but apparently does not elicit acid secretion (BERTACCINI et al. 1973). This species variation might be due to a more potent effect of bombesin on SS release in the fundic mucosa of rats. Gastrin secretion may be tonically suppressed by SS (SAFFOURI et al. 1979, 1980; CHIBA et al. 1981) and cholinergic stimulation of gastrin release may be mediated, at least in part, by a suppression of SS release (WOLFE et al. 1984).

Bombesin/GRP may also stimulate gastric acid secretion by mechanisms independent of gastrin release. For example, in vivo immunoneutralization of GRP in rats has been found to inhibit potently acid secretion in response to gastric distension under conditions where plasma gastrin levels are unchanged (MAILLIARD and WOLFE 1989). Under some experimental conditions, bombesin/GRP inhibits acid secretion in vivo (see Sect. B.V below).

2. Effects of Other Peptides

In general, the literature on the role of most gastric peptides in the stimulation of acid secretion is not well defined. There is increasing evidence suggesting involvement of several such peptides; however, the evidence for specific effects is often conflicting and species differences are evident. Morphine and methionine (Met)-enkephalin, which is detected by immunocytochemical methods in mucosal endocrine cells as well as myenteric plexus neurons and nerve fibers projecting to all layers of the stomach (SCHULTZBERG et al. 1980), reportedly stimulate gastric acid secretion under some conditions in dogs and humans (FELDMAN et al. 1980; KONTUREK et al. 1980b). Met-enkephalin releases bombesin-like immunoreactivity in the perfused rat stomach (Madaus et al. 1990). Met- and leucine (Leu)-enkephalin and substance P, which is present in low levels in the stomach with a neural distribution similar to Met-enkephalin (Schultzberg et al. 1980), have also been found to inhibit basal and stimulated release of somatostatin-like immunoreactivity or SLI (Chiba et al. 1980b; Kwok et al. 1985; McIntosh et al. 1983). In conscious fistula rats, however, Met-enkephalin infusion inhibits histamine but not pentagastrin-stimulated acid secretion (El Munshid et al. 1980). These different findings may reflect actions on different opioid receptor subtypes and/or conflicting central and peripheral responses.

With vasoactive intestinal peptide (VIP) there appears to be considerable species variation in responsiveness. Immunoreactive VIP nerve cell bodies are present mainly within the myenteric plexus of the stomach; however, VIP-containing fibers extend throughout all gastric layers (Schultzberg et al. 1980). In vivo infusion of VIP stimulates acid secretion in cats (Vagne et al. 1982), inhibits acid secretion in dogs (Makhlouf et al. 1978) and has no effect in humans (Holm-Bentzen et al. 1983). In the vascularly perfused rat stomach, VIP stimulates SS release (Chiba et al. 1980a; Saffouri et al. 1984a). Schubert (1991) recently demonstrated that VIP weakly and transiently increases acid secretion in the luminally perfused mouse stomach with the main effect being stimulation of a sustained release of SS. On the basis of these data it was proposed that VIP neurons may be at least partly responsible for tonic basal SS secretion in vivo. Additional experiments in other species are needed before this action of VIP can be generalized. It is probable that multiple intramural afferent/efferent mechanisms are involved and each may respond differently depending on prevailing conditions within the gastric lumen. For example, VIP and PHI (which is structurally similar to VIP and similarly distributed within specific species; Yiangou et al. 1985) stimulate release of bombesin-like immunoreactivity (BLI) in the luminally perfused rat stomach when luminal pH is maintained at 2 but SLI and not BLI is released at pH 7 (Schusdziarra et al. 1986).

Some peptide-containing neurons in the stomach may also release other mediators such as serotonin, dopamine or norepinephrine (Bech 1986a; Stevens et al. 1986). Both gamma-aminobutyric acid and serotonin can stimulate gastrin and inhibit SS release whereas dopamine stimulates SLI release in rats (Harty and Franklin 1983; Koop and Arnold 1984; Koop et al. 1983).

Pancreastatin-like immunoreactivity is present in endocrine cells in the gastric mucosa of pigs and appears to occur in relatively high levels in rat fundic mucosa (Schmidt et al. 1988; Johnston et al. 1987; Curry et al. 1990). Pancreastatin reportedly enhances meal-stimulated acid secretion in conscious dogs but apparently has no effect on basal secretion or that stimulated by cholinergic, histamine or pentagastrin mechanisms (Hashimoto et al. 1990). The site of action and physiologic relevance of this effect of pancreastatin is presently unknown.

IV. Intestinal Factors

Although distension and the presence of acid, fat and hyperosmolar solutions within the upper small intestine is known to inhibit gastric acid secretion, stimulatory components are also present. A stimulatory intestinal phase was documented as early as 1900 by LECONTE. Much like the gastro-intestinal phase, distension and infusion of protein digestion products into the intestine induce HCl secretion in the stomach (KONTUREK et al. 1978b). Extrinsic denervation does not block the stimulatory effect of food in the small intestine (GREGORY and IVY 1941) and crude intestinal extracts have been shown to stimulate acid secretion (ORLOFF et al. 1977). Furthermore, jejunoileal shunt operations in humans appear to reduce meal-stimulated acid secretion, particularly during late-phase secretion (HESSELFELDT et al. 1979).

The identity of the humoral factor(s) responsible for stimulation is controversial. Gastrin is present in intestinal mucosa in small amounts as is a putative stimulatory factor with sequence homology to rat liver fatty acid binding protein (WALZ et al. 1988), alternately referred to as intestinal phase hormone (ORLOFF et al. 1977), enterooxyntin (GROSSMAN 1974), porcine ileal peptide (WIDER et al. 1984) or gastrotropin (WALZ et al. 1988; DEFIZE et al. 1988; MOYER et al. 1988). GANTZ et al. (1989) recently isolated a full-length gastrotropin cDNA from a hog small intestinal cDNA library and tested biological activity of a gastrotropin sequence deduced from this cDNA as well as gastrotropin. Based on experiments with gastric fistula rats and isolated canine parietal cells, these authors concluded that neither type of gastrotropin stimulates acid secretion. LANDOR et al. (1980) have postulated that stimulatory effects of liver extract and amino acids in the intestine are mediated by increased levels of circulating amino acids. Thus the question of intestinal humoral regulation is not settled. The relative importance of this phase is also not clearly documented, with estimates ranging from 7% to 50% of maximal gastric acid secretion in response to liver extract (KAUFFMAN and GROSSMAN 1979; KONTUREK et al. 1978b).

V. Inhibitory Phases of Acid Secretion

1. Cephalic Influences

In parallel with the stimulatory phases of secretion, the inhibitory phases have classically been divided into cephalic, gastric and intestinal. GROSSMAN (1981) proposed that vagal (cephalic) excitation results in both stimulation and inhibition of gastric acid secretion because the maximal secretory response to physiological vagal stimulation is less than that which can be achieved with histamine or pentagastrin administration. This hypothesis is reinforced by the observation that immediately following vagotomy there is a marked rise in serum gastrin concentrations in the fed and fasted states (Feldman et al. 1979a; Feldman and Richardson 1981; Thompson et al. 1978; Hollinshead et al. 1985). Vagally induced release of an inhibitor is also supported by studies showing that gastrin-stimulated secretion is inhibited by sham feeding (Konturek 1989). SS has been suggested as a candidate inhibitor because it is released in response to sham feeding and other methods of physiologic vagal excitation (De Graef and Woussen-Colle 1985; Hollingshead et al. 1985; Ipp et al. 1982; Webb et al. 1984). However, other studies have not detected SS release during the cephalic phase (Schusdziarra and Schmid 1986).

2. Antral and Fundic Mechanisms

The antral mucosa plays a major role in modulation of gastric acid secretion. Woodward et al. (1954) originally observed that acidification of the antral mucosa reduces acid secretion. This effect has since been shown to be mainly the result of reduced gastrin release (Olbe 1964; Debas et al. 1974; Debas 1987; Walsh et al. 1976; Thompson and Swierczek 1977; Feldman and Walsh 1980). It is not clear whether decreased luminal pH directly affects G cell function or decreases antral gastrin release indirectly via either an increased release of SS from antral D cells or some other mechanism. The negative effects of acid on gastrin release have been a major consideration in the clinical treatment of peptic ulcer disease because both histamine H_2-receptor blockers and parietal cell H^+,K^+-ATPase inhibitors (see Sect. C.I) remove this negative influence resulting in elevation of serum gastrin levels. Prolonged elevation of serum gastrin in certain disease conditions such as Zollinger-Ellison syndrome stimulates growth of the gastric mucosa and may lead to precancerous conditions (Karnes and Walsh 1990).

Somatostatin, which was shown some years ago to inhibit acid secretion in vivo (Bloom et al. 1974), appears to be a major paracrine inhibitor as well as a possible endocrine inhibitor of acid secretion (Walsh 1987, 1988; Schubert and Shamburek 1990; Yamada and Chiba 1989). SS-containing D cells are present not only in the fundic and antral mucosa but also in other parts of the GI tract. D cells are closely associated with parietal cells in the fundus and G cells in the antrum and probably exert paracrine inhibitory influences in both regions of the stomach (Larsson et al. 1979). Somatostatin release from D cells in the fundus and antrum appears to be controlled by neural (cholinergic, noncholinergic, and adrenergic), paracrine and hormonal pathways, the relative importance of which may be species dependent (Schusdizarra and Schmidt 1986; Stevens et al. 1986). It is also probable that SS has multiple inhibitory sites of action within the gastric mucosa including, for example, the gastric parietal cell, G cells and histamine-containing ECL cells. In the gastric mucosa of dogs, rats and humans the tetradecapeptide SS (SS-14) is the predominant form and is considerably more potent than SS-28 in inhibiting acid secretion and gastrin release (KONTUREK et al. 1985; SEAL et al. 1982; FRANCIS et al. 1990). SS is an effective antagonist of pentagastrin- and carbachol-stimulated HCl secretion in vivo; however, in vivo effects on histamine-stimulated secretion are controversial (KONTUREK et al. 1976, 1985; ALBINUS et al. 1977; ROBEIN et al. 1979; EL MUNSHID et al. 1980; BECH 1986b). SS may tonically suppress HCl secretion in vivo and a functional linkage between gastrin and SS release has been proposed (SAFFOURI et al. 1980). For example, SS immunoneutralization increases basal gastrin release in perfused rat stomach (SAFFOURI et al. 1979) and elevates basal acid secretion in urethane-anesthetized rats (YANG et al. 1990) and in the isolated perfused rat stomach (SHORT et al. 1985; SEAL et al. 1988). Simultaneous infusion of SS antiserum along with GRP has also been found to increase gastrin release to a greater extent than does GRP infusion alone in the isolated, vascularly perfused rat stomach (SANDVIK et al. 1989).

Release of SS from fundic and antral D cells may be controlled by somewhat different mechanisms (OLSEN et al. 1987; SCHUBERT et al. 1991b); however, D cell activities in both parts of the stomach appear to be tonically suppressed by inhibitory cholinergic neurons. Cholinergic agonists decrease SS release in fundic and antral mucosal fragments. Pretreatment of isolated, luminally perfused mouse stomachs with pertussis toxin, an agent that prevents inhibitory effects of inhibitory GTP-binding (G_i) protein(s) on the enzyme adenylate cyclase by an adenosine-5′-diphosphate (ADP) ribosylation mechanism (UI 1990), blocks cholinergic inhibition and apparently unmasks a cholinergic stimulatory effect (SCHUBERT et al. 1989).

Under some experimental conditions, bombesin, which potently stimulates gastrin release, may also cause release of SS. HIRSCHOWITZ and MOLINA (1983) found that acid secretory rates in response to bombesin infusion in dogs were lower than those obtained by infusion of gastrin at concentrations similar to those released into plasma by bombesin infusion. Other studies suggest that SS may modulate bombesin inhibition of secretion. In addition to stimulating gastrin release, bombesin/GRP administration can increase SS release from the antrum and fundus in vivo (CHIBA et al. 1980b; SCHUSDZIARRA and SCHMIDT 1986), in vascularly perfused rat stomachs (DUVALL 1981; OLSEN et al. 1987; GUO et al. 1987), in isolated perfused, vagally intact pig antrum (HOLST et al. 1987b), in perifused rat antral and fundic mucosal fragments (SCHUBERT et al. 1988) and in perifused antral glands (RICHELSEN et al. 1983). Elimination of bombesin-induced gastrin release in the isolated, luminally perfused mouse stomach leads to bombesin-induced increases in SS secretion as well as inhibition of basal and histamine-stimulated acid secretion (SCHUBERT and HIGHTOWER 1989). Other studies have shown that in vivo administration of bombesin may also release other peptides (such as CCK from the duodenum; CANTOR et al. 1987). Therefore, effects of bombesin on intact tissue should be interpreted cautiously.

The presence of bombesin receptors on antral and/or fundic D cells has been the subject of some debate. In one study bombesin-binding sites were not detected on canine fundic mucosal cells although such sites were present in the antrum (VIGNA et al. 1987). In another study bombesin-binding sites were detected on fundic D cells as well as parietal cells (NAKAMURA et al. 1988). Based on data from isolated perfused rat stomachs, fundic D cells have been alternately proposed to possess and not to possess stimulatory bombesin receptors (SCHUBERT and HIGHTOWER 1989; SCHUBERT et al. 1991b). SCHUBERT et al. (1991b) have recently postulated that SS release from fundic D cells is mediated via bombesin-stimulated release of a non-cholinergic neurotransmitter. In the antrum, however, gastrin, which is released from G cells upon bombesin stimulation, was proposed as the modulator of SS release from D cells. This hypothesis is based on experiments with chambered rat antral and fundic segments. In antral segments the gastrin antagonist L365,260 blocked bombesin-stimulated SS release whereas TTX and atropine enhanced this response. In the fundus, L365,260 had no effect, TTX inhibited and atropine augmented SS release in response to bombesin/GRP. Studies by GUO et al. (1990) with isolated, vascularly perfused rat stomach also detected TTX inhibition of bombesin/GRP-stimulated SS release. TTX inhibition accounted for ~70% of SS release and gastrin antiserum blocked the remaining 30%. These authors concluded that bombesin-stimulated SS release is mediated mainly via an indirect mechanism involving a nicotinic cholinergic neural pathway with a minor, direct gastrin stimulation of SS release from antral D cells. Since paracrine mechanisms may involve higher local concentration than hormonal mechanisms, a paracrine effect may explain why HOLST et al. (1987b) did not detect gastrin-stimulated SS release in experiments with isolated, vagally perfused pig antrum and why gastrin is a relatively weak stimulator of SS release from D cell in vitro (Sect. C.III). However, there may also be species differences in these control mechanisms.

Prostaglandins (PGs) of the E_2 series are released following mucosal acidification (BEFRITS et al. 1984). Suppression of prostaglandin synthesis with the cyclooxygenase inhibitor, indomethacin, leads to increased HCl secretion in several species (EL-BAYAR et al. 1985; KONTUREK et al. 1979; LEVINE and SCHWARTZEL 1984). An initial report that inhibitory effects of SS are mediated by local release of PGs (LIGUMSKY et al. 1983) has not been confirmed in other studies (ALBINUS et al. 1985; MOGARD et al. 1985; KOOP et al. 1985).

Preganglionic stimulation of splanchnic nerves coupled with infusion of atropine increases release of SLI (MCINTOSH et al. 1981b) as does infusion of β-adrenergic agonists in isolated, vascularly perfused rat stomachs and in dogs with vagally denervated fundic pouches (KOOP et al. 1983; STEVENS et al. 1986). These data support a regulatory role for adrenergic modulation of acid secretion. More recently, in pylorus-ligated pigs, antral but not fundic SS release in response to vagal stimulation was inhibited in adrenalectomized animals (OLSEN et al. 1987). The mechanism by which adrenalectomy altered the vagal response was not characterized but it was speculated that the observed effects might be mediated via vagal sympathetic stimulation of the adrenals. There also appears to be a local nervous reflex originating in the oxyntic portion of the mucosa which may tonically suppress antral G cell gastrin secretion (SOON-SHIONG and DEBAS 1980; SJÖVALL et al. 1990). The mediator of this tonic inhibition is unknown.

Epidermal growth factor (EGF) has a well-known inhibitory effect on gastric acid secretion in vivo. Intravenous infusion of EGF inhibits HCl secretion in response to cholinergic agonists, pentagastrin, meals and sham feeding (KONTUREK et al. 1984; ELDER et al. 1982). Transforming growth factor-α (TGFα), which shares ~35% sequence homology with EGF and binds to the EGF receptor (CARPENTER 1987), has a similar potency to EGF in inhibiting HCl secretion in rats (GREGORY et al. 1988) and in chambered guinea pig gastric mucosa (FINKE et al. 1985; RHODES et al. 1986). EGF receptor-like immunoreactivity is reportedly associated with the basolateral membrane of parietal cells (MORI et al. 1987). Recent findings suggest high levels of TGFα/EGF and TGFα messenger RNA (mRNA) may be present in the gastric mucosa, particularly in parietal cells (CARTLIDGE and ELDER 1989; BEAUCHAMP et al. 1989). Since EGF also stimulates gastric mucosal growth (JOHNSON and GUTHRIE 1980; DEMBINSKI et al. 1982), its wide-ranging inhibitory effects on gastric HCl secretion are rather unusual. It may be that the intracellular mechanisms of acid secretory inhibition vs. growth stimulation follow different pathways. A preliminary report by GARNER et al. (1990) showing that EGF antibodies can reverse the mitogenic effect of EGF for as long as 10 h after the initial exposure but can only block the inhibitory effect on acid secretion if administered prior to EGF supports the concept of different activation mechanisms. In another recent study, ulceration of the mucosal epithelium in the GI tract was found to induce the development of a novel cell lineage from stem cells. This lineage produces and secretes abundant immunoreactive EGF (WRIGHT et al. 1990). Thus, it may be as WRIGHT et al. (1990) have proposed that the main in vivo role of EGF is to stimulate ulcer healing by inducing growth of new cells. These processes are discussed in detail in Chap. 12.

3. Intestinal Inhibitory Mechanisms

Negative feedback from the intestine has long been recognized as an effective means for controlling potentially damaging effects of rapid delivery of gastric contents to the small intestine. In 1930 KOSAKA and LIM postulated that an inhibitory factor, which they named enterogastrone, was released from small and large intestine in response to fat. In older physiology textbooks three intestinal peptides were often classified as major hormonal inhibitors of secretion or enterogastrones. These include secretin, gastric inhibitory polypeptide (GIP) and cholecystokinin (CCK), which are released in response to fat, acid and/or carbohydrate in the small intestine. Unfortunately, this distinction is not so clear cut in the literature (Konturek 1989; Walsh 1988; Schubert and Shamburek 1990).

a) Secretin

Johnson and Grossman (1969) concluded some years ago that secretin is a physiological enterogastrone and the only enterogastrone released by acid in the duodenum. Chey and Chang (1989) have continued to champion a physiological role for secretin. Data supporting such a role for secretin are based mainly on studies in which immunoneutralization of circulating secretin was found to increase gastrin release in dogs with vagally innervated fundic pouches. In these same studies, coinfusion of secretin and pentagastrin reduced the acid secretory response to pentagastrin (Chey et al. 1981). A putative secretin-releasing peptide has recently been isolated from rat small intestine (Li et al. 1990). Secretin infusion elevates plasma SS-like immunoreactivity in dogs (Rouiller et al. 1980); and inhibition of gastric acid secretion in rats by secretin and VIP, a member of the secretin family, is accompanied by increased release of SLI, a portion of which appears to be derived from the gastric antrum (Chiba et al. 1980a). Secretin also stimulates release of SS and inhibits gastrin release in response to carbachol in rat antral mucosa in short-term culture. This secretin inhibition of carbachol-stimulated gastrin release is blocked by SS antibodies (Wolfe et al. 1983b).

Other studies have questioned whether or not secretin is a physiological enterogastrone. Kleibeuker et al. (1984) were unable to demonstrate acid secretory inhibition in humans when secretin was infused at concentrations similar to circulating levels measured following peptone meal acidification. In studies in dogs fed liver extract meals, administration of antisecretin serum was found to block secretin-induced increases in pancreatic bicarbonate secretion but to have no significant effect on the acid secretory response or on the meal-stimulated rise in serum gastrin at meal pHs between 3 and 7 (Konturek et al. 1986). More recently, however, it has been reported that low, presumably physiologic doses of secretin can inhibit acid secretion in humans following duodenal acidification (Christiansen et al. 1988). An important consideration in such studies is the relative concentrations of related peptides needed to produce an effect in the species of interest. For example, comparisons of peptide effects on adenylyl cyclase activity in intestinal epithelial cell membranes indicated that secretin is ~500× less potent than VIP and ~10–20× less potent than PHI in rats; however, in humans, secretin is almost ineffective and PHI is ~500× less effective than VIP (Laburthe and Amiranoff 1989). Thus, species differences in secretin inhibition of gastric acid secretion might be related to relative potency of secretin action on another secretin family member's preferred receptor (see Chap. 5).

b) Cholecystokinin

There appears to be considerable variation in the mechanisms of CCK release and its in vivo actions. It is not yet clear whether divergent experimental findings represent true species differences or are the result of complications associated with in vivo and perfused in vitro models. CCK shares amino acid homology with gastrin but only weakly stimulates acid secretion by itself in most species (being a potent stimulant, however, in the cat (WAY 1971)). Duodenal acidification and fat in the duodenum and jejunum release CCK-like immunoreactivity in dogs and humans but apparently not in rats or cats (CHEN et al. 1985; KONTUREK 1989). CCK stimulates SS release in dogs in vivo (BEGLINGER et al. 1986; ROUILLER et al. 1980) but has no significant effect on SS release in rats (EISSELE et al. 1991). CCK inhibits pentagastrin-stimulated acid secretion in most species (WALSH 1987; MAYER et al. 1982; CORAZZAR et al. 1979). Initially CCK was thought to antagonize gastrin binding to the parietal cell receptor (GROSSMAN 1970). This mechanism is unlikely because CCK-8 and gastrin-17 bind and stimulate secretory-related activity equipotently in rabbit and canine parietal cells in vitro (HERSEY et al. 1983; SOLL et al. 1984b).

c) Gastric Inhibitory Polypeptide

Although gastric inhibitory peptide (GIP) was initially considered to be a good candidate for enterogastrone, more recent work suggests that CCK contamination may have contributed to initial positive findings (MAYER et al. 1982; SCHMIDT et al. 1987), with physiological concentrations of GIP being ineffective (YAMAGISHI and DEBAS 1980). Circulating GIP concentrations following fat ingestion do not appear to be sufficiently elevated to account for fat-induced inhibition of acid secretion (MAXWELL et al. 1980). In one study however, antibodies to GIP were found to enhance acid secretion in response to a peptone meal in dogs (WOLFE et al. 1983a). Moreover, GIP infusion in isolated perfused rat stomach produces a substantial increase in SLI release which is inhibited by vagal excitation (MCINTOSH et al. 1981a). Since the vagal inhibition of GIP-stimulated SLI release is only partially blocked by atropine, it has been suggested that noncholinergic neurotransmitters may also be involved (BROWN et al. 1989). This hypothesis has received some support from recent data showing that opioid peptides which act via μ- but not δ- or κ-type receptors can increase SLI release (MCINTOSH et al. 1990). Because GIP may be an effective inhibitor of HCl secretion under certain conditions (BROWN et al. 1989), the issue of its physiological role as an enterogastrone remains open.

Several other peptides with putative hormonal actions are released in response to fat in the intestine including neurotensin (NT), SS, peptide YY

(PYY, which is a member of the pancreatic polypeptide family), VIP and enteroglucagons [glucagon-like polypeptide-1 (GLP-1_{7-36}-NH_2), glicentin and oxyntomodulin (a 37 amino acid C-terminal fragment of glicentin (HOLST 1983)), which along with glucagon appear to be derived from the same translation pathway but are processed differently post-translationally].

d) Somatostatin

Somatostatin may serve an enterogastrone function in addition to its well-known paracrine actions in the gastric mucosa (COLTURI et al. 1984; UVNÄS-WALLENSTEN et al. 1981; SEAL et al. 1987, 1988). In human studies a comparison between effects of infused SS-14 and effects of monoclonal antibodies to SS on intestinal fat-induced inhibition of acid secretion suggested that the amount of SS released after eating might be sufficient to reduce acid secretion somewhat but not to suppress gastrin release (SEAL et al. 1987). In related studies a monoclonal antibody to SS blocked secretory inhibition induced by intraduodenal fat, suggesting a potential hormonal role for SS in duodenal control of HCl secretion (SEAL et al. 1988). Other studies, however, have questioned the role of SS in fat-induced inhibition of HCl secretion (MOGARD et al. 1988).

e) Neurotensin

Neurotensin (NT), present mainly in ileal mucosal N cells (CARRAWAY and LEEMAN 1976), is a potential enterogastrone. Its inhibitory action may or may not require gastric vagal innervation. ANDERSSON et al. (1980) showed that NT inhibited pentagastrin-stimulated acid secretion only in vagally innervated Heidenhain pouches in dogs. Proximal gastric vagotomy reportedly abolishes the inhibitory effect of intestinal oleic acid in humans (KIHL et al. 1981). However, other studies have found no effect of vagotomy on NT inhibition of pentagastrin-stimulated acid secretion in dogs (MATE et al. 1988). Plasma NT concentrations rise following a meal, but data from improved radioimmunoassays for NT suggest that circulating levels of the biologically active form are insufficient to account for fat-induced inhibition (MOGARD et al. 1987). In an immunoneutralization study, an opposite conclusion was drawn because monoclonal antibodies against NT blocked the inhibition of secretion produced by intraileal but not intraduodenal fat (SEAL et al. 1988).

f) Glucagon-Like Peptides

Glucagon-like peptide-1, a member of the enteroglucagon family, is also a potential candidate enterogastrone. Glucagon-like peptide-1 (GLP-1_{7-36}-NH_2) is found in high concentrations in the distal ileum (KREYMANN et al. 1988) and, following ingestion of a fat or mixed-component meal, plasma levels of GLP-1 are elevated (KREYMANN et al. 1987). GLP-1 inhibits gastrin secretion and stimulates SS release in the isolated, vascularly perfused rat stomach (EISSELE et al. 1990a). In humans, GLP-1 infusion at levels approximating those measured postprandially resulted in inhibition of pentagastrin-stimulated acid secretion (O'HALLORAN et al. 1990). Another enteroglucagon, glicentin, which inhibits acid secretion in rats (KIRKEGAARD et al. 1982), has also been proposed as a potentially relevant inhibitor because it is released from the intestine in response to hyperosmolar glucose solutions (PETERSEN et al. 1985). Glucagon and its gene-related peptide, oxyntomodulin, also inhibit acid secretion at pharmacological concentrations (DUBRASQUET et al. 1982; JARROUSSE et al. 1985; LOUD et al. 1988; SCHJOLDAGER et al. 1989a,b). These peptides appear not to be physiologic enterogastrones as their in vivo potency is too low. However, their effects may become important in certain pathologic conditions.

g) Peptide YY

Peptide YY (PYY), which is present in endocrine-like cells in the distal ileum and colon (LUNDBERG et al. 1982), is yet another candidate for a physiologic enterogastrone. This peptide is a member of the pancreatic polypeptide family that was isolated by TATEMOTO in 1982. Plasma levels of PYY rise following ingestion of fat-containing meals and the magnitude of the rise appears to be sufficient to inhibit acid secretion (PAPPAS et al. 1985; ADRIAN et al. 1985a,b). At present the mechanism of PYY inhibition of acid secretion is unclear. Unlike secretin, GLP-1, VIP, GIP and CCK, PYY has not been found to increase SS release (GREELEY et al. 1988; EISSELE et al. 1990b). PAPPAS et al. (1986) have suggested an action on the cephalic phase of secretion; however, others were unable to detect such an effect (HILL et al. 1991).

Since the inhibitory effects of several of the above-described peptides have not been correlated with circulating plasma levels following a meal, and since many must be administered in apparent pharmacological rather than physiological concentrations, their relevance as true enterogastrones is not clearly established. Another possibility is that several enterogastrones act in concent to inhibit acid secretion with the level of their release depending on the type of meal ingested. In this case, low circulating levels of several peptide hormones might act in concert to effect acid secretory inhibition. In support of this hypothesis, SOPER et al. (1990) recently found that neither ileal pouch infusion nor a mixed meal alone increased plasma concentrations of enteroglucagon, NT or PYY, but the combination of fat plus a mixed meal increased plasma levels of all three peptides. Clearly, much work remains to be done before physiological vs. pharmacological effects of these inhibitory peptides are established.

4. Neuropeptide Inhibitors

Central application of several neuropeptides including bombesin, CRF, CGRP, neuropeptide Y and opioid peptides inhibit acid secretion (TACHÉ and YANG 1990a,b). Intravenous administration of the neuropeptides CGRP I and II and calcitonin (BEGLINGER et al. 1988) as well as CRF and TRH also inhibit acid secretion. The site of action of these peptides and the physiologic relevance of their effect on acid secretion are not established (KONTUREK 1989; TACHÉ 1987, 1988). In the case of CGRP, infusion in isolated, perfused rat stomachs results in release of SS and a moderate decrease in gastrin release. Since these effects of CGRP are not altered by atropine, TTX or propranolol, KOOP et al. (1987) have suggested a direct action on D cells. As yet there is no in vitro evidence for such an action.

a) Substance P

Substance P (SP), a tachykinin which has been detected in the myenteric and submucosal plexuses as well as in the gastric mucosa (MINAGAWA et al. 1984; SCHULTZBERG et al. 1980), inhibits acid secretion stimulated by pentagastrin and peptone meals (KONTUREK et al. 1981; YOKOTANI and FUJIWARA 1985; MARTENSSON et al. 1984) but apparently does not inhibit secretion stimulated by the cholinergic agonist bethanechol (YOKOTANI and FUJIWARA 1985). Substance P also inhibits basal release of SLI as well as that stimulated by L-isoproterenol and GIP (CHIBA et al. 1980b; KWOK et al. 1985). Since atropine (a muscarinic-receptor blocker), hexamethonium (a ganglionic nicotinic-receptor blocker) and naloxone (a Met-enkephalin antagonist) did not block SP inhibition of SLI release and there was no concomitant increase in gastrin secretion in the presence of these antagonists, KWOK et al. (1985) have suggested that SP either acts directly on D cells or indirectly via noncholinergic, nonenkephalinergic neurons.

b) Galanin

Galanin is a recently discovered 29 amino acid peptide (TATEMOTO et al. 1983) that has been detected not only in the central nervous system (RATTAN 1991) but also in the intrinsic nervous system in the stomach and other parts of the GI tract (MELANDER et al. 1985; EKBLAD et al. 1985b; BISHOP et al. 1986). This peptide inhibits basal gastrin secretion (MADAUS et al. 1988; KWOK et al. 1988) and neuromedin C-induced gastrin release (MADAUS et al. 1988) independent of SS release (neuromedin C is a decapeptide with 90% sequence homology with the N-terminal end of bombesin; MCDONALD et al. 1979). Galanin has also been reported to inhibit basal, bombesin and pentagastrin, but not bethanechol, or histamine-stimulated gastric acid secretion (SOLDANI et al. 1988; ROSSOWSKI et al. 1989; YAGCI et al. 1990).

C. Peptide Effects on Isolated Cells

To date five different acid secretory-related cell types from the gastric mucosa have been studied either immediately following isolation or in primary culture. These include the gastric parietal cell, antral G and D cells, fundic D cells and histamine-containing (ECL/mast) cells. The best characterized of these is the parietal cell for which isolated cell models were first described by BERGLINDH et al. in 1976 and SOLL et al. in 1978–1979. The Berglindh preparation consists of small groups of parietal cells and chief cells in approximately equal proportions. A few endocrine-like cells are also contained in gastric glands. The preparation is unique in that cell polarity is preserved as is a glandular lumen into which acid is secreted (BERGLINDH et al. 1976). The disadvantages of the gland preparation are that paracrine influences cannot be ruled out and biochemical studies are complicated by the presence of other cell types. The retention of a paracrine influence may, however, be considered an advantage in studies aimed at comparing secretory-related functions in the more intact gland preparation with isolated parietal cells (BERQVIST and ÖBRINK 1979; CHEW 1983a; NYLANDER et al. 1985). From a historical perspective, it should be noted that enrichment of parietal cells in glands is similar to that in isolated parietal cell preparations obtained by a one-step centrifugal elutriation of gastric mucosal enzymatic digests. In many early studies, therefore, parietal cell content in glands and isolated parietal cells enriched by routine centrifugal elutriation was roughly comparable although this was not generally recognized by workers in the field at the time.

BERGLINDH et al. (1976; BERGLINDH and OBRINK 1976) were the first to demonstrate that agonist-stimulated acid secretory-related activity can be measured indirectly in glands using [^{14}C]-aminopyrine (AP), a weak base that accumulates in acidic spaces at a pH below 5, and oxygen consumption as indices of secretion. SOLL (1978a, 1980) later demonstrated that isolated parietal cells enriched by centrifugal elutriation also responded to secretory agonists with increased AP accumulation and oxygen consumption. The original Soll technique has been modified over the years to provide further enrichment of parietal cells. For example, BERGLINDH described a Nycodenz density gradient technique in 1985 which allowed enrichment of parietal cells to ~70%–80% purity in a single 10-min step. This technique was combined with the elutriation technique to yield cell enrichment approaching 100% purity (CHEW and BROWN 1986; CHEW 1990). Such highly purified parietal cell preparations were recently used to develop a primary parietal cell culture model that retains differentiated characteristics including agonist responsiveness (CHEW et al. 1989).

Fundic and antral D cells, antral G cells and histamine-containing cells have been studied immediately following enzymatic dispersion and in primary, short-term culture. These cell types are not as well characterized as the parietal cell mainly because they are present in very low numbers in the gastric mucosa and are not as readily enriched as the parietal cell due to their smaller size. Also, unlike the parietal cell, which is readily identified at the light microscopic level, these cells have fewer distinctive characteristics requiring immunohistochemical identification. Since enrichment of the cell type of interest in these preparation ranges from ~1% to 60% with the purity of most less than 20%, biochemical studies aimed at determining activities in individual cell types are performed at some frequency but are of questionable significance. Another problem with assigning a specific receptor type to cells in impure preparations is the potential for release of paracrines from contaminating cell types that may alter the function of the cell of interest. As yet, these problems have not been sufficiently addressed in studies of D, G and histamine-containing cells.

Experiments with isolated cells can provide important information on the sites of action and intracellular mechanisms of action of secretory agonists and antagonists. However, an important caveat in isolated cell studies is the potential for nonspecific effects, particularly those involving the use of agonists and antagonists acting through nonreceptor-mediated mechanisms. Moreover, with parietal cells any weak base with a pK_a similar to aminopyrine can compete with and thereby reduce AP accumulation. Also, agents that affect cellular morphology or cause cell swelling have the potential for effecting nonspecific decreases in AP accumulation because AP accumulates within intracellular compartments in isolated single cells and the lumen of isolated glands. Furthermore, AP accumulation is a steady-state measurement dependent not only on the activity of the H^+, K^+ ATPase but also on the relative leakiness of the cellular preparation. For these reasons, cellular mechanistic studies should be executed and interpreted with caution using as many different indices of secretory responsiveness as possible.

I. Parietal Cells

1. Peptide Activation Mechanisms

Parietal cell HCl secretion appears to be stimulated directly by histamine, acetylcholine and gastrin. Stimulation of parietal cells results in dramatic morphological transformations involving fusion of tubulovesicles within the cell to form intracellular canaliculi lined with elongated microvilli. This fusion process transports the H^+,K^+-ATPase or proton "pump" from a cytoplasmic tubulovesicular domain to the apical membrane leading to pump activation and HCl secretion (FORTE and SOLL 1989). The intracellular events leading to HCl secretion have been partially described. There is strong evidence for the presence of histamine H_2- and cholinergic M_3-type receptors on parietal cells (PFEIFFER et al. 1990; CHEW 1989; SOLL and BERGLINDH 1987; FORTE and SOLL 1989). Stimulatory effects of histamine are mediated by adenosine-3′,5′-cyclic monophosphate (cAMP)-dependent mechanisms with calcium potentially playing a minor role (CHEW 1989; SOLL and BERGLINDH 1987; FORTE and SOLL 1989; CHEW and PETROPOULOS 1991b; LJUNGSTRÖM and CHEW 1991). At least four different phosphoproteins of 27, 30, 40, and 80 kDa respond to histamine via cAMP-dependent mechanisms with increased phosphorylation (CHEW and BROWN 1987; URUSHIDANI et al.

1987; ODDSDOTTIR et al. 1988). With the exception of the 80-kDa protein, which has recently been identified as ezrin/cytovillin (FORTE, personal communication), the identity of these proteins is not yet established. Cholinergic agonists appear to activate parietal cells through intracellular release and extracellular influx of Ca^{2+} and possible protein kinase C (PKC)-dependent mechanisms (CHEW 1989; SOLL and BERGLINDH 1987; FORTE and SOLL 1989). Phosphorylation of proteins of 28, 36, and 66 kDa is increased in response to cholinergic stimulation. The 36- and 66-kDa proteins are also phosphorylated in response to PKC activator, 12-*O*-tetradecanoyl phorbol-13-acetate (TPA), whereas the 28-kDa protein is phosphorylated in the presence of the calcium ionophore ionomycin (BROWN and CHEW 1989). Chelation of intracellular and extracellular Ca^{2+} with the Ca^{2+} chelators EGTA and BAPTA has no effect on histamine-stimulated phosphorylation of the 27-and 40-kDa proteins or the 36- and 66-kDa proteins; however, carbachol-stimulated phosphorylation of the 28-kDa protein is abolished (BROWN and CHEW 1989). These data suggest that Ca^{2+} dependent and -independent mechanisms may be involved in cholinergic modulation of parietal cell activity.

a) Gastrin

Whether or not gastrin receptors are present on parietal cells in all species has been debated for a number of years (FORTE and SOLL 1989; SOLL and BERGLINDH 1987; CHEW 1989). In isolated parietal cells from dogs, gastrin weakly stimulates AP accumulation and there is apparent specific binding of ^{125}I-gastrin (SOLL 1978a,b, 1980, SOLL et al. 1984b). A putative gastrin receptor with a molecular weight of 78 kDa has also been detected on canine parietal cells (MU et al. 1987). Pentagastrin increases oxygen consumption and AP accumulation in isolated pig parietal cells (NORBERG et al. 1986; MARDH et al. 1987). In chambered frog mucosa, however, acid secretion stimulated by pentagastrin is blocked by the H_2-receptor antagonist cimetidine (EKBLAD 1985). In rat parietal cells, gastrin apparently does not increase AP accumulation (DIAL et al. 1981; PEREZ-REYES et al. 1983; GEHL et al. 1988). In this same species there is a recent report that no specific gastrin binding is detected (DOUGLAS et al. 1990); however, others have reported that labeled gastrin binds specifically to parietal cell membranes (SOUMARMON et al. 1977; NAKAMURA et al. 1987). Attempts to demonstrate gastrin-stimulated AP accumulation in human gastric glands have thus far proved unsuccessful (HAGLUND et al. 1982; LETH et al. 1991). Gastrin-stimulated AP accumulation does occur in rabbit glands (BERGLINDH et al. 1980; CHEW and HERSEY 1982); however, it has been argued that this effect is mediated solely by gastrin-induced histamine release from histamine-containing (ECL?) cells in the glands (BERGLINDH et al. 1980; SOLL and BERGLINDH 1987). It is generally agreed that gastrin stimulates histamine release from ECL cells in rabbit (Sect. C.IV). The argument against a direct effect of gastrin on rabbit parietal cells is based on data showing apparent complete inhibition of gastrin-stimulated AP accumulation in glands by the H_2-histamine antagonist, cimetidine (BERGLINDH et al. 1980). Other studies, however, suggest there is a small direct stimulation of AP accumulation in acutely isolated and cultured parietal cells (CHEW and HERSEY 1982; CHEW et al. 1989). Gastrin also potentiates AP accumulation in rabbit parietal cells (CHEW and HERSEY 1982) as it does in canine parietal cells (SOLL 1978b).

Additional support for the presence of gastrin receptors on parietal cells from rabbit and other species comes from experiments in which stimulatory effects of gastrin on intracellular Ca^{2+} metabolism and phosphoinositol (PI) turnover have been demonsrated. Both gastrin and CCK, which are equally effective albeit weak acid secretory agonists in isolated parietal cells from dog and rabbit (HERSEY et al. 1983; SOLL et al. 1984b), elevate intracellular $[Ca^{2+}]$ and stimulate PI turnover in parietal cell fractions from rabbits, dogs, guinea pigs and rat enriched to varying degrees of purity (CHEW and BROWN 1986; CHEW 1986; MUALLEM and SACHS 1984; CHIBA et al. 1988; ROCHE and MAGOUS 1989; ROCHE et al. 1990, 1991; PUURUNEN and SCHWABE 1987; TSUNODA 1986). ROCHE et al. (1991) have recently used two apparently specific CCK receptor subtype inhibitors developed by CHANG and coworkers (CHANG and LOTTI 1986; LOTTI and CHANG 1989) and BOCK et al. (1989) to characterize a gastrin-type receptor in rabbit parietal cells enriched to ~70% purity. The CCK-B receptor blocker, L365,260 [3-(acylamino)-benzodiazepine], which has a high affinity for gastrin and CCK receptors in the brain, potently inhibited gastrin-stimulated AP accumulation and [^{3}H]-inositol phosphate (InsP) production. In contrast, L365,718, a CCK-A receptor blocker with high affinity for pancreatic CCK receptors, was ~100× less potent. In related studies, ROCHE et al. (1990) have recently proposed that a pertussis toxin-sensitive guanine nucleotide binding (G) protein is involved in the action of gastrin on rabbit parietal cells. Pertussis toxin caused a rightward shift in gastrin dose-response curves for AP and [^{3}H]inositol trisphosphate ($InsP_3$) accumulation. However, the changes in AP accumulation were extremely small and cell purity was only ~70%. Therefore, these experiments will require additional confirmation.

In the initial studies of CHEW and BROWN (1986) in which parietal cells from rabbit were enriched to >95% purity, the Ca^{2+} response pattern as measured with the fluorescent intracellular Ca^{2+} indicator, fura-2, was different in parietal cell populations as compared to chief cell populations, with the chief cell response being more transient; thus, it was concluded that it was unlikely that the observed gastrin response was due to chief cell contamination. More recently in highly enriched parietal cells CCK, which is equipotent with gastrin in stimulating AP accumulation in rabbit parietal cells, was found to increase in situ phosphorylation of the same 36- and 66-kDa proteins that were phosphorylated in response to carbachol and the PKC actrivator TPA (BROWN and CHEW 1989). The identity and function of these phosphoproteins has not yet been elucidated. Parietal cells from rabbit in primary culture also exhibit a weak AP accumulation response to gastrin, which is only slightly improved over 3 days in culture (CHEW et al. 1989).

More direct evidence for a gastrin receptor on rabbit parietal cells was provided in recent experiments in which single, morphologically identifiable parietal cells in short-term primary culture were shown to respond to gastrin with an increase in intracellular Ca^{2+} concentration ($[Ca^{2+}]_i$), (CHEW and PETROPOULOS 1991a). A potentially interesting finding in these preliminary experiments was the apparent heterogeneity of the cellular Ca^{2+} response to gastrin. Only ~70% of cells sampled responded to gastrin with a rise in $[Ca^+]_i$ although all of these cells responded to a subsequent carbachol challenge. In addition, cells exhibited tachyphylaxis to gastrin but not carbachol, and the Ca^{2+}-signaling pattern with the two agonists was strikingly different. These data may explain previously observed potentiating interactions between histamine and gastrin in isolated glands and cells (SOLL 1978b; CHEW and HERSEY 1982). Whether or not gastrin receptors are also heterogeneously dispersed within parietal cell populations and the significance of the differential response to gastrin is as yet unknown. There is, however, general agreement that gastrin is a relatively weak stimulant of acid secretory responses in isolated cells from all species and it is probable that gastrin stimulation of histamine release from ECL cells is quantitatively more important than the direct action of gastrin on the parietal cell (LETH et al. 1988). It remains to be determined whether gastrin exerts other effects on parietal cells independent of its weak effect on acid secretion. The apparent absence of a gastrin response in rat, human and guinea pig parietal cells may be the result of damaging effects of digestive enzymes. The gastrin response is significantly more labile than that to histamine or carbachol and, in poorly responsive cells, is not detected (CHEW, unpublished observations). Clearly additional experiments are needed to resolve these issues.

b) Effects of Other Peptides

Over the past few years a number of other peptides that modulate acid secretion in vivo and in isolated perfused stomach preparations have been tested for effects on parietal cells and gastric glands. In contrast to in vivo findings in rats, the opioid peptides Met-enkephalin and Met-enkephalin-Arg^6-Phe^7 have been reported to weakly enhance histamine-stimulated AP accumulation in isolated rat parietal cells (SCHEPP et al. 1986). The bombesin-like peptide, gastrin-releasing peptide (GRP), which potently stimulates release of gastrin from isolated antral G cells (see below), has no effect on AP accumulation in human or rabbit gastric glands (LETH et al. 1991) or on rat parietal cells (GEHL et al. 1988). Bombesin also does not elevate $[Ca^{2+}]_i$ in single cultured parietal cells (CHEW, unpublished observations). Nanomolar concentrations of porcine ileal peptide preparations extracted from distal small intestine reportedly elevate $[Ca^{2+}]_i$ and increase

AP accumulation and pepsinogen release in isolated guinea pig parietal and chief cells (TSUNODA and WIDER 1987). WIDER and colleagues have proposed that porcine ileal peptide is a potent enteroxyntin; however, others have questioned whether this is so.

Effects of glucagon-like peptides on isolated parietal cells are, in some instances, in conflict with in vivo findings. Oxyntomodulin, which inhibits pentagastrin-stimulated acid secretion in vivo, increases cAMP production in rat fundic glands (BATAILLE et al. 1981); however, GEHL et al. (1988) were unable to detect any effect of oxyntomodulin on basal or histamine-stimulated AP accumulation in rat parietal cells enriched to ~20%. In related experiments, PERES-REYES et al. (1983) found no effect of secretin or glucagon on AP accumulation in canine parietal cells. Other glucagon-like peptides, including G-29 (pancreatic glucagon), GLP-1, GLP-2 and TGLP-1 (a truncated form of GLP-1) have been shown to elevate cAMP in rat fundic glands (GESPACH et al. 1982; SCHEPP and RUOFF 1984; HANSEN et al. 1988). Since all of these experiments were performed on mixed cell preparations, it is possible that glucagon-like receptors are present on a cell type other than the parietal cell. Indeed, it is difficult to explain why a presumed inhibitor of in vivo acid secretion (O'HALLORAN et al. 1990) would elevate cAMP in parietal cells as elevation of cAMP with either cAMP analogs or histamine stimulates secretion (GHEW 1989; SOLL and BERGDLINDH 1987; FORTE and SOLL 1989). It should be noted, however, that SCHMIDTLER et al. (1991) have recently reported that GLP-1 stimulates AP and cAMP accumulation in enriched rat parietal cell preparations in the presence of the phosphodiesterase inhibitor isobutylmethylxanthine (IBMX). Under these conditions, GLP-1 was more potent than histamine and nearly as efficacious. If confirmed in other species, these data will add a confusing dimension to the enteroglucagon story.

2. Peptide Inhibitory Mechanisms

To date only a few peptides have been found to inhibit AP accumulation in isolated parietal cells and gastric glands. None of the peptides tested thus far affect basal AP accumulation, which is significant in species such as the rabbit. Moreover, no peptide has thus far been found to produce complete inhibition in response to secretagogue stimulation. The physiological relevance of these in vitro effects is not established and results with different peptide inhibitors are often conflicting in the literature.

a) Somatostatin

In agreement with its weak effect on histamine-stimulated acid secretion in vivo, SS partially inhibits histamine-stimulated AP accumulation in isolated gastric glands and parietal cells from rabbit and parietal cells from dog, guinea pigs, humans (BATZRI 1982; CHEW 1983a; NYLANDER et al. 1985; BENGTSSON et al. 1989; LETH et al. 1991; ZDON et al. 1987; PARK et al. 1987).

Somatostatin also decreases histamine-stimulated cAMP accumulation in isolated rat gastric glands, guinea pig mucosal cells and canine parietal cells (GESPACH et al. 1980, 1983; BATZRI 1981; PARK et al. 1987), and decreases histamine-stimulated adenylyl cyclase activity (SCHEPP et al. 1983) and cAMP-dependent protein kinase activity in isolated rat mucosal preparations enriched in parietal cells (MANGEAT et al. 1982).

In rabbit gastric glands the IC_{50} for SS inhibition of histamine and gastrin-stimulated AP accumulation was found to be $\sim 3\,nM$ and inhibitory kinetics were noncompetitive; however, no direct SS inhibition of gastrin-stimulated AP accumulation was detected in enriched parietal cells. This led to the hypothesis that SS has a weak direct effect on the parietal cell with a more potent indirect inhibitory effect on histamine-containing cells (CHEW 1983a). Later studies by NYLANDER et al. (1985) demonstrated that SS inhibits not only gastrin-stimulated histamine release from rabbit glands but also carbachol-stimulated release. These experiments provided further support for the hypothesis that the major action of SS is to inhibit release of histamine from ECL cells. Since EL MUNSHID et al. (1980) were unable to detect SS inhibition of pentagastrin-stimulated histamine release in rats in vivo, there may be species differences in the mechanism of SS inhibition. Alternatively, in vivo measurements may not be sufficiently sensitive to detect a local inhibition.

In SS-binding studies with intact canine parietal cells, $[^{125}I]$-SS-14 and -28 were found to bind equally well to both high-($K_D \sim 3\,nM$) and low-affinity ($K_D \sim 0.2\,\mu M$) sites. Moreover, these analogs were equipotent in inhibition of agonist-stimulated AP accumulation (PARK et al. 1987). The K_D of the high-affinity receptor fits well with the IC_{50} for inhibition of AP accumulation; however, SS-14 is $\sim 10\times$ more potent than SS-28 in inhibiting HCl secretion in vivo (SEAL et al. 1982). These data suggest there may be more important sites of SS action in vivo (PARK et al. 1987). Also, there is a concern that these binding studies may not accurately reflect SS receptor binding as specific SS-14 and SS-28 receptor subtypes have been localized to a number of different cell types (PATEL et al. 1990).

In other studies with partially enriched canine and rabbit parietal cells, SS has been reported to inhibit AP accumulation stimulated by the adenylyl cyclase activator forskolin, the cAMP analog dibutyryl cAMP (dBcAMP), the cholinergic agonist carbachol, and pentagastrin (ZDON et al. 1987; PARK et al. 1987). Since SS did not inhibit carbachol-stimulated increases in phosphoinositol turnover or PKC activity in canine parietal cells, it was postulated that the inhibition occurred at an intracellular site (PARK et al. 1987). These data conflict with other studies in rabbit and dog in which SS was found not to inhibit dBcAMP- or carbachol-stimulated AP accumulation in glands or parietal cells (CHEW 1983a; SOLL 1989; LETH et al. 1991). The reason for these differences is not clear; however, ZDON et al. (1987) reported SS inhibition of forskolin-stimulated AP accumulation to be significantly less than the inhibition of the histamine response, a result which suggests the possibility of nonspecific effects. In the study of Park et al. (1987), the purity of the parietal cell preparation was ~70% and the phosphodiesterase inhibitor isobutylmethylxanthine (IBMX) was added along with pentagastrin. IBMX can enhance effects of endogenous histamine (Chew 1983b; Chew and Hersey 1982; Berglindh et al. 1980; Chen et al. 1988; Chuang et al. 1992). Thus, the apparent inhibition could have been the result of an interaction of endogenous histamine with other agonists.

At present, there is insufficient pharmacological data to define the kinetics of inhibition or the specificity of the inhibitory effects of SS on parietal cell function. There may be intracellular receptors for SS (Reyl and Lewin 1981; Guijarro et al. 1985). Reyl and Lewin (1981) have also postulated that SS activates a phosphoprotein phosphatase in parietal cells. This a potentially important observation because it may explain possible postreceptor actions of SS. Recently, SS analogs have been reported to stimulate a tyrosine phosphatase in pancreatic cancer cells, an action which was correlated with growth inhibition (Liebow et al. 1989). A dual inhibitory action of SS has been described in canine enteric endocrine cells (Barber et al. 1987). This effect may be related to cAMP-independent effects on Na/H exchange (Barber et al. 1989). Thus, a precedent for complex effects of SS on cellular functions exists but has not yet been demonstrated in parietal cells. It still may be, however, that the major action of SS is in the inhibition of secretagogue-induced histamine and gastrin release. Such an action would explain why SS is a potent inhibitor of in vivo pentagastrin and cholinergically stimulated secretion but not histamine-stimulated secretion.

b) Epidermal Growth Factor

Epidermal growth factor (EGF), which is a relatively potent inhibitor of acid secretion in vivo, partially inhibits AP accumulation in response to histamine in isolated gastric glands from rabbit and in parietal cells enriched to varying degrees of purity from several species (Dembinski et al. 1986; Shaw et al. 1987; Lewis et al. 1990a; Chew and Brown 1989). Similar effects have been obtained with TGFα (Lewis et al. 1990b; Chew and Brown 1989). Results with carbachol and EGF/TGFα are conflicting. Lewis et al. (1990a,b) reported that EGF but not TGFα inhibits carbachol-stimulated AP accumulation in rabbit parietal cells. Shaw et al. (1987) found no EGF inhibition of carbachol-stimulated AP accumulation in rat parietal cells; however, carbachol inhibition with both EGF and TGFα has been detected in rabbit parietal cells and glands (Chew and Brown 1989). Since the H_2-receptor antagonist cimetidine was used in these latter experiments, the inhibition could not be attributed to an indirect effect mediated by carbachol-induced release of histamine. There are also conflicting reports as to whether or not EGF inhibits AP accumulation stimulated by dBcAMP (Lewis et al. 1990a; Chew and Brown 1989; Shaw et al. 1987; Dembinski et al. 1986). These differences may be the result of a failure to block effects of endogenous histamine or may be dose related as very high concentrations of EGF were used in one study (DEMBINSKI et al. 1986). EGF causes PGE_2 release in perfused rat stomachs (CHIBA et al. 1982) and stimulates PGE_2 production in gastric mucosal cell preparations enriched to ~80% in parietal cells (SHAW et al. 1987). However, PGE_2 does not appear to be involved in EGF inhibition of acid secretory responses of isolated parietal cells because the PG synthesis inhibitor indomethacin does not block EGF inhibition of histamine-stimulated AP accumulation (SHAW et al. 1987).

Epidermal growth factor decreases histamine-stimulated cAMP production in rat and rabbit parietal cells (HATT and HANSEN 1988; LEWIS et al. 1990a). This action may involve a regulatory G protein (G_i?) since pretreatment of parietal cells with pertussis toxin overcomes EGF inhibition of histamine-stimulated AP accumulation (ATWELL and HANSON 1988; LEWIS et al. 1990a). Since carbachol-stimulated AP accumulation is not affected by pertussis toxin, EGF may have more than one site of action (LEWIS et al. 1990a). Although EGF has been found to activate a receptor-linked tyrosine kinase and to increase $[Ca^{2+}]_i$ and Pl metabolism in some cultured cell lines (CARPENTER 1987), there is presently no published evidence that these events occur in parietal cells. Furthermore, EGF has not been found to reduce carbachol-induced increases in parietal cell $[Ca^{2+}]_i$ (LEWIS et al. 1990a). Therefore, EGF inhibition of carbachol-stimulated AP accumulation may occur at an intracellular site beyond initial receptor-activated events.

Whether or not a direct inhibitory effect of EGF on parietal cell HCl secretion is a specific effect that also occurs in vivo is not definitely established. EGF has the ability to alter parietal cell cytoskeletal architecture (GONZALEZ et al. 1981) and to induce actin redistribution in carcinoma cells (SCHLESSINGER and GEIGER 1981), effects that might indirectly, and perhaps nonspecifically, reduce AP accumulation in isolated cells. Moreover, the IC_{50} for EGF inhibition of AP accumulation in rabbit gastric glands (LEWIS et al. 1990a; CHEW and BROWN 1989) and rat parietal cells (HATT and HANSON 1988; SHAW et al. 1987) is ~ 10- to 40-fold higher than the EC_{50} for the mitogenic effect of EGF (CARPENTER 1987) and the K_D for EGF binding to high-affinity sites on gastric glands and gastric mucosal membranes (FORGUE-LAFITTE et al. 1984; HORI et al. 1990). Preliminary results with cultured parietal cells suggest that EGF/TGFα may positively regulate acid secretory related functions at mitogenic concentrations (CHEW and BROWN 1989). If these data are correct, EGF/TGFα may act on different receptor subtypes. Alternatively, another yet undetected member of the EGF family with higher affinity for the inhibitory receptor may modulate EGF/TGFα inhibition.

c) Other Inhibitory Peptides

Pancreastatin, which apparently has no effect in vivo on basal or agonist-stimulated secretion and enhances meal-stimulated secretion (HASHIMOTO et al. 1990), partially inhibits histamine and carbachol-stimulated AP accumulation in rabbit parietal cells. As with EGF, pretreatment with pertussis toxin reversed pancreastatin inhibition of histamine but not carbachol, and carbachol-stimulated elevation in $[Ca^{2+}]_i$ was unaffected (Lewis et al. 1989). The relevance of pancreastatin inhibition of AP accumulation in isolated cells to effects on in vivo parietal cell activity has not been established.

The tachykinins, substance P and neurokinin A, have recently been reported to inhibit weakly AP accumulation in rat parietal cells stimulated with histamine, forskolin and dBcAMP but not those stimulated with carbachol (Schepp et al. 1990c). The specificity of these effects is unclear. Substance P stimulates pepsinogen release from chief cells and SP receptors have been detected on chief but not parietal cells from dogs (Vigna et al. 1989). Not surprisingly, the in vivo inhibitors VIP and PYY have also not been found to alter AP accumulation in human or rabbit parietal cells (Leth et al. 1991). VIP and secretin elevate cAMP in rat fundic and antral glands and in canine fundic mucosal cells (Wollin et al. 1979; Dupont et al. 1980; Gespach et al. 1983); however, as discussed by these authors it is unlikely that parietal cells are the target. Moreover, other studies in rabbit have shown secretin elevates cAMP in enriched chief cells but not parietal cells (Koeltz et al. 1982).

In parietal cells from guinea pigs enriched to ~60% purity, CGRP decreases AP accumulation stimulated by histamine, carbachol and pentagastrin (Umeda and Okada 1987). The specificity of this effect is unknown. CRF, which inhibits in vivo acid secretion following peripheral infusion or central microinjection, has no effect on AP accumulation in canine parietal cells stimulated with histamine, carbachol or pentagastrin (Todisco et al. 1987).

Calcitonin and GIP at micromolar concentrations reportedly have a weak, noncompetitive inhibitory effect that is not concentration related on histamine-stimulated AP accumulation in isolated rat parietal cells (Schepp et al. 1985). GIP apparently has no stimulatory effect on cAMP metabolism in gastric glands from guinea pigs (Gespach et al. 1982). It is unclear whether this peptide has an inhibitory effect on agonist-stimulated activity. Considering the concentrations of peptides used and the peculiar kinetics of inhibition, it is unlikely that these effects are physiologically relevant.

II. Antral G Cells

Based on isolated G cell studies, there is now reasonably good (but not conclusive) evidence that stimulatory bombesin/GRP and cholinergic receptors are present on antral G cells. Lichtenberger et al. (1980) were the first to demonstrate that it was possible to study G cell function in isolated mixed cell fractions from rodent mucosae. These cells released gastrin in response to cAMP analogs and peptone. In 1987 a method was developed for isolation and primary culture of canine G cells in which gastrin secretion could be evoked by receptor-mediated agonists and antagonists (GIRAUD et al. 1987; SUGANO et al. 1987). Approximately 12% of these cultured cells were G cells. Other cell types identified in the culture included mucus-containing cells (84%), and SLI and serotonin-immunoreactive cells (1–1.5%). Gastrin release was stimulated by bombesin and GRP analogs with a potency order of bombesin>$cGRP_{10}$>$cGRP_{23}$>$cGRP_{27}$. SS inhibited bombesin-stimulated release but not basal release with an IC_{50} of 10 p*M*. The EC_{50} for bombesin was 0.18 n*M* and that for $cGRP_{10}$ was 0.53 n*M*. Efficacies for the various analogs were similar. The SP analog spantide had no effect on basal or bombesin-stimulated gastrin release. Carbachol at a dose of 1 μ*M* stimulated significant release of gastrin but with lower efficacy than the maximal bombesin (10 n*M*) effect. In other studies SCHEPP et al. (1990a) compared a higher, and probably maximal, dose of carbachol (100 μ*M*) with a maximal stimulating dose (10 n*M*) of neuromedin C. Carbachol appeared to be almost equally efficacious with neuromedin C. More recently, specific, saturable and reversible binding of ^{125}I-[Tyr^4]-bombesin was detected in antral mucosal cell preparations enriched to ~2% G cells. Binding specificity of bombesin and GRP peptides correlated with that for gastrin-releasing activity and progressive enrichment for G cells from ~0.8% to 2% (VIGNA et al. 1990). CRF has no effect on gastrin release (TODISCO et al. 1987).

Althought a cAMP-dependent receptor mediator has not yet been identified, it appears that both cAMP and calcium-dependent pathways modulate G cell gastrin release. The cAMP-dependent agonists forskolin and dBcAMP, as well as the calcium ionophore A23187 and the PKC activator TPA, increase gastrin release (LICHTENBERGER et al. 1980; SUGANO et al. 1987; GIRAUD et al. 1987; SCHEPP et al. 1990a). Amino acids and amines also appear to release gastrin by different mechanisms. Amino acid but not amine-stimulated release is decreased by SS and enhanced by bombesin, dBcAMP and the cAMP phosphodiesterase inhibitor IBMX (DELVALLE and YAMADA 1990). In related studies, bombesin and GRP_{10} apparently increased gastrin release through a calcium/phosphatidyinositol (PI)-dependent mechanism (SUGANO et al. 1987). The order of potency for these peptides was similar for gastrin-releasing activity and PI breakdown: bombesin>GRP_{10}>neuromedin B (a decapeptide with 70% sequence identity with the N-terminal decapeptide of bombesin). Isolated G cells from rat appear to respond similarly to canine cells. SCHEPP et al. (1990a) have reported similar results in acutely isolated G cells from rats enriched to ~1.4% purity. Others have also demonstrated that human cells in short-term culture respond to bombesin, forskolin, A23187 and TPA (CAMPOS et al. 1990). SS suppressed forskolin and bombesin-stimulated gastrin release, a result which suggested a dual mode of action for this peptide on adenylate cyclase-dependent and -independent pathways. The biochemical mechanisms of gastrin release from G cells are discussed further in Chap. 4.

Galanin has recently been reported to inhibit potently (IC_{50} ~0.5 n*M*) gastrin release induced by neuromedin C (GRP_{10}) and carbachol as well as that induced by the receptor-independent agonists dBcAMP and TPA in acutely isolated G cells of ~1.4% purity (SCHEPP et al. 1990b). In these studies an intracellular site for galanin inhibition was proposed. Such an action is unusual but not without precedent. In pancreatic beta cells galanin reportedly induces membrane hyperpolarization by activating an ATP-dependent K^+ channel, thereby inhibiting actions of receptor-dependent and -independent secretagogues (DE WEILE et al. 1988). Moreover, SHARP et al. (1989) have suggested that, at least in RINm5F cells, galanin may work via two G protein-dependent mechanisms, one that hyperpolarizes the plasma membrane and a second that is closer to the exocytotic event. Of note in the experiments of SCHEPP and colleagues is that galanin inhibition occurred in the subnanomolar range and the inhibition was blocked by pertussis toxin. It is possible that galanin exerts a similar action on isolated G cells; however, the cellular mechanisms associated with G cell inhibition have not yet been determined nor has the physiologic relevance of this effect been defined.

III. Fundic and Antral D Cells

There appear to be two types of SS-containing cells in the gastric mucosa: antral D cells with their apical surface exposed to the gastric lumen (FUJITA and KOBAYASHI 1977) and fundic D cells which have no luminally exposed surface (LARSSON et al. 1979). Both fundic and antral D cells have recently been isolated and placed in primary culture. Of interest and potential relevance to in vivo studies are observations with D cells from canine fundus and human antrum in which gastrin has been shown to be a weak agonist as compared to CCK. A primary culture of canine fundic mucosal cells, a fraction of which contained SLI, was first described in 1984 (SOLL et al. 1984c). Cells containing SLI became progressively enriched from ~5% to 65% following 48 h of culture on a collagen matrix. These cells released SLI in response to dBcAMP, gastrin, epinephrine and β-adrenergic agonists. The gastrin response was potentiated by cAMP-elevating agonists including epinephrine. Bombesin apparently has no direct effect on SLI release from D cells (CHIBA and YAMADA 1987; CHIBA et al. 1987) nor does CRF (TODISCO et al. 1987). However, there is recent evidence that SS itself may regulate D cell function through an autocrine mechanism (PARK et al. 1989).

Binding of ^{125}I-[Leu^{15}]-gastrin has been correlated with progressive enrichment of SLI cells in elutriator fractions (SOLL et al. 1984a). A later study demonstrated that CCK stimulates SLI release much more potently than gastrin (SOLL et al. 1985). Similar observations have been made in fundic gastric glands from rabbit (BENGTSSON et al. 1989). These results suggest the possibility that fundic D cells possess CCK-A-type receptors rather than CCK-B-type receptors, which appear to be present on parietal cells. SOLL et al. (1985) have suggested that the potent stimulatory effect of CCK on SS release, which would be expected to inhibit HCl secretion, may offset its weak stimulatory effect on the parietal cell. Another possibility not yet addressed in the literature is that CCK also may be less potent than gastrin in releasing histamine from ECL cells. The net result may explain why CCK is such a weak in vivo stimulant of HCl secretion in some species. This may not be the case in rat, for example. EISSELE et al. (1991) recently found no effect of the potent CCK analog cerulein on SLI release in isolated, vascularly perfused stomachs. These data are different from findings in dogs where CCK infusion elevates SLI in plasma (BEGLINGER et al. 1986; SHUSDZIARRA and SCHMID 1986).

In related studies carbachol exerted a predominant inhibitory effect on SLI release that was blocked by pertussis toxin, suggesting muscarinic receptor linkage via a G_i-type protein. In the presence of pertussis toxin an apparent stimulatory effect of carbachol was unmasked. SCHUBERT and HIGHTOWER (1990) have reported similar results in isolated perfused rat stomach preparations. Carbachol also increased ^{3}H-IP$_3$ production independent of pertussis toxin (CHIBA and YAMADA 1987; CHIBA et al. 1987). These unusual effects of carbachol were characterized in more detail in a subsequent study by CHIBA et al. (1989), who found that carbachol inhibition of SS release was correlated with a decrease in cellular cAMP content, an increase in $[Ca^{2+}]_i$ and PI turnover. Since the cell preparations used in these experiments were impure, it is not clear whether one or more additional cell types contributed to the observed biochemical responses.

More recently, BUCHAN et al. (1990) found that human antral D cells in primary culture, like fundic D cells, do not release SS in response to bombesin. In contrast to the fundic preparation, CCK but not G-17 stimulated release of SLI. Since CCK is a more potent stimulant than gastrin in canine fundic D cells, the lack of gastrin effect on the antral cell preparation may have been related to the relative responsiveness of the two preparations. Of interest and potential concern is that the antral D cells were poorly responsive to forskolin but responded strongly to the Ca^{2+} ionophore A23187 plus the PKC activator TPA. These data suggest that antral D cells are controlled mainly via Ca^{2+}-dependent intracellular mechanisms. However, other interpretations such as reduced responsiveness are also possible.

The lack of a direct effect of bombesin on isolated D cells contrasts with in vitro studies of more intact preparations in which bombesin/GRP stimulates SS release (RICHELSEN et al. 1983; SCHUBERT et al. 1988, 1991a,b). BUCHAN et al. (1990) have suggested that this discrepancy may be a result of bombesin-induced release of another mediator which then stimulates SS secretion. As discussed above, similar hypotheses have been proposed based on studies with isolated perfused rat stomach models (GUO et al. 1990; SCHUBERT et al. 1991b). If the absence of gastrin receptors on antral D cells is confirmed, the GUO and SCHUBERT models will require modification.

Considering the interest in effects of putative enterogastrones on SS release in vivo, it is surprising that the effects of these agents apparently have not been examined in detail in cultured D cell preparations. A preliminary report in 1985 indicated that VIP and glucagon inhibit release of SLI in canine D cell cultures (Chiba et al. 1985) but a full-length publication has not been forthcoming. There are also no reports of effects or lack thereof with secretin, GIP and other glucagon-like peptides.

IV. Histamine-Containing Cells

At one time, histamine was proposed to be the final common mediator of acid secretion (Code 1965). Cells containing histamine are present in relatively large numbers in rat and mouse gastric mucosa with lesser numbers in humans and dogs. In rats, rabbits and frogs histamine appears to be localized to enterochromaffin-like (ECL) cells whereas in dogs, humans and pigs much of the histamine in the gastric mucosa is contained in mast cells in the lamina propria with a relatively small fraction possibly possessing endocrine-like characteristics (Lorenz et al. 1973; Aures et al. 1968; Hankanson et al. 1986; Thunberg 1967; Soll et al. 1979, 1981; Simonsson et al. 1988). Although the subject is still somewhat controversial (Soll 1989), gastrin and/or acetylcholine have been shown to release histamine in several species including man and dogs in vivo, in chambered frog gastric mucosa, and in isolated perfused rat stomachs (Gerber and Barnes 1987; Ekblad 1985; Sandvik et al. 1987; Code 1965; Peden et al. 1982; Man et al. 1981; Rangachari 1975; Lundell et al. 1986). Bergqvist and Öbrink (1979) first showed that histamine is released from isolated gastric glands from rabbit in response to gastrin. Nylander et al. (1985) confirmed these findings and showed that histamine was also released in response to acetylcholine. A gastrin receptor on rat ECL cells has also recently been detected with a biotinylated gastrin antagonist, biotinyl-L-Trp-L-Leu-β-Ala (Douglas et al. 1990).

Some years ago Soll et al. (1979) isolated a mast cell-like population of cells from dog gastric mucosa that had similar morphology to classic mast cells but failed to release histamine in response to compound 48/80, a potent releaser of histamine from peritoneal mast cells. Mast-like cells from the canine mucosa were later placed in short-term primary culture and functions compared with a mast cell-enriched fraction from liver (Soll et al. 1988). Both cell types released histamine in response to immunoglobulin E (IgE) and adenosine; however, the fundic mast cells did not release histamine in response to either gastrin or acetylcholine. These data suggested that the mast cell fraction isolated by Soll and colleagues was not the histamine-containing cell involved in control of gastric acid secretion. Recent data indicate that, at least in humans, histamine in the gastric mucosa is contained in two cell populations with ~22%–35% of the total in endocrine-like cells. These ECL cells appear to be identical with ECL cells from other species and are apparently confined to the oxyntic gland portion of the stomach (Lönroth et al. 1990; Simonsson et al. 1988). There is also now preliminary evidence that gastrin-responsive cells are also present in canine fundic mucosa (Chuang et al. 1991). As yet no mechanistic data on ECL cells have been published. Tielemans et al. (1990) have recently provided evidence that ECL cells in all areas of the mouse gastric mucosal pits proliferate by mitotic division. This is potentially an important finding because, if ECL cells do divide with any regularity, it may be possible to establish a long-term culture of these cells which will then allow detailed studies of, for example, mitogenic effects of gastrin. Such studies could have important implications in the ability to characterize potential interrelationships between hypergastrinemia resulting from Zollinger-Ellison syndrome and certain forms of anti-ulcer therapy and ECL carcinoid tumors.

Acknowledgements. I wish to acknowledge colleagues who have worked in my laboratory on different aspects of parietal cell control mechanism including Magnus Ljungström, Milton Brown and Anne Petropoulos. I am grateful to the National Institutes of Health for funding this research for many years through PHS Grant DK31900. I am especially grateful to Frank Chew, Jr., for preparing the illustrations for this chapter. I also wish to thank Cynthia Allen and De'Anna Stevens for help in compilation of the bibliography.

References

Adrian T, Ferri G, Bacarese-Hamilton A, Fussel H, Polak JM, Bloom SR (1985a) Human distribution and release of a putative new gut hormone, peptide YY. Gastroenterology 89:1070–1077

Adrian T, Savage A, Sagor G, Allen J, Bacarese-Hamilton A, Tatemoto K, Polak JM, Bloom SR (1985b) Effect of peptide YY on gastric, pancreatic, and biliary function in humans. Gastroenterology 89:494–499

Albinus M, Reed JD, Gomez-Pan A, Schally AV, Coy DH (1977) Comparison of the inhibition of histamine- and pentagastrin-stimulated gastric acid secretion by somatostatin in the cat. Agents Actions 7:501–506

Albinus M, Gomez-Pan A, Hirst B, Shaw B (1985) Evidence against prostaglandin-mediation of somatostatin-inhibition of gastric secretions. Regul Pept 10: 259–266

Andersson S, Rosell S, Sjodin J, Folkers K (1980) Inhibition of acid secretion from vagally innervated and denervated gastric pouches by (gln^4)-neurotensin. Scand J Gastroenterol 15:253–256

Atwell MM, Hanson PJ (1988) Effect of pertussis toxin on the inhibition of secretory activity by prostaglandin E_2, somatostatin, epidermal growth factor and 12-*O*-tetradecanoylphorbol 13-acetate in parietal cells from rat stomach. Biochim Biophys Acta 971:282–288

Aures D, Håkanson R, Owman CH, Sporrong B (1968) Cellular stores of histamine and monoamines in the dog stomach. Life Sci 7:1147–1153

Barber DL, Gregor M, Soll AH (1987) Somatostatin and muscarinic inhibition of canine enteric endocrine cells: cellular mechanisms. Am J Physiol 253: G684–G689

Barber DL, McGuire ME, Ganz MB (1989) Beta-adrenergic and somatostatin receptors regulate Na-H exchange independent of cAMP. J Biol Chem 264: 21038–21042

Basso N, Improta G, Melchiorri P, Sopranzi N (1974) Gastrin release by bombesin in the antral pouch dog. Rend Gastroenterol 6:95-98

Bataille D, Gespach C, Coudray A, Rosselein G (1981) "Enteroglucagon": a specific effect on gastric glands isolated from the rat fundus. Evidence for an "oxyntomodulin" action. Biosci Rep 1:151–155

Batzri S (1981) Direct action of somatostatin on dispersed mucosal cells from guinea pig stomach. Biochim Biophys Acta 677:521–524

Beauchamp RD, Barnard JA, McCutchen CM, Cherner JA, Coffey RJ (1989) Localization of TGFα and its receptor in gastric mucosal cells: implications for a regulatory role in acid secretion and mucosal renewal. J Clin Invest 84: 1017–1023

Bech K (1986a) Effect of serotonin on bethanechol-stimulated gastric acid secretion and gastric antral motility in dogs. Scand J Gastroenterol 21:655–661

Bech K (1986b) Effect of somatostatin on histamine-stimulated gastric acid and pepsin secretion in dogs. Scand J Gastroenterol 21:662–668

Befrits R, Samuelsson K, Johansson C (1984) Gastric acid inhibition by antral acidification mediated by endogenous prostaglandins. Scand J Gastroenterol 7:899–904

Beglinger C, Ribes G, Whitehouse I, Loubatieres-Mariani MM, Grötzinger U, Gyr K (1986) Effect of exocrine pancreatic secretagogues on circulating somatostatin in dogs. Am J Physiol 250:G15–G20

Beglinger C, Born W, Hilderbrand P, Ensinck JW, Burkhardt J, Fisher JA, Gyr K (1988) Calcitonin gene related peptides I and II and calcitonin: distinct effects on gastric acid secretion in humans. Gastroenterology 95:958–965

Bengtsson P, Lundqvist G, Nilsson G (1989) Inhibition of acid formation and stimulation of somatostatin release by cholecystokinin-related peptides in rabbit gastric glands. J Physiol (Lond) 419:765–774

Berglindh T (1985) Improved one-step purification of isolated gastric parietal cell from rabbit and dog. Fed Proc 440:1203A

Berglindh T, Öbrink KJ (1976) A method for preparing isolated glands from the rabbit gastric mucosa. Acta Physiol Scand 96:150–159

Berglindh T, Helander HF, Öbrink KJ (1976) Effects of secretagogues on oxygen consumption, aminopyrine accumulation and morphology on isolated gastric glands from the rabbit. Acta Physiol Scand 97:401–414

Berglindh T, Sachs G, Takeguchi N (1980) Ca^{2+}-dependent secretagogue stimulation in isolated rabbit gastric glands. Am J Physiol 239:G90–G94

Bergqvist E, Öbrink KJ (1979) Gastrin-histamine as a normal sequence in gastric acid stimulation in the rabbit. Ups J Med Sci 84:145–154

Bertaccini G, Erspamer V, Impicciatore M (1973) The actions of bombesin on gastric secretion of the dog and the rat. Br J Pharmacol 49:437–444

Bishop AE, Polak JM, Bauer FE, Christofides ND, Carlei F, Bloom SR (1986) Occurrence and distribution of a newly discovered peptide, galanin, in the mammalian enteric nervous system. Gut 27:894–897

Black JW (1990) Neurochemical control of oxyntic cell secretion. Hepatogastroenterology Suppl 37:31

Black JW, Duncan WAM, Durant GJ, Ganellin CR, Parsons ME (1972) Definition and antagonism of histamine H_2-receptors. Nature 385–390

Bloom SR, Mortimer CH, Thorner MO, Besser GM, Hall R, Gomez-Pan A, Roy VM, Russel RCG, Coy DH, Kastin AJ, Scally AV (1974) Inhibition of gastrin and gastric acid secretion by growth-hormone release-inhibiting hormone. Lancet 2:1106–1109

Bock MG, Di Paredo RM, Evans BE, Rittle KE (1989) Benzodiazepine gastrin and brain cholecystokinin receptor ligands: L-365,260. J Med Chem 32:13–16

Bonfils S, Mignon M, Roze C (1979) Vagal control of gastric secretion. In: Crane RK (ed) Gastrointestinal physiology III. University Park Press, Baltimore, pp 59–106 (International review of physiology)

Brown JC, Buchan AMJ, McIntosh CHS, Pederson RA (1989) Gastric inhibitory polypeptide. In: Schultz SG, Makhlouf GM (eds) The gastrointestinal system. Oxford University Press, New York, pp 403–430 (Handbook of physiology, vol 2: Neural and endocrine biology, sect 6)
Brown MR, Chew CS (1989) Carbachol-induced protein phosphorylation in parietal cells: regulation by $[Ca^{2+}]_i$. Am J Physiol 257:G99–G110
Buchan AMJ, Curtis SB, Meloche RM (1990) Release of somatostatin immunoreactivity from human antral D cells in culture. Gastroenterology 99:690–696
Bunnett NW, Clark B, Debas HT, del Milton RC, Kovacs TO, Orloff MS, Pappas TN, Reeve JR Jr, Rivier JE, Walsh JH (1985) Canine bombesin-like gastrin releasing peptides stimulate gastrin release and acid secretion in the dog. J Physiol (Lond) 365:121–130
Campos RV, Buchan AMJ, Meloche RM, Pederson RA, Kwok YN, Coy DH (1990) Gastrin secretion from human antral G cells in culture. Gastroenterology 99: 36–44
Cantor P, Holst JJ, Knuthsen S, Rehfeld JF (1987) Effect of neuroactive agents on cholecystokinin release from the isolated, perfused porcine duodenum. Acta Physiol Scand 130:627–632
Carpenter G (1987) Receptors for epidermal growth factor and other polypeptide mitogens. Annu Rev Biochem 56:881–914
Carraway R, Leeman SE (1976) Characterization of radio-immunoassayable neurotensin in the rat. Its differential distribution in the central nervous system, small intestine, and stomach. J Biol Chem 251:7045–7052
Cartlidge SA, Elder JB (1989) Transforming growth factor α and epidermal growth factor levels in normal human gastrointestinal mucosa. Br J Cancer 60:657–660
Chang RSL, Lotti VJ (1986) Biochemical and pharmacological characterization of an extremely potent and selective nonpeptide cholecystokinin antagonist. Proc Natl Acad Sci USA 83:4923–4926
Chen MCY, Amirian DA, Toomey M, Sanders MJ, Soll AH (1988) Prostanoid inhibition of canine parietal cells: mediation by the inhibitory guanosine triphosphate-binding protein of adenylate cyclase. Gastroenterology 94: 1121–1129
Chen YF, Chey WY, Chang T-M, Lee KY (1985) Duodenal acidification releases cholecystokinin. Am J Physiol 249:G29–G33
Chew CS (1983a) Inhibitory action of somatostatin on isolated gastric glands and parietal cells. Am J Physiol 245:G221–G229
Chew CS (1983b) Forskolin stimulation of acid and pepsinogen secretion in isolated gastric glands. Am J Physiol 245:C371–C380
Chew CS (1986) Cholecystokinin, carbachol, gastrin, histamine and forskolin increase $[Ca^{2+}]_i$ in gastric glands. Am J Physiol 250:G814–G823
Chew CS (1989) Intracellular activation events for parietal cell HCl secretion. In: Forte JG (ed) The gastrointestinal system. Oxford University Press, New York, pp 255–266 (Handbook of physiology, sect 6)
Chew CS (1990) cAMP technologies, functional correlates in gastric parietal cells. Methods Enzymol 191:640–661
Chew CS, Brown MR (1986) Release of intracellular Ca^{2+} and elevation of inositol trisphosphate by secretagogues in parietal and chief cell isolated from rabbit gastric mucosa. Biochim Biophys Acta 888:116–125
Chew CS, Brown MR (1987) Histamine increases phosphorylation of 27 and 40 kDa parietal cell proteins. Am J Physiol 253:G823–G829
Chew CS, Brown MR (1989) TGFα and EGF enhance acid secretion in cultured parietal cells. FASEB J 3:A1152
Chew CS, Hersey SJ (1982) Gastrin stimulation of isolated gastric glands. Am J Physiol 242:G504–G512
Chew CS, Petropoulos A (1991a) Gastrin elevates calcium in single cultured parietal cells. FASEB J 5:A1060

Chew CS, Petropoulos A (1991b) Thapsigargin potentiates histamine-stimulated HCl secretion in gastric parietal cells but does not mimic cholinergic agonists. Cell Regul 2:27–39

Chew CS, Ljungström M, Smolka A, Brown MR (1989) Primary culture of secretagogue-responsive parietal cells from rabbit gastric mucosa. Am J Physiol 256:G254–G263

Chey WY, Chang T-M (1989) Secretin. In: Schultz SG, Makhlouf GM (eds) The gastrointestinal system. Oxford University Press, New York, pp. 359–402 (Handbook of physiology, vol 2: neuronal and endocrine biology, sect 6)

Chey WY, Kim MS, Lee KY, Chang TM (1981) Secretin is an enterogastrone in the dog. Am J Physiol 240:G239–G244

Chiba T, Yamada T (1987) Mechanisms for the stimulatory and inhibitory effects of carbamylcholine on canine gastric D-cells. Biochem Biophys Res Commun 147:140–144

Chiba T, Taminato T, Kadowaki S, Abe H, Chihara K, Goto Y, Seino Y, Fujita T (1980a) Effects of glucagon, secretin and vasoactive intestinal polypeptide on gastric somatostatin and gastrin release from isolated perfused rat stomach. Gastroenterology 79:67–71

Chiba T, Taminato T, Kadowaki S, Inoue Y, Mori K, Seino Y, Abe H, Chihara K, Matsukura S, Fujita T, Goto Y (1980b) Effects of various gastrointestinal peptides on gastric somatostatin release. Endocrinology 106:145–149

Chiba T, Kadowaki S, Taminato T, Chihara K, Seino Y, Matsukura S, Fujita T (1981) Effect of antisomatostatin γ-globulin on gastrin release in rats. Gastroenterology 81:321–326

Chiba T, Hirata Y, Taminato T, Kadowaki S, Matsukura S, Fujita T (1982) Epidermal growth factor stimulates prostaglandin E release from isolated perfused rat stomach. Biochem Biophys Res Commun 105:370–374

Chiba T, Park J, Yamada T (1985) Glucagon and vasoactive intestinal polypeptide stimulate somatostatin secretion from isolated canine fundic mucosal cell cultures (Abstr). Gastroenterology 88:1348

Chiba T, Raffoul K, Yamada T (1987) Divergent stimulatory and inhibitory actions of carbamylcholine on gastric D-cells. J Biol Chem 252:8467–8469

Chiba T, Fisher SK, Park J, Seguin EB, Agranoff BN, Yamada T (1988) Carbamylcholine and gastrin induce inositol lipid turnover in canine gastric parietal cells. Am J Physiol 255:G99–G105

Chiba T, Fujita T, Yamada T (1989) Carbachol inhibits stimulant-induced increases in fundic D-cell cytosolic Ca^{2+} concentration. Am J Physiol 257:G308–G312

Christiansen J, Hansen B, Schaffalitzky de Muckadell OB (1988) Effect of low-dose exogenous secretin on pentagastrin- and meal-stimulated gastric acid secretion in man. Dig Dis Sci 33:1277–1281

Chuang C-N, Chen MCY, Davidson S, Tanner M, Soll AH (1992) Gastrin induced of histamine release from canine fundic mucosal non-mast cells in primary culture. Am J Physiol (in press)

Code CF (1965) Histamine and gastric secretion: a later look, 1955–1965. Fed Proc 24:1311

Colturi TJ, Unger RH, Feldman M (1984) Role of circulating somatostatin in regulation of gastric acid secretion, gastrin release, and islet cell function. J Clin Invest 74:417–423

Corazziari E, Solomon TE, Grossman MI (1979) Effect of ninety-five percent pure cholecystokinin on gastrin stimulated acid secretion in man and dog. Gastroenterology 77:91–95

Curry WJ, Johnston CF, Shaw C, Buchanan KD (1990) Distribution and partial characterisation of immunoreactivity to the putative C-terminus of rat pancreastatin. Regul Pept 30:207–219

Debas HT (1987) Peripheral regulation of gastric acid secretion. In: Johnson LR (ed) Physiology of the gastrointestinal tract, vol 2, 2nd edn. Raven, New York, pp 931–945

Debas HT, Konturek SJ, Walsh JH, Grossman MI (1974) Proof of a pyloro-oxyntic reflex for stimulation of acid secretion. Gastroenterology 66:526–532

Defize J, Wider MD, Walz D, Hunt RH (1988) Isolation and partial characterization of gastrotropic from canine ileum: further studies of the parietal and chief cell response. Endocrinology 123:2578–2584

De Graef J, Woussen-Colle MC (1985) Effects of sham feeding, bethanechol, and bombesin on somatostatin release in dogs. Am J Physiol 248:G1–G7

Del Valle J, Yamada T (1990) Amino acids and amines stimulate gastrin release from canine antral G-cells via different pathways. J Clin Invest 85:139–143

Dembinski A, Gregory H, Konturek SJ, Polanski M (1982) Trophic action of epidermal growth factor on the pancreas and gastroduodenal mucosa in rats. J Physiol (Lond) 325:35–42

Dembinski A, Drozdowicz D, Gregory S, Konturek SJ, Warzecha Z (1986) Inhibition of acid formation by epidermal growth factor in isolated rabbit gastric glands. J Physiol (Lond) 378:347–357

De Weile J, Schmith-Antomarchi H, Fosset M, Lazdunski M (1988) ATP-sensitive K^+ channels that are blocked by hypoglycemia-inducing sulfonylureas in insulin-secretin cells are activated by galanin, a hyperglycemia-inducing hormone. Proc Natl Acad Sci USA 85:1312–1316

Dial E, Thompson WJ, Rosenfeld GC (1981) Isolated parietal cells: histamine response and pharmacology. J Pharmacol Exp Ther 219:585–590

Dockray GJ, Vaillant C, Walsh JH (1979) The neuronal origin of bombesin-like immunoreactivity in the rat gastrointestinal tract. Neuroscience 4:1561–1568

Douglas AJ, Walker B, Johnston CF, Murphy RF (1990) Visualisation of the gastrin receptor within rat mucosa using a biotinylated gastrin antagonist. Int J Pept Protein Res 35:306–309

Dubrasquet M, Bataille D, Gespach C (1982) Oxyntomodulin (glucagon-37 or bioactive enteroglucagon): a potent inhibitor of pentagastrin-stimulated acid secretion in rats. Biosci Rep 2:391–395

Dupont C, Gespach C, Chenut B, Rosselin G (1980) Regulation of vasoactive intestinal peptide of cyclic AMP accumulation in gastric epithelial cells. FEBS Lett 113:25–28

Duval JW, Saffouri B, Weir GC, Walsh JH, Arimura A, Makhlouf GM (1981) Stimulation of gastrin and somatostatin secretion from the isolated rat stomach by bombesin. Am J Physiol 241:G242–G247

Eissele R, Koop H, Arnold R (1990a) Effect of glucagon-like peptide-1 on gastric somatostatin and gastrin secretion in the rat. Scand J Gastroenterol 25:449–454

Eissele R, Koop H, Arnold R (1990b) Effect of peptide YY on gastric acid secretion, gastrin and somatostatin release in the rat. Gastroenterology 28:129–131

Eissele R, Koop I, Schaar M, Koop H, Arnold R (1991) Role of cholecystokinin in the control of gastric somatostatin in the rat: in vivo and in vitro studies. Regul Pept 32:333–340

Ekblad E, Ekelund M, Graffner H, Håkanson R, Sundler F (1985a) Peptide containing nerve fibers in stomach wall of rat and mouse. Gastroenterology 89:73–85

Ekblad E, Rökaeus A, Håkanson R, Sundler F (1985b) Galanin nerve fibers in the rat gut: distribution, origin and projections. Neuroscience 16:355–363

Ekblad MEB (1985) Histamine: the sole mediator of pentagastrin-stimulated acid secretion. Acta Physiol Scand 125:135–143

Eklund M, Håkanson R, Vallgren S (1987) Effects of cimetidine, atropine and pirenzepine on basal and stimulated gastric acid secretion in the rat. Eur J Pharmacol 138:225–232

El-Bayer H, Steel L, Montcalm E, Danquechin-Dorval E, Dubois A, Shea-Donohue T (1985) The role of endogenous prostaglandin in the regulation of gastric acid secretion in rhesus monkeys. Prostaglandins 30:401–417

Elder JB, Ganguli PC, Gillespie IE, Gerring EL, Gregory H (1982) Effect of urogastrone on gastric secretion and plasma gastrin levels in normal subjects. Gut 16:887–893

El Munshid HA, Håkanson R, Liedberg G, Sundler F (1980) Effects of various gastrointestinal peptides on parietal cells and endocrine cells in the oxytic mucosa of rat stomach. J Physiol (Lond) 305:249–265

Feldman M (1984) Inhibition of gastric acid secretion by selective and nonselective anticholinergics. Gastroenterology 86:361–366

Feldman M, Richardson CT (1981) Gastric acid secretion in humans. In: Johnson LR (ed) Physiology of the gastrointestinal tract, vol 1. Raven, New York, pp 693–707

Feldman M, Richardson CT (1986) Role of thought, sight, smell, and taste of food in the cephalic phase of gastric acid secretion in humans. Gastroenterology 90: 428–433

Feldman M, Walsh JH (1980) Acid inhibition of sham feeding-stimulated gastrin release and gastric acid secretion: effect of atropine. Gastroenterology 78: 772–776

Feldman M, Dickerman RM, McClelland RN, Cooper KA, Walsh JH, Richardson CT (1979a) Effect of selective proximal vagotomy on food-stimulated gastric acid secretion and gastrin release in patients with duodenal ulcer. Gastroenterology 76:926–931

Feldman M, Richardson CT, Taylor IL, Walsh JH (1979b) Effect of atropine on vagal release of gastrin and pancreatic polypeptide. J Clin Invest 63:294–298

Feldman M, Walsh JH, Taylor IL (1980) Effect of naloxone and morphine on gastric acid secretion and on serum gastrin and pancreatic polypeptide concentrations in humans. Gastroenterology 79:294–298

Feldman M, Unger RH, Walsh JH (1985) Effect of atropine on plasma gastrin and somatostatin concentrations during sham-feeding in man. Regul Pept 12: 345–352

Finke U, Rutten M, Murphy RA, Silen W (1985) Effects of epidermal growth factor on acid secretion from guinea pig gastric mucosa: in vitro analysis. Gastroenterology 88:1175–1182

Forgue-Lafitte ME, Kobari L, Gespach C, Chamblier MC, Rosselin G (1984) Characterization and repartition of epidermal growth factor-urogastrone receptors in gastric glands isolated from young and adult guinea pigs. Biochim Biophys Acta 798:192–198

Forte JG, Soll AH (1989) Cell biology of hydrochloric acid secretion. In: Forte JG (ed) The gastrointestinal system. Oxford University Press, New York, pp 207–228 (Handbook of physiology, sect 6)

Francis BH, Baskin DG, Saunders DR, Ensinck JW (1990) Distribution of somatostatin-14 and somatostatin-28 gastrointestinal-pancreatic cells of rats and humans. Gastroenterology 99:1283–1291

Fujita T, Kobayashi S (1977) Structure and function of gut endocrine cells. Int Rev Cytol 6:187–233

Gantz I, Nothwehr SF, Lucey M, Sacchettini JC, DelValle J, Banaszak LJ, Naud M, Gordon JI, Yamada T (1989) Gastrotropin: not an enterooxyntin but a member of a family of cytoplasmic hydrophobic ligand binding proteins. J Biol Chem 264:20248–20254

Garner A, Gregory H, Hampson SE, Stanier AM, Willshire IR, Young JA (1990) Epidermal growth factor: comparison of antisecretory and mitogenic activities. Digestion 46 Suppl 2:254

Gehl J, Jeppesen JL, Poulsen SS, Holst JJ (1988) The gastric acid secretagogue gastrin-releasing peptide and the inhibitor oxyntomodulin do not exert their effect directly on the parietal cell in the rat. Digestion 40:144–151

Gerber JG, Barnes JS (1987) Histamine release in vivo by pentagastrin from the canine stomach. J Pharmacol Exp Ther 243:887–892

Gespach C, Dupont C, Bataille D, Rosselin G (1980) Selective inhibition by somatostatin of cyclic AMP production in rat gastric glands. Demonstration of a direct effect on the parietal cell function. FEBS Lett 114:247–252

Gespach C, Bataille D, Dutrillaux M-C, Rosselin G (1982) The interaction of glucagon, gastric inhibitory peptide and somatostatin with cyclic AMP production systems present in rat gastric glands. Biochim Biophys Acta 720:7–16

Gespach C, Bataille D, Dutrillaux MC, Rosselin G (1983) Regulation by vasoactive intestinal peptide, histamine, somatostatin-14 and -28 of adenosine monophosphate levels in gastric glands isolated from the guinea pig fundus or antrum. Endocrinology 112:1597–1606

Giraud AS, Soll AH, Cuttitta F, Walsh JH (1987) Bombesin stimulation of gastrin release from canine gastrin cells in primary culture. Am J Physiol 252: G413–G420

González A, Garrido J, Vial JD (1981) Epidermal growth factor inhibits cytoskeleton-related changes in the surface of parietal cells. J Cell Biol 88: 108–114

Goyal RK (1983) Neurology of the gut. In: Sleisinger MH, Fordtran JS (eds) Gastrointestinal disease: pathophysiology, diagnosis, management, 3rd edn. Saunders, Philadelphia, pp 97–115

Greeley GH Jr, Guo Y-S, Gomez G, Lluis F, Singh P, Thompson JC (1988) Inhibition of gastric acid secretion by peptide YY is independent of gastric somatostatin release in the rat. Proc Soc Exp Biol Med 189:325–328

Gregory H, Thomas CE, Young JA, Willshire IR, Garner A (1988) The contribution of the C-terminal undecapeptide sequence of urogastrone-epidermal growth factor to its biological action. Regul Pept 22:217–226

Gregory RA, Ivy AC (1941) The humoral stimulation of gastric secretion. Q J Exp Physiol 31:111–128

Grossman MI (1970) Gastrin and its activities. Nature 228:1147–1150

Grossman MI (1974) Candidate hormones of the gut. Gastroenterology 67: 1016–1019

Grossman MI (1981) Regulation of gastric acid secretion. In: Johnson LR (ed) Physiology of the gastrointestinal tract. Raven, New York, pp 659–671

Guijarro LG, Arilla E, Lopex-Ruiz MP, Prieto JC, Whitford C, Hirst BH (1985) Somatostatin binding sites in cytosolic fraction isolated from rabbit antral and fundic gastric mucosa. Regul Pept 10:207–215

Guo Y-S, Mok L, Cooper CW, Greeley GH Jr, Thompson JC, Singh P (1987) Effect of gastrin-releasing peptide analogues on gastrin and somatostatin release from isolated rat stomach. Am J Physiol 253:G206–G210

Guo Y-S, Thompson JC, Singh P (1990) Role of gastrin in bombesin-stimulated somatostatin release. Gastroenterology 99:1297–1302

Gwyn DG, Leslie RA, Hopkins DA (1985) Observations on the afferent and efferent organization of the vagus nerve and the innervation of the stomach in the squirrel monkey. J Comp Neurol 239:163–175

Haglund U, Elander B, Fellenius E, Leth R, Rehnberg O, Olbe L (1982) The effects of secretagogues on isolated human gastric glands. Scand J Gastroenterol 17: 455–460

Håkanson R, Böttcher G, Ekblad E (1986) Histamine in endocrine cells in the stomach. Histochemistry 86:5–17

Hansen AB, Gespach CP, Rosselin GE, Holst JJ (1988) Effect of truncated glucagon-like peptide 1 on cAMP in rat gastric glands and HGT-1 human gastric cancer cells. FEBS Lett 236:119–122

Harty RF, Franklin PA (1983) GABA affects the release of gastrin and somatostatin from rat antral mucosa. Nature 303:623–624

Hashimoto T, Kogire M, Lluis F, Gomez G, Tatemoto K, Greeley GH Jr, Thompson JC (1990) Stimulatory effect of pancreastatin in gastric acid secretion in conscious dogs. Gastroenterology 99:61–65

Hatt JF, Hanson PJ (1988) Inhibition of gastric acid secretion by epidermal growth factor: effects on cyclic AMP and on prostaglandins production in rat isolated parietal cells. Biochem J 255:789–794

Hersey SJ, May D, Schyberg D (1983) Stimulation of pepsinogen release from isolated gastric glands by cholecystokinin-like peptides. Am J Physiol 244: G192–G197

Hesselfeldt P, Christiansen J, Reyfeld JF, Backer O (1979) Meal stimulated gastric acid and gastrin secretion before and after jejuno-ileal shunt operation in obese patients. Scand J Gastroenterol 14:13–16

Hill FLC, Zhang T, Gomez G, Greeley GH Jr (1991) Peptide YY, a new gut hormone (a mini-review). Steroids 56:77–82

Hirschowitz BI (1989) Neural and hormonal control of gastric secretion. In: Forte JG (ed) The gastrointestinal system. Oxford University Press, New York, pp 127–158 (Handbook of physiology, sect 6)

Hirschowitz BI, Fong J (1990) Effect of antral vagotomy in dogs on gastrin and gastric secretion with various stimuli. Am J Physiol 258:G919–G925

Hirschowitz BI, Molina E (1983) Relation of gastric acid and pepsin secretion to serum gastrin levels in dogs given bombesin and gastrin-17. Am J Physiol 244:G546–551

Hoffman HH, Schnitzlein HN (1961) The number of nerve fibers in the vagus nerve of man. Anat Rec 139:429–436

Hollinshead JW, Debas HT, Yamada T, Elashoff J, Osadchey B, Walsh JH (1985) Hypergastrinemia develops within 24 hours of truncal vagotomy in dogs. Gastroenterology 88:35–40

Holm-Bentzen M, Christiansen J, Kirkegaard P, Olson PS, Petersen B, Fahrenkrug JJ (1983) The effect of vasoactive intestinal polypeptide on meal-stimulated gastric acid secretion in man. Scand J Gastroenterol 18:659–661

Holst JJ (1983) Molecular heterogeneity of glucagon in normal subjects and in patients with glucagon-producing tumours. Diabetologia 24:359–365

Holst JJ, Knuhtsen S, Orskov C, Skak-Nielsen T, Poulsen SP, Nielsen OV (1987a) GRP nerves in pig antrum: role of GRP in vagal control of gastrin secretion. Am J Physiol 253:G643–G649

Holst JJ, Knuhtsen S, Orskov C, Skak-Nielsen T, Poulsen S, Nielsen OV (1987b) GRP-producing nerves control antral somatostatin and gastrin secretion in pigs. Am J Physiol 253:G767–G774

Hori R, Nomura H, Iwakawa S, Okumura K (1990) Characterization of epidermal growth factor receptors on plasma membranes isolated from rat gastric mucosa. Pharmacol Res 7:665

Impicciatore M, Debas H, Walsh JH, Grossman MI, Bertaccini G (1974) Release of gastrin and stimulation of acid secretion by bombesin in dog. Gastroenterology 6:99–101

Improta G, Broccardo M (1990) Tachykinins: effects on gastric secretion and emptying in rats. Pharmacol Res 22:605

Ipp E, Piran U, Richter H, Garberoglio C, Moossa A, Rubinstein AH (1982) Central control of peripheral circulating somatostatin in dogs: effect of 2-deoxyglucose. Am J Physiol 243:E213–E216

Ippoliti AF, Maxwell V, Isenberg JI (1976) The effect of various forms of milk on gastric-acid secretion. Studies in patients with duodenal ulcer and normal subjects. Ann Intern Med 84:286–289

Jain DK, Wolfe MM, McGuigan JE (1985) Functional and anatomical relationships between antral gastrin cells and gastrin-releasing peptide neurons. Histochemistry 82:463–467

Johnson LR, Grossman MI (1969) Characteristics of the inhibition of gastric secretion by secretin. Am J Physiol 217:1401–1404

Johnson LR, Guthrie PD (1980) Stimulation of rat oxyntic gland mucosal growth by epidermal growth factor. Am J Physiol 238:G45–G49

Johnston CF, McGrath SJ, McCollum AP, Sloan JM, Murphy RF, Buchanan KD (1987) C-terminal pancreastatin in normal human gastrointestinal tract and in apudomas. Regul Pept 18:381

Jarrousse C, Audousset-Puech M-P, Dubrasquet M, Niel H, Martinez J, Bataille D (1985) Oxyntomodulin (glucagon-37) and its C-terminal octapeptide inhibit gastric acid secretion. FEBS Lett 188:81–84

Karnes WE Jr, Walsh JH (1990) The gastrin hypothesis. J Clin Gastroenterol Suppl 12:S7–S12

Kauffman GL Jr, Grossman MI (1979) Serum gastrin during intestinal phase of acid secretion in dogs. Gastroenterology 77:26–30

Kihl B, Rokaeus A, Rosell S, Olbe L (1981) Fat inhibition of gastric acid secretion in man and plasma concentrations of neurotensin-like immunoreactivity. Gastroenterology 16:513–526

Kirkegaard PA, Moody A, Holst J, Loud F, Olsen P, Christiansen J (1982) Glicentin inhibits gastric acid secretion in the rat. Nature 297:156–157

Kleibeuker JH, Eysselein VE, Maxwell VE, Walsh JH (1984) Role of endogenous secretin in acid-induced inhibition of human gastric function. J Clin Invest 73:526–532

Knutson U, Olbe L (1974) The effect of exogenous gastrin on the acid sham feeding response in antrum-bulb-resected duodenal ulcer patients. Scand J Gastroenterol 9:231–238

Koeltz HR, Hersey SJ, Sachs G, Chew CS (1982) Pepsinogen release from isolated gastric glands. Am J Physiol 243:G218–G225

Konturek SJ (1989) Inhibition of gastric acid secretion. In: Forte JG (ed) The gastrointestinal system. Oxford University Press, New York, pp 159–184 (Handbook of physiology, sect 9)

Konturek SJ, Tasler J, Cieszkowski M, Coy DH, Schally AV (1976) Effect of growth hormone release inhibiting hormone on gastric secretion, mucosal blood flow, and serum gastrin. Gastroenterology 70:737–741

Konturek SJ, Kwiecien N, Obtulowicz W, Mikos E, Sito E, Olesky J, Popiela T (1978a) Cephalic phase of gastric secretion in healthy subjects and duodenal ulcer patients: role of vagal innervation. Gut 20:875–881

Konturek SJ, Radecki T, Kwiecien N (1978b) Stimuli for intestinal phase of gastric secretion in dogs. Am J Physiol 234:E64–E69

Konturek SJ, Mikos E, Pawlik W, Walus K (1979) Direct inhibition of gastric secretion and mucosal blood flow by arachidonic acid. J Physiol (Lond) 286: 15–28

Konturek SJ, Obtulowicz W, Kwiecien N, Oleksy J (1980a) Effects of pirenzepine and atropine on gastric secretory and plasma hormonal responses to sham feeding in patients with duodenal ulcer. Scand J Gastroenterol Suppl 66:63–69

Konturek SJ, Tasler J, Cieszkowski M, Mikos E, Coy DH, Schally AV (1980b) Comparison of methionine-enkephalin and morphine in the stimulation of gastric acid secretion in the dog. Gastroenterology 78:294–300

Konturek SJ, Jaworek J, Tasler J, Cieszkowski M, Pawlik W (1981) Effect of substance P and its C-terminal hexapeptide on gastric and pancreatic secretion in the dog. Am J Physiol 241:G74–G81

Konturek SJ, Cieszkowski M, Jaworek J, Konturek K, Brozozowskii T, Gregory H (1984) Effects of epidermal growth factor on gastrointestinal secretions. Am J Physiol 246:G580–G586

Konturek SJ, Kwiecien N, Obtulowicz W, Bielanski W, Oleksy J, Schally AV (1985) Effects of somatostatin 14 and somatostatin 28 on plasma hormonal and gastric secretory response to cephalic and gastrointestinal stimulation in man. Scand J Gastroenterol 20:31–38

Konturek SJ, Konturek JW, Domschke S, Domschke W, Varga L, Halter F (1986) Effects of secretin antibodies on gastric H^+ inhibition and pancreatic HCO_3 stimulation by acidified liver extract meal in dogs. Hepatogastroenterology 33:170–175

Koop H, Arnold R (1984) Serotoninergic control of somatostatin and gastrin release from the isolated rat stomach. Regul Pept 9:101–108

Koop H, Behrens I, Bothe E, Koschwitz H, Mclntosh CHS (1983) Adrenergic control of rat gastric somatostatin and gastrin release. Scand J Gastroenterol 18:65–71

Koop H, Schwarz C, Koop I, Dionysius J, Arnold R (1985) Role of endogenous prostaglandins in somatostatin-induced inhibition of gastrointestinal hormone secretion. Hepatogastroenterology 32:288–292

Koop H, Eissele R, Kuhlkamp V, Bothe E, Dionysius J, Arnold R (1987) Calcitonin gene-related peptide stimulates rat gastric somatostatin release in vitro. Life Sci 40:541–546

Kosaka T, Lim RKS (1930) Demonstration of the humoral agent in fat inhibition of gastric secretion. Proc Soc Exp Biol 27:890–891

Kovacs TOG, Walsh JH, Maxwell V, Wong HC, Azuma T, Katt E (1989) Gastrin is a major mediator of the gastric phase of acid secretion in dogs: proof by monoclonal antibody neutralization. Gastroenterology 97:1406–1413

Kreymann B, Williams G, Ghatei MA, Bloom SR (1987) Glucagon-like peptide-1 (7-36): a physiological incretin in man. Lancet 2:1300–1304

Kreymann B, Yiangou Y, Kanse S, Williams G, Ghatei MA, Bloom SR (1988) Isolation and characterization of GLP-1 7-36 amide from rat intestine. FEBS Lett 242:167–170

Kwok YN, McIntosh CHS, Pederson RA, Brown JC (1985) Effect of substance P on somatostatin release from the isolated perfused rat stomach. Gastroenterology 88:90–95

Kwok YN, Verchere CB, McIntosh CHS, Brown JC (1988) Effect of galanin on endocrine secretion from the isolated perfused rat stomach and pancreas. Eur J Pharmacol 145:49–54

Laburthe M, Amiranoff B (1989) Peptide receptors in intestinal epithelium. In: Schultz SG, Makhlouf GM (eds) The gastrointestinal system. Oxford Universtiy Press, New York, pp 215–243 (Handbook of physiology, vol 2: Neuronal and endocrine biology, sect 6)

Landor JH, Gough AL, Rai VS, Lim MK (1980) Amino acids as possible mediators of the intestinal phase of gastric secretion. Surg Gynecol Obstet 150:203–207

Larsson L-I, Goltermann N, de Magistris L, Rehfeld JF, Schwartz TW (1979) Somatostatin cell processes as pathways for paracrine secrtion. Science 205: 1393–1395

Laughton WB, Powley TL (1987) Localization of efferent function in the dorsal motor nucleus of the vagus. Am J Physiol 252:R13–R25

Leconte P (1900) Fonctions gastro-intestinales. Cellule 17:285–318

Lenz HJ, Ferrari-Taylor J, Isenberg JI (1983) Wine and five percent ethanol are potent stimulants of gastric secretion in humans. Gastroenterology 85: 1082–1087

Leth R, Olbe L, Haglund U (1988) The pentagastrin-induced gastric acid response in humans. Scand J Gastroenterol 23:224–228

Leth R, Lundell L, Olbe L (1991) Effects of some gastrointestinal peptides on isolated human and rabbit gastric glands. Scand J Gastroenterol 26:89–96

Levine RA, Schwartzel EH Jr (1984) Effect of indomethacin on basal and histamine stimulated human gastric acid secretion. Gut 25:718–722

Lewis JJ, Goldenring JR, Asher VA, Modlin IM (1989) Pancreastatin: a novel peptide inhibitor of parietal cell signal transduction. Biochem Biophys Res Comm 163:667–673

Lewis JJ, Goldenring JR, Asher VA, Modlin IM (1990a) Effects of epidermal growth factor on signal transduction in rabbit parietal cells. Am J Physiol 258:G476–G483

Lewis JJ, Goldenring JR, Modlin IM, Coffey RJ (1990b) Inhibition of parietal cell H^+ secretion by transforming growth factor alpha: a possible autocrine regulatory mechanism. Surgery 108:220–227

Li P, Lee Y, Chang T-M, Chey WY (1990) Mechanism of acid-induced release of secretin in rats (presence of a secretin-releasing peptide). J Clin Invest 86: 1474–1479

Lichtenberger LM, Forssmann WG, Ito S (1980) Functional responsiveness of an isolated and enriched fraction of rodent gastrin cells. Gastroenterology 79: 447–459

Liebow C, Reilly C, Serrano M (1989) Somatostatin analogues inhibit growth of pancreatic cancer by stimulating tyrosine phosphatase. Proc Natl Acad Sci USA 86:2003–2007

Ligumsky M, Goto Y, Debas H, Yamada T (1983) Prostaglandins mediate inhibition of gastric acid secretion by somatostatin in the rat. Science 219:301–303

Ljungström M, Chew CS (1991) Calcium oscillations in single cultured gastric parietal cells. Am J Physiol 260:C67–C78

Lönroth H, Håkanson R, Lundell L, Sundler F (1990) Histamine containing endocrine cells in the human stomach. Gut 31:383–388

Lorenz W, Matejka E, Schmal A, Seidal W, Rieman HJ, Uhlig R, Mann S (1973) A phylogenetic study on the occurrence and distribution of histamine in the gastrointestinal tract and other tissues of man and various animals. Comp Gen Pharmacol 4:229–250

Lotti VJ, Chang RSL (1989) A new potent and selective nonpeptide gastrin and brain cholecystokinin receptor (CCK-B) ligand:L-365, 260. Eur J Pharmacol 162:273–280

Loud FB, Holst JJ, Rehfeld JF, Christiansen J (1988) Inhibition of gastric acid secretion in humans by glucagon during euglycaemia, hyperglycaemia and hypoglycaemia. Dig Dis Sci 33:530–534

Lundberg JM, Tatemoto K, Terenius I, Hellstrom PM, Mutt V, Hökfelt T (1982) Localization of peptide YY (PYY) in gastrointestinal endocrine cells and effects on intestinal blood flow and motility. Proc Natl Acad Sci USA 79: 4471–4475

Lundell L, Forssell H, Lonroth H, Rosengren E, Wingren U (1986) Histamine storage and formation in canine gastric mucosa-effect of pentagastrin stimulation. Acta Physiol Scand 128:587–595

Madaus S, Schusdziarra V, Seufferlein T, Classen M (1988) Effect of galanin on gastrin and somatostatin release from the rat stomach. Life Sci 42:2381–2387

Madaus S, Schusdziarra V, Erberl T, Seufferlein T, Classen M (1990) Effect of met-enkephalin and met-enkephalin-ARG^{6}-PHE^{7} on bombesin-like immunoreactivity (BLI), somatostatin and gastrin secretion from the perfused rat stomach. Hepatogastroenterology 37:201–203

Mailliard ME, Wolfe MM (1989) Effect of antibodies to the neuropeptide GRP on distention-induced gastric acid secretion in the rat. Regl Pept 26:287–296

Makhlouf GM, Zfass AM, Said SI, Schebalin M (1978) Effects of synthetic vasoactive intestinal peptide (VIP), secretin and their partial sequences on gastric secretion. Proc Soc Exp Biol Med 157:565–568

Man WK, Saunders JH, Ingoldby C, Spencer J (1981) Effect of pentagastrin on histamine output from the stomach in patiens with duodenal ulcer. Gut 22: 916–922

Mangeat C, Gespach C, Marchis-Mouren G, Rosselin G (1982) Differential effects of histamine, vasoactive intestinal peptide, prostaglandin E_2 and somatostatin on cyclic AMP-dependent protein kinase activation in gastric glands isolated from the guinea pig fundus and antrum. Regul Pept 3:155–168

Mårdh S, Song Y-H, Carlsson C, Björkman T (1987) Mechanisms of stimulation of acid production in parietal cells isolated from the pig gastric mucosa. Acta Physiol Scand 131:589–598

Martensson HG, Akande B, Yeo C, Jaffe BM (1984) The role of substance P in the control of gastric acid secretion. Surgery 95:567–571

Martindale R, Kauffman GL, Levin S, Walsh JH, Yamada T (1982) Differential regulation of gastrin and somatostatin secretion from isolated perfused rat stomachs. Gastroenterology 83:240–244
Mate L, Doyle HR, Sakamoto T, Townsend CM Jr (1988) The mechanism of the inhibitory action of neurotensin on pentagastrin-stimulated gastric secretion in dogs. Surg Gynecol Obstet 166:206–210
Maxwell V, Shulkes A, Brown JC, Solomon TE, Walsh JH, Grossman MI (1980) Effect of gastric inhibitory polypeptide on pentagastrin stimulated acid secretion in man. Dig Dis Sci 25:113–116
Mayer EA, Elashoff J, Mutt V, Walsh J (1982) Reassessment of gastric acid inhibition by cholecystokinin and gastric inhibitory polypeptide in dogs. Gastroenterology 83:1047–1050
McArthur K, Hogan D, Isenberg JI (1982) Relative stimulatory effects of commonly ingested beverages on gastric acid secretion in humans. Gastroenterology 83: 199–203
McDonald TJ, Jornvall H, Nilsson G (1979) Characterization of a gastrin releasing peptide from porcine non-antral gastric tissue. Biochem Biophys Res Commun 90:227–233
McGuigan JE, Trudeau WJ (1968) Immunochemical measurement of elevated levels of gastrin in the serum of patients with pancreatic tumors of the Zollinger-Ellison variety. N Engl J Med 278:1308–1313
McIntosh CHS, Pederson RA, Koop H, Brown JC (1981a) Gastric inhibitory polypeptide stimulated secretion of somatostatin-like immunoreactivity from the stomach: inhibition by acetylcholine or vagal stimulation. Can J Physiol Pharmacol 59:468–472
McIntosh CHS, Pederson RA, Muller M, Brown JC (1981b) Autonomic nervous control of gastric somatostatin secretion from the isolated perfused rat stomach. Life Sci 29:1477–1483
McIntosh CHS, Kwok YN, Mordhorst T, Nishimura E, Pederson RA, Brown JC (1983) Enkephalinergic control of somatostatin secretion from the perfused rat stomach. Can J Physiol Pharmacol 61:657–663
McIntosh CHS, Jia X, Kwok YN (1990) Characterization of the opioid receptor type mediating inhibition of rat gastric somatostatin secretion. Am J Physiol 259: G922–G927
Melander T, Hökfelt T, Rökaeus A, Fahrenkrug J, Tatemoto K, Mutt V (1985) Distribution of galanin-like immunoreactivity in the gastrointestinal tract of several mammalian species. Cell Tissue Res 239:253–270
Minagawa H, Shiosaka S, Inoue H (1984) Origins and three-dimensional distribution of substance P-containing structures on the rat stomach using whole-mount tissue. Gastroenterology 86:51–59
Mogard MH, Kauffman GL Jr, Pehlevanian M, Golanska E, Elashoff JD, Walsh JH (1985) Prostaglandins may not mediate inhibition of gastric acid secretion by somatostatin in the rat. Regul Pept 10:231–236
Mogard MH, Maxwell V, Sytnik B, Walsh JH (1987) Regulation of gastric acid secretion by neurotensin in man. J Clin Invest 80:1064–1067
Mogard MH, Maxwell V, Wong H, Reedy TJ, Sytnik B, Walsh JH (1988) Somatostatin may not be a hormonal messenger of fat-induced inhibition of gastric functions. Gastroenterology 94:405–408
Mori S, Morishita Y, Sakai S, Kurimoto S, Okamoto M, Kawamoto T, Kuroki T (1987) Electron microscopic evidence for epidermal growth receptor (EGF-R)-like immunoreactivity associated with the basolateral surface of gastric parietal cells. Acta Pathol Jpn 37:1909–1917
Moyer MP, Dixon PS, Aust PB, Wider WD (1988) Effects of a new gastrointestinal (GI) peptide, gastrotropin, on growth and differentiation of human GI epithelial cells. Gastroenterology 94:312

Mu F-T, Baldwin G, Weinstock J, Stockman D, Toh BH (1987) Monoclonal antibody to the gastrin receptor on parietal cells recognizes a 78-kDa protein. Proc Natl Acad Sci USA 84:2698–2702
Muallem S, Sachs G (1984) Changes in cytosolic free Ca^{2+} in isolated parietal cells: differential effects of secretagogues. Biochim Biophys Acta 805:181–185
Nakamura M, Oda M, Kaneko K, Yonei Y, Tsukada N, Komatsu H, Tsugu M, Tsuchiya M (1987) Autoradiographic demonstration of gastrin binding sites in rat gastric mucosa. Peptides 8:391–398
Nakamura M, Oda M, Kaneko K, Akaiwa Y, Tsukada N, Komatsu H, Tsuchiya M (1988) Autoradiographic demonstration of gastrin-releasing peptide binding sites in rat gastric mucosa. Gastroenterology 94:968–976
Nilsson G, Simon J, Yalow RS, Berson SA (1972) Plasma gastrin and gastric acid responses to sham feeding and feeding in dogs. Gastroenterology 63:51–59
Norberg L, Ljungström M, Vega F, Mårdh S (1986) Stimulation of acid formation by histamine, carbachol and pentagastrin in isolated parietal cells. Acta Physiol Scand 126:385–390
Nylander O, Bergqvist E, Obrink KJ (1985) Dual inhibitory actions of somatostatin on isolated gastric glands. Acta Physiol Scand 125:111–119
Oddsdottir M, Goldenring JR, Adrian TE, Zdon MJ, Zucker KA, Modlin IM (1988) Identification and characterization of a cytosolic 30 kDa histamine stimulated phosphoprotein in parietal cell cytosol. Biochem Biophys Res Commun 154: 489–496
O'Halloran DJ, Nikou GC, Kreymann B, Ghatei MA, Bloom SR (1990) Glucagon-like peptide-1 (7-36)-NH_2: a physiological inhibitor of gastric acid secretion in man. J Endocrinol 126:169–173
Olbe L (1964) Potentiation of sham feeding response in Pavlov pouch dogs by subthreshold amounts of gastrin with and without acidification of denervated antrum. Acta Physiol Scand 61:244–254
Olsen M, Holst JJ, Sottimano C, Nielsen OV (1987) Autonomic nervous control of fundic secretion of somatostatin and antral secretion of gastrin and somatostatin in pigs. Digestion 36:24–35
Orloff MJ, Guillemin RCL, Nakaji NT (1977) Isolation of the hormone responsible for the intestinal phase of gastric secretion. Gastroenterology 72:820
Pappas TN, Debas H, Goto Y, Taylor I (1985) Peptide YY inhibits meal-stimulated pancreatic and gastric secretion. Am J Physiol 248:G118–G123
Pappas TN, Debas HT, Taylor IL (1986) Enterogastrone-like effect of peptide YY is vagally mediated in the dog. J Clin Invest 77:49–53
Park J, Chiba T, Yamada T (1987) Mechanisms for direct inhibition of canine gastric parietal cells by somatostatin. J Biol Chem 262:14190–14196
Park J, Chiba T, Yokotani K, Delvalle J, Yamada T (1989) Somatostatin receptors on canine fundic D-cells: evidence for autocrine regulation of gastric somatostatin. Am J Physiol 257:G235–G241
Patel YC, Murthy KK, Escher EE, Banville D, Spiess J, Srikant CB (1990) Mechanism of action of somatostatin: an overview of receptor function and studies of the molecular characterization and purification of somatostatin receptor proteins. Metabolism 39:63–69
Peden NR, Boyd EJS, Callahan H, Shepherd DM, Wormsley KG (1982) The effects of impromidine and pentagastrin on gastric output of histamine, acid and pepsin in man. Hepatogastroenterology 29:30–34
Perez-Reyes E, Payne NA, Gerber JG (1983) Effect of somatostatin, secretin, and glucagon on secretagogue stimulated aminopyrine uptake in isolated canine parietal cells. Agents Actions 13:265–268
Petersen B, Christiansen J, Holst JJ (1985) A glucose-dependent mechanism in jejunum inhibits gastric acid secretion: a response mediated through enteroglucagon? Scand J Gastroenterol 20:193–197

Peterson WL, Barnett C, Walsh JH (1986) Effect of intragastric infusions of ethanol and wine on serum gastrin concentration and gastric acid secretion. Gastroenterology 91:1390–1395

Pfeiffer A, Hanack C, Kopp R, Tacke R, Moser U, Mutschler E, Lambrecht G, Herawi M (1990) Human gastric mucosa expresses glandular M_3 subtype of muscarinic receptors. Dig Dis Sci 35:1468–1472

Preshaw RM (1970) Gastric acid output after sham feeding and during release or infusion of gastrin. Am J Physiol 219:1409–1416

Puurunen J, Schwabe U (1987) Effect of gastric secretagogues on the formation of IPs in isolated gastric cells of the rat. Br J Pharmacol 90:479–490

Radke R, Stach W, Weiss R (1980) Innervation of the gastric wall related to acid secretion: a light and electron microscopy study on rats, rabbits and guinea pigs. Acta Biol Med Ger 39:687–696

Rangachari PK (1975) Histamine release by gastric stimulants. Nature 253:53–55

Rattan S (1991) Role of galanin in the gut. Gastroenterology 100:1762–1768

Reyl F, Lewin MJM (1981) Intracellular receptor for somatostatin in gastric mucosal cells: decomposition and reconstruction of somatostatin-stimulated phosphoprotein phosphatases. Biochim Biophys Acta 675:295–300

Rhodes JA, Tam JP, Finke U, Saunders M, Bernanke J, Silen W, Murphy RA (1986) Transforming growth factor alpha inhibits secretion of gastric acid. Proc Natl Acad Sci USA 83:3844–3846

Richardson CT, Walsh JH, Cooper KA, Feldman M, Fordtran JS (1977) Studies on the role of cephalic-vagal stimulation in the acid secretory response to eating in normal human subjects. J Clin Invest 60:435–441

Richelsen B, Rehfeld JF, Larsson L-I (1983) Antral gland cell column: a method for studying release of gastric hormones. Am J Physiol 245:G463–G469

Robein MJ, de la Mare MC, Dubrasquet JM, Bonfils S (1979) Utilization of the perfused stomach in anaesthetized rats to study the inhibitory effect of somatostatin in gastric acid secretion. Agents Actions 9:415–421

Roche S, Magous R (1989) Gastrin and CCK-8 induce inositol 1,4,5-triphosphate in rabbit gastric parietal cells. Biochim Biophys Acta 1014:313–318

Roche S, Bali J-P, Magous R (1990) Involvement of a pertussis toxin-sensitive G protein in the action of gastrin on gastric parietal cells. Biochim Biophys Acta 1055:287–294

Roche S, Bali J-P, Galleyrand J-C, Magous R (1991) Characterization of a gastrin-type receptor on rabbit gastric parietal cells using L365, 260 and L364, 718. Am J Physiol 260:G182–G188

Rossowski WJ, Coy DH (1989) Inhibitory action of galanin on gastric acid secretion in pentobarbital-anesthetized rats. Life Sci 44:1807–1813

Rouiller D, Schusdziarra V, Harris V, Unger RH (1980) Release of pancreatic and gastric somatostatin-like immunoreactivity in response to the octapeptide of cholecystokinin, secretin, gastric inhibitory polypeptide and gastrin-17 in dogs. Endocrinology 107:524–529

Saffouri B, Weir G, Bitar K, Makhlouf G (1979) Stimulation of gastrin secretion from the perfused rat stomach by somatostatin antiserum. Life Sci 25: 1749–1754

Saffouri B, Weir GC, Bitar KN, Makhlouf GM (1980) Gastrin and somatostatin secretion by perfused rat stomach: functional linkage of antral peptides. Am J Physiol 238:495–501

Saffouri B, Duvall JW, Arimura A, Makhlouf GM (1984) Effects of vasoactive intestinal peptide and secretin on gastrin and somatostatin secretion from the perfused rat stomach. Gastroenterology 86:839–842

Saffouri B, DuVal JW, Makhlouf GM (1984b) Stimulation of gastrin secretion in vitro by intraluminal chemicals: regulation by intramural cholinergic and non-cholinergic neurons. Gastroenterology 87:557–561

Sandvik AK, Waldum HL, Kleveland PM, Sognen BS (1987) Gastrin produces an immediate and dose-dependent histamine release preceding acid secretion in the totally isolated, vascularly perfused rat stomach. Scand J Gastroenterol 22: 803–808

Sandvik AK, Holst JJ, Waldrum HL (1989) The effect of gastrin-releasing peptide on acid secretion and the release of gastrin, somatostatin, and histamine in totally isolated, vascularly perfused rat stomach. Scand J Gastroenterol 24:9–15

Schepp W, Ruoff H-J (1984) Adenylate cyclase and H^+ production of isolated rat parietal cells response to glucagon and histamine. Eur J Clin Pharmacol 98:9–18

Schepp W, Heim H-K, Ruoff H-J (1983) Comparison of the effect of PGE_2 and somatostatin on histamine-stimulated ^{14}C-aminopyrine uptake and cyclic AMP formation in isolated rat gastric mucosal cells. Agents Actions 13:200–206

Schepp W, Miederer SE, Ruoff H-J (1985) Effects of hormones (calcitonin, GIP) and pharmacological antagonists (ranitidine and famotidine) on isolated rat parietal cells. Regul Pept 12:297–308

Schepp W, Schneider J, Schusdziarra V, Classen M (1986) Naturally occurring opioid peptides modulate H^+ production by isolated parietal cells. Peptides 7:885–890

Schepp W, Prinz C, Håkanson R, Schusdziarra V, Classen M (1990a) Bombesin-like peptides stimulate gastrin release from isolated rat G-cells. Regul Pept 28: 241–253

Schepp W, Prinz C, Tatge C, Håkanson R, Schusdziarra V, Classen M (1990b) Galanin inhibits gastrin release from isolated rat gastric G-cells. Am J Physiol 258:G596–G602

Schepp W, Schmidtler J, Tatge C, Schusdziarra V, Classen M (1990c) Effect of substance P and neurokinin A on rat parietal cell function. Am J Physiol 259:G646–G654

Schjoldager B, Mortensen PE, Myhre J, Christiansen J, Holst JJ (1989a) Oxyntomodulin from distal gut. Role in regulation of gastric and pancreatic functions. Dig Dis Sci 34:1411–1419

Schjoldager BT, Mortensen PE, Christiansen J, Orskov C, Holst JJ (1989b) GLP-1 (glucagon-like peptide 1) and truncated GLP-1, fragments of human proglucagon, inhibit gastric acid secretion in humans. Dig Dis Sci 34:703–708

Schlessinger J, Geiger B (1981) Epidermal growth factor induces redistribution of actin and actinin in human epidermal carcinoma cells. Exp Cell Res 134: 273–279

Schmidt WE, Siegel EG, Kummel H, Gallwitz BW, Creutzfeldt W (1987) Commercially available preparations of porcine glucose-dependent insulinotropic polypeptide (GIP) contain a biologically inactive GIP-fragment and cholecystokinin-33/-39. Endocrinology 120:835–837

Schmidt WE, Siegel EG, Lamberts R, Gallwitz B, Creutzfeldt W (1988) Pancreastatin: molecular and immunocytochemical characterization of a novel peptide in porcine and human tissues. Endocrinology 123:1395–1404

Schmidtler J, Schepp W, Janczewska I, Weigert N, Fürlinger C, Schusdziarra V, Classen M (1991) GLP-1-(7-36) amide, -(1-37), and -(1-36) amide: potent cAMP-dependent stimuli of rat parietal cell function. Am J Physiol 260: G940–G950

Schubert ML (1991) The effect of vasoactive intestinal polypeptide on gastric acid secretion is predominantly mediated by somatostatin. Gastroenterology 100: 1195–1200

Schubert ML, Hightower J (1989) Inhibition of acid secretion by bombesin is partly mediated by release of fundic somatostatin. Gastroenterology 97:561–567

Schubert ML, Hightower J (1990) Functionally distinct muscarinic receptors on gastric somatostatin cells. Am J Physiol 258:G982–G987

Schubert ML, Shamburek RD (1990) Control of acid secretion. Gastroenterol Clin North Am 19:1–24

Schubert ML, Saffouri B, Walsh JH, Makhlouf GM (1985) Inhibition of neurally mediated gastrin secretion by bombesin antiserum. Am J Physiol 248: G456–G462

Schubert ML, Saffouri B, Makhlouf GM (1988) Identical patterns of somatostatin secretion from isolated antrum and fundus of rat stomach. Am J Physiol 254: G20–G24

Schubert ML, Hightower J, Makhlouf GM (1989) Linkage between somatostatin and acid secretion: evidence from use of pertussis toxin. Am J Physiol 256: G418–G422

Schubert ML, Hightower J, Coy DH, Makhlouf GM (1991a) Regulation of acid secretion by bombesin/GRP neurons of the gastric fundus. Am J Physiol 260: G156–G160

Schubert ML, Jong MJ, Makhlouf GM (1991b) Bombesin/GRP-stimulated somatostatin secretion is mediated by gastrin in the antrum and by intrinsic neurons in the fundus. Am J Physiol 261:G885–G889

Schultzberg M, Hökfelt T, Nilsson G, Terenius L, Rehfeld JF, Brown M, Elde R, Goldstein M, Said S (1980) Distribution of peptide- and catecholamine-containing neurons in the gastrointestinal tract of rat and guinea pig: immunohistochemical studies with antisera to substance P, vasoactive intestinal polypeptide, enkephalins, somatostatin, gastrin/cholecystokinin, neurotensin and dopamine β-hydroxylase. Neuroscience 5:689–744

Schusdziarra V, Schmid R (1986) Physiological and pathophysiological aspects of somatostatin. Scand J Gastroenterol 21 Suppl 119:29–41

Schusdziarra V, Schmid VR, Bender H, Schusdziarra M, Rivier J, Vale W, Classen M (1986) Effect of vasoactive intestinal peptide, peptide histidine isoleucine and growth hormone-releasing factor-40 on bombesin-like immunoreactivity, somatostatin and gastrin release from the perfused rat stomach. Peptides 7: 127–133

Seal AM, Yamada T, Debas H, Hollinshead J, Osadchey B, Aponte G, Walsh J (1982) Somatostatin-14 and -28: clearance and potency on gastric function in dogs. Am J Physiol 243:97–102

Seal AM, Meloche RM, Liu YQE, Buchan AMJ, Brown JC (1987) Effects of monoclonal antibodies to somatostatin on somatostatin-induced and intestinal fat-induced inhibition of gastric acid secretion in the rat. Gastroenterology 92:1187–1192

Seal AM, Liu E, Buchan A, Brown J (1988) Immunoneutralization of somatostatin and neurotensin: effect on gastric acid secretion. Am J Physiol 255:G40–G45

Shapiro R, Miselis R (1985) The central organization of the vagus nerve innervating the stomach of the rat. J Comp Neurol 238:473–488

Sharp GWG, LeMarchand-Brustel Y, Yada T (1989) Galanin can inhibit insulin release by a mechanism other than membrane hyperpolarization or inhibition of adelate cyclase. J Biol Chem 264:7302–7309

Shaw GP, Hatt JF, Anderson NG, Hanson PJ (1987) Action of epidermal growth factor on acid secretion by rat isolated parietal cells. Biochem J 244:699–704

Short GM, Doyle JM, Wolfe MM (1985) Effect of antibodies to somatostatin on acid secretion and gastrin release by the isolated perfused rat stomach. Gastroenterology 88:984–988

Simonsson M, Eriksson S, Håkanson R (1988) Endocrine cells in the human oxyntic mucosa. A histochemical study. Scand J Gastroenterol 23:1089–1099

Sjövall M, Lindstedt G, Olbe L, Lundell L (1990) Effect of parietal cell vagotomy and cholinergic blockade on gastrin release in man induced by gastrin-releasing peptide. Digestion 46:114–120

Soldani G, Mengozzi G, Della LA, Intorre L, Martelli F, Brown DR (1988) An analysis of the effects of galanin on gastric acid secretion and plasma levels of gastrin in the dog. Eur J Pharmacol 154:313–318

Soll AH (1978a) The actions of secretagogues on oxygen uptake by isolated mammalian parietal cells. J Clin Invest 61:370–380

Soll AH (1978b) The interaction of histamine with gastrin and carbamylcholine on oxygen uptake by isolated mammalian parietal cells. J Clin Invest 61:381–389

Soll AH (1980) Secretagogue stimulation of ^{14}C-aminopyrine accumulation by isolated canine parietal cells. Am J Physiol 238:G366–G375

Soll AH (1989) Gastric mucosal receptors. In: Schultz SG, Makhlouf GM (eds) The gastrointestinal system. Oxford University Press, New York, pp 193–214 (Handbook of physiology, vol 2, Neuronal and endocrine biology, sect 6)

Soll AH, Berglindh T (1987) Physiology of isolated gastric glands and parietal cells: receptors and effectors regulating function. In: Johnson LR (ed) Physiology of the gastrointestinal tract, 2nd edn. Raven, New York, pp 883–909

Soll AH, Lewin K, Beaven MA (1979) Isolation of histamine-containing cells from canine fundic mucosa. Gastroenterology 77:1283–1290

Soll AH, Lewin KJ, Beaven MA (1981) Isolation of histamine-containing cells from rat gastric mucosa: biochemical and morphologic differences from mast cells. Gastroenterology 80:717–727

Soll AH, Amirian DA, Thomas LP, Park J, Elashoff JD, Beaven MA, Yamada T (1984a) Gastrin receptors on non-parietal cells isolated from canine fundic mucosa. Am J Physiol 247:G715–G723

Soll AH, Amirian DA, Thomas LP, Reedy TJ, Elashoff JD (1984b) Gastrin receptors on isolated canine parietal cells. J Clin Invest 73:1434–1447

Soll AH, Yamada T, Park J, Thomas LP (1984c) Release of somatostatin-like immunoreactivity from canine fundic mucosal cells in primary culture. Am J Physiol 247:558–566

Soll AH, Amirian DA, Park J, Elashoff JD, Yamada T (1985) Cholecystokinin potently releases somatostatin from canine fundic mucosal cells in short-term culture. Am J Physiol 248:569–573

Soll AH, Toomey M, Culp D, Shanahan F, Beaven MA (1988) Modulation of histamine release from canine fundic mucosal mast cells. Am J Physiol 254: G40–G48

Soon-Shiong B, Debas HT (1980) Fundic inhibition of acid secretion and gastrin release in the dog. Gastroenterology 79:867–872

Soper NJ, Chapman NJ, Kelly KA, Brown ML, Phillips SF, Cao VL (1990) The "ileal brake" after ileal pouch-anal anastomosis. Gastroenterology 98:111–116

Soumarmon A, Cheret AM, Lewin MJM (1977) Localization of gastrin receptors in intact isolated and separated rat fundic cells. Gastroenterology 73:900–903

Stenquist B (1979) Studies on vagal activation of gastric acid secretion in man. Acta Physiol Scand Suppl 465:1–31

Stenquist B, Nilsson G, Rehfeld JF, Olbe L (1979) Plasma gastrin concentrations following sham feeding in duodenal ulcer patients. Gastroenterology 14:305–312

Stevens MH, Thirlby RC, Richardson CT, Fredrickson MA, Unger RH, Feldman M (1986) Inhibitory effects of β-adrenergic agonists on gastric acid secretion in dogs. Am J Physiol 251:G453–G459

Strunz UT, Grossman MI (1978) Effect of intragastric pressure on gastric emptying and secretion. Am J Physiol 235:E552–E555

Strunz UT, Walsh JH, Grossman MI (1978) Stimulation of gastrin release in dogs by individual amino acids. Proc Soc Exp Biol Med 157:440–441

Sugano K, Park J, Soll AH, Yamada T (1987) Stimulation of gastrin release by bombesin and canine gastrin releasing peptides. Studies with isolated canine G-cells in primary culture. J Clin Invest 79:935–942

Taché Y (1987) Central nervous system regulation of gastric acid secretion. In: Johnson LR (ed) Physiology of the gastrointestinal tract, 2nd edn. Raven, New York, pp 911–930

Taché Y (1988) CNS peptides and regulation of gastric acid secretion. Annu Rev Physiol 50:19–39

Taché Y, Yang H (1990a) Brain regulation of gastric acid secretion by peptides. Sites and mechanisms of action. Ann NY Acad Sci 597:128–145

Taché Y, Yang H (1990b) Hypothalamic, medullary, and spinal sites of action of peptides influencing gastric acid secretion. Ann NY Acad Sci 597:130–145

Tatemoto K (1982) Isolation and characterization of peptide YY (PYY), a candidate gut hormone that inhibits pancreatic exocrine secretion. Proc Natl Acad Sci USA 79:2514–2518

Tatemoto K, Rokaeus A, Jornvall H, McDonald TJ, Mutt V (1983) Galanin-novel biologically active peptide from porcine intestine. FEBS Lett 164:124–128

Tatsuta M, Itoh T, Okunda S, Tamura H, Baba M, Yamamura H (1981) Inhibition of gastrin secretion by pirenzepine (LS 519) in treatment of gastric ulcer. Scand J Gastroenterol 16:269–271

Taylor IL, Byrne WJ, Christie DL, Ament ME, Walsh JH (1982) Effect of individual L-amino acids on gastric acid secretion and serum gastrin and pancreatic polypeptide release in humans. Gastroenterology 83:272–278

Tepperman BL, Walsh JH, Preshaw RM (1972) Effect of antral denervation on gastrin release by sham feeding and insulin hypoglycemia in dogs. Gastroenterology 63:973–980

Thompson JC, Swierczek JS (1977) Acid and endocrine responses to meals varying in pH in normal and duodenal ulcer subjects. Ann Surg 186:541–548

Thompson JC, Lowder WS, Peurifoy JT, Swierczek JS, Rayford PL (1978) Effect of selective proximal vagotomy and truncal vagotomy on gastric acid and serum gastrin responses to a meal in duodenal ulcer patients. Ann Surg 188: 431–438

Thunberg R (1967) Localization of cells containing and forming histamine in the gastric mucosa of the rat. Exp Cell Res 47:108–115

Tielemans Y, Willems G, Sundler F, Håkanson R (1990) Self-replication of enterochromaffin-like cells in the mouse stomach. Digestion 45:138–146

Todisco A, Park J, Lezoche E, Debas H, Tache Y, Yamada T (1987) Peripheral acid inhibitory action of corticotropin releasing factor: mediation by nongastric mechanisms. Gastroenterology 92:919–924

Tsunoda Y (1986) Gastrin induces intracellular Ca^{2+} release and acid secretion regulation by the microtubular-microfilamentous system. Biochim Biophy Acta 855:186–188

Tsunoda Y, Wider MD (1987) Porcine ileal polypeptide causes an increase in cytoplasmic Ca^{2+} in both parietal and chief cells resulting in acid and pepsinogen secretion. Biochim Biophys Acta 905:118–124

Ui M (1990) G proteins identified as pertussis toxin substrates. In: Naccache PH (ed) G proteins and calcium signaling. CRC, Boca Raton, pp 3–27

Umeda Y, Okada T (1987) Inhibition of gastric acid secretion by human calcitonin gene-related peptide with picomolar potency in guinea-pig parietal cell preparations. Biochem Biophys Res Commun 146:430–436

Urushidani T, Hanzel DK, Forte JG (1987) Protein phosphorylation associated with stimulation of rabbit gastric glands. Biochim Biophys Acta 930:209–219

Uvnäs B (1942) The part played by the pyloric region in the cephalic phase of gastric secretion. Acta Physiol Scand Suppl 13 4:1–85

Uvnäs-Wallensten K, Efendic S, Johannsson C, Sjodin L, Cranwell PD (1981) Effect of intraantral and intrabulbar pH on somatostatin-like immunoreactivity in peripheral venous blood of conscious dogs. The possible function of somatostatin as an inhibitory hormone of gastric acid secretion and its possible identity with bulbogastrone and antral chalone. Acta Physiol Scand 111:397–408

Vagne M, Konturek SJ, Chayvialle JA (1982) Effect of vasoactive intestinal peptide on gastric secretion in the cat. Gastroenterology 83:250–255

Varner AA, Modlin IM, Walsh JH (1981) High potency of bombesin for stimulation of human gastrin release and gastric acid secretion. Regul Pept 1:289–296

Vigna SR, Mantyh CR, Giraud AS, Soll AH, Walsh JH, Mantyh PW (1987) Localization of specific binding sites for bombesin in the canine gastrointestinal tract. Gastroenterology 93:1287–1295

Vigna SR, Mantyh CR, Soll AH, Maggio JE, Mantyh PW (1989) Substance P receptors on canine chief cells: localization, characterization, and function. J Neurosci 9:2878–2876

Vigna SR, Giraud AS, Mantyh PW, Soll AH, Walsh JH (1990) Characterization of bombesin receptors on canine antral gastrin cells. Peptides 11:259–264
Walsh JH (1987) Gastrointestinal hormones. In: Johnson LR (ed) Physiology of the gastrointestinal tract, 2nd edn. Raven, New York, pp 181–253
Walsh JH (1988) Peptides as regulators of gastric acid secretion. Annu Rev Physiol 50:41–63
Walsh JH, Richardson CT, Fordtran JS (1976) pH dependence of acid secretion and gastrin release in normal and ulcer patients. J Clin Invest 57:1125–1131
Walsh JH, Maxwell V, Ferrari J, Varner AA (1981) Bombesin stimulates human gastric function by gastrin-dependent and independent mechanisms. Peptides 2:193–198
Walsh JH, Kovacs TOG, Maxwell V, Cuttitta F (1988) Bombesin-like peptides as regulators of gastric function. Ann NY Acad Sci 547:217–224
Walz DA, Wider MD, Snow JW, Dass C, Desiderio DM (1988) The complete amino acid sequence of porcine gastrotropin, an ileal protein which stimulates gastric acid and pepsinogen secretion. J Biol Chem 263:14189–14195
Way L (1971) Effect of cholecystokinin and caerulein on gastric secretion in cats. Gastroenterology 60:560–565
Webb S, Levy I, Wass JA, Llorens A, Penman E, Casamitjana R, Wu P, Gaya J, Martinez MJ, Riviera F (1984) Studies on the mechanisms of somatostatin release after insulin induced hypoglycemia in man. Clin Endocrinol (Oxf) 21: 667–675
Wider MD, Vinik AL, Heldsinger A (1984) Isolation and partial characterization of an entero-oxyntin from porcine ileum. Endocrinology 115:1484–1491
Wolfe MM, Hocking M, Maico D, McGuigan J (1983a) Effects of antibodies to gastric inhibitory peptide on gastric acid secretion and gastrin release in the dog. Gastroenterology 84:941–948
Wolfe MM, Reel GM, McGuigan JE (1983b) Inhibition of gastrin release by secretin is mediated by somatostatin in cultured rat antral mucosa. J Clin Invest 72: 1586–1593
Wolfe MM, Jain DK, Reel GM, McGuigan JE (1984) Effects of carbachol on gastrin and somatostatin release in rat antral tissue culture. Gastroenterology 87:86–93
Wollin A, Soll AH, Samloff IM (1979) Actions of histamine, secretin, and PGE_2 on cyclic AMP production by isolated canine fundic mucosal cells. Am J Physiol 237:E437–E443
Woodward ER, Lyon ES, Landor J, Dragstedt LR (1954) The physiology of the gastric antrum. Gastroenterology 27:766–785
Wright NA, Pike C, Elia G (1990) Induction of a novel epidermal growth factor-secreting cell lineage by mucosal ulceration in human gastrointestinal stem cells. Nature 343:82–85
Yagci RV, Alptekin N, Rossowski WJ, Brown A, Coy DH, Ertan A (1990) Inhibitory effect of galanin on basal and pentagastrin-stimulated gastric acid secretion in rats. Scand J Gastroenterol 25:853–858
Yamada T, Chiba T (1989) Somatostatin. In: Schultz SG, Makhlouf GM (eds) The gastrointestinal system. American Physiological Society, Bethesda, pp 431–453 (Handbook of physiology, vol 2: Neuronal and endocrine biology, sect 6)
Yamagishi T, Debas H (1980) Gastric inhibitory polypeptide (GIP) is not the primary mediator of the enterogastrone action of fat in the dog. Gastroenterology 78:931–936
Yanagisawa K, Yang H, Walsh JH, Tache Y (1990) Role of acetylcholine, histamine and gastrin in the acid response to intracisternal injection of TRH analog, RX 77368, in the rat. Regul Pept 27:161–170
Yang H, Wong H, Walsh JH, Tache Y (1989) Effect of gastrin monoclonal antibody 28.2 on acid response to chemical vagal stimulation in rats. Life Sci 45:2413–241
Yang H, Wong H, Wu V, Walsh JH, Taché Y (1990) Somatostatin monoclonal antibody immunoneutralization increases gastrin and gastric acid secretion in urethane-anethetized rats. Gastroenterology 99:659–665

Yiangou Y, Christofides ND, Blank MA, Yanaihara N, Tatemoto K, Bishop AE, Polak JM, Bloom SR (1985) Molecular forms of peptide histidine isoleucine-like immunoreactivity in the gastrointestinal tract. Gastroenterology 89:516–524
Yokotani K, Fujiwara M (1985) Effects of substance P on cholinergically stimulated gastric acid secretion and mucosal blood flow in rats. J Pharmacol Exp Ther 232:826–830
Zdon MJ, Zucker KA, Adrian TE, Modlin IM (1987) Somatostatin analogue inhibition of isolated parietal cell secretion. Surgery 102:967–973

CHAPTER 8

Peptides and Enteric Neural Activity

J.J. GALLIGAN

A. Introduction

There have been many advances in recent years in our understanding of the neurophysiology and neurochemistry of enteric neurons. The mechanisms by which neurotransmitters, including some peptides, affect gastrointestinal (GI) function have, in some cases, been described in great detail. There is a large number of peptides contained in and released from enteric nerves and endocrine cells in the GI tract and the localization and characterization of these peptides and their precursors are the subjects of other chapters in this volume. This chapter will focus on the actions of GI peptides on single neurons found in the intramural plexuses of the GI tract, including nerve cells in the gall bladder and pancreas. The focus will be on the receptors present on neurons, the ionic mechanisms by which peptides alter neuronal activity and the transduction systems, if known, by which peptide receptors are coupled to ion channels in enteric nerves.

B. Electrophysiology of Enteric Nerves

The classification scheme of HIRST et al. (1974) will be used when describing subtypes of enteric neurons. There are two types of neuron when using this scheme. "S" neurons receive fast synaptic input and the action potential is blocked by tetrodotoxin (TTX). "AH" neurons do not receive fast synaptic input and the action potential contains a significant calcium component and is only partly blocked by TTX. The action potential in AH neurons is followed by a long-lasting (2–10 s) afterhyperpolarization that is due to a potassium conductance activated by calcium ($G_{K(Ca)}$) entering the neuron during the action potential (NISHI and NORTH 1973; HIRST et al. 1974, 1985; MORITA et al. 1982). The afterhyperpolarization limits the firing rate of AH neurons.

C. Actions of Peptides on Enteric Nerves

I. Tachykinin Peptides

1. Tachykinin Peptides in Enteric Nerves

There are three tachykinin peptides contained in the nervous system: substance P (SP), neurokinin A (NKA, previously called substance K) and neurokinin B (NKB, neuromedin K) (see reviews by EMSON et al. 1987; MAGGIO and MANYTH 1989; Chap. 1, this volume). SP and NKA, but not NKB, have been localized in enteric nerves (DEACON et al. 1987). In addition to multiple tachykinin peptides, there are also multiple sites of action as tachykinins have direct actions on effectors in the GI tract (smooth muscle, blood vessels, epithelial cells) and on neurons. This discussion will focus on the actions of tachykinins on enteric neurons.

2. Tachykinins Depolarize Enteric Neurons

Tachykinins applied by superfusion, by pressure ejection from a micropipette or by ionophoresis depolarize most neurons in guinea pig ileum myenteric plexus (KATAYAMA and NORTH 1978; KATAYAMA et al. 1979; NEMETH et al. 1983; HANANI et al. 1988; JOHNSON et al. 1981; TOKIMASA and NORTH 1984; GALLIGAN et al. 1987; PALMER et al. 1987b), in myenteric neurons from guinea pig cecum (HANANI and BURNSTOCK 1985), in myenteric neurons of guinea pig distal colon (WADE and WOOD 1988), in submucous neurons of guinea pig ileum and cecum (MIHARA et al. 1985; SURPRENANT et al. 1987; AKASU and TOKIMASA 1989) and in myenteric neurons of rat ileum (BROOKES et al. 1988; WILLARD and NISHI 1985; WILLARD 1990a). S and AH neurons are excited by tachykinins (KATAYAMA et al. 1979; GALLIGAN et al. 1987; HANANI et al. 1988).

3. Tachykinins Decrease Resting Potassium Conductance (G_K)

Tachykinin-induced depolarizations are associated with an increase in membrane input resistance indicating that tachykinins close membrane ion channels normally open at rest (KATAYAMA and NORTH 1978; KATAYAMA et al. 1979; NEMETH et al. 1983; HANANI and BURNSTOCK 1985; MIHARA et al., 1985; GALLIGAN et al. 1987; PALMER et al. 1987; HANANI et al. 1988; WADE and WOOD 1988; WILLARD 1990a). When recording responses in normal solutions (5 mM KCl), tachykinin-induced depolarizations reverse polarity near -90 mV, the potassium equilibrium potential (E_K). When the extracellular K^+ concentration was altered, the reversal potential for tachykinin responses changed according to the Nernst equation prediction for responses mediated by changes in potassium conductance (G_K) (KATAYAMA and NORTH 1978; KATAYAMA et al. 1979; WADE and WOOD 1988; GALLIGAN et al. 1987; HANANI et al. 1988; WILLARD 1990a). In submucous S neurons, voltage

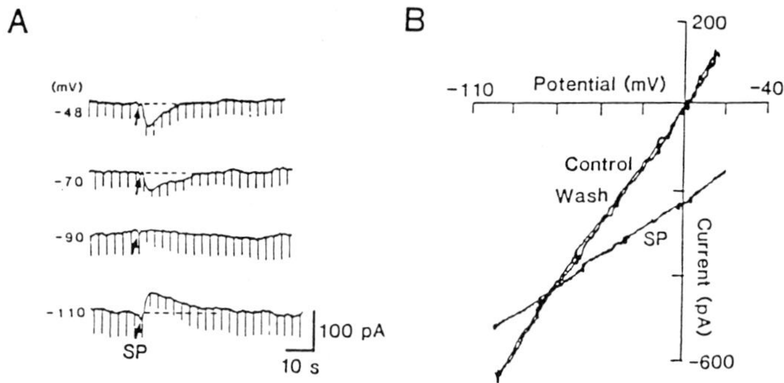

Fig. 1A,B. Reversal of substance P (SP)-induced inward currents. **A** Current recordings obtained during single-electrode voltage clamp experiment in one submucous plexus neuron. Holding potentials are indicated to the left of each current recording: resting potential was −48 mV. A 10-ms SP pressure pulse was applied (*arrows*) at each holding potential; the inward current became smaller with hyperpolarization and reversed to an outward current beyond E_K (approximately −95 mV in this experiment). A decreased membrane conductance is apparent during the SP-induced current responses in the current recordings (downward deflections) produced by 20-mV step hyperpolarizations. **B** Current-voltage (I-V) relations obtained from a different neuron before, during and after a 2-min superfusion with 60 n*M* SP. A slow 2-mV/s ramp command was applied between −50 and −100 mV in control solution and during the peak of the SP current. The SP current intersected the control curve at −90 mV. Resting potential was −58 mV. (From SURPRENANT et al. 1987)

clamp studies of SP-induced inward currents showed that these currents also reversed polarity near E_K and SP currents were voltage independent between −40 and −110 mV (Fig. 1A,B) (SURPRENANT et al. 1987). In addition, SP response amplitude was proportional to the resting membrane potential and input resistance. The higher the initial input resistance, the less negative the resting membrane potential and SP responses were smaller. These data led to the conclusion that SP closes a voltage-independent resting G_K (SURPRENANT et al. 1987).

Tachykinins depolarize myenteric and submucous AH neurons and inhibit the spike afterhyperpolarization ($G_{K(Ca)}$; see above). Depolarization and inhibition of the afterhyperpolarization both contribute to neuronal excitation as the membrane potential is closer to action potential threshold during depolarization and inhibition of the afterhyperpolarization permits repetitive firing. The membrane potential of myenteric and submucous AH neurons is determined, in part, by a resting $G_{K(Ca)}$. This conclusion is based on the results of studies showing that treatments that block calcium channels, such as low Ca^{2+}/high Mg^{2+} solutions or cobalt-containing solutions, depolarize AH neurons and increase input resistance (GRAFE et al. 1980; HIRST et al. 1985; NORTH and TOKIMASA 1987). The resting $G_{K(Ca)}$ is kept open by a persistent inward calcium current. The calcium current is small (100–200 picoamperes) near resting potentials and is insensitive

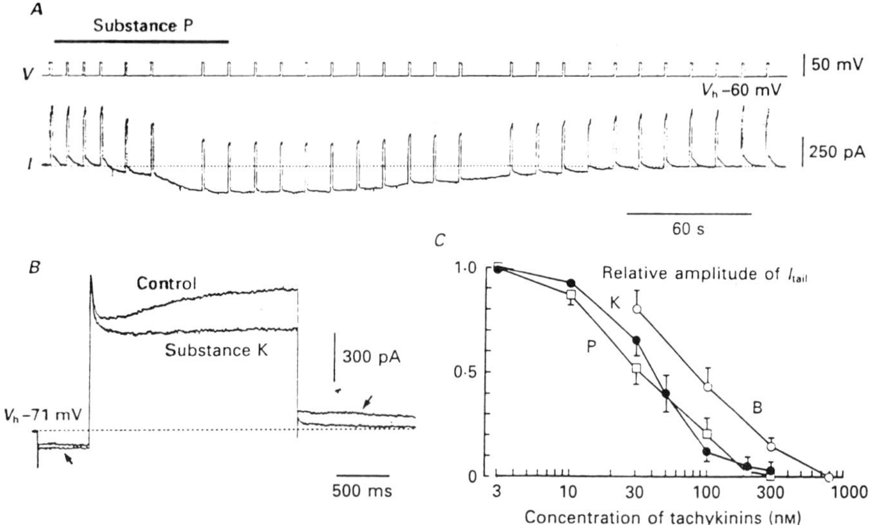

Fig. 2A,B. Tachykinin actions on cation currents ($I_{K(Ca)}$). Recordings in **A** and **B** were obtained from different AH cells. **A** Substance P (200 n*M*) was added to the superfusate during the time *indicated by a bar*. Holding potential, −60 mV. Depolarizing step commands were 30 mV for 1 s. **B** The cell was clamped from a holding potential of −71 mV to −86 mV for 500 ms and then to −11 mV for 2 s. Clamping currents in control and in the presence of neurokinin A (NKA) or substance K, 200 n*M*) are superimposed. A current trace *indicated by arrows* denotes the clamp current in control. Net inward current produced by NKA (40 pA at −71 mV) is not shown. **C** Dose inhibition curves for SP (*open squares*, *P*), NKA (*filled circles*, *K*) and NKB (*open circles*, *B*). The peak amplitude of the outward tail current in control was taken as 100%. Vertical bars are standard errors of mean (n = 3 − 5). (From AKASU and TOKIMASA 1989)

to dihydropyridine antagonists and to omega-conotoxin, blockers of some neuronal calcium channels (AKASU and TOKIMASA 1989). There is evidence that the resting $G_{K(Ca)}$ is the same conductance activated by the action potential (NORTH and TOKIMASA 1987) and it is likely that peptides that inhibit the spike afterhyperpolarization would inhibit resting $G_{K(Ca)}$. This appears to be the case for tachykinin peptides acting on submucous AH neurons. When AH neurons were held at membrane potentials near rest (−50 to −60 mV) and then depolarized to more positive potentials, a slowly activating outward current was observed. Following repolarization, a slowly declining tail current was recorded (Fig. 2A,B). The slowly activating outward current and declining tail current represent activation and deactivation of the persistent $G_{K(Ca)}$ (HIRST et al. 1985; NORTH and TOKIMASA 1987; AKASU and TOKIMASA 1989). In submucous AH neurons, tachykinins induce an inward current and the slowly developing outward current activated by a depolarizing command and tail currents flowing on repolarization are suppressed (Fig. 2A,B). Tachykinins are therefore inhibiting resting $G_{K(Ca)}$ (AKASU and TOKIMASA 1989).

These data do not discriminate between a reduction in calcium entry, alterations in intracellular calcium disposition or a direct effect of tachykinins on the potassium channel as the mechanism by which $G_{K(Ca)}$ is reduced. The actions of muscarine and tachykinins on enteric neurons are very similar (TOKIMASA and NORTH 1984) and data from studies of muscarinic suppression of the spike afterhyperpolarization may provide some insight. Acetylcholine (ACh), acting at muscarinic receptors, and muscarine depolarize myenteric AH neurons and depress the spike afterhyperpolarization (NORTH and TOKIMASA 1983; GALLIGAN et al. 1989). The mechanism by which these drugs suppress the AH was investigated by applying ACh ionophoretically onto myenteric AH neurons *after* the afterhyperpolarization or aftercurrent (when using voltage clamp) had been evoked; that is after calcium had entered the cell and had activated $G_{K(Ca)}$. These studies demonstrated that the time course of the afterhyperpolarization or aftercurrent was shortened by ACh applied after $G_{K(Ca)}$ had been activated. Furthermore, when the calcium-dependent action potential amplitude and duration were measured or when calcium currents were studied directly, it was found that ACh or muscarine did not inhibit calcium entry (NORTH and TOKIMASA 1983; GALLIGAN et al. 1989). Therefore, suppression of the afterhyperpolarization wat not solely dependent on inhibition of calcium entry and other processes (i.e., calcium sequestration, alterations in the K-channel, etc.) are involved.

In addition to $G_{K(Ca)}$, the background or "leak" conductance in enteric AH neurons is reduced by tachykinins. In calcium-free solutions (when $G_{K(Ca)}$ is unavailable), tachykinins produce an inward current. This is a direct demonstration that tachykinin receptors are coupled to a decrease in the background G_K as well as to $G_{K(Ca)}$. The A current or transient potassium current and the inwardly rectifying potassium current were unaffected by tachykinins (AKASU and TOKIMASA 1989).

In summary, tachykinin receptor agonists excite most enteric neurons by decreasing resting G_K. In S neurons, in which resting $G_{K(Ca)}$ does not contribute significantly to the resting membrane potential, only the background G_K is inhibited. In AH neurons, tachykinins suppress resting $G_{K(Ca)}$ and background G_K. Also in AH neurons, the spike afterhyperpolarization is suppressed. Both of these actions in AH neurons lead to neuronal excitation.

4. Tachykinin-Induced Depolarizations Associated with a Conductance Increase

In guinea pig gallbladder neurons (MAWE 1990) and in some myenteric neurons of guinea pig and rat ileum (GALLIGAN et al. 1987; WILLARD and NISHI 1985), SP induces a depolarization associated with a fall in input resistance. In gallbladder neurons, the SP response decreased in amplitude at potentials positive to rest and increased in amplitude at more hyperpolarized potentials (MAWE 1990). In guinea pig myenteric neurons, the reversal potential of this response was estimated to be near −15 mV (GALLIGAN et al. 1987). These data are consistent with either a tachykinin-induced increase in a nonspecific cation conductance or a chloride conductance.

In cultured myenteric neurons of guinea pig cecum and rat ileum, and in myenteric neurons of guinea pig duodenum, SP or senktide (a tachykinin agonist analog, see below) produce a fast depolarization (WILLARD and NISHI 1985; HANANI and BURNSTOCK 1985; HANANI et al. 1988). The ionic basis of the fast response has not been studied.

5. Tachykinin Receptors on Enteric Neurons

Mammalian tachykinin receptors have been designated as neurokinin-1 (NK-1), neurokinin-2 (NK-2) and neurokinin-3 (NK-3) receptors (HENRY 1987). Each receptor type is believed to be the specific receptor for one of the endogenous tachykinin peptides. According to this hypothesis, SP, NKA and NKB are endogenous ligands for the NK-1, NK-2 and NK-3 receptors respectively (BUCK et al. 1984; HENRY 1987).

Some synthetic analogs of SP possess antagonist properties and exhibit selectivity for tachykinin receptor subtypes. [D-Pro^{2}, D-Trp^{7,9}]SP was shown to be a specific antagonist of SP-induced and noncholinergic neurogenic contractions of guinea pig taenia coli (LEANDER et al. 1981) and these initial studies were followed by the development of a number of antagonist peptides with greater potency and reduced agonist activity (REGOLI et al. 1985). In GI tissues, the antagonists were effective at blocking the direct excitatory actions of tachykinins on target cells, i.e., intestinal muscle (KILBINGER et al. 1986), submucosal arterioles (GALLIGAN et al. 1990) and intestinal epithelium (KEAST et al. 1985); the receptor on these target cells is likely to be the NK-1 subtype (formerly SP-P; LEE et al. 1982). However, these analogs were poor antagonists of tachykinin actions on neurons as indicated by data from studies of SP-induced release of [^{3}H]ACh from guinea pig myenteric neurons. The antagonists [D-Arg^{1}, D-Pro^{2}, D-Trp^{7,9}, Leu^{11}]SP, [D-Pro^{2}, D-Trp^{7,9}]SP, [Arg^{5}, D-Trp^{7,9}, Nle^{11}]SP, [D-Arg^{1}, D-Trp^{7,9}, Leu^{11}]SP, and [D-Pro^{4}, D-Trp^{7,9,10}]SP$_{(4-11)}$ either did not block SP-induced ACh release or did so with low affinity (KILBINGER et al. 1986; FEATHERSTONE et al. 1986). Furthermore, it was found that [D-Arg^{1}, D-Pro^{2}, D-Trp^{7,9}, Leu^{11}]SP, [D-Arg^{1}, D-Trp^{7,9}, Leu^{11}]SP, [D-Pro^{2}, D-Trp^{7,9}]SP and [D-Pro^{2}, D-Phe^{7}, D-Trp^{9}]SP did not block SP-induced depolarizations of myenteric or submucous neurons (NEMETH et al. 1983; SURPRENANT et al. 1987; GALLIGAN et al. 1987; WILLARD 1990a). These data indicate that the receptor on nonneural target cells is different from the neuronal receptor, which is unlikely to be the NK-1 receptor.

As described above, SP, NKA and NKB are believed to be endogenous ligands for NK-1, NK-2 and NK-3 receptors respectively. While each peptide is somewhat selective for its proposed receptor, the peptides retain significant activity at other receptors. When the actions of the naturally occurring mammalian tachykinin peptides were studied on myenteric neurons of guinea pig ileum it was found that concentration-response curves for these peptides were similar with a rank order potency of SP > NKA > NKB (GALLIGAN et al. 1987). A similar result was obtained when measuring suppression of $G_{K(Ca)}$ in submucous neurons (Fig. 2C) (AKASU and TOKIMASA 1989). The differences in potency among the naturally occurring peptides (two- to tenfold) were not sufficient to permit firm conclusions regarding the nature of the receptor present on enteric neurons. However, studies of neurogenic contractions of guinea pig ileum after selective desensitization of NK-1 receptors using SP methyl ester (NK-1 selective agonist), or block of the NK-1 receptor with [Arg^6, D-Trp^{7,9}, MePhe^8]SP$_{(6-11)}$, showed that NKB was 30- to 100-fold more potent than SP in eliciting longitudinal muscle contraction. The actions of NKB were neurogenic as they were blocked by TTX (LAUFER et al. 1985). WORMSER et al. (1986) synthesized a series of tachykinin analogs with enhanced selectivity for subtypes of tachykinin receptor. Succinyl-[Asp^6, MePhe^8]SP$_{(6-11)}$ (senktide) was 70000 times more potent at NK-3 versus NK-1 receptors in guinea pig ileum. Senktide applied to myenteric neurons produced responses identical to those of SP; neurons were depolarized, input resistance increased and senktide responses reversed polarity at E_K (HANANI et al. 1988). When SP and senktide were applied to myenteric neurons by pressure ejection from a pipette positioned near impaled neurons, it was found that longer pulses of SP were needed to produce responses equivalent to those elicited by senktide. It was estimated that senktide was 20–100 times more potent than SP at depolarizing myenteric neurons and therefore NK-3 receptors are localized to myenteric neurons in guinea pig ileum (HANANI et al. 1988).

6. Tachykinin Receptors Coupled to Polyphosphoinositide Hydrolysis

Forskolin is a diterpine that stimulates adenylate cyclase and increases intracellular cAMP (SEAMON and DALY 1981). In myenteric neurons, forskolin mimics the action of tachykinins; it depolarizes neurons, increases input resistance and depolarizations reverse polarity at E_K (NEMETH et al. 1986). Adenosine inhibits forskolin-induced depolarizations but not those induced by intracellularly applied adenosine-3′,5′-cyclic monophosphate (cAMP) or membrane permeable analogs of cAMP and it was concluded that adenosine acted directly on adenylate cyclase to inhibit cAMP formation (PALMER et al. 1987a). Adenosine did not affect responses to SP, and forskolin and SP seemed to produce their actions via different intracellular pathways (PALMER et al. 1987b). Biochemical studies of SP actions on guinea pig and rat ileum showed that SP promoted breakdown of phosphoinositol (PI) and accumulation of InsP, $InsP_2$ and to a lesser extent $InsP_3$ (WATSON and DOWNES 1983; HOLZER and LIPPE 1985) but increases in cAMP levels were not observed (WATSON 1984). These results were obtained in preparations containing nerves and muscle and a proportion of inositol phosphate measured would result from tachykinin stimulation of muscle cells. More recently, it has been shown that SP, NKA, NKB, SP-methyl ester and senktide stimulated the accumulation of inositol phosphates, with senktide being 100-fold more potent than SP. The actions of maximally effective concentrations of senktide and SP-methyl ester were additive, indicating two different mechanisms or sites of action while the effects of SP and NKB were not additive, indicating a common site of action. The NK-1 antagonist [D-Pro^2, D-$Trp_{7,9,10}$]$SP_{(4-11)}$ blocked inositol phosphate accumulation stimulated by SP-methyl ester but not by senktide. Therefore, NK-1 receptors on muscle and NK-3 receptors on enteric neurons may both be coupled to PI hydrolysis as an intracellular transduction system (GUARD et al. 1988). Some phorbol esters mimic the actions of diacylglycerol (DAG) in stimulating protein kinase C; DAG is a product of PI hydrolysis (CASTAGNA et al. 1982). Phorbol esters applied to enteric neurons mimicked the actions of tachykinins; neurons were depolarized, input resistance was increased and spike afterhyperpolarizations in AH neurons were suppressed (MIHARA et al. 1987b; NORTH et al. 1987). These observations are consistent with PI hydrolysis being an intermediate step in the actions of tachykinins on enteric nerves.

7. Synaptic Activation of Tachykinin Receptors

The evidence that SP and other tachykinins act as neurotransmitters released from enteric nerves has been reviewed previously (BARTHO and HOLZER 1985). During intracellular recordings from S and AH neurons, repetitive stimulation of interganglionic nerve strands produces a noncholinergic slow excitatory synaptic potential (slow EPSP) that is due to a decrease in G_K (KATAYAMA and NORTH 1978; WOOD and MAYER 1978; JOHNSON et al. 1980; GRAFE et al. 1980; SURPRENANT 1984; MIHARA et al. 1985; BROOKES et al. 1988). The slow EPSP is identical to responses produced by tachykinins (see above) and the data available favor the hypothesis that SP is a mediator of slow synaptic transmission between enteric neurons. Specific blockade of slow EPSPs with an NK-3 antagonist would provide the most direct evidence in support of this hypothesis. Such an antagonist has not been used in studies of enteric nerves up to this time, but alternative studies have provided some convincing data. In addition to identity of action, SP is released from enteric neurons in a calcium-dependent manner on electrical stimulation (BARON et al. 1983; HOLZER 1984). Proteolytic enzymes, such as chymotrypsin and carboxypeptidase A, reduce in a reversible manner the amplitude of some slow EPSPs recorded from myenteric neurons, suggesting that a peptide may be a mediator. Furthermore, SP responses and slow EPSPs cross-desensitize (JOHNSON et al. 1981). SP is contained in myenteric neurons with short (<0.5 mm) axonal projections (COSTA et al. 1981). When recordings were made from myenteric neurons under conditions when long axonal projections had been cut, slow EPSPs could still be evoked by electrical stimulation of interganglionic nerve strands. Therefore, some neurons releasing the mediator(s) of slow EPSPs have short projections, as do SP-containing neurons (BORNSTEIN et al. 1984). In cultured myenteric neurons it is possible to stimulate a single neuron and to record intracellularly from a second neuron synaptically connected to the first cell (WILLARD 1990a). Repetitive stimulation (5 Hz and greater) of the presynaptic neuron evoked a slow EPSP in the second cell. The slow EPSP was mimicked by SP and the slow EPSP was blocked reversibly by SP antisera. Immunohistochemical studies of presynaptic neurons revealed that most cells contained immunoreactivity for SP (WILLARD 1990a). Finally it is interesting to note that, while NKB may be a NK-3 receptor agonist, NKB has not been detected in enteric neurons. It is possible that SP and/or NKA present in and released from enteric neurons are the ligands acting at enteric NK-3 receptors (DEACON et al. 1987).

II. Vasoactive Intestinal Peptide

1. Vasoactive Intestinal Peptide Receptors Coupled to G_K Decrease

In myenteric plexus of guinea pig small intestine, vasoactive intestinal peptide (VIP) increased the firing rate of neurons during extracellular recordings (WILLIAMS and NORTH 1979). Intracellular recordings from myenteric neurons of guinea pig and rat small intestine and submucous neurons of guinea pig cecum showed that VIP depolarized neurons and depolarizations were accompanied by an increased input resistance. Spike afterhyperpolarizations in AH neurons were suppressed by VIP and VIP responses reversed polarity at E_K (MIHARA et al. 1985; ZAFIROV et al. 1985; PALMER et al. 1987b; WILLARD 1990b).

2. Vasoactive Intestinal Peptide Receptors May Be Coupled to Adenylate Cyclase in Enteric Nerves

In guinea pig myenteric plexus, the actions of forskolin, membrane permeable analogs of cAMP and phosphodiesterase inhibitors were identical to those of VIP (NEMETH et al. 1986; PALMER et al. 1986b, 1987b). Adenosine pretreatment blocked the actions of forskolin and VIP (PALMER et al. 1987b; see also Sect. I.6). These observations are consistent with, but do not prove, the hypothesis that cAMP is an intracellular signal for the actions of VIP in enteric nerves.

3. Synaptic Potentials Mediated by Vasoactive Intestinal Peptide

As with other peptides, the availability of potent and selective receptor antagonists has hindered the study of synaptic events that may be mediated by VIP. In cultured rat myenteric neurons, stimulation of some presynaptic neurons evoked both fast and slow EPSPs in postsynaptic cells. Fast EPSPs were blocked by hexamethonium and the presynaptic neurons were concluded to be cholinergic. Slow EPSPs were not blocked by atropine and were concluded to be noncholinergic. The presynaptic neurons releasing ACh and the noncholinergic mediator of the slow EPSP were called dual function neurons as they released two transmitters (WILLARD 1990b). Immunohistochemical studies of presynaptic dual function neurons revealed that many of these cells contained VIP immunoreactivity (WILLARD and NISHI 1987; WILLARD 1990b). VIP antisera reversibly blocked slow EPSPs evoked by stimulation of dual function neurons without affecting the resting membrane potential or fast EPSPs recorded from postsynaptic neurons. SP antisera did not block slow EPSPs recorded from the postsynaptic neurons (WILLARD 1990b). These data indicate that some dual function neurons in cultured rat myenteric neurons may use VIP or a closely related substance as a noncholinergic excitatory transmitter.

III. Calcitonin Gene-Related Peptide

1. Calcitonin Gene-Related Peptide Receptors Coupled to G_K Decrease

Calcitonin gene-related peptide (CGRP) applied to myenteric AH neurons in guinea pig ileum produced a slow depolarization associated with an increased input resistance and inhibition of spike afterhyperpolarizations; CGRP depolarization reversed polarity at E_K (PALMER et al. 1986a,b, 1987b). These data suggest that CGRP depolarizes myenteric AH neurons by decreasing resting G_K.

2. Transduction Mechanism

As described above, treatments which directly or indirectly raise intracellular cAMP mimic slow EPSPs and mimic the actions of CGRP (PALMER et al. 1986a,b, 1987b). CGRP-induced depolarizations were not blocked by adenosine pretreatment, suggesting that CGRP-induced excitation utilizes a nonadenylate cyclase-dependent pathway (see Sect. C.I.6). However, studies of CGRP-induced ACh release from cultured myenteric neurons of guinea pig taenia coli have shown that this stimulatory action is blocked by dideoxyadenosine, an inhibitor of adenylate cyclase (MULHOLLAND and JAFFER 1990). These conflicting results could be attributed to differences in experimental methods but it is clear that more work is needed to define intracellular signaling pathways used by CGRP.

IV. Cholecystokinin

1. Cholecystokinin Receptors Coupled to a G_K Decrease

In vitro, cholecystokinin (CCK) excites myenteric and submucous neurons (SATO et al. 1973; NEMETH et al. 1985; NGU 1983). Intracellular recordings from guinea pig myenteric neurons have shown that CCK applied either by superfusion or by pressure ejection from a pipette positioned near the neuron depolarized more than half of the S and AH neurons tested (NEMETH et al. 1985). In the same study, 20% of myenteric neurons were inhibited by CCK but this response was not studied in detail. CCK-induced depolarizations were associated with an increase in input resistance and reversed polarity at E_K. The spike afterhyperpolarization in AH cells was suppressed. These data indicate that CCK inhibits resting G_K of myenteric neurons, resulting in a slow membrane depolarization and an increased neuronal excitability (NEMETH et al. 1985).

2. Transduction Mechanism

In guinea pig myenteric neurons, CCK-induced depolarizations are blocked by adenosine, a treatment which antagonizes adenylate cyclase-dependent depolarizations of these neurons (PALMER et al. 1987a,b). However, further study is required before definitive conclusions can be reached concerning the transduction pathway utilized by CCK in enteric nerves.

3. Receptors on Enteric Neurons

In addition to depolarizing enteric neurons, CCK is an effective stimulant of nerve-mediated GI contractions and of ACh release from enteric nerves. The nonpeptide CCK antagonist L-364,718 binds to peripheral CCK receptors but not to those in the CNS (CHANG and LOTTI 1986). L-364,718 partially blocked CCK-induced ACh release from guinea pig myenteric plexus (ZELLES et al. 1990; LU et al. 1991), indicating that the CCK receptor on enteric neurons is of the peripheral subtype (CCK_A). In addition, L-365,260, an antagonist of CCK receptors in the brain (CCK_B), also partly blocked CCK-induced release of ACh from myenteric neurons. When tissues were incubated with both antagonists, CCK-induced release of ACh was inhibited completely (LU et al. 1991). These data indicate that CCK_A and CCK_B receptors are localized to myenteric neurons in guinea pig small intestine.

4. Cholecystokinin Facilitates Neurotransmission in Pancreatic and Gallbladder Ganglia

Stimulation of preganglionic nerve trunks entering intramural ganglia in cat pancreas (MA and SZURSZEWSKI 1991), guinea pig (MAWE 1991) and opossum gallbladder (BAUER et al. 1991) evoked fast nicotinic EPSPs during intracellular recordings from ganglion cells. EPSP amplitude was increased during treatment with CCK octapeptide (CCK-8). In all three studies, CCK-8 did not change the resting properties of postsynaptic cells. In pancreatic ganglia, CCK-8 induced increases in EPSP amplitude were due to an increase in the number of ACh quanta released by each action potential and also to an increase in the sensitivity of the nicotinic receptor to ACh (MA and SZURSZEWSKI 1991). In guinea pig and opossum gallbladder neurons, ACh sensitivity of postsynaptic cells was unaffected by CCK-8 (BAUER et al. 1991; MAWE 1991). In guinea pig gallbladder, the facilitating action of CCK-8 was mediated at CCK_A receptors (MAWE 1991).

V. Gastrin-Releasing Peptide/Bombesin

Gastrin-releasing peptide (GRP) or bombesin depolarized guinea pig ileum myenteric AH neurons and submucous neurons of guinea pig cecum (ZAFIROV et al. 1985; MIHARA et al. 1985). In AH neurons, this response was associated with increased input resistance and inhibition of spike after-hyperpolarizations. Depolarizations reversed polarity at E_K, suggesting that GRP and bombesin depolarize myenteric AH neurons by suppressing resting G_K (ZAFIROV et al. 1985). The actions of GRP were antagonized by adenosine pretreatment, suggesting that GRP may act via an adenylate cyclase dependent pathway (PALMER et al. 1987b).

VI. Motilin

Porcine motilin depolarizes myenteric neurons in guinea pig antrum (TACK et al. 1991). Almost one-half of neurons tested were depolarized by motilin. The depolarization was associated with an increase in input resistance and the reversal potential was between −80 and −90 mV (near E_K).

VII. Opioid Peptides

1. Opioid Peptides Hyperpolarize Enteric Neurons and Increase G_K

Extracellular recordings showed that [Met^5]- and [Leu^5]-enkephalin inhibit spontaneous firing of myenteric neurons in guinea pig small intestine (NORTH and WILLIAMS 1976). Studies using intracellular recording methods showed that enkephalins hyperpolarized guinea pig and rat myenteric neurons and hyperpolarizations were associated with a fall in input resistance (NORTH et al. 1979; WILLARD and NISHI 1985). The amplitude of opioid-induced hyperpolarizations changed as a function of extracellular K^+ concentration and responses reversed polarity at E_K (MORITA and NORTH 1982). These findings were extended to submucous neurons of guinea pig intestine, where opioid-induced hyperpolarizations were associated with a decrease in input resistance and reversed polarity at E_K (MIHARA and NORTH 1986; NORTH et al. 1987; TATSUMI et al. 1990). Opioid-induced hyperpolarizations of myenteric and submucous neurons were most often observed in S neurons (SURPRENANT and NORTH 1985; MIHARA and NORTH 1986; GALLIGAN and NORTH 1991; PILLAI and JOHNSON 1991).

2. Opioid Peptides Activate an Inwardly Rectifying G_K

The G_K activated by opioids in submucous neurons of guinea pig intestine is an inwardly rectifying conductance. The current-voltage (I-V) relationship of opioid-induced currents was steeper as membrane potentials approached and became negative to E_K (NORTH et al. 1987; TATSUMI et al. 1990). In normal [K^+], there was a threefold increase in opioid-induced conductance as cells were hyperpolarized from potentials near rest to E_K. The opioid conductance was blocked by rubidium ions and inward movement of K^+ through the opioid-activated channel was blocked by extracellular cesium ions (NORTH et al. 1987).

3. Transduction Mechanism for Opioid-Activated G_K

The mechanism by which opioid receptors couple to G_K has been studied in guinea pig submucous neurons (NORTH et al. 1987; TATSUMI et al. 1990). Opioid-induced currents were not altered by forskolin, dibutyryl cAMP, or phorbol esters. These data indicated that the opioid-induced conductance was not activated by protein kinase A or C dependent pathways. When recordings were made with electrodes containing the nonhydrolyzable guanosine triphosphate (GTP) analog GTPγS, opioid-induced currents or hyperpolarizations did not recover or recovered slowly after agonist washout. Furthermore, in tissues pretreated with pertussis toxin (PTX), to inactivate some guanine nucleotide-binding (G) proteins, opioid-induced hyperpolarizations were not observed. When using whole cell patch clamp methods to record from PTX-treated neurons, opioid-induced outward currents could be restored if neurons were dialyzed with activated G_i or G_o proteins (TATSUMI et al. 1990). These data indicate that opioid receptors couple to inwardly rectifying potassium channels via a G protein.

4. Opioid Receptors on Enteric Neurons

The availability of selective agonists and antagonists for opioid receptor subtypes has permitted characterization of receptors localized to enteric neurons. In guinea pig myenteric neurons, μ-opioid receptors are coupled to G_K. Myenteric neurons were hyperpolarized by normorphine and Met^5-enkephalin and the actions of these agonists were blocked by low concentrations (30 nM) of naloxone (SURPRENANT and NORTH 1985). Although normorphine and [Met^5]-enkephalin show only moderate selectivity for μ-receptors, hyperpolarizations of myenteric neurons were also brought about by the selective μ-agonist, [D-Ala^2, N-$MePhe^4$, Gly^5-ol]enkephalin (DAGOL) (PILLAI and JOHNSON 1991).

The δ-opioid receptor is coupled to G_K in guinea pig submucous neurons (MIHARA and NORTH 1986; NORTH et al. 1987; TATSUMI et al. 1990). Normorphine and DAGOL do not change membrane potential while, in the same neurons, the δ-selective agonist [D-Pen^2, D-Pen^5]enkephalin (DPDPE) hyperpolarized most neurons (NORTH et al. 1987). Opioid-induced hyperpolarizations were blocked by naloxone and the δ-selective antagonist [*N*-bisallyl(aminoisobutyrate)Leu^5]enkephalin (ICI 174,864) (MIHARA and NORTH 1986).

5. Opioid Receptors Coupled to Inhibition of Calcium Channels

Whole cell calcium currents recorded from acutely dissociated submucosal neurons of guinea pig ileum and cecum were not blocked by dihydropyridine antagonists but were blocked by omega-conotoxin (SURPRENANT et al. 1990). These calcium currents showed some inactivation and were similar to N-type calcium currents described by NOWYCKY et al. (1985). [Met^5]enkephalin produced a concentration-dependent (0.1–10 μ*M*) decrease in calcium currents that was blocked by naloxone and ICI 174,864 and was mimicked by DPDPE. DAGOL and dynorphin (κ-agonist) did not inhibit calcium currents in submucous neurons. Enkephalin-induced inhibition of calcium currents was not observed in PTX-pretreated cells but inhibition was restored by dialysis of these cells with purified G_i and/or G_o. Dialysis of neurons with GTPγS mimicked opioid-induced inhibition of calcium currents. These data indicate that δ-opioid receptors couple via a G protein (G_i/G_o) to inhibition of calcium currents in guinea pig submucous neurons (SURPRENANT et al. 1990).

In myenteric neurons of guinea pig and rat intestine, opioid receptors may be coupled to inhibition of calcium currents. This tentative conclusion is supported by the observation that opioid peptides decreased the duration of calcium-dependent action potentials in AH neurons (CHERUBINI and NORTH 1985; WILLARD and NISHI 1985). In guinea pig myenteric neurons, inhibition of calcium-dependent action potentials was mediated at κ-receptors (CHERUBINI and NORTH 1985).

6. Functional Implications of Opioid Peptide Action on Enteric Nerves

One of the most extensively studied actions of opioids has been inhibition of ACh release from myenteric neurons in guinea pig ileum (PATON 1957; SCHAUMAN 1957). The cellular mechanism by which opioids inhibit neurotransmitter release has not been established unequivocally. Hyperpolarization of the soma or nerve fibers preventing action potential initiation or propagation, hyperpolarization of nerve terminals preventing action potential invasion of terminals or direct inhibition of voltage-dependent calcium entry are all potential mechanisms for inhibition of transmitter release (MORITA and NORTH 1982; CHERUBINI et al. 1985; SHEN and SURPRENANT 1990). In guinea pig intestine, opioids inhibit neurogenic contractions in vitro by inhibiting ACh release from enteric motoneurons but in other species, such as human and dog, opioids stimulate contractions (BURKS et al. 1982; BURKS 1986). Opioid-induced contractions are non-propulsive accounting, in part, for the antidiarrheal actions of opioids (BURKS 1986). However, the apparent stimulatory action of opioids on intestinal contractions is inconsistent with the known inhibitory actions of opioid peptides on enteric neurons. Recently it has been shown that opioids suppress inhibitory neuromuscular transmission in dog small intestine and human colon in vitro (BAUER and SZURSZEWSKI 1991; HOYLE et al. 1990). This was established by measuring the amplitude of inhibitory junction potentials (IJPs) recorded from circular muscle. In dog intestine both μ-and δ-agonists inhibited IJPs while in human colon this action was mediated at δ-receptors. In both studies, it was concluded that opioid receptors were localized to nerve terminals as opioids inhibited IJPs in preparations in which the myenteric plexus had been removed. Suppression of inhibitory neuromuscular transmission could contribute to the uncoordinated, non-propulsive motility observed in dogs and humans following opioid treatment (BAUER and SZURSZEWSKI 1991; HOYLE et al. 1990).

VIII. Somatostatin

1. Somatostatin Hyperpolarizes Enteric Neurons by Increasing an Inwardly Rectifying G_K

Extracellular recordings from myenteric neurons of guinea pig small intestine showed that somatostatin (SS) inhibited spontaneous firing (WILLIAMS and NORTH 1978). When using intracellular methods, SS applied by ionophoresis or by superfusion depolarized some myenteric S and AH neurons. Depolarization was more commonly observed following ionophoretic application of SS from electrodes containing high concentrations of SS (KATAYAMA and NORTH 1980). SS-induced depolarizations were also observed in a subsequent study using pressure application from a pipette containing high concentrations (>1 m*M*) of SS (MIHARA et al. 1987b). Depolarization has been attributed to a nonspecific action of high peptide concentrations (MIHARA et al. 1987b).

Somatostatin hyperpolarized 90% of neurons in guinea pig ileum and cecum submucous plexus (MIHARA et al. 1987b). SS hyperpolarizations were recorded in low-Ca^{2+}/high-Mg^{2+} solutions, indicating a direct action of SS on neurons from which recordings were obtained. Hyperpolarizations were associated with decreased input resistance and reversed polarity at E_K. Under voltage clamp, SS-currents reversed polarity at E_K (Fig. 3A) and the SS conductance increased as the membrane potential approached E_K (Fig. 3B). That is, the SS conductance rectified inwardly. SS-outward currents were blocked by rubidium ions and inward currents (recorded negative to

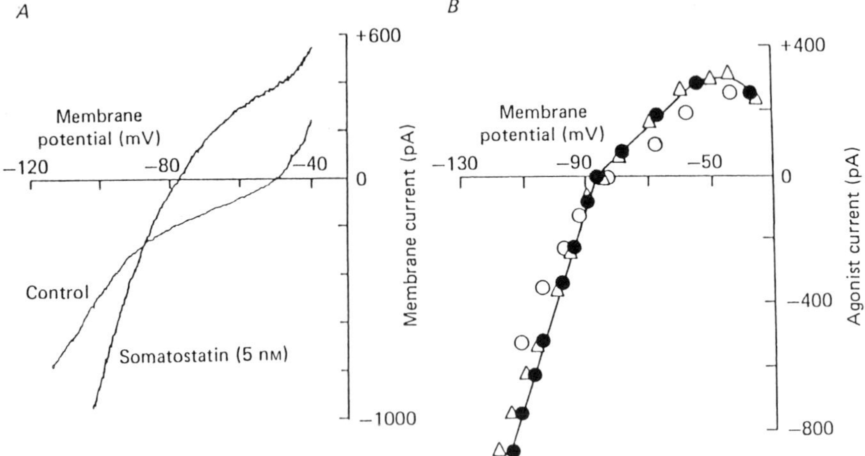

Fig. 3A,B. Submucous neurons show inward rectification in the resting state and in the presence of somatostatin. **A** Steady state I-V relationships obtained in control and at the peak of outward current produced by 5 n*M* somatostatin. Control I-V curve shows increased slope at potentials negative to −86 mV; I-V curve obtained in the presence of somatostatin shows increased membrane conductance, inward rectification beyond E_K and reversal from outward to inward current at −86 mV. **B** I-V relations obtained by subtracting control I-V curves from I-V curves obtained in the presence of somatostatin (5 n*M*, *filled circles*), the δ-opioid agonist DPDPE (200 n*M*, *open circles*) and the α_2-adrenoceptor agonist UK 14304 (200 n*M*, *open triangles*). Agonist-induced currents show inward rectification at potentials negative to −86 mV. (From MIHARA et al. 1987b)

E_K) were blocked in a voltage-dependent manner by external cesium ions (MIHARA et al. 1987b).

2. AG-Protein Couples the Somatostatin Receptor to the Potassium Channel

Somatostatin-induced hyperpolarizations and outward currents were unaffected by forskolin or phorbol esters (MIHARA et al. 1987b). When recordings were made with electrodes containing GTPγS, SS hyperpolarizations were maintained after agonist washout. Furthermore, whole cell outward currents induced by SS were blocked when the pipette contained PTX (TATSUMI et al. 1990). These data point to the involvement of a G protein in coupling the SS receptor to potassium channels (MIHARA et al. 1987b; NORTH 1989; TATSUMI et al. 1990).

3. Nonadrenergic Inhibitory Postsynaptic Potential Mimicked by Somatostatin

In extrinsically denervated submucosal preparations of guinea pig ileum, repetitive stimulation of interganglionic nerve strands evoked a slow IPSP in some neurons. The IPSP was not blocked by α_2-adrenergic antagonists and reversed polarity E_K. At this time, SS is the only known substance contained in submucosal neurons which mimics the IPSP (MIHARA et al. 1987a). However, definitive proof that SS is a mediator of slow IPSPs awaits studies done with specific SS receptor blockers.

4. Somatostatin Inhibits Calcium Currents

Whole cell calcium currents recorded from acutely dissociated submucosal neurons of guinea pig ileum and cecum were described above. SS produced a concentration-dependent (1–100 n*M*) decrease in peak calcium current. When the holding potential was −70 mV and the command potential was 10 mV, SS (10 n*M*) reduced peak current by 50%. SS-induced inhibition of calcium currents was not observed in PTX-pretreated cells but inhibition was partly restored by intracellular dialysis with purified G_i and G_o. In addition, SS-induced inhibition of calcium currents was mimicked by intracellular dialysis with GTPγS. In submucous neurons, SS receptors are therefore coupled to inhibition of calcium currents via a G protein (SURPRENANT et al. 1990).

5. Functional Consequences of Somatostatin Action on Enteric Neurons

Somatostatin inhibits intestinal contractions via an action on cholinergic nerves (GUILLEMIN 1976) and inhibits intestinal secretion by inhibiting secretomotor nerves (KEAST et al. 1986). These functional responses have a basis in the actions of SS on single enteric neurons. SS inhibits fast nicotinic synaptic transmission between enteric nerves. Inhibition of synaptic transmission is attributed to hyperpolarization of the nerve cell body which blocks action potential initiation (SHEN and SURPRENANT 1990). SS-induced inhibition of neuronal calcium currents would also contribute to inhibition of synaptic or neuroeffector transmission. Hyperpolarization of myenteric motor nerves and submucous secretomotor nerves and/or inhibition of calcium currents would therefore account for SS-induced inhibition of nerve-mediated contractions and secretomotor responses (see also Chaps. 9, 10).

IX. Galanin

Intracellular recordings from myenteric neurons of guinea pig ileum showed that galanin hyperpolarized AH neurons. Hyperpolarizations were associated with a decreased input resistance and reversed polarity near E_K (TAMURA et al. 1988). Calcium-dependent action potentials in AH neurons (potassium channels blocked by extracellular tetraethylammonium and 4-aminopyridine and intracellular cesium chloride) were shortened in duration by galanin (TAMURA et al. 1988). In addition fast EPSPs recorded from myenteric S neurons were inhibited by galanin. It is unclear if this was a presynaptic action or a reduction in the sensitivity of postsynaptic cells to ACh as responses to locally applied ACh were reduced by galanin (TAMURA et al. 1987). Inhibition of calcium currents and membrane hyperpolarization both could contribute to the inhibitory effects of galanin on synaptic transmission.

X. Neuropeptide Y

The actions of NPY on whole cell and single calcium currents have been studied in cultured rat myenteric neurons (HIRNING et al. 1990). In these neurons two types of calcium currents were examined. In the whole cell mode, there were inactivating and non-inactivating calcium currents. The inactivating component had the characteristics of "N" type channels while the non-inactivating calcium current was blocked by dihydropyridines and was called an "L"-type current (NOWYCKY et al. 1985). NPY ($0.1\,\mu M$) reduced peak calcium current by 40% and was more effective at reducing N than L currents (Fig. 4A–C). Measurements of intracellular calcium were accomplished using microfluorimetric methods and the calcium indicator dye fura-2. Exposing neurons to $50\,mM$ K^+ produced an increase in intracellular calcium that was blocked by NPY (Fig. 4D). NPY-mediated inhibition of the rise in intracellular calcium was not observed in PTX-pretreated cells, suggesting the involvement of a G protein (HIRNING et al. 1990).

Single-channel calcium currents recorded in the cell-attached mode revealed two types of calcium channel; these corresponded to N (14-pico Siemens (pS) single-channel conductance) and L (27-pS single-channel conductance) type calcium channels. NPY applied outside the patch pipette did not affect calcium channel activity, indicating that NPY does not utilize a diffusible second messenger to inhibit neuronal calcium currents. Instead there must be close receptor-channel coupling apparently mediated by a PTX-sensitive G protein (HIRNING et al. 1990).

D. Conclusions

While there are a large number of neuropeptides and peptide hormones, the ion channels and transduction mechanisms by which peptides can act on enteric neurons are limited. In general, peptides have been shown to act via three principal mechanisms on enteric neurons. Peptides act at: (1) receptors coupled to a decrease in resting G_K; (2) receptors coupled to an increase in G_K or (3) receptors coupled to an inhibition of voltage-activated calcium channels. Peptide receptors are coupled to ion channels via three potential intracellular transduction pathways. The available data are indirect but indicate that peptide receptors coupled to a decrease in G_K may utilize either cAMP-dependent mechanisms or a pathway dependent on PI hydrolysis and protein kinase C activation. Peptide receptors coupled to increases in G_K do not use cAMP-dependent or PI/protein kinase C-dependent pathways. The coupling of some peptide receptors to the inwardly rectifying potassium channel in enteric nerves is similar to that occurring

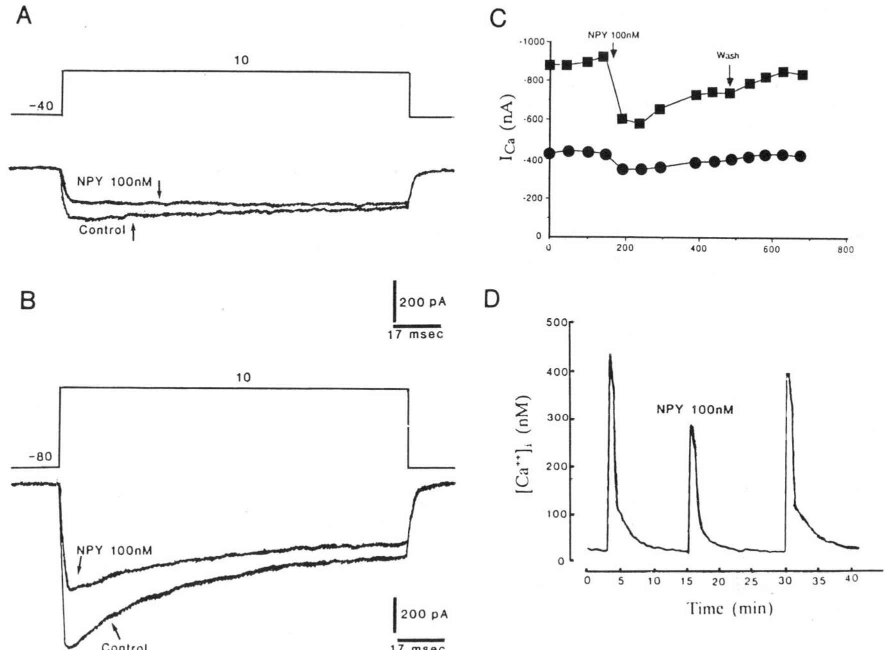

Fig. 4A–D. Inhibition of N-type Ca^{2+} channel currents by neuropeptide Y (NPY). **A,B** Currents elicited from holding potential of −40 mV (**A**) and holding potential of −80 mV, **B** before and after addition of 100 n*M* NPY. **C** Time course of NPY application. NPY (100 n*M*) was added *at arrow* and washed out *at arrow labeled WASH*. *Squares* plot the peak current response; *circles* plot the effects of NPY on sustained currents at the end of the depolarization. **D** Effects of 100 n*M* NPY on 50 m*M* K^+-induced intracellular calcium transients assayed by fura-2 spectrofluorometry. Middle peak obtained in the presence of NPY (100 n*M*). NPY was applied 60 s before high [K^+]. (From HIRNING et al. 1990)

between muscarinic receptors and potassium channels in cardiac tissue (PFAFFINGER et al. 1985). In the heart, the muscarinic receptor is coupled directly to the potassium channel via the activated α-subunit of a G protein (YATANI et al. 1987). It is possible that a similar mechanism is operative in enteric nerves (NORTH et al. 1987; MIHARA et al. 1987b; TATSUMI et al. 1990). This mechanism may also couple peptide receptors to calcium channels in enteric neurons (HIRNING et al. 1990).

References

Akasu T, Tokimasa T (1989) Potassium currents in submucous neurones of guinea pig caecum and their synaptic modification. J Physiol (Lond) 416:571–588

Baron SA, Jaffe BM, Gintzler AR (1983) Release of substance P from the enteric nervous system: direct quantitation and characterization. J Pharmacol Exp Ther 227:365–368

Bartho L, Holzer P (1985) Search for a physiological role for substance P in gastrointestinal motility. Neuroscience 16:1–32

Bauer AJ, Szurszewski JH (1991) Effect of opioid peptides on circular muscle of canine duodenum. J Physiol (Lond) 434:409–422

Bauer AJ, Hanani M, Muir TC, Szurszewski JH (1991) Intracellular recordings from gall bladder ganglia of opossums. Am J Physiol 260:299–306

Bornstein JC, North RA, Costa M, Furness JB (1984) Excitatory synaptic potentials due to activation of neurons with short projections in the myenteric plexus. Neuroscience 11:723–731

Brookes SJH, Ewart WR, Wingate DL (1988) Intracellular recordings from cells in the myenteric plexus of the rat duodenum. Neuroscience 24:297–307

Buck SH, Burcher E, Shults C, Lovenberg W, O'Donohue TL (1984) Novel pharmacology of substance K binding sites: a third type of tachykinin receptor. Science 226:987–989

Burks TF (1986) Actions of drugs on gastrointestinal motility. In: Johnson LR, Christensen J, Jackson MJ, Jacobson ED, Walsh JH (eds) Physiology of the digestive tract, vol 1, 2nd edn. Raven, New York, p 723

Burks TF, Hirning LD, Galligan JJ, Davis TP (1982) Motility effects of opioid peptides in dog intestine. Life Sci 31:2237–2240

Castagna M, Takai Y, Kaibuchi K, Sano K, Kikkawa U, Nishizuka Y (1982) Direct activation of calcium activated phospholipid-dependent protein kinase by tumor promoting phorbol esters. J Biol Chem 257:7847–7851

Chang RSL, Lotti VJ (1986) Biochemical and pharmacological characterization of an extremely potent and selective nonpeptide cholecystokinin antagonist. Proc Natl Acad Sci USA 83:4923–4926

Cherubini E, North RA (1985) Mu and kappa opioids inhibit transmitter release by different mechanisms. Proc Natl Acad Sci USA 82:1860–1863

Cherubini E, Morita K, North RA (1985) Opioid inhibition of synaptic transmission in the guinea pig myenteric plexus. Br J Pharmacol 85:805–817

Costa M, Furness JB, Llewellyn-Smith IJ, Cuello AC (1981) Projections of substance P-containing neurons within the guinea pig small intestine. Neuroscience 6: 411–424

Deacon CF, Agoston DV, Nau R, Conlon JM (1987) Conversion of neuropeptide K to neurokinin A and vesicular co-localization of neurokinin A and substance P in neurons of the guinea pig small intestine. J Neurochem 48:141–136

Emson PC, Diez-Guerra FJ, Hirai H (1987) Mammalian tachykinins: neurochemistry and pharmacology. In: Turner AJ (ed) Neuropeptides and their peptidases. VCH, Weinheim, p 87

Featherstone RL, Fosbraey P, Morton IKM (1986) A comparison of the effect of three substance P antagonists on tachykinin stimulated [^{3}H]acetylcholine release in the guinea-pig ileum. Br J Pharamcol 87:73–78

Galligan JJ, North RA (1991) Opioid, 5-HT_{1A} and α_2 receptors localized to subsets of guinea pig myenteric neurons. J Auton Nerv Syst 32:1–12

Galligan JJ, North RA, Tokimasa T (1989) Muscarinic agonists and potassium currents in guinea-pig myenteric neurones. Br J Pharmacol 96:193–203

Galligan JJ, Tokimasa T, North RA (1987) Effects of three mammalian tachykinins on single enteric neurons. Neurosci Lett 82:167–171

Galligan JJ, Jiang M-M, Shen K-Z, Surprenant A (1990) Substance P mediates neurogenic vasodilation in extrinsically denervated guinea-pig submucosal arterioles. J Physiol (Lond) 420:267–280

Grafe P, Mayer CJ, Wood JD (1980) Synaptic modulation of calcium-dependent potassium conductance in myenteric neurones in the guinea-pig. J Physiol (Lond) 305:235–248

Guard S, Watling KJ, Watson SP (1988) Neurokinin$_3$ receptors are linked to inositol phospholipid hydrolysis in the guinea-pig ileum longitudinal muscle myenteric plexus preparation. Br J Pharmacol 94:148–154

Guillemin R (1976) Somatostatin inhibits the release of acetylcholine induced electrically in the myenteric plexus. Endocrinology 99:1653–1654

Hanani M, Burnstock G (1984) Substance P evokes fast and slow responses in cultured myenteric neurons of the guinea pig. Neurosci Lett 48:19–23

Hanani M, Burnstock G (1985) The actions of substance P and serotonin on myenteric neurons in culture. Brain Res 358:276–281

Hanani M, Chorev M, Gilon C, Selinger Z (1988) The actions of receptor selective substance P analogs on myenteric neurons: an electrophysiological investigation. Eur J Pharmacol 153:247–253

Henry JL (1987) Discussion of nomenclature for tachykinins and tachykinin receptors. In: Henry JL, Couture R, Cuello AC, Pelletier G, Quirion R, Regoli D (eds) Substance P and neurokinins. Springer, Berlin Heidelberg New York, p 17

Hirning LD, Fox AP, Miller RJ (1990) Inhibition of calcium currents in cultured myenteric neurons by neuropeptide Y: evidence for direct receptor channel coupling. Brain Res 532:120–130

Hirst GDS, Holman ME, Spence I (1974) Two types of neurones in the myenteric plexus of duodenum in the guinea pig. J Physiol (Lond) 236:303–326

Hirst GDS, Johnson SM, van Helden DF (1985) The slow calcium-dependent potassium current in a myenteric neurone of the guinea pig ileum. J Physiol (Lond) 361:315–337

Holzer P (1984) Characterization of the stimulus induced release of immunoreactive substance P of the guinea pig small intestine. Brain Res 297:127–136

Holzer P, Lippe IT (1985) Substance P action on phosphoinositides in guinea pig intestinal muscle: a possible transduction mechanism. Naunyn Schmiedebergs Arch Pharmacol 329:50–55

Hoyle CHV, Kamm MA, Burnstock G, Lennard-Jones JE (1990) Enkephalins modulate inhibitory neuromuscular transmission in circular muscle of human colon via delta opioid receptors. J Physiol (Lond) 431:465–478

Johnson SM, Katayama Y, North RA (1980) Slow synaptic potentials in neurones of the myenteric plexus. J Physiol (Lond) 301:505–516

Johnson SM, Katayama Y, Morita K, North RA (1981) Mediators of slow synaptic potentials in the myenteric plexus of the guinea pig ileum. J Physiol (Lond) 320:176–186

Katayama Y, North RA (1978) Does substance P mediate slow synaptic excitation within the myenteric plexus? Nature 274:387–388

Katayama Y, North RA (1980) The action of somatostatin on neurones of the myenteric plexus of the guinea-pig ileum. J Physiol (Lond) 303:315–323

Katayama Y, North RA, Williams JT (1979) The action of substance P on neurons of the myenteric plexus of the guinea pig small intestine. Proc Soc Lond [B] 206:191–208

Keast JR, Furness JB, Costa M (1985) Different substance P receptors are found on mucosal epithelial cells and submucous neurons of the guinea pig small intestine. Naunyn Schmiedebergs Arch Pharmacol 318:281–287

Keast JR, Furness JB, Costa M (1986) Effects of noradrenaline and somatostatin on basal and stimulated mucosal ion transport in the guinea pig small intestine. Naunyn Schmiedebergs Arch Pharmacol 337:393–399

Kilbinger H, Stauss P, Erlhof I, Holzer P (1986) Antagonist discrimination between subtypes of tachykinin receptors in the guinea pig ileum. Naunyn Schmiedebergs Arch Pharmacol 334:181–187

Laufer R, Wormser U, Friedman ZY, Gilon C, Chorev M, Selinger Z (1985) Neurokinin B is a preferred agonist for a neuronal substance P receptor and its action is antagonized by enkephalin. Proc Natl Acad Sci USA 82:7444–7448

Leander S, Hakanson R, Rosell S, Folkers K, Sundler F, Tornqvist K (1981) A specific substance P antagonist blocks smooth muscle contractions induced by non-cholinergic, non-adrenergic nerve stimulation. Nature 294:467–469

Lee CM, Iversen LL, Hanley MR, Sandberg B (1982) The possible existence of multiple receptors for substance P. Naunyn Schmiedebergs Arch Pharmacol 318:281–287

Lu Y, Louie DS, Owyang C (1991) Characterization of neuronal cholecystokinin receptors in the myenteric plexus: role of CCK A and CCK B receptors in cholinergic transmission. Gastroenterology 100:A653

Ma RC, Szurszewski JH (1991) Quantitative analysis of facilitating effect of sulfated cholecystokinin (S-CCK-8) on nicotinic transmission in cat intrapancreatic ganglia. Gastroenterology 100:A653

Maggio JE, Manyth PW (1989) Gut tachykinins. In: Schultz SG, Makhlouf GM, Rauner BB (eds) Neural and endocrine biology. Oxford University Press, New York, p 661 (Handbook of physiology sect 6: The gastrointestinal system, vol 2)

Mawe GM (1990) Intracellular recording from neurones if the guinea pig gall bladder. J Physiol (Lond) 429:323–338

Mawe GM (1991) The role of cholecystokinin in ganglionic transmission in guinea-pig gall-bladder. J Physiol (Lond) 439:89–102

Mihara S, North RA (1986) Opioids increase potassium conductance in submucous neurones of guinea-pig caecum by activating δ-receptors. Br J Pharmacol 88: 315–322

Mihara S, Katayama Y, Nishi S (1985) Slow synaptic potentials in neurones of submucous plexus of guinea pig caecum and their mimicry by noradrenaline and various peptides. Neuroscience 16:1057–1068

Mihara S, Nishi S, North RA, Surprenant A (1987a) A non-adrenergic, non-cholinergic slow inhibitory post synaptic potential in neurones of the guinea pig submucous plexus. J Physiol (Lond) 390:357–365

Mihara S, North RA, Surprenant A (1987b) Somatostatin increases an inwardly rectifying potassium conductance in guinea-pig submucous plexus neurones. J Physiol (Lond) 390:335–355

Morita K, North RA (1982) Opiate activation of potassium conductance in myenteric neurons: inhibition by calcium ion. Brain Res 242:145–190

Morita K, North RA, Tokimasa T (1982) The calcium-activated potassium conductance in guinea pig myenteric neurones. J Physiol (Lond) 329:341–354

Mulholland MW, Jaffer S (1990) Stimulation of acetylcholine release in myenteric plexus by calcitonin gene-related peptide. Am J Physiol 259:G934–G939

Nemeth PR, Ewart WR, Wood JD (1983) Effects of the putative substance P antagonists (D-Pro^2, D-Phe^7, D-Trp^9 and D-Pro^2, D-Trp^7, D-Trp^9) on electrical activity of myenteric neurons. J Auton Nerv Syst 8:165–169

Nemeth PR, Zafirov DH, Wood JD (1985) Effects of cholecystokinin, caerulein and pentagastrin on electrical behavior of myenteric neurons. Eur J Pharmacol 116:263–285

Nemeth PR, Palmer JM, Wood JD, Zafirov DH (1986) Effects of forskolin on electrical behaviour of myenteric neurones in guinea-pig small intestine. J Physiol (Lond) 376:439–450

Ngu MC (1983) The effect of cholecystokinin octapeptide on guinea pig enteric neurons. Aust N Z J Med 13:316

Nishi S, North RA (1973) Intracellular recording from the myenteric plexus of the guinea pig ileum. J Physiol (Lond) 231:471–491

North RA (1989) Drug receptors and the inhibition of nerve cells. Br J Pharmacol 98:13–28

North RA, Tokimasa T (1983) Depression of calcium dependent potassium conductance of guinea pig myenteric neurones by muscarinic agonists. J Physiol (Lond) 342:253–266

North RA, Tokimasa T (1987) Persistent calcium sensitive potassium current and the resting properties of guinea-pig myenteric neurones. J Physiol (Lond) 386: 333–353

North RA, Williams JT (1976) Enkephalin inhibits firing of myenteric neurones. Nature 264:460–461

North RA, Katayama Y, Williams JT (1979) On the mechanism and site of action of enkephalin on single myenteric neurones. Brain Res 165:67–77

North RA, Williams JT, Surprenant A, Christie MJ (1987) Mu and delta receptors belong to a family of receptors that are coupled to potassium channels. Proc Natl Acad Sci USA 84:5487–5491

Nowycky MC, Fox AP, Tsien RW (1985) Three types of neuronal calcium channel with different calcium agonist sensitivity. Nature 316:440–443

Palmer JM, Scheman M, Tamura K, Wood JD (1986a) Calcitonin gene related peptide excites myenteric plexus neurons. Eur J Pharmacol 132:163–170

Palmer JM, Wood JD, Zafirov DH (1986b) Elevation of adenosine 3′,-5′ phosphate mimics slow synaptic excitation in myenteric neurones of the guinea pig. J Physiol (Lond) 376:451–460

Palmer JM, Wood JD, Zafirov DH (1987a) Purinergic inhibition in the small intestinal myenteric plexus of the guinea-pig. J Physiol (Lond) 387:357–369

Palmer JM, Wood JD, Zafirov DH (1987b) Transduction of aminergic and peptidergic signals in enteric neurones of the guinea pig. J Physiol (Lond) 387:371–383

Paton WDM (1957) The action of morphine and related substances on contraction and acetylcholine output of coaxially stimulated guinea pig ileum. Br J Pharmacol 11:119–127

Pfaffinger PJ, Martin JM, Hunter DD, Nathanson NM, Hille B (1985) GTP-binding proteins couple cardiac muscarinic receptors to a K channel. Nature 303: 250–253

Pillai NP, Johnson SM (1991) The electrophysiological effects of [D-Ala^2, *N*-Me-Phe^4, Gly^5-ol]enkephalin on guinea pig myenteric neurons. Eur J Pharmacol 192:227–233

Regoli D, Orleans -Juste PD, Drapeau G, Dion S, Escher E (1985) Pharmacological characterization of substance P antagonists. In: Hakanson R, Sundler F (eds) Tachykinin antagonists. Elsevier, Amsterdam, p 277

Sato TI, Takayanagi K, Takagi K (1973) Pharmacological properties of electrical activities obtained from neurons in Auerbach's plexus. Jpn J Pharmacol 23: 665

Schauman W (1957) Inhibition by morphine of acetylcholine release from the intestine of guinea pig. Br J Pharmacol 12:115–118

Seamon KB, Daly JW (1981) Forskolin: a unique diterpine activator of cyclic AMP-generating systems. J Cyclic Nucleotide Res 1:201–224

Shen K-Z, Surprenant A (1990) Mechanisms underlying presynaptic inhibition through α_2-adrenoceptors in guinea pig submucosal neurones. J Physiol (Lond) 431:609–628

Surprenant A (1984) Slow excitatory synaptic potentials recorded from neurones of guinea pig submucous plexus. J Physiol (Lond) 351:343–361

Surprenant A, North RA (1985) μ and α_2 adrenoceptors coexist on myenteric but not on submucous neurones. Neuroscience 16:425–430

Surprenant A, North RA, Katayama Y (1987) Observations on the actions of substance P and [D-Arg^1, D-Pro^2, D-Trp^{7,9}, Leu^{11}]substance P in single neurons of the guinea pig submucous plexus. Neuroscience 20:189–199

Surprenant A, Shen K-Z, North RA, Tatsumi H (1990) Inhibition of calcium currents by noradrenaline, somatostatin and opioids in guinea pig submucosal neurones. J Physiol (Lond) 431:585–608

Tack JF, Janssens W, Janssens J, Vantrappen G, Wood JD (1991) Motilin and erythromycin excite myenteric neurons in gastric antrum of the guinea pig. Gastroenterology 100:A500

Tamura K, Palmer JM, Wood JD (1987) Galanin suppresses nicotinic synaptic transmission in the myenteric plexus of guinea pig small intestine. Eur J Pharmacol 136:445–446

Tamura K, Palmer JM, Winkelman CK, Wood JD (1988) Mechanism of action of galanin on myenteric neurons. J Neurophysiol 60:966–979

Tatsumi H, Costa M, Schimerlik M, North RA (1990) Potassium conductance increased by noradrenaline, opioids, somatostatin, and G-proteins: whole cell recording from guinea pig submucous neurons. J Neurosci 10:1675–1682
Tokimasa T, North RA (1984) Many transmitters, few channels. In: Chan-Palay V, Chan-Palay SL (eds) Coexistence of neuroactive substances in neurons. Wiley, New York, pp 217–224
Wade PR, Wood JD (1988) Actions of serotonin and substance P on myenteric neurons of guinea pig distal colon. Eur J Pharmacol 148:1–8
Watson SP (1984) The action of substance P on contraction, inositol phospholipids and adenylate cyclase in rat small intestine. Biochem Pharmacol 33:3733–3737
Watson SP, Downes CP (1983) Substance P induced hydrolysis of inositol phospholipids in guinea pig ileum and rat hypothalamus. Eur J Pharmacol 93:245–253
Willard AL (1990a) Substance P mediates synaptic transmission between rat myenteric neurones in cell culture. J Physiol (Lond) 426:453–471
Willard AL (1990b) A vasoactive intestinal peptide-like cotransmitter at cholinergic synapses between rat myenteric neurons in cell culture. J Neurosci 10:1025–1034
Willard AL, Nishi R (1985) Neurons dissociated from rat myenteric plexus retain differentiated properties when grown in cell culture. II. Electrophysiological properties and responses to neurotransmitter candidates. Neuroscience 16:201–211
Willard AL, Nishi R (1987) Neuropeptides mark functionally distinguishable cholinergic enteric neurons. Brain Res 422:163–167
Williams JT, North RA (1978) Inhibition of firing of myenteric neurones by somatostatin. Brain Res 155:165–168
Williams JT, North RA (1979) VIP excites neurones of the myenteric plexus. Brain Res 175:174–177
Wood JD, Mayer CJ (1978) Slow synaptic excitation mediated by serotonin in Auerbach's plexus. Nature 276:836–837
Wormser U, Laufer R, Hart Y, Chorev M, Gilon C, Selinger Z (1986) Highly selective agonists for substance P receptor subtypes. EMBO J 5:2805–2808
Yatani A, Codina J, Brown AM, Birnbaumer L (1987) Direct activation of mammalian atrial muscarinic potassium channels by GTP regulatory protein G_K. Science 235:207–211
Zafirov DH, Palmer JM, Nemeth PR, Wood JD (1985) Bombesin, gastrin releasing peptide and VIP excite myenteric neurons. Eur J Pharmacol 115:103–107
Zelles T, Harsing LG, Vizi ES (1990) Characterization of neuronal cholecystokinin receptor by L-364,718 in Auerbach's plexus. Eur J Pharmacol 178:101–104

CHAPTER 9

Peptidergic Regulation of Smooth Muscle Contractility

J.R. Grider

A. Introduction

The main functions of the smooth muscle of the gut are to mix and propel intraluminal contents in a coordinated manner in order to facilitate the digestion and absorption of nutrients and expulsion of unabsorbed residue. These functions are determined by intrinsic electrical and mechanical properties of the muscle and their regulation by hormones and neurotransmitters. The myenteric plexus, one of the two main components of the enteric nervous system, is particularly important in the regulation of smooth muscle activity. This chapter will discuss peptide hormones and transmitters that are important in the regulation of smooth muscle activity in the gut.

Smooth muscle of the gut is organized into two muscle layers: an inner circular and an outer longitudinal layer, named for the orientation of muscle cells within each. The myenteric plexus is located between them and innervates both muscle layers (Furness and Costa 1987). In addition, neurons in this plexus project fibers to neurons in the same plexus as well as to neurons in the submucosal plexus and in extrinsic paravertebral ganglia. In some species (dogs, humans), the innermost layer of circular muscle receives additional innervation from the submucosal plexus (Furness et al. 1990). Submucosal neurons also project axons to the muscularis mucosa, a thin band of muscle located between the lamina propria and the submucosa.

B. Regulation of Smooth Muscle by Peptide Hormones

I. Cholecystokinin

Cholecystokinin (CCK) is secreted from endocrine cells of the proximal intestine and is released into the circulation in various molecular forms (CCK-58, CCK-33, CCK-8) which possess the same active C-terminal moiety containing a sulfated tyrosine necessary for physiological activity. Depending on the region of the gut, the action of CCK on smooth muscle can be direct, neurally mediated or both. The direct effect is contractile whereas the neurally mediated effects can be either contractile or relaxant depending on the nature of transmitter released by the action of CCK.

Neuronal receptors are more sensitive to CCK than are muscle receptors (Cox et al. 1990; Grider and Makhlouf 1987c). Thus, at low physiological concentrations (1–10 pM), CCK causes contraction of gallbladder muscle by stimulating the release of acetylcholine from cholinergic neurons (Hanyu et al. 1990; Takahashi et al. 1991). In the sphincter of Oddi, CCK acts on noncholinergic neurons to cause relaxation by stimulating the release of the relaxant neuropeptide VIP. Blockade of the neurally mediated action of CCK by tetrodotoxin unmasks a direct contractile effect on sphincteric muscle (Wiley et al. 1988). A similar pattern is observed in the lower esophageal sphincter but its physiological significance in this region is uncertain. The sensitivity of neural receptors is also evident in the action of CCK on intestinal muscle where atropine inhibits the greater part of the response to low and moderate concentrations of CCK (Grider and Makhlouf 1987c; Lucaities et al. 1991).

Smooth muscle receptors for CCK are found in all regions of the gut (Biancani et al. 1987; Bitar and Makhlouf 1982b; Bitar et al. 1982; Cox et al. 1990; Menozzi et al. 1989; Morini et al. 1990). In the gallbladder and intestine, CCK receptors on muscle cells are of the CCK-A type; these receptors are characteristically activated by sulfated CCK much more potently than by desulfated CCK or sulfated and desulfated gastrin, and are blocked by the benzodiazepine derivative L364,718 (Grider and Makhlouf 1990; Morini et al. 1990). In the stomach, gastrin and CCK are equipotent agonists and the receptors are sensitive to the inhibitory effect of both L364,718 and the putative gastrin antagonist L365,260 (Bitar et al. 1982; Grider and Makhlouf 1990; Menozzi et al. 1989); these receptors are best labeled gastrin/CCK receptors to distinguish them from CCK-B receptors in the brain, where CCK remains slightly more potent than gastrin.

The binding of CCK to muscle receptors produces an increase in intracellular calcium ($[Ca^{2+}]_i$) and results in contraction. In circular muscle of the stomach, gallbladder, intestine, and lower esophageal and anal sphincters, the increase in $[Ca^{2+}]_i$ is mediated by inositol 1,4,5-trisphosphate-dependent release of Ca^{2+} from intracellular stores. The increase in $[Ca^{2+}]_i$ in longitudinal muscle cells is mediated by Ca^{2+} influx and is inhibited by Ca^{2+} channel blockers (Biancani et al. 1987; Grider and Makhlouf 1988a; Lee et al. 1989; Murthy and Makhlouf 1991).

Three physiological actions of CCK on the musculature of the gut have been identified: these include neurally mediated (and probably direct) contraction of the gallbladder, neurally mediated relaxation of the sphincter of Oddi and neurally mediated relaxation of the stomach resulting in delay in gastric emptying (Liddle et al. 1989). The last involves activation of CCK receptors located on primary afferent fibers originating in the stomach. Activation of these receptors triggers an extragastric reflex that leads to relaxation of gastric muscle by inhibiting the activity of cholinergic neurons and/or causing the release of a relaxant neurotransmitter from intragastric (i.e., postganglionic) neurons (Raybould and Tache 1988; Dockray 1990).

II. Motilin

Motilin is secreted by endocrine cells in the proximal intestine and released into the circulation in a cyclical fashion. Motilin causes direct contraction of smooth muscle of the stomach and proximal intestine. The direct contractile effect of motilin, like that of CCK, has been confirmed by studies using isolated muscle cells (LOUIE and OWYANG 1988; STRUNZ et al. 1975). Radioligand binding studies using crude membranes derived from the musculature of the stomach and intestine suggest the existence of only one type of motilin receptor (PEETERS et al. 1988). Erythromycin and its derivatives interact with this receptor, eliciting similar actions on the musculature. This property has led to the designation of these antibiotics as motilides (ITOH et al. 1984). Their therapeutic potential as prokinetic agents is currently under investigation (c.f. Chap. 15, this volume).

In human and dog, motilin appears to be responsible for initiating phase 3 of the interdigestive migrating motor complex (MMC) in the stomach and duodenum. A role for motilin in the regulation of this motor complex, which originates in the stomach and propagates caudad, was first postulated by ITOH et al. (1975). Evidence in favor of this notion includes the following: (1) infusion of motilin into the circulation initiates activity within the stomach that corresponds to phase 3 of the MMC (CHEY and LEE 1980; VANTRAPPEN et al. 1979); (2) motilin concentrations fluctuate in a cyclical manner with peaks that coincide with the start of phase 3 activity (PEETERS et al. 1980; VANTRAPPEN et al. 1979); and (3) neutralization of circulating motilin with motilin antiserum interrupts ongoing MMC activity and/or prevents the occurrence of phase 3 activity (LEE et al. 1983; POITRAS 1984). Somatostatin and pancreatic polypeptide can cause a decrease in circulating levels of motilin, but they do so only at supraphysiological concentrations (PEETERS et al. 1983).

III. Peptide YY and Neuropeptide Y

Peptide YY (PYY), pancreatic polypeptide (PP), and neuropeptide Y (NPY) are members of a family of peptides with partial sequence homology. PP is present in endocrine cells of the pancreas and PYY in endocrine cells of the distal small intestine and colon; NPY is present in neurons of the myenteric and submucosal plexuses and in extrinsic adrenergic neurons that innervate gastrointestinal blood vessels.

Both PYY and NPY interact with two types of receptors designated Y_1 and Y_2. The receptors have been identified in central and peripheral neurons, in intestinal epithelial cells and in vascular smooth muscle cells (SHEIKH et al. 1989, 1991; SHEIKH 1991; LABURTHE et al. 1986). Y_1 receptors are selectively activated by [Leu^{13},Pro^{34}]NPY and Y_2 receptors by NPY_{13-16} (SHEIKH 1991). There are currently no selective antagonists for either receptor type.

Evidence that one or both receptor type is present on smooth muscle cells of the gut is based on the ability of NPY to cause contraction of isolated gastric and intestinal muscle cells (GRIDER, unpublished observations). Based on the location of these receptors, the action of PYY and NPY on the musculature is likely to be both direct and neurally mediated. Both NPY and PYY inhibit acetylcholine release from enteric cholinergic neurons (WILEY et al. 1991). Accordingly the action of PYY and NPY on the musculature is likely to reflect the net effect of direct contraction of muscle cells and inhibition of acetylcholine release.

Evidence for a physiological role of PYY in the regulation of smooth muscle function is based on studies in which the peptide was infused into the circulation in amounts designed to reproduce postprandial concentrations. Under these conditions, PYY delays gastric emptying in humans and dogs (PAPPAS et al. 1986; ADRIAN et al. 1985). When infused into the circulation, PYY and NPY suppress the migrating motor complex (HELLSTROM 1987). PYY could act in this capacity as a hormone whereas NPY could act as a neurotransmitter regulating the components of this complex. However, the physiological significance of these effects remains uncertain.

Corelease of NPY and norepinephrine from extrinsic adrenergic fibers appears to be responsible for the early (norepinephrine) and sustained (NPY) phases of vasoconstriction in numerous viscera including the gut (SHEIKH 1991; see also Chap. 11, this volume).

IV. Neurotensin

Neurotensin is present in endocrine cells of the distal small intestine and colon. In some species (dog and human), neurotensin is also found in neurons of the myenteric and submucosal plexuses (BARBER et al. 1989).

The existence of neurotensin receptors has been examined by autoradiography and by radioligand binding in synaptosomes and in crude membranes derived from smooth muscle of the intestine (AHMAD and DANIEL 1991; KITABGI and FREYCHET 1979; KITABGI et al. 1984; SEYBOLD et al. 1990). These studies suggest that neurotensin receptors are present on muscle cells and enteric neurons.

The effects of neurotensin on gut smooth muscle are complex and variable depending on the species and the region of the gut examined. Direct and neurally mediated contractile and relaxant effects occur depending on the concentration of neurotensin. Thus, in guinea pig colon, neurotensin at low concentrations (0.1 n*M*) causes direct (tetrodotoxin-insensitive) relaxation (KITABGI and VINCENT 1981). In the mouse colon, however, even lower concentrations (0.01 n*M*) cause contraction mediated by release of substance P; higher concentrations of neurotensin cause transient relaxation followed by sustained contraction (FONTAINE and LEBRUN 1985). Similar biphasic effects have been reported in canine intestine (FOX et al. 1987; AHMAD and DANIEL 1991).

The dual location of neurotensin in endocrine cells and neurons of the intestine and the complexity of its actions have made it difficult to identify a physiological role for this peptide. In vivo studies suggest that neurotensin may have a role in the regulation of the gastrocolic reflex. Infusion of neurotensin in human and rat delays gastric emptying and suppresses the interdigestive motor complex but it enhances colonic motility (BLACKBURN et al. 1980; HELLSTROM et al. 1982).

C. Regulation of Smooth Muscle by Neural Peptides

The organization of neurons in the myenteric plexus is well conserved and has been studied extensively in several species, including human, dog, guinea pig and rat. These neurons have been characterized electrophysiologically as S and AH neurons, morphologically as Dogiel type I, II and III neurons, and neurochemically in terms of their content of peptide and nonpeptide transmitters. Correlations have emerged from these classifications. Thus, the majority of S neurons are morphologically Dogiel type I neurons whereas AH neurons are Dogiel type II neurons. Immunostaining has proved to be most useful and has resolved myenteric neurons into two major populations each comprising about 40%–45% of neurons in the myenteric plexus without overlap between them. The first consists of neurons containing vasoactive intestinal peptide (VIP) and its homolog, PHI (Peptide with N-terminal histidine and C-terminal isoleucine), which is cosynthesized with it in the same precursor (or PHM for peptide with N-terminal histidine and C-terminal methionine in humans); the second consists of neurons containing substance P (SP) and its homolog, substance K, also known as neurokinin A or NKA; it is possible that acetylcholine is also present in SP/NKA neurons. The two main populations of neurons correspond to major roles for SP/NKA/acetylcholine as excitatory transmitters and VIP/PHI or PHM as inhibitory (i.e., relaxant) transmitters. Subgroups in the two major populations of neurons contain one or more of the following peptides: dynorphin, [Met]enkephalin, mammalian bombesin (also known as gastrin-releasing peptide or GRP), NPY, and galanin. Small groups of neurons distinct from the two main populations contain one of the following: somatostatin (SS), serotonin and gamma-aminobutyric acid (GABA).

I. Vasoactive Intestinal Peptide, Peptide Histidine Isoleucine, and Peptide Histidine Methionine

Vasoactive intestinal peptide neurons in the myenteric plexus project their fibers caudad to neurons within the plexus, to neurons in the submucosal plexus and extrinsic paravertebral ganglia, and to muscle cells in the adjacent circular and longitudinal muscle layers. The topography of VIP

Table 1. Regions of gut in which neurally induced relaxation is inhibited by VIP antiserum or VIP antagonist

Region of gut	Species	Reference
Sphincters		
Lower esophageal (LES)	Cat	BIANCANI et al. (1984, 1986)
	Opossum	GOYAL et al. (1980)
Oddi (SO)	Cat	WILEY et al. (1988)
Internal anal (IAS)	Rabbit	BIANCANI et al. (1985)
	Opossum	NURKO et al. (1989)
Stomach		
	Rat	KAMATA et al. (1988)
	Guinea pig	GRIDER et al. (1985a)
		GRIDER and RIVIER (1990)
	Cat	D'AMATO et al. (1988)
	Dog	ANGEL et al. (1983)
Intestine		
	Guinea pig	GRIDER and MAKHLOUF (1986)
	Human	GRIDER (1989a)
Colon		
	Guinea pig	GRIDER et al. (1985b)
		GRIDER and RIVIER (1990)
	Rat	GRIDER and MAKHLOUF (1986)

neurons suggests that VIP can influence the activity of other neurons in addition to having a direct effect on smooth muscle cells.

Vasoactive intestinal peptide and PHI (or PHM) cause relaxation of vascular and visceral smooth muscle, including muscle from all regions of the gut. Radioligand binding studies using freshly isolated or cultured gastric or intestinal smooth muscle cells demonstrate the presence of one receptor type with greater affinity for VIP than for PHI or PHM (BITAR and JENSEN 1983; CHIJIIWA et al. 1989). Studies of relaxation using freshly isolated muscle cells confirm the location of this receptor on muscle cells (BITAR and MAKHLOUF 1982a). Radioligand studies using synaptosomal membranes show that VIP-binding sites are present also on nerve terminals (MAO et al. 1991). VIP receptors on smooth muscle cells are coupled to adenylate cyclase and produce relaxation by increasing intracellular levels of adenosine-3′,5′-cyclic monophosphate (cAMP) (BITAR and MAKHLOUF 1982a; ITO et al. 1990; MCHENRY et al. 1991).

The presence of VIP in a major population of neurons that innervate the smooth muscle layers coupled with its ability to cause direct relaxation of smooth muscle cells in all regions of the gut suggests that VIP is the transmitter in motor neurons responsible for relaxation. Thus, VIP is released by electrical field stimulation of gastric and intestinal smooth muscle strips (GRIDER et al. 1985a,b; GRIDER and MAKHLOUF 1987a). The release is abolished by tetrodotoxin and by withdrawal of Ca^{2+} from the medium. A close stoichiometry exists between VIP release and relaxation in that the magnitude of relaxation is proportional to the amount of VIP released by electrical (i.e., neural) stimulation (GRIDER and MAKHLOUF 1987a). Evidence that VIP release is coupled to muscle relaxation is based on the ability of specific VIP antisera and selective VIP antagonists to inhibit neurally mediated relaxation in various regions of the gut in several species (Table 1). These include guinea pig, rat, dog and pig stomach (FAHRENKRUG et al. 1978a; GRIDER et al. 1985a; ITO et al. 1988; KAMATA et al. 1988), guinea pig tenia coli (GRIDER et al. 1985b) and feline lower esophageal sphincter (BIANCANI et al. 1984). PHI (or PHM), which is cosynthesized and coreleased with VIP, could participate in neurally mediated relaxation. However, because PHI/PHM is more extensively processed than VIP and is a weaker agonist, the extent of its contribution is unknown (AGOSTON et al. 1989; FAHRENKRUG et al. 1985).

A recently developed technique which enables selective protection of VIP receptors on muscle cells provides strong evidence for the role of VIP as the relaxant transmitter in myenteric motor neurons (GRIDER 1990a). VIP receptors were protected by incubation of gastric and tenial muscle strips with VIP or the VIP antagonist VIP_{10-28}, followed by addition of *N*-ethylmaleimide to inactivate all unprotected residual receptors. After washing, tissues treated in this fashion retained their ability to relax in response to VIP or to electrical field stimulation but lost their ability to relax in response to other agonists (e.g., isoproterenol or adenosine-5′-triphosphate). The tissues also retained their ability to respond to forskolin, an activator of adenylate cyclase, and to dibutyryl cAMP, implying that signal transduction pathways mediating VIP-induced relaxation were intact. Of importance, tissues in which ATP was used as protective agent retained their ability to relax in response to ATP, but lost their ability to relax in response to neural stimulation (Fig. 1). This finding suggests that VIP, but not ATP, is a mediator of relaxation induced by neural stimulation.

The participation of VIP in the regulation of a physiological function has been examined in relation to its role in stretch-induced relaxation (FAHRENKRUG et al. 1978b; NURKO et al. 1989) and in the regulation of the peristaltic reflex in rat, guinea pig, cat and human intestine (GRIDER and MAKHLOUF 1986; GRIDER 1989a; SJÖQVIST and FAHRENKRUG 1987). Studies using an isolated segment of intestine or colon that enables relaxation and VIP release to be measured simultaneously show that VIP is responsible for the descending relaxation component of the peristaltic reflex. In rat, guinea pig and human intestine, VIP release occurs only during descending relaxation. Neutralization of VIP with VIP antiserum or inhibition of its action by VIP antagonists inhibits descending relaxation (GRIDER and MAKHLOUF 1986; GRIDER 1989a; GRIDER and RIVIER 1990) (Fig. 2). Relaxation induced by low grades of stretch was abolished whereas relaxation induced by high grades of stretch was inhibited by about 40%–80%.

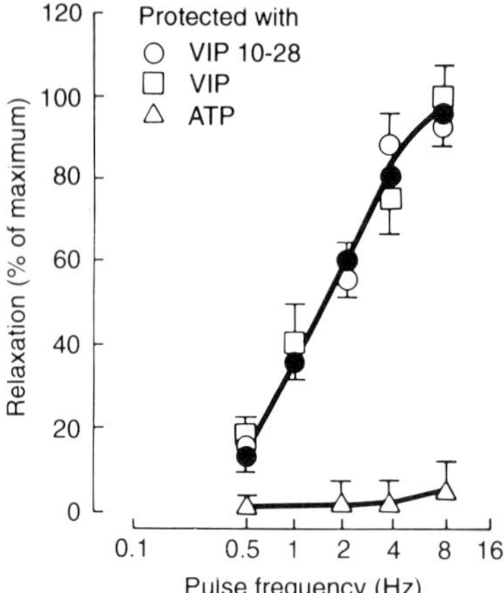

Fig. 1. Relaxation of guinea pig gastric muscle strips in response to field stimulation before (*closed symbols*) and after (*open symbols*) selective protection of either VIP receptors or ATP receptors. When VIP receptors were preserved with VIP (*open squares*) or VIP antagonist VIP_{10-28} (*open circles*), relaxation in response to neural stimulation was fully preserved. When ATP receptors were preserved the strips failed to relax in response to neural stimulation (*open triangles*). (From GRIDER 1990a)

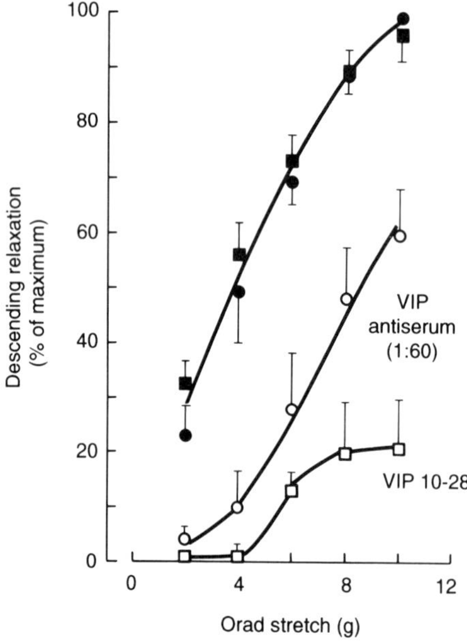

Fig. 2. Inhibition of descending relaxation component of the peristaltic reflex in rat colon by VIP antiserum (final dilution 1:60) and the VIP antagonist VIP_{10-28} ($10\,\mu M$). (From GRIDER and MAKHLOUF 1986; GRIDER and RIVIER 1990)

II. Tachykinins (Neurokinins)

Tachykinin (neurokinin) neurons in the myenteric plexus project their fibers to other neurons within the plexus as well as to smooth muscle cells in the adjacent muscle layers. As noted above, the neurons contain both SP and NKA, which are cosynthesized within the same precursor (β-preprotachykinin), as well as acetylcholine (STERNINI et al. 1989). The probable colocalization of acetylcholine and tachykinins in the same neurons raises the possibility of differential release of peptide and nonpeptide excitatory transmitters. Neurokinin B (NKB), which is derived from a separate precursor, is not found in neurons of the myenteric plexus.

Tachykinins cause direct and neurally mediated contraction of gastric and intestinal smooth muscle. These effects are mediated by receptors located on cholinergic neurons and smooth muscle cells. Three receptor types have been identified: NK-1 receptors for which SP and SP methylester are preferred ligands (WORMSER et al. 1986), NK-2 receptors for which NKA and [β-Ala^{8}]NKA$_{4-10}$ are preferred ligands (MAGGI et al. 1990; REGOLI et al. 1990), and NK-3 receptors for which NKB and senktide are preferred ligands (GUARD et al. 1990). NKB acting presumably on NK-3 receptors is most potent in eliciting acetylcholine release; its absence from the gut, however, precludes a physiological function for this peptide. SP is more effective than NKA in eliciting acetylcholine release, whereas NKA appears to be the more effective direct agonist of muscle contraction.

Tachykinins and acetylcholine are the most likely transmitters in motor neurons responsible for contraction (COSTA et al. 1985). Both acetylcholine and tachykinins are released from muscle strips by field stimulation, the former at low frequencies, the latter at high frequencies. Contraction induced by low stimulus frequencies is inhibited by atropine whereas contraction induced by higher frequencies is inhibited by tachykinin antagonists.

The location and properties of tachykinins are consistent with a role for these peptides in the regulation of the ascending contraction component of the peristaltic reflex (GRIDER and MAKHLOUF 1986; GRIDER 1989a,b). SP and NKA release occurs only during ascending contraction in both human and rat intestine. Ascending contraction elicited by low grades of stretch is abolished by atropine but is minimally inhibited by tachykinin antagonists or antisera, indicating that it is largely mediated by release of acetylcholine from motor neurons (Fig. 3). In contrast, ascending contraction elicited by high grades of stretch is strongly inhibited by tachykinin antagonists. A combination of atropine and tachykinin antagonists abolishes ascending contraction at all grades of stretch, implying that release of these transmitters accounts fully for ascending contraction. The pattern of inhibition by atropine and tachykinin antagonists is consistent with differential release of these transmitters from the same population of neurons.

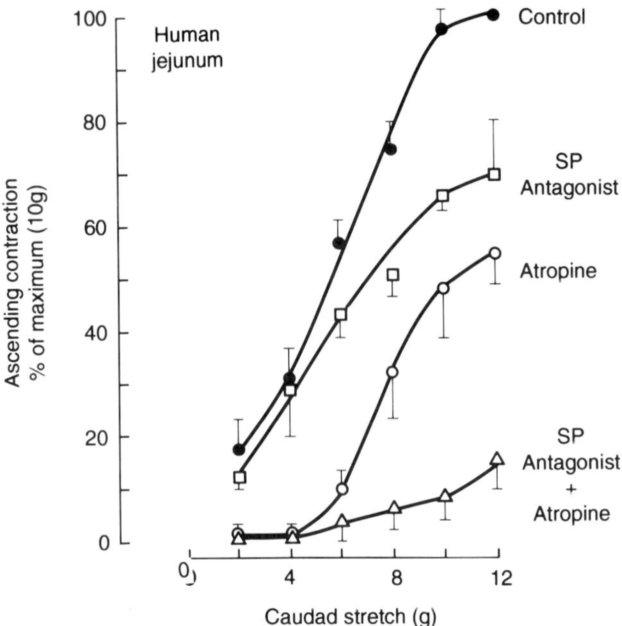

Fig. 3. Inhibition of ascending contraction component of the peristaltic reflex in human intestine by SP antagonist [D-Pro^2, D-Trp^{7,9}]substance P, the muscarinic cholinergic antagonist atropine, or both. An identical pattern of inhibition occurred in rat colonic segments. (From GRIDER 1989a)

III. Opioid Peptides

Opioid peptides in neurons of the myenteric plexus are derived from two precursors: preproenkephalin and preprodynorphin (KAKIDANI et al. 1982; NODA et al. 1982). The first yields several copies of [Met^5]enkephalin and C-terminally extended derivatives of [Met^5]enkephalin. The second yields [Leu^5]enkephalin-containing peptides including α- and β-neoendorphin, dynorphin-13 and dynorphin-17. Further processing of dynorphin yields [Leu^5]enkephalin. The extent of processing influences the response to opioid peptides by generating fragments with greater affinity for specific opioid receptor types. Derivatives of proenkephalin are present in much greater abundance than derivatives of prodynorphin in the myenteric plexus (CORBETT et al. 1988).

Dynorphin and α-neoendorphin are ligands of kappa(κ)-opioid receptors whereas [Met^5]enkephalin and its C-terminally extended forms are ligands of delta(δ)-opioid receptors (CHAVKIN et al. 1882; WUSTER et al. 1981). In the stomach and intestine, [Leu^5]enkephalin is a ligand of mu(μ)-opioid receptors on cholinergic neurons and smooth muscle cells and of δ-receptors on noncholinergic neurons. The location and properties of these receptors determine the effects of opioid peptides in the gut. These consist of: (1) direct contraction of smooth muscle cells (BITAR and MAKHLOUF 1982c, 1985; GRIDER and MAKHLOUF 1991), (2) presynaptic inhibition of acetylcholine release from myenteric cholinergic neurons (WATERFIELD et al. 1977) and (3) inhibition of release of VIP from myenteric neurons (GRIDER and MAKHLOUF 1987b).

Radioligand binding and contractile studies on freshly isolated muscle cells have demonstrated the presence of three types of opioid receptors (μ, δ and κ) capable of mediating contraction. In several species (human, rabbit and guinea pig), the receptors were found only in muscle cells isolated from the circular but not longitudinal muscle layer (BITAR and MAKHLOUF 1985; GRIDER and MAKHLOUF 1991; KUEMMERLE et al. 1991). The presence of opioid receptors mediating inhibition of acetylcholine from cholinergic neurons is well known (WATERFIELD et al. 1977). The rank order of potency for opioid peptides in inhibiting acetylcholine release (dynorphin > [Met^5]enkephalin > [Leu^5]enkephalin) is similar to that for direct contraction of muscle cells. The presence of opioid receptors on noncholinergic neurons is evident from the inhibitory effect of opioid peptides on VIP release (GRIDER and MAKHLOUF 1987b). VIP neurons are tonically active and are responsible for a dominant inhibitory (relaxant) neural tone that normally masks myogenic phasic activity in the intestine. Neutralization of VIP with VIP antiserum, inhibition of VIP's effect with VIP antagonists or inhibition of VIP release with opioid agonists unmasks the intrinsic myogenic phasic activity. In their effect on VIP neurons, [Met^5]enkephalin and [Leu^5]enkephalin are equipotent and interact with δ-receptors; dynorphin is a weak, partial agonist of VIP release from VIP neurons in contrast to its potent effect on cholinergic neurons and muscle cells.

Opioid neurons participate as modulatory neurons in the regulation of the descending relaxation component of the peristaltic reflex (GRIDER and MAKHLOUF 1987d). The neurons are tonically active and exert a continuous restraint on VIP release from VIP motor neurons. Their effect can be mimicked by addition of opioid agonists which inhibit VIP release and descending relaxation. The activity of opioid neurons decreases during descending relaxation, thereby eliminating the restraint exerted by opioid neurons on VIP neurons. Consistent with this notion, addition of the opiate antagonist naloxone augments VIP release and descending relaxation; this result confirms the participation of opioid neurons.

IV. Somatostatin

Somatostatin (SS) is found in neurons of the myenteric and submucosal plexuses as well as in paracrine cells of the stomach and endocrine cells of the intestine. It is not known whether endocrine release of SS can influence neuromuscular activity. Release of SS from myenteric neurons, however, has an important regulatory influence on propulsive activity (see below).

Somatostatin is present in a small population of myenteric neurons that project caudad within the plexus but not into the adjacent muscle layers (Keast et al. 1984). The topography of these neurons makes it unlikely that SS has a direct effect on smooth muscle cells. Consistent with this notion, SS has no contractile or relaxant effect in muscle cells isolated from guinea pig or human intestine (McHenry et al. 1991).

The effects of SS on smooth muscle function are neurally mediated. Somatostatin inhibits the release of acetylcholine and opioid peptides from myenteric neurons (Guillemin 1976; Teitelbaum et al. 1984; Furness and Costa 1979) but it stimulates the release of GABA (Takeda et al. 1989) and VIP. These effects of SS are reflected in its ability to regulate the descending relaxation phase of the peristaltic reflex (Grider et al. 1987). Somatostatin is released concomitantly with VIP only during descending relaxation. Addition of SS antiserum inhibits VIP release and descending relaxation implying that SS release exerts a stimulatory influence on VIP release during peristalsis. Consistent with this notion, exogenous SS augments VIP release and descending relaxation. More recent studies (Grider 1991) show that the effect of SS on VIP neurons is mediated via GABA and opioid neurons. The release of SS inhibits the activity of opioid neurons, thereby eliminating their restraint on VIP neurons. In addition, the release of SS stimulates the activity of GABA neurons which have a stimulatory influence on VIP neurons. Thus, during descending relaxation, activation of SS-containing neurons eliminates the influence of inhibitory opioid neurons and increases the influence of stimulatory GABA neurons leading to increase in VIP release and relaxation.

V. Gastrin-Releasing Peptide and Neuromedin B

Gastrin-releasing peptide (GRP) and its homolog, neuromedin B (NMB), are the mammalian counterparts of amphibian bombesins. GRP neurons project caudad within the myenteric plexus and into the adjacent muscle layers (Costa et al. 1984). Autoradiographic studies combined with the use of selective antagonists have revealed the presence of two receptor types on muscle cells for which GRP and NMB, respectively, are the preferred ligands (Von Schrenck et al. 1989, 1990; Severi et al. 1991). Pharmacological studies using muscle strips show additionally the presence of bombesin/GRP receptors on cholinergic and noncholinergic neurons of the myenteric plexus.

The effect of GRP-related peptides reflects the presence of receptors on muscle cells and neurons. GRP/bombesin causes contraction of gastric, intestinal and gallbladder muscle by acting directly on muscle cells and by stimulating the release of acetylcholine and/or SP/NKA (Micheletti et al. 1988; Severi et al. 1988, 1991; Mayer et al. 1982; Angel et al. 1984; Zetler 1980). The role of bombesin/GRP in regulation of smooth muscle function is unknown. Its only physiological role appears to be regulation of gastrin secretion.

VI. Galanin

Galanin is present in a small population of neurons of the myenteric plexus which project caudad within the plexus as well as into the adjacent muscle layers (EKBLAD et al. 1985). Galanin receptors are present on smooth muscle cells isolated from the stomach and intestine. The receptors act to augment relaxation induced by VIP and appear to do so by opening K^+ channels (GRIDER and MAKHLOUF 1988b). The augmentatory effect of galanin is inhibited by the K^+ channel blocker apamin. In innervated muscle strips, galanin can also act to stimulate the release of acetylcholine from myenteric neurons (YAU et al. 1986). Consistent with these effects, galanin causes contraction of the lower esophageal sphincter and relaxation of the internal anal sphincter (RATTAN and GOYAL 1988; CHAKDER and RATTAN 1991). The physiological role of galanin, however, remains uncertain.

VII. Calcitonin Gene-Related Peptide

Neurons containing calcitonin gene-related peptide (CGRP) are present in the myenteric plexus of the small intestine and colon but are absent from the plexus in the proximal gut (stomach and esophagus) (FURNESS and COSTA 1987). In the stomach, nerve fibers containing CGRP are of extrinsic origin and represent primary afferent fibers sensitive to the sensory neurotoxin capsaicin (MULDERRY et al. 1988; HOLZER 1988). CGRP-containing sensory afferents are also present in the intestine and cannot be distinguished immunohistochemically from intrinsic CGRP fibers.

Calcitonin gene-related peptide receptors are present on gastric smooth muscle cells, where they mediate cAMP-dependent relaxation. In this region, however, CGRP fibers are of extrinsic origin and the physiological significance of this property is unknown (MATON et al. 1988). In innervated smooth muscle, CGRP has been reported to cause direct relaxation of the internal anal sphincter (CHAKDER and RATTAN 1990) and neurally mediated contraction of intestinal muscle (HOLZER et al. 1989). Studies using selective CGRP receptor antagonists suggest that CGRP may be the transmitter in sensory neurons that mediates the stretch-induced peristaltic reflex (GRIDER 1990b).

References

Adrian TE, Savage AP, Sagor GR, Allen JM, Bacarese-Hamilton AJ, Tatemoto K, Polak JM, Bloom SR (1985) Effect of peptide YY on gastric, pancreatic and biliary function in humans. Gastroenterology 89:494–499

Agoston DV, Fahrenkrug J, Mikkelesen JD, Whittaker VP (1989) A peptide with N-terminal histidine and C-terminal isoleucine amide (PHI) and vasoactive intestinal peptide (VIP) are copackaged in myenteric neurones of the guinea pig ileum. Peptides 10:571–573

Ahmad S, Daniel EE (1991) Receptors for neurotensin in canine small intestine. Peptides 12:623–629

Angel F, Go VLW, Schmalz PF, Szurszewski JH (1983) Vasoactive intestinal polypeptide. A putative neurotransmitter in the canine gastric muscularis mucosa. J Physiol (Lond) 341:641–645

Angel F, Go VLW, Szurszewski JH (1984) Innervation of the muscularis mucosa of canine proximal colon. J Physiol (Lond) 357:93–108

Barber DL, Buchan AMJ, Leeman SE, Soll AH (1989) Canine enteric submucosal cultures: transmitter release from neurotensin-immunoreactive neurons. Neuroscience 32:245–253

Biancani P, Walsh JH, Behar J (1984) Vasoactive intestinal polypeptide. A neurotransmitter for lower esophageal sphincter relaxation. J Clin Invest 73:963–967

Biancani P, Walsh JH, Behar J (1985) Vasoactive intestinal peptide: a transmitter for relaxation of the rabbit internal anal sphincter. Gastroenterology 89:867–874

Biancani P, Coy DH, Hillemeier C, Behar J (1986) [NAc-Tyr^1,D-Phe^2]-GRF(1–29)-NH_2 and [D-Phe^2]VIP act as VIP antagonists and reduce neurally mediated lower esophageal sphincter relaxation. Gastroenterology 91:1046

Biancani P, Hillemeier C, Bitar KN, Makhlouf GM (1987) Contraction mediated by Ca^{2+} influx in esophageal muscle and by Ca^{2+} release in the LES. Am J Physiol 253:G760–G766

Bitar KN, Jensen RT (1983) Binding of ^{125}I-VIP to isolated gastric smooth muscle cells. Gastroenterology 84:1107

Bitar KN, Makhlouf GM (1982a) Relaxation of isolated gastric smooth muscle cells by vasoactive intestinal peptide. Science 216:531–533

Bitar KN, Makhlouf GM (1982b) Receptors on smooth muscle cells: characterization by contraction and specific antagonists. Am J Physiol 242:G400–407

Bitar KN, Makhlouf GM (1982c) Specific opiate receptors on isolated mammalian gastric smooth muscle cells. Nature 297:72–74

Bitar KN, Makhlouf GM (1985) Selective presence of opiate receptors on intestinal circular muscle cells. Life Sci 37:1545–1550

Bitar KN, Saffouri B, Makhlouf GM (1982) Cholinergic and peptidergic receptors on isolated human antral smooth muscle. Gastroenterology 82:832–837

Blackburn AM, Bloom SR, Long RG, Fletcher DR, Christofides ND, Fitzpatrick ML, Baron JH (1980) Effect of neurotensin on gastric function in man. Lancet 1:987–989

Chakder S, Rattan S (1990) [Tyr^0]-calcitonin gene related peptide28–37 (rat) as a putative antagonist of calcitonin gene related peptide responses on opossum internal anal sphincter smooth muscle. J Pharmacol Exp Ther 253:200–206

Chakder S, Rattan S (1991) Effect of galanin on the opossum internal anal sphincter: structure-activity relationship. Gastroenterology 100:711–718

Chavkin C, James IF, Goldstein A (1982) Dynorphin is a specific endogenous ligand of the κ opioid receptor. Science 215:413–415

Chey W, Lee KY (1980) Motilin. Clin Gastroenterol 9:645–656

Chijiiwa Y, Mader C, Grider JR, Makhlouf GM (1989) Characterization of VIP receptors in cultured gastric smooth muscle cells. Gastroenterology 96:A86

Corbett AD, McKnight AT, Kosterlitz HW (1988) Tissue content of opioid peptides in the myenteric plexus-longitudinal muscle of guinea pig small intestine. J Neurochem 51:32–37

Costa M, Furness JB, Yanaihara N, Yanaihara C, Moody TW (1984) Distributions and projections of neurons with gastrin releasing peptide/bombesin-like immunoreactivity in the guinea pig small intestine. Cell Tissue Res 235:285–293

Costa M, Furness JB, Pullin CO, Bornstein J (1985) Substance P enteric neurons mediate non-cholinergic transmission of the circular muscle of the guinea pig intestine. Naunyn Schmiedebergs Arch Pharmacol 328:446–453

Cox KL, von Schrenck T, Moran TH, Gardner JD, Jensen RT (1990) Characterization of cholecystokinin receptors on the sphincter of Oddi. Am J Physiol 259:G873–G881

D'Amato M, de Beurme FA, Lefebvre RA (1988) Comparison of the effect of vasoactive intestinal polypeptide and non-adrenergic, non-cholinergic neuron stimulation in the cat gastric fundus. Eur J Pharmacol 152:71–82

Dockray GJ (1990) Neuronal actions of cholecystokinin. In: Thompson J (ed) Gastrointestinal endocrinology: receptors and postreceptor mechanisms. Academic, New York, pp 321–332

Ekblad E, Rokaeus A, Häkanson R, Sundler S (1985) Galanin nerve fibers in rat gut: distribution, origin, and projections. Neuroscience 16:355–363

Fahrenkrug J, Galbo H, Holst JJ, Schaffalitzky de Muckadell OB (1978a) Influence of the autonomic nervous system on the release of vasoactive intestinal polypeptide from the porcine gastrointestinal tract. J Physiol (Lond) 280:405–422

Fahrenkrug J, Haglund U, Jodal M, Lundgren O, Olbe L, Schaffalitzky de Muckadell OB (1978b) Nervous release of vasoactive intestinal polypeptide in the gastrointestinal tract of cats: possible physiological implications. J Physiol (Lond) 284:405–422

Fahrenkrug J, Bek T, Lundberg JM, Hökfelt T (1985) VIP and PHI in cat neurons: co-localization but variable tissue content possible due to differential processing. Regul Pept 12:21–34

Fontaine J, Lebrun P (1985) Effects of neurotensin on the isolated mouse distal colon. Eur J Pharmacol 107:141–147

Fox JET, Kostolanska F, Daniel EE, Allescher HD, Hanke T (1987) Mechanism of excitatory actions of neurotensin on canine small intestine circular muscles in vivo and in vitro. Can J Physiol Pharmacol 65:2254–2259

Furness JB, Costa M (1979) Actions of somatostatin on excitatory and inhibitory nerves in the intestine. Eur J Pharmacol 56:69–74

Furness JB, Costa M (1987) The enteric nervous system. Churchill Livingstone, Edinburgh

Furness JB, Lloyd KCK, Sternini C, Walsh JH (1990) Projections of substance P, vasoactive intestinal peptide and tyrosine hydroxylase immunoreactive nerve fibers in the canine intestine, with special reference to the innervation of the circular muscle layer. Arch Histol Cytol 53:129–140

Goyal RK, Rattan S, Said S (1980) VIP as a possible neurotransmitter of non-cholinergic, non-adrenergic inhibitory neurons. Nature 288:378–380

Grider JR (1989a) Identification of neurotransmitters regulating intestinal peristaltic reflex in humans. Gastroenterology 97:1414–1419

Grider JR (1989b) Tachykinins as transmitters of ascending contractile component of the peristalsis reflex. Am J Physiol 257:G709–G714

Grider JR (1990a) Identification of neurotransmitters by selective protection of postjunctional receptors. Am J Physiol 258:G103–G106

Grider JR (1990b) Identification of calcitonin gene related peptide (CGRP) as a sensory neurotransmitter of stretch-induced peristaltic reflex. Gastroenterology 98:A355

Grider JR (1991) Somatostatin (SS) neurons regulate the descending limb of the peristaltic reflex by modulating the activity of opioid and GABA neurons. Gastroenterology 100:A445

Grider JR, Makhlouf GM (1986) Colonic peristaltic reflex: identification of vasoactive intestinal peptide as mediator of descending relaxation. Am J Physiol 251:G40–G45

Grider JR, Makhlouf GM (1987a) Prejunctional inhibition of vasoactive intestinal peptide release. Am J Physiol 253:G7–G12

Grider JR, Makhlouf GM (1987b) Suppression of inhibitory neural input to colonic circular muscle by opioid peptides. J Pharmacol Exp Ther 243:205–210

Grider JR, Makhlouf GM (1987c) Regional and cellular heterogeneity of cholecystokinin receptors mediating muscle contraction in the gut. Gastroenterology 92:175–180

Grider JR, Makhlouf GM (1987d) Role of opioid neurons in the regulation of intestinal peristalsis. Am J Physiol 253:G226–G231

Grider JR, Makhlouf GM (1988a) Contraction mediated by Ca^{2+} release in circular and Ca^{2+} influx in longitudinal intestinal muscle cells. J Pharmacol Exp Ther 244:432–437

Grider JR, Makhlouf GM (1988b) The modulatory function of galanin: potentiation of VIP-induced relaxation in isolated smooth muscle cells. Gastroenterology 94:A157

Grider JR, Makhlouf GM (1990) Distinct receptors for cholecystokinin and gastrin on muscle cells of stomach and gallbladder. Am J Physiol 259:G184–G190

Grider JR, Makhlouf GM (1991) Identification of opioid receptors on gastric muscle cells by selective receptor protection. Am J Physiol 260:G103–G107

Grider JR, Rivier JR (1990) Vasoactive intestinal peptide (VIP) as transmitter of inhibitory motor neurons of the gut: evidence from the use of selective VIP antagonists and VIP antiserum. J Pharmacol Exp Ther 253:738–742

Grider JR, Cable MB, Said SI, Makhlouf GM (1985a) Vasoactive intestinal peptide (VIP) as neural mediator of gastric relaxation. Am J Physiol 248:G73–G78

Grider JR, Cable MB, Bitar KN, Said SI, Makhlouf GM (1985b) Vasoactive intestinal peptide. Relaxant transmitter in tenia coli of the guinea pig. Gastroenterology 89:36–42

Grider JR, Arimura A, Makhlouf GM (1987) Role of somatostatin neurons in intestinal peristalsis: facilitatory interneurons in descending pathways. Am J Physiol 253:G434–G438

Guard S, Watson SP, Maggio JE, Too HP, Watling KJ (1990) Pharmacological analysis of [^{3}H]-senktide binding to NK_3 tachykinin receptors in guinea pig ileum longitudinal muscle-myenteric plexus and cerebral cortex membranes. Br J Pharmacol 99:767–773

Guillemin R (1976) Somatostatin inhibits the release of acetylcholine induced electrically in the myenteric plexus. Endocrinology 99:1653–1654

Hanyu N, Dodds WJ, Layman RD, Hogan WJ, Chey WY, Takahashi I (1990) Mechanism of cholecystokinin-induced contraction of the opossum gallbladder. Gastroenterology 98:1299–1306

Hellstrom PM (1987) Mechanisms involved in colonic vasoconstriction and inhibition of motility induced by neuropeptide Y. Acta Physiol Scand 129:549–556

Hellstrom PM, Nylander G, Rosell S (1982) Effects of neurotensin on the transit of gastrointestinal contents in the rat. Acta Physiol Scand 115:239–243

Holzer P (1988) Local effector functions of capsaicin-sensitive sensory nerve endings: involvement of tachykinins, calcitonin gene-related peptide and other neuropeptides. Neuroscience 24:739

Holzer P, Bartho L, Matusak O, Bauer V (1989) Calcitonin gene-related peptide action on intestinal circular muscle. Am J Physiol 256:G546–G552

Ito S, Ohga A, Ohta T (1988) Gastric relaxation and vasoactive intestinal peptide output in response to reflex vagal stimulation in the dog. J Physiol (Lond) 404:683–693

Ito S, Krokawa A, Ohga A, Ohta T, Swabe K (1990) Mechanical, electrical, and cyclic nucleotide responses to peptide VIP and inhibitory nerves stimulation in rat stomach. J Physiol (Lond) 430:337–353

Itoh Z, Aizawa R, Takeuchi S, Couch EF (1975) Hunger contractions and motilin. In: Vantrappen G (ed) Proceedings of the 5th international symposium on gastrointestinal motility, Typoff Press, Herenthals, Belgium. pp 48–55

Itoh Z, Nakayama M, Suzuki T, Arai H, Wakabayashi K (1984) Erythromycin mimics exogenous motilin in gastrointestinal contractile activity in the dog. Am J Physiol 247:G688–G694

Kakidani H, Furutani Y, Takahashi H, Noda M, Morimoto Y, Hirose T, Asai M, Inayama S, Nakanishi S, Numa S (1982) Cloning and sequence analysis of cDNA for porcine β-neo-endorphin/dynorphin precursor. Nature 298:245–249

Kamata K, Sakamoto A, Kasuya Y (1988) Similarities between the relaxations induced by vasoactive intestinal peptide and by stimulation of the non-adrenergic, non-cholinergic neurons in the rat stomach. Naunyn Schmiedebergs Arch Pharmacol 338:401–406

Keast JR, Furness JB, Costa M (1984) Somatostatin in human enteric nerves. Distribution and characterization. Cell Tissue Res 237:299–308

Kitabgi P, Freychet P (1979) Neurotensin: contractile activity, specific binding, and lack of effect on cyclic nucleotides in intestinal smooth muscle. Eur J Pharmacol 55:35–42

Kitabgi P, Vincent J-P (1981) Neurotensin is a potent inhibitor of guinea pig colon contractile activity. Eur J Pharmacol 74:311–318

Kitabgi P, Kwan CY, Fox JET, Vincent JP (1984) Characterization of neurotensin binding to rat gastric smooth muscle receptor sites. Peptides 5:917–923

Kuemmerle JF, Jin J-G, Grider JR, Makhlouf GM (1991) Characterization of distinct μ, δ, and κ opioid receptors on intestinal muscle cells with selective radioligands and receptor protection. Gastroenterology 100:A650

Kuemmerle JF, Makhlouf GM (1992) Characterization of opioid receptors in intestinal muscle cells by selective radioligands and receptor protection. Am J Physiol 263:G269–G276

Laburthe M, Chenut B, Rouyer-Fessard C, Tatemoto K, Clouvineau A, Servin S, Amiranoff B (1986) Interaction of peptide YY with rat intestinal epithelial plasma membranes: binding of the radioiodinated peptide. Endocrinology 118:1910–1917

Lee KY, Chang TM, Chey WY (1983) Effect of rabbit antimotilin serum on myoelectric activity and plasma motilin concentration in fasting dog. Am J Physiol 245:G547–G553

Lee KY, Biancani P, Behar J (1989) Calcium sources utilized by cholecystokinin and acetylcholine in the cat gallbladder muscle. Am J Physiol 256:G785–G788

Liddle RA, Gertz BJ, Kanayama S, Beccaria L, Coker LD, Turnbull TA, Morita T (1989) Effects of a novel cholecystokinin (CCK) receptor antagonist, MK-329, on gallbladder contraction and gastric emptying in humans. J Clin Invest 84:1220–1225

Louie DS, Owyang C (1988) Motilin receptors on isolated gastric smooth muscle cells. Am J Physiol 254:G210–G216

Lucaities VL, Mendelsohn LG, Mason NR, Cohen ML (1991) CCK-8, CCK-4 and gastrin-induced contractions in guinea pig ileum: evidence for differential release of acetylcholine and substance P by CCK-A and CCK-B receptors. J Pharmacol Exp Ther 56:695–703

Maggi CA, Patacchini R, Giachetti A, Meli A (1990) Tachykinin receptors in the circular muscle of the guinea pig ileum. Br J Pharmacol 101:996–1000

Mao YK, Barnett W, Coy DH, Tougas G, Daniel EE (1991) Distribution of vasoactive intestinal polypeptide (VIP) binding in circular muscle and characterization of VIP binding in canine small intestine. J Pharmacol Exp Ther 258:986–991

Maton PN, Sutliff VE, Zhou ZC, Collins SM, Gardner JD, Jensen RT (1988) Characterization of receptors for calcitonin gene-related peptide on gastric smooth muscle cells. Am J Phsyiol 254:G789–G794

Mayer EA, Elashoff J, Walsh JH (1982) Characterization of bombesin effects on canine gastric muscle. Am J Physiol 243:G141–G147

McHenry L, Murthy KS, Grider JR, Makhlouf GM (1991) Inhibition of muscle cell relaxation by somatostatin: tissue-specific, cAMP-dependent, pertussis toxin-sensitive. Am J Physiol 261:G45–G49

Menozzi D, Gardner JD, Jensen RT, Maton P (1989) Properties of receptors for gastrin and CCK on gastric smooth muscle cells. Am J Physiol 257:G73–G79

Micheletti R, Grider JR, Makhlouf GM (1988) Identification of bombesin receptors on isolated muscle cells from human intestine. Regul Pept 21:219–226

Morini G, Barocelli E, Impicciatore M, Grider JR, Makhlouf GM (1990) Receptor type for cholecystokinin on isolated muscle cells of the guinea pig. Regul Pept 28:313–321

Mulderry PK, Ghatel MA, Spookes RA, Jones PM, Pierson AM, Hamid QA, Kanse S, Amara SG, Burrin JM, Legon S, Polak JM, Bloom SR (1988) Differential expression of α-CGRP and β-CGRP by primary sensory neurons and enteric autonomic neurons of the rat. Neuroscience 25:195–205

Murthy KS, Makhlouf GM (1991) Phosphoinositide metabolism in intestinal smooth muscle: preferential production of Ins(1,4,5)P_3 in circular muscle cells. Am J Physiol 261:G945–G951

Noda M, Furutani Y, Takahashi H, Toyosato M, Hirose T, Inayama T, Nakanishi S, Numa S (1982) Cloning and sequence analysis of cDNA for bovine adrenal preproenkephalin. Nature 295:202–206

Nurko S, Dunn BM, Rattan S (1989) Peptide histidine isoleucine and vasoactive intestinal polypeptide cause relaxation of opossum internal anal sphincter via two distinct receptors. Gastroenterology 96:403–413

Pappas TN, Debas HT, Chang AM, Taylor IL (1986) Peptide YY release by fatty acids is sufficient to inhibit gastric emptying in dogs. Gastroenterology 89:494–499

Peeters TL, Vantrappen G, Janssens J (1980) Fasting motility levels are related to the interdigestive motility complex. Gastroenterology 79:716–719

Peeters TL, Janssens J, Vantrappen GR (1983) Somatostatin and the interdigestive migrating motor complex in man. Regul Pept 5:209–217

Peeters TL, Bormans V, Vantrappen G (1988) Comparison of motilin binding to crude homogenates of human and canine gastrointestinal smooth muscle tissue. Regul Pept 23:171–182

Poitras P (1984) Motilin is a digestive hormone in the dog. Gastroenterology 87:909–913

Rattan S, Goyal RK (1988) Effects of galanin on the opossum lower esophageal sphincter. Life Sci 41:2783–2790

Raybould HE, Tache Y (1988) Cholecystokinin inhibits gastric motility and emptying via a capsaicin-sensitive vagal pathway in rats. Am J Physiol 255:G242–G246

Regoli D, Rhaleb N-E, Dion S, Tousignant C, Rouissi N, Jukic D, Drapeau G (1990) Neurokinin A. A pharmacological study. Pharmacol Res 22:1–14

Severi C, Grider JR, Makhlouf GM (1988) Identification of separate bombesin and substance P receptors on isolated muscle cells from canine gallbladder. J Pharmacol Exp Ther 245:195–198

Severi C, Jensen RT, Erspamer V, d'Arpino L, Coy DH, Torsoli A, delle Fave G (1991) Different receptors mediate the action of bombesin-related peptides on gastric smooth muscle cells. Am J Physiol 260:G683–G690

Seybold VS, Treder BG, Aanonsen LM, Parsons A, Brown DR (1990) Neurotensin binding sites in porcine jejunum: biochemical characterization and intramural localization. Synapse 6:81–90

Sheikh P (1991) Neuropeptide Y and peptide YY: major modulators of gastrointestinal blood flow and function. Am J Physiol 261:G701–G715

Sheikh SP, Hakanson R, Schwartz TW (1989) Y_1 and Y_2 receptors for neuropeptide Y. FEBS Lett 245:209–214

Sheikh SP, Roach E, Fuhlendorff J, Williams JA (1991) Localization of Y_1 receptors for NPY and PYY on vascular smooth muscle cells in rat pancreas. Am J Physiol 260:250–257

Sjöqvist A, Fahrenkrug J (1987) Release of vasoactive intestinal polypeptide anally of a local distension of the feline small intestine. Acta Physiol Scand 130:433–438

Sternini C, Anderson K, Frantz G, Krause JE, Brecha N (1989) Expression of substance P/neurokinin A-encoding preprotachykinin messenger ribonucleic acids in the rat enteric nervous system. Gastroenterology 97:348–356

Strunz U, Domschke W, Mitznegg P, Domschke S, Schubert E, Wuensch E, Jaerger E, Demling L (1975) Analysis of the motor effect of 13-norleucine motilin on the rabbit, guinea pig, rat and human alimentary tract in vitro. Gastroenterology 68:1485–1491

Takahashi T, May D, Owyang C (1991) Cholinergic dependence of gallbladder response to cholecystokinin in the guinea pig in vivo. Am J Physiol 261:G565–G569

Takeda T, Taniyama K, Baba S, Tanaka C (1989) Putative mechanisms involved in excitatory and inhibitory effects of somatostatin on intestinal motility. Am J Physiol 257:G532–G538

Teitelbaum DH, O'Dorisio TM, Perkins WE, Gaginella TS (1984) Somatostatin modulation of peptide-induced acetylcholine release in guinea pig ileum. Am J Physiol 246:G509–G514

Vantrappen G, Janssens J, Peeters TL, Bloom SR, Christofides N, Hellemans J (1979) Motilin and the interdigestive migrating motor complex in man. Am J Dig Dis 24:497–500

Von Schrenck T, Heinz-Erian P, Moran T, Mantey SA, Gardner JD, Jensen RT (1989) Neuromedin B receptor in esophagus: evidence for subtypes of bombesin receptors. Am J Physiol 256:G747–G758

Von Schrenck T, Wang LH, Coy DH, Villanueva ML, Mantey S, Jensen RT (1990) Potent bombesin receptor antagonists distinguish receptor subtypes. Am J Physiol 259:G468–G473

Waterfield AA, Smokcum RWS, Hughes J, Kosterlitz HW, Henderson G (1977) In vitro pharmacology of the opioid peptides, enkephalins and endorphins. Eur J Pharmacol 43:107–116

Wormser U, Laufer R, Hart Y, Choren M, Coilon C, Selinger Z (1986) Highly selective agonists for substance P receptor. EMBO J 5:2805–2808

Wiley JW, O'Dorisio TM, Owyang C (1988) Vasoactive intestinal polypeptide mediates cholecystokinin-induced relaxation of the sphincter of Oddi. J Clin Invest 81:1920–1924

Wiley JW, Lu Y, Owyang C (1991) Mechanism of action of peptide YY to inhibit gastric motility. Gastroenterology 100:865–872

Wuster M, Rubini P, Schultz R (1981) The preference of putative pro-enkephalins for different types of opiate receptors. Life Sci 29:1219–1227

Yau WM, Dorsett JA, Youther ML (1986) Evidence for galanin as an inhibitory neuropeptide on cholinergic neurons in the guinea pig small intestine. Neurosci Lett 72:305–308

Zetler G (1980) Antagonism of the gut contracting effects of bombesin and neurotensin by opioid peptides, morphine, atropine or tetrodotoxin. Pharmacology 21:348–354

CHAPTER 10

Peptidergic Regulation of Intestinal Electrolyte Transport

S.M. O'Grady

A. Introduction

The intestinal epithelium plays an essential role in the control of extracellular fluid volume, electrolyte composition and acid-base balance in vertebrate animals (Powell 1987; Halm and Frizzell 1991). Active transport of monovalent ions across the epithelium establishes an osmotic driving force for the absorption or secretion of fluid (Finkelstein 1987; Spring 1991). Many studies have shown that the transport pathways involved in net electrolyte movements are regulated by neuropeptides and peptide hormones (Brown and Miller 1991). Much of our understanding of the second messenger systems and epithelial transport pathways affected by these peptides is incomplete. This is particularly true for mammalian intestine and colon due to the variety of cell types within the epithelium and to secondary effects of peptide neurohormones on submucosal nerves, resident leukocytes, and enteroendocrine cells. The first objective of this review is to provide a general overview of the mechanisms of sodium (Na^+), potassium (K^+), chloride (Cl^-) and bicarbonate (HCO_3^-) transport across the small intestine and colon. More detailed reviews of intestinal and colonic ion transport can be found in the following references (Powell 1987; Donowitz and Welsh 1987; Sullivan and Field 1991; Halm and Frizzell 1991). The second objective is to summarize our present understanding of how various peptide neurohormones affect the transport properties of intestinal and colonic epithelia. The peptides discussed in this review have been divided into two categories depending on whether they stimulate electrolyte and fluid secretion or promote absorption.

B. Transport Models for Absorption and Secretion

The mammalian intestinal epithelium possesses a crypt-villus type organization (Madara 1991). The villous region appears to be the principal site of electrolyte, nutrient and fluid absorption whereas cells in the crypts primarily secrete anions and fluid (Welsh et al. 1982; Sullivan and Field 1991). Ileal villous cells may also secrete Cl. This conclusion is based on the observation that prostaglandin E_2 and serotonin produce Cl-dependent decreases in apical membrane fractional resistance consistent with an increase in Cl secretion (STEWART and TURNBERG 1989). The primary active transport mechanism responsible for transepithelial monovalent ion transport is the Na-K ATPase (GLYNN and KARLISH 1975). This enzyme is located in the basolateral membrane of both absorptive and secretory cell types (ALBIN and GUTMAN 1979; SCHULTZ and HUDSON 1991). Inhibition of the Na-K ATPase by compounds such as ouabain abolishes transepithelial ion transport (DIBONA and MILLS 1979). Sodium ion is actively absorbed along the entire length of the intestine and colon. Other solutes and electrolytes such as glucose, amino acids, Cl and HCO_3 are indirectly coupled to Na transport. The term "secondary active transport" describes the movement of solutes against their electrochemical gradients through indirect coupling to Na transport (SCHULTZ and HUDSON 1991). The specific ion transport pathways responsible for Na absorption and secondary active transport of monovalent anions are presented in the next sections.

I. Sodium, Chloride and Bicarbonate Absorption

A general model for Na, Cl and HCO_3 absorption in mammalian jejunum is presented in Fig. 1A. In the jejunum, electrogenic Na absorption is mediated by Na-glucose and Na-amino acid cotransporters located in the apical membrane (HOPFER 1987). Electroneutral Na absorption is mediated by Na-proton (H^+) exchange (MURER et al. 1976; GUNTHER and WRIGHT 1983; KNICKELBEIN et al. 1983; MONTROSE and KIMMICH 1985). The Na-H exchanger has been identified in both apical and basolateral membranes of villous cells and in the basolateral membrane of crypt cells (CASSANO et al. 1984; BARROS et al. 1986; KNICKELBEIN et al. 1988; TASCO et al. 1988). Apical membrane Na-H exchange activity produces acidification of the intestinal lumen. The secreted protons combine with luminal HCO_3 ions to produce CO_2 and H_2O (TURNBERG et al. 1970a). Absorption of HCO_3 results from the diffusion of CO_2 across the apical membrane and conversion to HCO_3 within the cell by the enzyme carbonic anhydrase (TURNBERG et al. 1970b; HEINTZE et al. 1982; POWELL 1987). The mechanism of HCO_3 efflux across the basolateral membrane is unknown but may involve an Na-$(HCO_3)_x$ cotransport pathway similar to that proposed for renal proximal tubule cells (GRASSL et al. 1987; LOPES et al. 1987) or perhaps a conductive pathway (TABCHARANI et al. 1989; WHITE 1989). In contrast to the transcellular absorption of Na and HCO_3, a major portion of Cl absorption appears to be mediated by the paracellular pathway (TURNBERG et al. 1970b). The driving force for paracellular Cl movement comes from the transepithelial potential difference produced by the electrogenic absorption of Na. The existence of an HCO_3-insensitive, sulfate-dependent Cl exchange pathway in the apical membrane of rat jejunum suggests that transcellular Cl absorption may also occur in some species (VAANDRAGER and DEJONGE 1988). The mechanism of Cl efflux across the basolateral membrane is unknown.

A

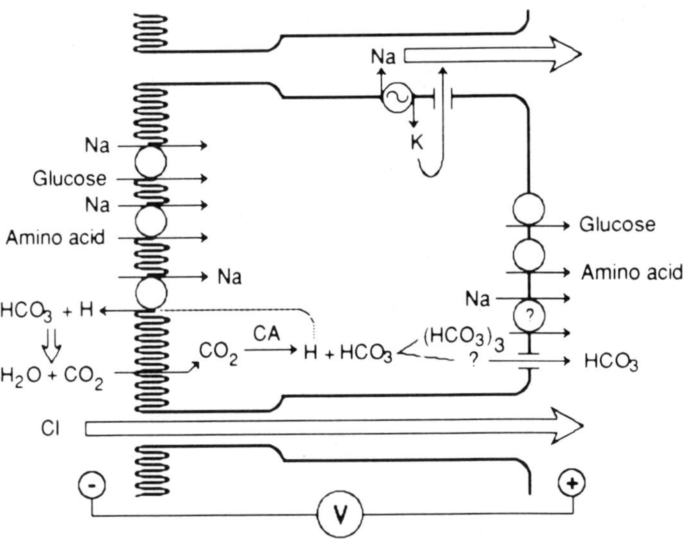

B

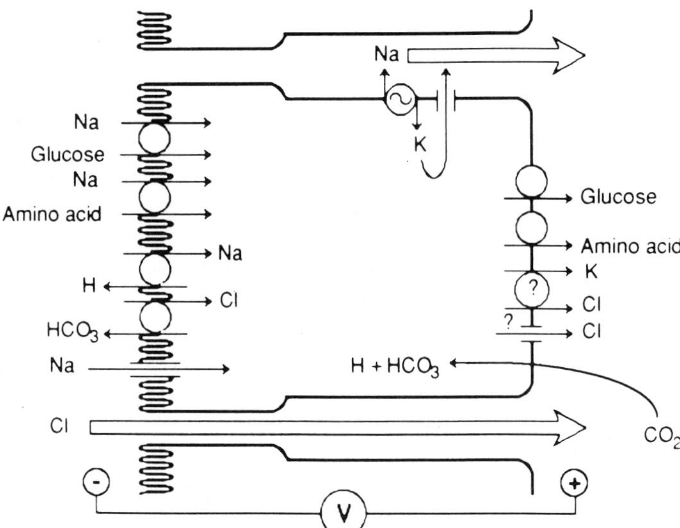

Fig. 1. A Transport model of Na, Cl and HCO_3 absorption across mammalian jejunum. **B** Absorption of Na and Cl across the mammalian ileum

The proposed mechanism for Na and Cl absorption in mammalian ileum is presented in Fig. 1B. Electrogenic Na absorption is similar to that described for the jejunum and primarily involves Na-coupled nutrient cotransport processes located in the apical membrane. There is also evidence for a phenamil-sensitive Na conductive pathway in rabbit ileum that contributes to electrogenic Na absorption (SELLIN et al. 1989). Several studies using brush-border membrane vesicles suggest that Na and Cl uptake across the apical membrane is mediated by Na-H and Cl-HCO_3 exchange pathways (KNICKELBEIN et al. 1985; HOPFER and LINEDTKE 1987). These antiporters are coupled by cell pH such that inhibition of either process ultimately limits the availability of intracellular H or HCO_3 for exchange by the other transport pathway. Experiments with rabbit (KNICKELBEIN et al. 1988) and rat (VAANDRAGER and DEJONGE 1988) ileal brush border membrane vesicles (BBMV) suggest that oxalate and SO_4 may also be exchanged for Cl across the apical membrane by separate anion exchange mechanisms other than the Cl-HCO_3 antiporter. Na and Cl uptake across the apical membrane may also involve a furosemide-sensitive Na-Cl cotransport mechanism. Loop diuretic compounds such as bumetanide and furosemide have been shown to block an Na-K-2Cl cotransport system in the apical membrane of the winter flounder intestine (O'GRADY et al. 1987). Experiments using rat ileum in vivo showed that furosemide inhibits NaCl absorption and suggested that Na and Cl transport are directly coupled (HUMPHREYS 1976). Experiments with rabbit and human ileal BBMVs also showed that Na-dependent Cl uptake could be blocked by furosemide (FAN et al. 1983; STOLL et al. 1987). It is important to note, however, that the effects of high concentrations of furosemide on mammalian intestine could be due to a nonselective effect on Cl-HCO_3 exchange activity (STOLL et al. 1987). Thus a role for Na-Cl cotransport in mammalian intestine is unclear at this time. In either case, anion exchange and Na-Cl cotransport activity at the apical membrane both produce increases in intracellular Cl concentration above electrochemical equilibrium. Thus a driving force for Cl efflux exists across the basolateral membrane (O'GRADY et al. 1987; HALM and FRIZZELL 1991). The exact mechanism for Cl transport out of the cell is presently unknown but could involve a conductive pathway or a KCl cotransport mechanism. Experiments with the *Necturus* gallbladder epithelium indicate that a KCl cotransport system is responsible for electroneutral Cl efflux across the basolateral membrane (REUSS 1983).

Electrogenic Na uptake from the lumen of the colon is mediated by amiloride-sensitive Na channels in the apical membrane (Fig. 2). Estimates of single-channel current and amiloride-blocking kinetics of these channels are similar to Na channels observed in frog skin, toad bladder and distal tubule (WILLS et al. 1984; GARTY and BENOS 1988; KLEYMAN and CRAGOE 1988). Coupled Na and Cl absorption also takes place in the colon. In rat colon, net Na and Cl absorption is dependent on carbonic anhydrase activity and mutually dependent on the presence of both ions in the bathing solution

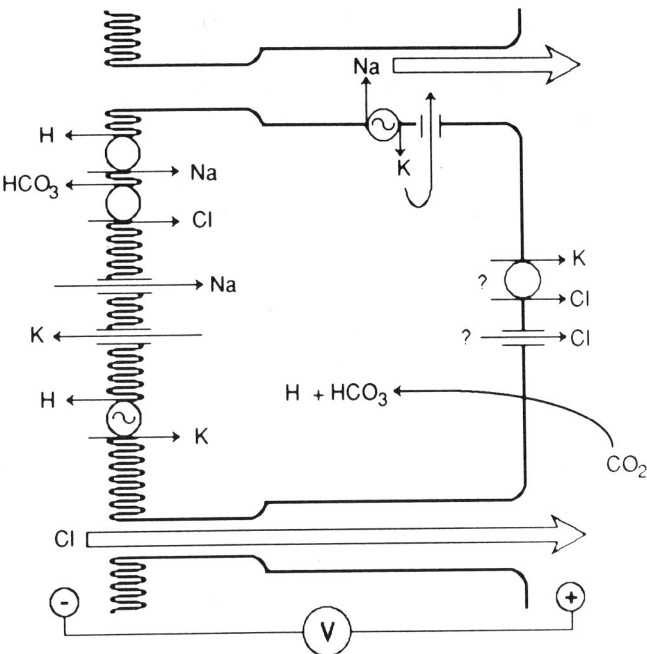

Fig. 2. Model of Na and Cl absorption across mammalian distal colon epithelium

(BINDER 1978). In contrast, Na absorption in rabbit proximal colon is unaffected by Cl replacement but Cl absorption is dependent on the presence of Na (SELLIN and DESOIGNIE 1984). This result suggests that Na and Cl transport are not tightly coupled as in rat colon and that an alternative mechanism for HCO_3 efflux may be active under conditions where Cl is absent from the bathing solution. Amiloride at high concentrations (0.5–1 m*M*) blocks Na absorption in the colon, consistent with involvement of Na-H exchange. Measurements of intracellular pH (pH_i) in rabbit proximal colon have shown that pH_i decreases in the presence of amiloride and when Na is removed from the bathing solution (AHN et al. 1985). This result is also consistent with a role for Na-H exchange in Na absorption. Absorption of Cl in rabbit distal colon is blocked by the disulfonic stilbene derivative 4-acetamido-4′-isothiocyanostilbene-2,2′-disulfonate (SITS) and by inhibition of carbonic anhydrase activity with acetazolamide (HATCH et al. 1984; DUFFEY 1984). These results suggest that Cl-HCO_3 exchange mediates Cl uptake across the apical membrane. Cl efflux across the basolateral membrane is presumably mediated by an electroneutral KCl cotransport system and by a Cl channel. KCl cotransport and K channels in the basolateral membrane provide a mechanism for regulating intracellular K activity in cells where coupled NaCl absorption takes place.

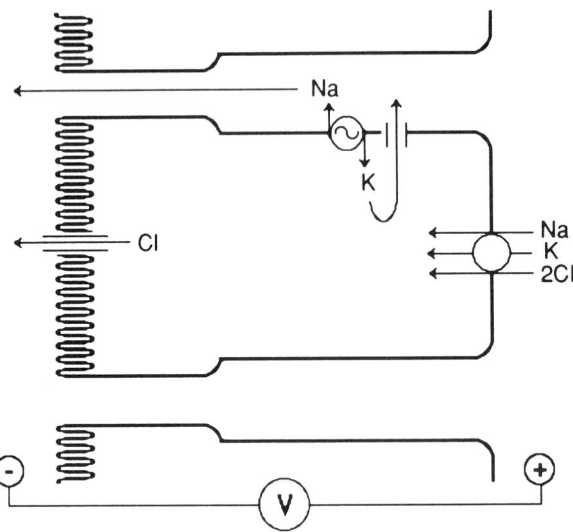

Fig. 3. General model for Cl secretion in a variety of secretory epithelia

II. Chloride and Bicarbonate Secretion

The mechanism for active Cl secretion in a variety of secretory epithelia is presented in Fig. 3. In general, Cl uptake across the basolateral membrane is mediated by a bumetanide-sensitive, electroneutral Na-K-2Cl cotransport mechanism in the basolateral membrane (LIEDTKE 1989; HAAS 1989). The cotransporter has been identified in human colonic cell lines (T_{84} and HT-29 cell lines) and in basolateral membrane vesicles from rabbit distal colon epithelium (DHARMSATHAPHORN et al. 1985; FRANKLIN et al. 1989; WIENER and VAN OS 1989) but not in mammalian small intestine. It is worth noting that experiments with furosemide and bumetanide on crypt cells from rat and human intestine suggest that the cotransporter does not significantly contribute to basolateral Cl uptake under secretory conditions (STEWART and TURNBERG 1989). An alternative pathway for Cl uptake in small intestine has not been proposed. Bumetanide-insensitive Cl secretion has also been observed in porcine distal colon (TRAYNOR and O'GRADY 1991). Chloride uptake by this pathway is inhibited by HCO_3 removal and by inhibition of carbonic anhydrase activity with acetazolamide. This result suggests a possible role for Cl-HCO_3 exchange in Cl transport across the basolateral membrane of secretory cells in porcine colon.

In epithelia where a basolateral Na-K-2Cl cotransport system exists, Na that enters the cell with Cl is recycled across the basolateral membrane in exchange for K by the Na-K ATPase. Potassium, which enters the cell via the Na-K ATPase and Na-K-2Cl cotransport system, exits the cell across the basolateral membrane through K channels (HALM and FRIZZELL 1991).

Chloride, which is present above electrochemical equilibrium within the cell (WELSH 1983; SHOROFSKY et al. 1983), exits across the apical membrane through Cl channels in response to the negative intracellular potential. Under conditions of active secretion, electrogenic K efflux across the basolateral membrane maintains the electrical driving force necessary for Cl exit across the apical membrane (McCANN and WELSH 1990; DAWSON 1991).

The mechanism of intestinal HCO_3 secretion is poorly understood. Bicarbonate secretion in the ileum is dependent on serosal HCO_3 and is not inhibited when carbonic anhydrase activity is blocked with acetazolamide (SMITH et al. 1985). This result indicates that a mechanism for transporting HCO_3 across the basolateral membrane is involved in the secretory process. An Na-$(HCO_3)_x$ cotransport system has recently been identified in the basolateral membrane of parietal cells (TOWNSLEY and MACHEN 1989) and in proximal tubule cells (LOPES et al. 1987). Na-$(HCO_3)_x$ cotransport with a stoichiometry of 1:1 has been proposed as a pathway for transporting HCO_3 into secretory cells above electrochemical equilibrium (SMITH et al. 1985; TOWNSLEY and MACHEN 1989). Efflux of HCO_3 across the apical membrane is presumably electrogenic and could involve anion channels that also conduct Cl, or an outwardly directed Na-$(HCO_3)_x$ cotransport mechanism in the apical membrane. Apical Na-$(HCO_3)_x$ cotransport may also explain Na secretion which has previously been reported in rabbit ileum and porcine proximal jejunum (FIELD 1971; CHANDAN et al. 1991).

III. Potassium Transport

Potassium absorption in small intestine is thought to be mediated by the paracellular pathway (TURNBERG 1971). Absorption of monovalent ions and H_2O results in an increase in K concentration in the lumen of the small intestine relative to the plasma. When the transepithelial K concentration gradient exceeds the transepithelial potential difference produced by electrogenic Na absorption, paracellular K transport from the lumen-to-serosa exceeds paracellular K movement in the opposite direction resulting in net K absorption. In the colon, transcellular K absorption is electroneutral and is independent of Na-K ATPase activity in the basolateral membrane (McCABE et al. 1982; HALM and DAWSON 1984). Colonic K absorption also appears to be linked to H secretion since removal of K reduces luminal acidification in guinea pig colon (SUZUKI and KANEKO 1987). Experiments with rabbit distal colon BBMV have shown that an H-K ATPase is present which is insensitive to ouabain (GUSTIN and GOODMAN 1981; KAUNITZ and SACHS 1986). In contrast to rabbit colon, K absorption in guinea pig and turtle colon is blocked by mucosal ouabain (McCABE et al. 1982; HALM and DAWSON 1984). The effect of ouabain appears to be on a transport mechanism other than the Na-K ATPase since Na replacement does not significantly alter the rate of K absorption in guinea pig colon. Thus the pharmacology of the H-K ATPase shows some species differences in ouabain sensitivity. Efflux of K across the basolateral membrane appears to be mediated by a Ba^{2+}-sensitive K channel in rabbit distal colon (Halm and Frizzell 1986). Since K uptake across the apical membrane is linked to H efflux, a mechanism for electrogenic HCO_3 efflux or H influx must exist to sustain H secretion and to balance electrogenic K efflux from the cell. A transport mechanism responsible for acid loading and charge balance has not been identified.

Active K secretion has also been characterized in colonic epithelia. Potassium secretion linked to Na absorption has been observed in turtle colon (Halm and Dawson 1984) but is not associated with Na absorption in mammalian colon (Sullivan and Smith 1986). Active K secretion across guinea pig, rabbit and rat colon is linked to electrogenic Cl secretion and is insensitive to mucosal amiloride or aldosterone stimulation (McCabe and Smith 1985). In Cl secreting cells, Na uptake by the Na-K-2Cl cotransporter augments K uptake by Na-K ATPase resulting in accumulation of K within the cell. Inhibition of cotransport activity with bumetanide inhibits net K secretion in mammalian colon (Halm and Frizzell 1986). Apical K channels provide a pathway for electrogenic K efflux and net K transport across the epithelium. Potassium secretion in turtle colon is blocked by Ba; however, Ba does not inhibit K secretion in rabbit colon (Dawson 1991). Triethylammonium and cesium produce a partial block of K secretion in rabbit colon. The basic properties and regulation of K channels responsible for secretion in mammalian colon have not been characterized.

IV. Paracellular Pathway and Solute Transport

Active electrolyte transport across the intestinal epithelium establishes the driving force for a significant amount of paracellular ion transport (Madara 1991). The occluding junction or "tight junction" provides a barrier to macromolecular diffusion and possesses varying degrees of permselectivity to monovalent ions along the length of the intestine (i.e., the paracellular permeability of the jejunum is greater than that of the ileum). The regional differences are presumably dependent on the number of junctional strands between adjacent cells (Madara et al. 1980). The strands appear as tightly spaced chains of intramembrane particles (Knutton et al. 1978). It has been suggested that the spaces between particles represent junctional pores through which movement of monovalent ions can occur. Tight junctions in mammalian intestine have greater permeability to cations relative to anions (Davis et al. 1982). Just below the occluding junction, there exists a band of actin and myosin filaments that surround the apical border of the cell. Contraction of the perijunctional band of filaments occurs following stimulation of Na-dependent glucose absorption (Madara et al. 1987). This contraction produces an increase in paracellular permeability and a decrease in cation selectivity of the junctional complex (Madara and Pappenheimer 1987). The result is to increase paracellular water and nutrient permeability and to facilitate the paracellular absorption of nutrients under postprandial conditions. It is worth noting, however, that the amount of nutrient and fluid transport across the epithelium by this mechanism is limited relative to transcellular transport since the total junctional surface area is much less than the apical membrane surface area of the epithelium.

C. Regulation of Ion Transport by Peptide Neurohormones

I. Peptides That Stimulate Intestinal Secretion or Inhibit Absorption

1. Vasoactive Intestinal Peptide and Related Peptides

Vasoactive intestinal peptide (VIP) is a 28 amino acid peptide that is structurally related to the glucagon-secretin family of peptides (BROWN and MILLER 1991). Peptide histidine isoleucine (PHI) and peptide histidine methionine (PHM) have been isolated from porcine and human intestine respectively and are highly homologous to VIP. Indeed, these peptides are components of the VIP prohormone and are colocalized with VIP (ITOH et al. 1983). VIP and related peptides have been shown to be effective secretagogues throughout the intestinal tract. Studies in rat duodenum in vivo have shown that VIP and PHI inhibit fluid absorption and stimulate HCO_3 secretion (ISENBERG et al. 1984). In ileum and jejunum, VIP and secretin stimulate fluid secretion, inhibit net Na, K, Cl and HCO_3 absorption and increase anion secretion following their intravenous infusion (BARBEZAT 1973; DAVIS et al. 1981). Studies in rat colon in vivo have shown that VIP and PHI stimulate fluid and anion secretion (WU et al. 1979). The transport-related actions of VIP in vivo primarily result from a direct effect on epithelial cells. Secondary effects have also been reported on submucosal nerves in rat jejunum where inhibition of neuronal activity with tetrodotoxin (TTX) decreases VIP-induced secretion (COUPAR 1986). Secondary effects on intestinal blood flow may also contribute to the decrease in electrolyte absorption observed in vivo (MAILMAN 1978; see Chap. 11, this volume).

Experiments in vitro have shown that VIP stimulates Cl secretion and inhibits Na and Cl absorption (SCHWARTZ et al. 1974; RACUSEN and BINDER 1977; MCCULLOCH et al. 1987). Similar effects have been reported for secretin and glucagon (WALDMAN et al. 1977). Receptors for VIP have been identified in several human colon adenocarcinoma cell lines including T_{84} cells, HT-29 cells and Caco-2 cells (ZWEIBAUM et al. 1991; see Chap. 5, this volume). In cultured T_{84} cells, VIP stimulates adenylate cyclase activity resulting in an increase in intracellular adenosine-3′,5′-cyclic monophosphate (cAMP) concentration (MCCABE and DHARMSATHAPHORN 1988). Cyclic AMP subsequently stimulates bumetanide-sensitive Cl influx across the basolateral membrane and accelerates the rate of Cl efflux across the apical membrane (DHARMSATHAPHORN et al. 1985; MANDEL et al. 1986). VIP stimulation also produces an increase in basolateral rubidium (Rb) efflux (a measure of K secretion) that is blocked by Ba and quinidine (MCROBERTS et al. 1985). Whole-cell patch clamp experiments with T_{84} cells have shown that cAMP activates a Cl channel with a linear current-voltage relationship over a range of voltages from -100 to $+100$ mV. In contrast, ionomycin (a Ca^{2+} ionophore) activates an outwardly rectifying Cl channel over the same range of voltages. The Cl currents produced by cAMP and ionomycin were additive, suggesting that cAMP and Ca stimulate Cl secretion by activating separate Cl conductive pathways in the apical membrane (CLIFF and FRIZZELL 1990). Cyclic AMP activation of Cl channels in T_{84} cells is dependent on cAMP-dependent protein kinase activity. T_{84} cells, transfected with an expression vector that contains a mutant form of the regulatory subunit of cAMP-dependent protein kinase, do not respond to VIP stimulation even though adenylate cyclase activation in mutant cells is not significantly different from control cells (ROGERS et al. 1990). Protein phosphorylation experiments with intact T_{84} cells following VIP stimulation indicate that at least four soluble proteins (ranging from 17 to 37 kDa) are phosphorylated. The function of these phosphoproteins is unknown but it seems unlikely that they represent ion transport proteins (COHN 1987). VIP-stimulated Cl secretion has also been documented in HT-29 cells and in Caco-2 cells. In HT-29 cells, VIP stimulates bumetanide-sensitive K uptake across the basolateral membrane, reflecting an increase in Na-K-2Cl cotransport activity (TURNER et al. 1990a). Elevations in intracellular cAMP also produce decreases in apical membrane fractional resistance and an increase in short-circuit current (Isc, a measure of active electrogenic ion transport) consistent with Cl channel activation and net Cl secretion (ZWEIBAUM et al. 1991). In Caco-2 cells, VIP and other compounds that increase intracellular cAMP increase apical membrane Cl permeability (GRASSET et al. 1985). In contrast to HT-29 cells and T_{84} cells, however, cAMP does not activate Na-K-2Cl cotransport activity or any other known Cl influx pathway in the basolateral membrane. Thus Caco-2 cells are unable to sustain high rates of Cl secretion despite activation of a relatively large apical membrane Cl conductance (ZWEIBAUM et al. 1991).

In a human small intestine epithelial cell line (intestine 407 cells), VIP receptors linked to adenylate cyclase activation have been identified (YADA et al. 1989). VIP produces an increase in intracellular Ca^{2+} concentration $[Ca^{2+}]_i$ which results in K channel activation and hyperpolarization of 407 cells (YADA and OKADA 1984). Stimulation with dibutyryl-cAMP also increases $[Ca^{2+}]_i$ and hyperpolarizes these cells, suggesting that cAMP stimulates the release of Ca from intracellular stores (YADA et al. 1989). Similar results were observed in an earlier study in chicken enterocytes using 8-bromo cAMP (SEMRAD and CHANG 1987). However, VIP has only a minor effect on $[Ca^{2+}]_i$ in HT-29 cells (TURNER et al. 1990a). Thus the effect of cAMP on Ca mobilization appears to vary between cell types. It is worth noting that peptides related to VIP such as secretin and glucagon have been shown to stimulate inositol phospholipid breakdown in pancreatic acinar cells (TRIMBLE et al. 1986) and hepatocytes (WAKELAM et al. 1986). Hence, VIP-induced changes in intracellular Ca could potentially involve coupling to phospholipase C activity and phosphoinositol (PI) turnover.

2. Gastrin-Releasing Peptide and Related Peptides

Gastrin-releasing peptide (GRP) and neuromedin B (NMB) are mammalian peptides that are homologous to the amphibian dermal peptides bombesin (BB) and ranatensin (RT). BB-like immunoreactivity has been localized in submucosal nerve cell bodies and ganglia from canine (DANIEL et al. 1985) and rat (MOGHIMZADEH et al. 1983) small intestine. Studies in canine jejunum in vivo have shown that BB inhibits NaCl and fluid absorption (BARBEZAT and REASBECK 1983). Ussing chamber studies with rat and guinea pig small intestine have shown that BB produces a transient increase in Isc consistent with a stimulation of Cl secretion (KACHUR et al. 1982a; CHANG et al. 1986). In guinea pig ileum, pretreatment of the tissue with TTX had no effect on the increase in Isc produced by BB, suggesting that the transport-related actions of this peptide are not mediated by submucosal nerves (KACHUR et al. 1982a). In contrast, studies with porcine jejunum indicate that 40% of the Isc response to GRP is blocked by TTX but not by atropine or hexamethonium (CHANDAN et al. 1988). This result suggests that GRP and amphibian homologs stimulate the release of noncholinergic neurotransmitters from submucosal nerves in addition to their direct actions on the epithelium. GRP stimulation of Isc in porcine jejunum is significantly inhibited by serosal furosemide and by Cl replacement. Transepithelial flux experiments indicate that GRP stimulates unidirectional serosal-to-mucosal Na and Cl fluxes and increases net Cl secretion over 100%. GRP, NMB, BB and RT have the following order of potency in porcine jejunum: GRP≈RT≥NMB>BB. The amphibian peptides BB and RT, however, are more effective in elevating Isc than their mammalian homologs (CHANDAN et al. 1988). In addition to its effects on ion transport, GRP also stimulates crypt cell proliferation, suggesting the possibility that changes in ion transport activity may be associated with intestinal growth (WOLL and ROZENGURT 1989; see Chap. 12, this volume).

3. Substance P and Neurokinins

Substance P (SP) is a member of a large family of peptides with homologous C-terminal regions called tachykinins (ERSPAMER 1981). Neurokinin A (NKA) and neurokinin B (NKB) are mammalian tachykinins that may also play a role in the regulation of intestinal ion transport (TATEMOTO et al. 1985). Studies in vivo have shown that intravenous administration of SP decreases the absorption (or stimulates secretion) of Na, K, Cl and fluid in the canine jejunum (McFADDEN et al. 1986). Experiments on small in intestinal mucosa from rats, guinea pigs, chinchillas and chickens in vitro have shown that SP produces a transient increase in Isc (Kachur et al. 1982; Chang et al. 1986; Keast et al. 1985). Radioisotopic flux studies and Cl substitution experiments in guinea pig ileum suggest that the SP -induced Isc change results from an increase in Cl secretion (Perdue et al. 1987.) In rat ileum, SP stimulates anion secretion and inhibits Na absorption (Walling et al. 1977). Inhibition of Na-H exchange has been reported in isolated chicken villous cells, suggesting that SP reduces net Na absorption by blocking this transport pathway (Chang et al. 1986). NKA and NKB produce effects similar to SP on ion transport in porcine proximal jejunum (Brown et al. 1992). In this tissue, their order of potency is SP>NKA>NKB, which suggests that the SP effect is mediated by a neurokinin-1 receptor (Parsons et al. 1992). In guinea pig and rabbit intestine, a portion of the SP effect is blocked by TTX (Hubel et al. 1984; Keast et al. 1985). Acetylcholine is at least one of the neurotransmitters that is released followed SP stimulation (Perdue et al. 1987). The SP antagonist [D-Arg^{1},D-Pro^{2},D-Trp^{7,9},Leu^{11}] SP blocks the direct but not indirect actions of SP on the epithelium. This result suggests that different NK receptors are involved in the SP-induced stimulation of submucosal nerves (Keast et al. 1985).

The anion secretion produced by SP appears to be mediated by increases in intracellular Ca. Chelation of extracellular Ca blocks the effects of SP on Na influx in isolated chicken enterocytes (Chang and Musch 1990). Removal of extracellular Ca from the serosal surface of guinea pig ileal epithelium inhibits the SP-stimulated Isc response, suggesting that SP receptors are linked to a mechanism for Ca influx in the basolateral membrane (Brown 1987). Calcium channel blockers such as verapamil, D-600 and loperamide also reduce the SP-stimulated Isc observed in rabbit, guinea pig, chinchilla and chicken intestine, suggesting a role for basolateral Ca channels (Donowitz et al. 1982; Chang et al. 1986). The TTX-sensitive actions of SP are also mediated to some degree by increases in intracellular Ca. Acetylcholine, released from submucosal nerve terminals in response to SP stimulation, binds to muscarinic receptors that are presumably coupled to inositol phospholipid turnover and release of Ca from intracellular stores (Chandan et al. 1991). The increase in intracellular Ca is presumably linked to protein phosphorylation and regulation of Cl secretion but the details of these events with regard to SP stimulation are poorly understood.

4. Neurotensin

In mammals, neurotensin (NT) is a 13 amino acid peptide that is structurally homologous to the mammalian hexapeptide neuromedin N (NMN) and the amphibian dermal peptide xenopsin (XP) (Carraway and Ferris 1983; Minamino et al. 1984). In vitro experiments in small intestine from rodents, chickens and pigs have shown that NT increases Isc and that this effect is attenuated under anion replacement conditions (Kachur et al. 1982b; Brown and Treder 1989; Chang et al. 1986). In porcine distal jejunum, NT inhibits

Cl absorption; the relative potency of NT and related peptides is NT > XP > NMN with NT as the most effective peptide in stimulating Isc (BROWN and TREDER 1989). Structure-activity experiments with NT in isolated guinea pig ileum show that amino acid substitutions in the C-terminal portion of the peptide produce large changes in potency and efficacy (KACHUR et al. 1982b). In both guinea pig ileum and porcine distal jejunum (which is functionally equivalent to the ileum of other mammals), the effects of NT are completely blocked by TTX (KACHUR et al. 1982b; BROWN and TREDER 1989). In guinea pig ileum, the effects of NT may be mediated by SP release from submucosal nerves. In porcine distal jejunum, neurotransmitters other than SP appear to be involved. Direct actions of NT have been reported in HT-29 cells (MORRIS et al. 1990; TURNER et al. 1990b). NT stimulates an outwardly rectifying Cl channel with properties that are similar to Ca-activated Cl channels in T_{84} cells. NT also activates a K conductance in these cells.

The effects of NT on intestinal ion transport appear to be mediated by increases in $[Ca^{2+}]_i$ occurring either in epithelial cells themselves or in mucosal neurons. In rodent, rabbit, chicken and pig intestine, the effects of NT are dependent on extracellular Ca and blocked by verapamil and related compounds that inhibit SP-mediated increases in Isc (CHANG et al. 1986). In HT-29 cells, however, NT increases intracellular Ca through activation of inositol phospholipid turnover (AMAR et al. 1986) and by increasing Ca uptake from the bathing solution (MORRIS et al. 1990). The initial increases in Cl current produced by NT appears to be due to the release of Ca from intracellular stores near the membrane. As $[Ca^{2+}]_i$ rises, inactivation of the Cl channel takes place which is not observed when cells are stimulated with ionomycin. This result suggests that NT also activates an inhibitory mechanism not involving Ca that may serve as a feedback control mechanism to regulate cell volume and $[Cl]_i$. Neurotensin stimulation of Ca follows intracellular Ca release and is blocked by La^{3+} (MORRIS et al. 1990). This influx pathway provides for a sustained increase in intracellular Ca which may play some role in replenishing intracellular Ca stores.

5. Natriuretic Peptides

Atrial natriuretic peptide (ANP) and brain natriuretic peptide (BNP) are homologous peptides originally isolated from rat heart and porcine brain respectively (SUDOH et al. 1988). Immunohistochemical localization studies have also identified ANP in the submucosa of human large intestine (GERBES et al. 1990), rat ileum and colon (VUOLTEENAHO et al. 1988) and guinea pig jejunum and colon (VOLLMAR et al. 1988). BNP has been localized in nerve fibers throughout the wall of the stomach and jejunum of the rat (SHARKEY et al. 1991). ANP and atrial extracts have been shown to inhibit fluid absorption in mammalian intestine in anesthetized animals (MARTINEZ et al. 1986). In vitro experiments with rat intestine mounted in Ussing chambers show that ANP stimulates Na and Cl secretion in the jejunum and presumably Cl secretion in the colon (Patrick et al. 1987; Moriarty et al. 1990). The ANP effects on rat colon are blocked by pretreatment with TTX and atropine, suggesting that the ANP effect on secretion is secondary to stimulation of acetylcholine release from submucosal nerves. BNP also appears to stimulate Cl secretion in rat jejunum through stimulation of submucosal nerves (Sharkey et al. 1991). In porcine distal colon the effects of BNP and ANP are resistant to TTX and to 5,8,11,14-eicosatetraynoic acid (ETYA, an inhibitor of arachidonic acid metabolism), suggesting a direct effect on the epithelium (Traynor and O'Grady 1991). ANP and BNP stimulate K and Cl secretion in porcine distal colon. The effects of both peptides are blocked by HCO_3 replacement and are unaffected by serosal addition of bumetanide. Thus BNP-stimulated Cl secretion does not appear to involve Na-K-2Cl cotransport as a mechanism for basolateral Cl uptake; rather, Cl influx may be dependent on Cl-HCO_3 exchange. Analysis of concentration-response curves for BNP and ANP show that BNP is more potent and efficacious than ANP in stimulating electrogenic Cl secretion in porcine colon. This result suggests that the receptor for natriuretic peptides is likely to be an ANP-B receptor which possesses a higher affinity for BNP relative to ANP.

Atrial natriuretic peptide has been shown to inhibit Na absorption in chicken enterocytes (Semrad et al. 1991) and to block Na and Cl absorption in flounder intestine (O'Grady et al. 1985). Inhibition of Na absorption in chicken intestine results from blocking Na-H exchange activity in the apical membrane. This effect of ANP is mimicked by 8-bromoguanosine-3′,5′-cyclic monophosphate (8-Br cGMP). ANP-induced decreases in Na and Cl absorption in flounder intestine result from blockade of an apical membrane Na-K-2Cl cotransport system that is also inhibited by 8-Br cGMP. ANP produces a fourfold increase in cGMP content of the flounder intestinal epithelium without affecting levels of cAMP (O'Grady 1989). In rat enterocytes in culture (CRL 1589 cells), ANP activates particulate guanylate cyclase (five- to tenfold) with an $EC_{50} = 6\,nM$ (Crane et al. 1990). Accumulation of cGMP within these cells following ANP treatment was 25-fold greater than in control cells. In chicken and flounder intestine, the effects of ANP and 8-Br cGMP could be reversed by the isoquinoline sulfonamide derivative H-8, which is known to block cGMP-dependent protein kinase activity in a variety of cells (O'Grady et al. 1988; Semrad et al. 1991). Similar effects of H-8 have been reported in intermedullary collecting duct cells where ANP is known to inhibit Na absorption by blocking amiloride-sensitive Na channels in the apical membrane (Light et al. 1989). Thus, in intestinal epithelia and mammalian distal nephron, ANP action appears to be mediated by the cGMP second messenger system. This may not be true in porcine colon, however, since 8-Br cGMP produces an inhibition of Na absorption in vitro which is not observed when tissues are stimulated with BNP or ANP (DuVall et al., unpublished observations). Qualitative differences in the flux data suggest that another second messenger pathway is involved in the transport-related actions of BNP in porcine colon.

II. Peptides That Stimulate Intestinal Absorption or Inhibit Secretion

1. Angiotensin

Angiotensin II (ANG II) and angiotensin III (ANG III) both enhance intestinal electrolyte and fluid absorption in response to decrease in extracellular fluid volume (LEVENS 1984). ANG II stimulates Na and fluid absorption across everted sacs from rat small intestine and colon (DAVIES et al. 1970). In vivo studies with rat jejunum have shown that infusion of ANG II at doses below those required to alter blood pressure results in electroneutral solute and fluid absorption that can be blocked by α_1-adrenoceptor blockade or by depletion of catecholamines from sympathetic nerves (BOLTON et al. 1975). The effects of ANG II on fluid absorption are also mimicked by norepinephrine (NE), suggesting that ANG II stimulates the release of NE from submucosal sympathetic nerves (LEVENS et al. 1979).

Ussing chamber experiments using rat jejunum have shown that ANG II increases the basal Isc presumably by increasing Cl secretion (COX et al. 1987). ANG II was also found to inhibit the Isc in rat descending colon but the ionic basis for this effect was not characterized. The effects of ANG II in both segments were resistant to TTX treatment but blocked by inhibitors of prostaglandin synthesis. Prostanoids such as prostaglandin E_2 (PGE_2) are effective secretagogues in mammalian small intestine and colon (POWELL 1991). Dehydration, Na depletion and other conditions that reduce extracellular volume presumably increase jejunal electrolyte and fluid absorption through the release of ANG II at low concentrations (LEVENS 1984). Angiotensin II also increases fluid and electrolyte absorption in distal colon by a mechanism that is independent of the actions of aldosterone (LEVENS et al. 1981b).

2. Vasopressin

Vasopressin is known to have important effects on fluid absorption in the distal nephron under conditions of dehydration. In mammalian descending colon, vasopressin increases NaCl absorption and inhibits active Cl secretion (BRIDGES et al. 1983). The effects of vasopressin are attenuated in dexamethasone-pretreated tissues where electrogenic Na absorption is maximal (BRIDGES et al. 1984). This result suggests that the increase in Na uptake following vasopressin stimulation is mediated by Na channels in the apical membrane. The increase in Cl absorption is presumably due to an increase in Cl-HCO_3 exchange activity but no direct evidence to support this conclusion is available. The effects of vasopressin cannot be blocked by opiate or adrenergic receptor blockers (HORNYCH et al. 1973). It has been suggested that vasopressin produces its effects on colonic ion transport by decreasing intracellular Ca concentration. However, direct experimental evidence for this hypothesis is lacking at this time.

3. Neuropeptide Y

Neuropeptide Y (NPY) has been identified in submucosal neurons in mammalian small intestine (Allen et al. 1987). Peptide YY is a closely related peptide that is found in enteroendocrine cells in the small intestine and colon (Tatemoto 1982). In rabbit ileum, PYY is more potent than NPY in reducing Isc (Cox et al. 1988). Repeated treatment of the tissue with NPY produces a tachyphylaxis response (Hubel and Renquist 1986). Cross-tachyphylaxis to PYY has not been determined. Ussing chamber experiments with rat and rabbit ileum indicate that NPY increases net Cl absorption with inconsistent effects on Na transport (Friel et al. 1986; Hubel and Renquist 1986). Neuropeptide Y also inhibits VIPergic and cholinergic stimulation of ion transport in guinea pig distal colon (McCulloch et al. 1987). In rabbit and porcine ileum, NPY stimulates the mucosal-to-serosal Cl flux whereas, in rat ileum, NPY increases net Cl absorption by reducing the serosal-to-mucosal Cl flux (Cox et al. 1988; Brown et al. 1990a). In porcine ileum, the effects of NPY on Cl absorption are blocked by the disulfonic stilbene derivative 4,4′-diisothiocyanostilbene-2,2′-disulfonic acid (DIDS), suggesting that NPY stimulates a Cl-HCO_3 exchange pathway located in the apical membrane. Neuropeptide Y also produces a significant increase in tissue conductance in porcine ileum, suggesting that the peptide increases the paracellular permeability of the epithelium (Brown et al. 1990a).

The effects of NPY in rat and rabbit ileum are not blocked by pretreatment of the tissues with TTX (Hubel and Renquist 1986; Cox et al. 1988). This suggests that NPY directly interacts with receptors on the epithelial cells to produce changes in ion transport (see Chap. 5). In contrast, the effects of NPY in porcine ileum are blocked by TTX, indicating that neurotransmitters released from submucosal nerves are responsible for the transport-related actions of NPY in this tissue (Brown et al. 1990a). Pretreatment of tissues with phentolamine (an α-adrenoceptor blocker) attenuates the effects of NPY on Isc. This result suggests that part of the NPY effect may be mediated by NE. The effects of NPY on colonic Isc are also blocked by ETYA, indicating that NPY-induced neurotransmitter release may in turn stimulate prostaglandin or leukotriene release within the tissue.

4. Somatostatin

Somatostatin (SS) is a peptide that is present in enteroendocrine cells (D cells contain the 28 amino acid form of SS) and nerve fibers (containing the 14 amino acid peptide) in small intestine and colon (Buchan et al. 1985).

D-cells and SS-containing nerve fibers are closely associated with crypt cells and are more numerous in the small intestine relative to the colon (see Chap. 1). In vivo studies with SS-14 and the metabolically stable synthetic analog octreotide have shown that SS inhibits VIP-induced secretion of fluid and electrolytes in human jejunum and reduces secretory diarrhea in patients with VIP-producing tumors (Davis et al. 1980; Maton et al. 1985). In Ussing chamber studies of rabbit ileum, porcine distal jejunum and rat colon, SS-14 produces long-term decreases in basal Isc. This effect in rabbit ileum results from inhibition of anion secretion and stimulation of electroneutral Na and Cl absorption (Dharmsathaphorn 1985). In porcine distal jejunum SS-14 increases HCO_3-dependent Cl absorption without significantly affecting Na transport. SS-14 also blocks the HCO_3-dependent residual flux in porcine jejunum which suggests inhibition of active HCO_3 secretion (Brown et al. 1990b). In T_{84} cells, SS-14 inhibits VIP- and prostaglandin-induced Cl secretion, providing functional evidence for a direct effect of the peptide on epithelial cells (Dharmsathaphorn et al. 1984). Additional evidence supporting a direct action of SS-14 on the epithelium comes from experiments with rabbit ileum where blockade of opiate and α-adrenergic receptors did not alter the effects of SS-14 on ion transport (Guandalini et al. 1980).

The intracellular mechanisms responsible for the transport-related actions of SS-14 in mammalian intestine have not been identified. SS-14 is known to block the effects of substances that increase intracellular cAMP and cGMP but the peptide does not appear to block agonist-induced stimulation of these second messenger systems (Dharmsathaphorn et al. 1980a; Guandalini et al. 1980). SS-14 and related analogs also do not alter secretory responses of the epithelium to agents that increase intracellular Ca (Dharmsathaphorn et al. 1980b). Thus the antisecretory and proabsorptive actions of SS appear to occur at a site distinct from the activation of second messenger pathways that regulate intestinal ion transport.

5. Opioid Peptides

The three families of opiate peptides include: (1) the enkephalin pentapeptides, (2) dynophin/α-neoendorphin and (3) β-endorphin (Miller and Hirning 1989). Although peptides from each of these families have been detected in mammalian intestine, only the enkephalins possess a distribution that suggests a role in ion transport (see Chap. 1). In vivo studies with synthetic opiate compounds and enkephalin peptides have shown that systemic opiate administration leads to increases in Na, Cl and fluid absorption in the jejunum and ileum (Warhurst et al. 1983; Fogel and Kaplan 1984). Enkephalins also inhibit fluid secretion induced by VIP and other secretagogues that increase intracellular cAMP in mammalian small intestine and colon (Turnberg 1983; Coupar 1983). In vitro studies with β-casomorphins and enkephalin derivatives have shown that these peptides produce sustained decreases in Isc with little or no effect on tissue conductance (DOBBINS et al. 1980; MCKAY et al. 1981; BINDER et al. 1984; BERSCHNEIDER et al. 1988). These changes in Isc appear to be associated with inhibition of anion secretion and stimulation of Na and Cl absorption. The transport-related actions of opioid peptides are not mediated by receptors present on enterocytes (DOBBINS et al. 1980). The observation that TTX blocks the effects of enkephalin derivatives in rabbit ileum, porcine distal jejunum and mouse jejunum indicates that submucosal nerves are the principal site of action for these peptides (DOBBINS et al. 1980; BINDER et al. 1984; SHELDON et al. 1990; QUITO and BROWN 1991). Pretreatment of porcine distal jejunum with phentolamine did not alter the effects of enkephalin derivatives on Isc, suggesting that NE is not the neurotransmitter responsible for the proabsorptive actions of these peptides (QUITO and BROWN 1991). The nonadrenergic, noncholinergic neurotransmitter substance that mediates opioid peptide action in the intestine has not been identified in any epithelial preparation to date.

The cellular mechanism of action of opioid peptides on epithelial ion transport is poorly understood. Morphine and methionine-enkephalin are known to block PGE_2- and VIP-stimulated secretion but have no effect on increases in cAMP content of the epithelium produced by these substances (COUPAR 1983). Moreover, the effects of enkephalin derivatives reverse the effects of 8-Br cAMP on Cl absorption in porcine jejunum, suggesting that enkephalin derivatives do not act by altering cAMP production (QUITO and BROWN 1991).

III. Other Peptides That Regulate Intestinal Electrolyte Transport

The list of peptides that affect intestinal electrolyte transport continues to expand. Recent studies have shown that neuromedin U (NMU), galanin (GAL), interleukin (IL-1 and IL-3) and cholecystokinin (CCK) all modulate the transport properties of the small intestine and colon. The octapeptide form of neuromedin U increases the Isc in porcine distal jejunum with an EC_{50} of 0.7 n*M*. Its effects are dependent on the presence of Cl in the bathing solution and are blocked by TTX. Atropine and hexamethonium have no effect on NMU-mediated increases in Isc, suggesting that NMU stimulates intestinal Cl secretion through interactions with noncholinergic enteric neurons (BROWN and QUITO 1988). Galanin is another gut peptide that has been shown to produce small decreases in basal Isc but to potently attenuate transmural electric stimulation of the Isc in porcine distal jejunum (IC_{50} = 13 n*M*). Cl replacement experiments suggest that the decrease in Isc produced by GAL is the result of a decrease in Cl secretion (BROWN et al. 1990c). The results indicate that GAL may affect electrically evoked ion transport in the porcine distal jejunum. The peptide has also been reported to inhibit transmural electric stimulation of Isc in the guinea pig colon in vitro but its efficacy in this tissue was much lower than that in porcine jejunum (MCCULLOCH et al. 1987). A third gut peptide that affects intestinal ion transport through stimulation of enteric nerves is cholecystokinin (CCK). CCK octapeptide stimulation of anion secretion in guinea pig ileum was shown to be completely blocked by TTX pretreatment (KACHUR et al. 1991). Approximately 50% of the increase in Isc produced by CCK was blocked by atropine, suggesting that acetylcholine release from submucosal nerves is responsible for a large fraction of CCK-induced anion secretion. The effects of CCK were also blocked by the CCK antagonists proglumide and lorglumide. This result indicates that the effect of CCK on enteric nerves is mediated through specific CCK receptors.

Recent studies with interleukins 1 and 3 (IL-1 and IL-3) in chicken intestine indicate that cytokines may affect intestinal electrolyte transport (CHANG et al. 1990). The increase in Isc produced by IL-1 and IL-3 in chicken intestine was blocked by Cl or HCO_3 replacement, suggesting that these peptides stimulate anion secretion. Pretreatment of tissues with piroxicam (an inhibitor of cyclooxygenase activity) blocked the effects of IL-1 and IL-3 in this preparation. Interleukins increase anion secretion by stimulating PGE_2 synthesis and release from cells in the lamina propria and submucosa. Thus, interleukins produced by macrophages, monocytes and activated T lymphocytes may play a role in inflammatory responses that activate intestinal anion secretion. Whether IL-1 and IL-3 produce similar effects in mammalian intestine remains to be determined.

D. Concluding Remarks

Our mechanistic understanding of peptide neurohormone regulation of intestinal ion transport is remarkably incomplete. Several peptides produce effects on intestinal transport function by stimulating the release of undefined neurotransmitter or arachidonic acid metabolites, making it difficult to trace the signal transduction pathways that are relevant to the control of ion transport mechanisms. In addition, the receptors that mediate peptide actions in a variety of intestinal epithelia are poorly characterized. Thus, even in tissue preparations where connections have been made between second messenger systems and ion transport regulation, little can be said about the mechanism of peptidergic control of ion transport. However, important progress has been made with peptides (i.e., VIP and SP) that have direct actions on intestinal epithelial cells in culture. Patch clamp and fluorescence imaging techniques have been successfully used in cultured cells to measure the activities of specific transport proteins in response to hormone or second messenger stimulation. These cells have also been used to investigate receptor-activated signaling pathways in greater detail than in native epithelia. Thus cultured cells will continue to serve as valuable tools for understanding the cascade of events that connect hormone-receptor interactions at the membrane with the regulation of specific transport proteins involved in electrolyte absorption or secretion. It is important to note, however, that the actions of specific neurohormones on transformed cells may differ significantly from those of native epithelial cells. Moreover, the secondary effects of peptide neurohormones on intrinsic neurons and inflammatory cells are important factors to consider in establishing their role in the physiology of absorption and secretion. Thus experiments that define these indirect modulatory actions are valuable contributions to our understanding of the communication that takes place between various cell types within the intestine. The results of these investigations have already led to the development of useful pharmacotherapies for the treatment of diarrheal diseases. Further therapeutic developments will depend on a greater mechanistic understanding of peptide neurohormones in health and disease.

References

Ahn J, Chang EB, Field M (1985) Phorbol ester inhibition of Na-H exchange in rabbit proximal colon. Am J Physiol 249:C527–C530

Albin D, Gutman Y (1979) Tritiated ouabain binding and dissociation in rabbit colon: effects of ions and drugs. Biochem Pharmacol 28:3181–3188

Allen JM, Hughes J, Bloom SR (1987) Presence, distribution and pharmacological effects neuropeptide Y in mammalian gastrointestinal tract. Dig Dis Sci 32:506–512

Amar S, Kitabgi P, Vincent JP (1986) Activation of phosphoinositol turnover by neurotensin receptors in the human colonic adenocarcinoma cell line HT29. FEBS Lett 201:31–36

Barbezat GO (1973) Stimulation of intestinal secretion by peptide hormones. Scand J Gastroenterology 8:3–21

Barbezat GO, Reasbeck PG (1983) Effects of bombesin, calcitonin and enkephalin on canine jejenual water and electrolyte transport. Dig Dis Sci 28:273–277

Barros F, Dominguez P, Velasco G, Lazo PS (1986) Na/H exchange is present in the basolateral membranes from rabbit small intestine. Biochem Biophys Res Commun 134:827–834

Berschneider HM, Martens H, Powell DW (1988) Effect of BW942C, an enkephalin-like pentapeptide, on sodium and chloride transport in the rabbit ileum. Gastroenterology 94:127–136

Binder HJ (1978) Sodium and chloride transport across colonic mucosa in the rat. In: Hoffman JF (ed) Membrane transport processes. Raven, New York, p 309

Binder HJ, Laurenson JP, Dobbins JW (1984) Role of opiate receptors in the regulation of enkephalin stimulation of active sodium and chloride absorption. Am J Physiol 247:G432–G436

Bolton JE, Munday KA, Parsons BJ, York BG (1975) Effects of angiotensin II on fluid transport, transmural potential difference and blood flow by rat jejunum in vivo. J Physiol (Lond) 253:411–428

Bridges RJ, Nell G, Rummel W (1983) Influence of vasopressin and calcium on electrolyte transport across isolated colonic mucosa of the rat. J Physiol (Lond) 338:463–475

Bridges RJ, Rummel W, Wallenberg P (1984) Effects of vasopressin on electrolyte transport across isolated colon from normal and dexamethasone-treated rats. J Physiol (Lond) 355:11–23

Brown DR (1987) Intracellular mediators of peptide action in the intestine and airways: focus on ion transport function. Am Rev Respir Dis 136:S43–S48

Brown DR, Miller RJ (1991) Neurohormonal control of fluid and electrolyte transport in intestinal mucosa. In: Schultz SG, Field M, Frizzell RA (eds)

Intestinal absorption and secretion. The gastrointestinal system IV. American Physiological Society, Bethesda MD, p 527 (Handbook of physiology, Sect. 6)
Brown DR, Quito FL (1988) Neuromedin U octapeptide alters ion transport in porcine jejunum. Eur J Pharmacol 155:159–162
Brown DR, Treder BG (1989) Neurohumoral regulation of ion transport in the porcine distal jejunum. Actions of neurotensin and its natural homologs. J Pharmacol Exp Ther 249:348–357
Brown DR, Boster SL, Overend MF, Parsons AM, Treder BG (1990a) Actions of neuropeptide Y on basal, cyclic AMP-induced and neurally evoked ion transport in porcine distal jejunum. Regul Pept 29:31–47
Brown DR, Overend MF, Treder BG (1990b) Neurohormonal regulation of ion transport in the porcine distal jejunum. Actions of somatostatin-14 and its natural and synthetic homologs. J Pharmacol Exp Ther 252:126–134
Brown DR, Hildebrand KR, Parsons AM, Soldani G (1990c) Effects of galanin on smooth muscle and mucosa of porcine jejunum. Peptides 11:497–500
Brown DR, Parsons AM, O'Grady SM (1992) Substance P produced sodium and bicarbonate secretion in porcine jejunal mucosa through an action on enteric neurons. J Pharmacol Exp Ther 261:1206–1212
Buchan AMJ, Sikora KJ, Levy JG, McIntosh CHS (1985) An immunocytochemical investigation with monoclonal antibodies to somatostatin. Histochemistry 83:175–180
Carraway RE, Ferris CF (1983) Isolation, biological and chemical characterization and synthesis of a neurotensin-related hexapeptide from chicken intestine. J Biol Chem 258:2475–2479
Cassano G, Stieger B, Murer H (1984) Na/H and Cl/HCO_3 exchange in rat jejunal and rat proximal tubular brush border membrane vesicles. Studies with acridine orange. Pflugers Arch 400:309–317
Chandan R, Newell S, Brown DR (1988) Actions of gastrin-releasing peptide and related mammalian and amphibian peptides on ion transport in the porcine proximal jejunum. Regul Pept 23:1–14
Chandan R, Hildebrand KR, Seybold VS, Soldani G, Brown DR (1991) Cholinergic neurons and muscarinic receptors mediate anion secretion in pig distal jejunum. Eur J Pharmacol 193:265–273
Chang EB, Musch MW (1990) Calcium mediated neurohumoral inhibition of chicken enterocyte Na influx: role of phosphatidylinositol metabolites. Life Sci 46:1913–1921
Chang EB, Brown DR, Wang NS, Field M (1986) Secretagogue-induced changes in membrane calcium permeability in chicken and chinchilla ileal mucosa: selective inhibition by loperamide. J Clin Invest 78:281–287
Chang EB, Musch MW, Mayer L (1990) Interleukins 1 and 3 stimulate anion secretion in chicken intestine. Gastroenterology 98:1518–1524
Cliff WH, Frizzell RA (1990) Separate Cl conductances activated by cAMP and Ca in Cl secreting epithelial cells. Proc Natl Acad Sci USA 87:4956–4960
Cohn J (1987) Vasoactive intestinal peptide stimulates protein phosphorylation in a colonic epithelial cell line. Am J Physiol 253:G420–G424
Coupar IM (1983) Characterization of the opiate receptor population mediating inhibition of VIP-induced secretion from the small intestine of the rat. Br J Pharmacol 80:371–376
Coupar IM (1986) Tetrodotoxin inhibits directly acting stimulants of intestinal fluid secretion. J Pharm Pharmacol 38:553–555
Cox HM, Cuthbert AW, Munday KA (1987) The effect of angiotensin II upon electrogenic ion transport in rat intestinal epithelia. Br J Pharmacol 90:393–401
Cox HM, Cuthbert AW, Håkanson R, Wahlestedt C (1988) The effect of neuropeptide Y and peptide YY on electrogenic ion transport in rat intestinal epithelia. J Physiol (Lond) 398:65–80
Crane MR, O'Hanley P, Waldman SA (1990) Rat intestinal cell atrial natriuretic peptide receptor coupled to guanylate cyclase. Gastroenterology 99:125–131

Daniel EE, Costa M, Furness JB, Keast JR (1985) Peptide neurons in the canine small intestine. J Comp Neurol 237:227–238
Davies NT, Munday KA, Parsons BJ (1970) The effect of angiotensin on rat intestinal fluid transfer. J Endocrinol 48:39–46
Davis GR, Camp RC, Raskin P, Krejs GJ (1980) Effect of somatostatin infusion on jejunal water and electrolyte transport in a patient with secretory diarrhea due to malignant carcinoid syndrome. Gastroenterology 78:346–349
Davis GR, Santa-Ana CA, Morawski SG, Fordtran JS (1981) Effect of vasoactive intestinal polypeptide on active and passive transport in the human jejunum. J Clin Invest 67:1687–1694
Davis GR, Santa-Ana CA, Morawski SG, Fordtran JS (1982) Permeability characteristics of human jejunum, ileum, proximal colon and distal colon: results of potential difference measurements and unidirectional fluxes. Gastroenterology 83:844–850
Dawson DC (1991) Ion channels in colonic salt transport. Annu Rev Physiol 53:321–339
Dharmsathaphorn K (1985) Intestinal somatostatin function. Adv Exp Biol Med 188:463–473
Dharmsathaphorn K, Binder HJ, Dobbins JW (1980a) Somatostatin stimulates sodium and chloride absorption in the rabbit ileum. Gastroenterology 78:1559–1565
Dharmsathaphorn K, Racusen L, Dobbins JW (1980b) Effects of somatostatin on ion transport in the rat colon. J Clin Invest 66:813–820
Dharmsathaphorn K, McRoberts JA, Mandel KG, Tisdale LD, Masui H (1984) A human colonic tumor cell line that maintains vectorial electrolyte transport. Am J Physiol 246:G204–G208
Dharmsathaphorn K, Mandel KG, Masui H, McRoberts JA (1985) Vasoactive intestinal polypeptide-induced Cl secretion by a colonic epithelial cell line: direct participation of a basolaterally localized Na-K-Cl cotransport system. J Clin Invest 75:462–471
DiBona DR, Mills JW (1979) Distribution of Na pump sites in transporting epithelia. Fed Proc 38:134–143
Dobbins JW, Racusen L, Binder HJ (1980) Effect of D-alanine methionine enkephalin amide on ion transport in rabbit ileum. J Clin Invest 66:19–28
Donowitz M, Welsh MJ (1987) Regulation of mammalian small intestinal electrolyte secretion. In: Johnson LR (ed) Physiology of the gastrointestinal tract. Raven, New York, p 1351
Donowitz M, Fogel R, Battisti L, Asarkof N (1982) The neurohumoral secretagogues carbachol, substance P and neurotensin increase Ca^{+2} influx and calcium content in rabbit ileum. Life Sci 31:1929–1937
Duffey ME (1984) Intracellular pH and bicarbonate activities in rabbit colon. Am J Physiol 246:C558–C561
Erspamer V (1981) The tachykinin peptide family. Trends Neurosci 4:267–269
Fan CC, Faust RG, Powell DW (1983) Coupled sodium chloride transport by rabbit ileal brush border membrane vesicles. Am J Physiol 244:G375–G385
Field M (1971) Ion transport in rabbit ileal mucosa. II. Effects of cyclic 3′,5′-AMP. Am J Physiol 221:992–997
Finkelstein A (1987) Water movement through lipid bilayers, pores, and plasma membranes, vol 4. Wiley, New York
Fogel R, Kaplan RB (1984) Role of enkephalins in regulation of basal intestinal water and ion absorption in the rat. Am J Physiol 246:G386–G392
Franklin CC, Turner JT, Kim HD (1989) Regulation of Na-K-Cl cotransport and [^{3}H]bumetanide binding site density by phorbolesters in HT29 cells. J Biol Chem 264:6667–6673
Friel DD, Miller RJ, Walker MW (1986) Neuropeptide Y: a powerful modulator of epithelial ion transport. Br J Pharmacol 88:425–431

Garty H, Benos DJ (1988) Characteristics and regulatory mechanisms of the amiloride-blockable Na channel. Physiol Rev 68:309–373

Gerbes AL, Denecke H, Nathrath W (1990) Presence of atrial natriuretic factor prohormone in enterochromaffin cells of the human large intestine. Gastroenterology 98:A495

Glynn IM, Karlish JD (1975) The sodium pump. Annu Rev Physiol 27:13–55

Grasset E, Bernabeu J, Pinto M (1985) Epithelial properties of human colonic carcinoma cell line Caco-2: effect of secretagogues. Am J Physiol 248:C410–C418

Grassl SM, Holohan PD, Ross CR (1987) HCO_3 transport in basolateral membrane vesicles isolated from rat renal cortex. J Biol Chem 262:2682–2687

Guandalini S, Kachur JF, Smith PL, Miller RJ, Field M (1980) In vitro effects of somatostatin on ion transport in rabbit intestine. Am J Physiol 238:G67–G74

Gunther RD, Wright EM (1983) Na, Li, Cl transport by brush border membranes from rabbit jejunum. J Membr Biol 74:85–94

Gustin MC, Goodman DBP (1981) Isolation of brush border membrane from rabbit descending colon epithelium: partial characterization of a unique potassium activated ATPase. J Biol Chem 256:10651–10656

Haas M (1989) Properties and diversity of (Na-K-Cl) cotransporters. Annu Rev Physiol 51:443–457

Halm DR, Dawson DC (1984) Potassium transport by turtle colon: active secretion and active absorption. Am J Physiol 246:C315–C322

Halm DR, Frizzell JA (1986) Active K transport across rabbit distal colon: relation to Na absorption and Cl secretion. Am J Physiol 251:C252–C267

Halm DR, Frizzell RA (1991) Ion transport across the large intestine. In: Schultz SG, Field M, Frizzell RA (eds) The gastrointestinal system IV. American Physiological Society, New York, p 257 (Handbook of physiology)

Hatch M, Freel RW, Goldner AM, Earnest DL (1984) Oxalate and chloride absorption by the rabbit colon: sensitivity to metabolic and anion transport inhibitors. Gut 25:232–237

Heintze K, Petersen KU, Wood JR (1982) Effects of bicarbonate on fluid and electrolyte transport by guinea pig and rabbit gallbladder: stimulation of absorption. J Membr Biol 62:175–181

Hopfer U (1987) Membrane transport mechanisms for hexoses and amino acids in the small intestine. In: Johnson LR (ed) Physiology of the gastrointestinal tract. Raven, New York, p 1499

Hopfer U, Liedtke CM (1987) Proton and bicarbonate transport mechanisms in the intestine. Annu Rev Physiol 49:51–67

Hornych A, Meyer P, Milliez P (1973) Angiotensin, vasopressin and cyclic AMP: effects on sodium and water fluxes in rat colon. Am J Physiol 224:1223–1229

Hubel KA, Renquist KS (1986) The effect of neuropeptide Y on ion transport by the rabbit ileum. J Pharmacol Exp Ther 238:167–169

Hubel KA, Renquist KS, Shirazi S (1984) Neural control of ileal ion transport: role of substance P in rabbit and man. Gastroenterology 86:1118

Humphreys MH (1976) Inhibition of Na and Cl absorption from perfused rat ileum by furosemide. Am J Physiol (Lond) 230:1571–1523

Isenberg JI, Wallin B, Johansson C, Smedfors B, Mutt V, Tatemoto K, Emas S (1984) Secretin, VIP and PHI stimulate rat proximal duodenal surface epithelial bicarbonate secretion in vivo. Regul Pept 8:315–320

Itoh S, Obata K, Yanaihara N, Okamoto H (1983) Human preprovasoactive intestinal polypeptide contains a novel PHI-27 like peptide, PHM-27. Nature 304:547–549

Kachur JF, Miller RJ, Field M, Rivier J (1982a) Neurohumoral control of ileal electrolyte transport. I. Bombesin and related peptides. J Pharmacol Exp Ther 220:449–455

Kachur JF, Miller RJ, Field M, Rivier J (1982b) Neurohumoral control of ileal electrolyte transport. II. Neurotensin and substance P. J Pharmacol Exp Ther 220:456–463

Kachur JF, Phillips GS, Gaginella TS (1991) Neuromodulation of guinea pig intestinal electrolyte transport by cholecystokinin octapeptide. Gastroenterology 100:344–349

Kaunitz JD, Sachs G (1986) Identification of a vanadate-sensitive, potassium-dependent proton pump from rabbit colon. J Biol Chem 261:14005–14010

Keast JR, Furness JB, Costa M (1985) Different substance P receptors are found on the mucosal epithelial cells and submucous neurons of the guinea pig small intestine. Naunyn Schmiedebergs Arch Pharmacol 329:382–387

Kleyman TR, Cragoe EJ (1988) Amiloride and its analogs as tools in the study of ion transport. J Membr Biol 105:1–21

Knickelbein RG, Aronson PS, Atherton W, Dobbins JW (1983) Sodium and chloride transport across rabbit ileal brush border. I. Evidence for Na-H exchange. Am J Physiol 245:G504–G510

Knickelbein RG, Aronson PS, Dobbins JW (1985) Substrate and inhibitor specificity of anion exchangers on the brush border membrane of rabbit ileum. J Membr Biol 88:199–204

Knickelbein RG, Aronson PS, Dobbins JW (1988) Membrane distribution of sodium-hydrogen and chloride-bcarbonate exchangers in crypt and villus cell membranes from rabbit ileum. J Clin Invest 82:2158–2163

Knutton S, Limbrick AR, Robertson JD (1978) Structure of occluding junctions in ileal epithelial cells of suckling rats. Cell Tissue Res 191:449–462

Levens NR (1984) Modulation of jejunal ion and water absorption by endogenous angiotensin after dehydration. Am J Physiol 246:G700–G709

Levens NR, Munday KA, Parsons BJ, Poat JA, Stewart CP (1979) Noradrenaline as a possible mediator of the actions of angiotensin on fluid transport by rat jejunum in vivo. J Physiol (Lond) 286:351–360

Levens NR, Peach MJ, Carey RM, Poat JA, Munday KA (1981a) Response of rat jejunum to angiotensin II: role of norepinephrine and prostaglandins. Am J Physiol 240:G17–G24

Levens NR, Peach MJ, Carey RM, Poat JA, Munday KA (1981b) Changes in an electroneutral transport process mediated by angiotensin II in rat distal colon in vivo. Endocrinology 108:1497–1504

Liedtke CM (1989) Regulation of Cl transport in epithelia. Annu Rev Physiol 51:143–160

Light DB, Corbin JD, Stanton BA (1989) Dual ion channel regulation by cGMP and cGMP-dependent protein kinase. Nature 344:336–339

Lopes AG, Siebens AW, Giebisch G, Boron W (1987) Electrogenic Na/HCO_3 cotransport across basolateral membrane of isolated perfused *Necturus* proximal tubule. Am J Physiol 253:F340–F350

Madara JL (1991) Functional morphology of epithelium of the small intestine. In: Schultz SG, Field M, Frizzell RA (eds) The gastrointestinal system IV. American Physiological Society, New York, p 83 (Handbook of physiology)

Madara JL, Pappenheimer JR (1987) The structural basis for physiological regulation of paracellular pathways in intestinal epithelia. J Membr Biol 100:149–164

Madara JL, Trier JS, Neutra MR (1980) Structural changes in the plasma membrane accompanying differentiation of epithelial cells in human and monkey small intestine. Gastroenterology 78:963–975

Madara JL, Moore R, Carlton S (1987) Alteration in intestinal tight junction structure and permeability by cytoskeletal contraction. Am J Physiol 253:C854–C861

Mailman D (1978) Effects of vasoactive intestinal polypeptide on intestinal absorption and blood flow. J Physiol (Lond) 279:121–132

Mandel KG, Dharmsathaphorn K, McRoberts JA (1986) Characterization of a cAMP-activated Cl transport pathway in the apical membrane of a human colonic epithelial cell line. J Biol Chem 261:704–712

Martinez SA, Vidal NA, Carchio SM, Karara AL (1986) Inhibition of water-sodium absorption by an atrial extract. Can J Physiol Pharmacol 64:244–247
Maton PN, O'Dorisio TM, Howe BA, McArthur KE, Howard JM, Cherner JA, Malarkey TB, Collen MJ, Gardner JD, Jensen RT (1985) Effect of a long acting somatostatin analog (SMS201,995) in a patient with pancreatic cholera. N Engl J Med 312:17–21
McCabe RD, Dharmsathaphorn K (1988) Mechanism of VIP stimulated chloride secretion by intestinal epithelial cells. Ann NY Acad Sci 527:326
McCabe RD, Smith PL (1985) Colonic potassium and chloride secretion: role of cAMP and calcium. Am J Physiol 248:G103–109
McCabe R, Cooke HJ, Sullivan LP (1982) Potassium transport by rabbit descending colon. Am J Physiol 242:C81–C86
McCann JD, Welsh MJ (1990) Regulation of Cl and K channels in airway epithelium. Annu Rev Physiol 52:115–135
McCulloch CR, Kuwahara A, Condon CD, Cooke HJ (1987) Neuropeptide modification of Cl secretion in guinea pig distal colon. Regul Pept 19:35–43
McFadden D, Zinner MJ, Jaffe BM (1986) Substance P induced intestinal secretion of water and electrolytes. Gut 27:267–272
McKay JS, Linaker BD, Turnberg LA (1981) Influence of opiates on ion transport across rabbit ileal mucosa. Gastroenterology 80:279–284
McRoberts JA, Beuerlein G, Dharmsathaphorn K (1985) Cyclic AMP and calcium activated K transport in a human colonic epithelial cell line. J Biol Chem 260:14163–14172
Miller RJ, Hirning LD (1989) Opioid peptides of the gut. In: Makhlouf GM (ed) Neural and endocrine biology. American Physiological Society, Bethesda, p 631 (Handbook of physiology, sect 6: The gastrointestinal system)
Minamino N, Kangawa K, Matsuo H (1984) Neuromedin N: a novel neurotensin-like peptide identified in porcine spinal cord. Biochem Biophys Res Commun 122:542–549
Moghimzadeh E, Ekman R, Hakanson R, Yanaihara N, Sundler F (1983) Neuronal gastrin releasing peptide in the mammalian gut and pancreas. Neuroscience 10:553–563
Montrose MH, Kimmich GA (1985) Relative rates of Na/H and Cl/OH exchange reactions in isolated intestinal cells. Ann NY Acad Sci 456:232–234
Moriarty KJ, Higgs NB, Lees M, Tonge A, Wardle TD, Warhurst G (1990) Influence of atrial natriuretic peptide on mammalian large intestine. Gastroenterology 98:647–653
Morris AP, Kirk KL, Frizzell RA (1990) Simultaneous analysis of cell Ca^{+2} and Ca^{+2}-stimulated chloride conductance in colonic epithelial cells (HT29). Cell Regul 1:951–963
Murer H, Hopfer U, Kinne R (1976) Sodium/proton antiport in brush border membrane vesicles isolated from rat small intestine and kidney. Biochem J 154:597–604
O'Grady SM (1989) Cyclic nucleotide-mediated effects of ANF and VIP on flounder intestinal ion transport. Am J Physiol 256:C142–C146
O'Grady SM, Field M, Nash NT, Rao MC (1985) Atrial natriuretic factor inhibits Na-K-Cl cotransport in teleost intestine. Am J Physiol 249:C531–C534
O'Grady SM, Palfrey HC, Field M (1987) Characteristics and functions of Na-K-Cl cotransport in epithelial tissues. Am J Physiol 253:C177–C192
O'Grady SM, DeJonge HR, Vaandrager AB, Field M (1988) Cyclic nucleotide-dependent protein kinase inhibition by H-8: effects on ion transport. Am J Physiol 254:C115–C121
Parsons AM, Seybold VS, Chandan R, Vogt J, Larson AA, Murray CR, Soldani G, Brown DR (1992) Neurokinin receptors and mucosal ion transport in porcine jejunum. J Pharmacol Exp Ther 261:1213–1221
Patrick MK, Chow O, O'Loughlin EV, Gall DG (1987) The effect of atrial natriuretic factor (ANF) on small intestinal electrolyte transport. Gastroenterology 92:1568

Perdue MH, Galbrith R, Davison JS (1987) Evidence for substance P as a functional neurotransmitter in guinea pig small intestinal mucosa. Regul Pept 18:63–74

Powell DW (1987) Ion and water transport in the intestine. In: Andreoli TE, Fanestil DD, Hoffman JF, Schultz SG (eds) Membrane transport processes in organized systems. Plenum, New York, p 175

Powell DW (1991) Immunophysiology of intestinal electrolyte transport. In: Schultz SG, Field M, Frizzell RA (eds) The gastrointestinal system IV. American Physiological Society, New York, p 591 (Handbook of physiology)

Quito FL, Brown DR (1991) Neurohormonal regulation of ion transport in the porcine distal jejunum. Enhancement of sodium and chloride absorption by submucosal opiate receptors. J Pharmacol Exp Ther 256:833–840

Racusen LC, Binder HJ (1977) Alteration of large intestinal electrolyte transport by vasoactive intestinal polypeptide in the rat. Gastroenterology 73:790–796

Reuss L (1983) Basolateral KCl cotransport in a NaCl absorbing epithelium. Nature 305:723–725

Rogers KV, Goldman PS, Frizzell RA, McKnight GS (1990) Regulation of Cl transport in T_{84} cell clones expressing a mutant regulatory subunit of cAMP-dependent protein kinase. Proc Natl Acad Sci USA 87:8975–8979

Schultz SG, Hudson RL (1991) Sodium-coupled transport mechanisms in epithelia. In: Schultz SG, Field M, Frizzell RA (eds) The gastrointestinal system IV. American Physiological Society, New York, p 45 (Handbook of physiology)

Schwartz CJ, Kimberg DV, Sheerin HE, Field M, Said SI (1974) Vasoactive intestinal peptide stimulation of adenylate cyclase and active electrolyte secretion in intestinal mucosa. J Clin Invest 54:536–544

Sellin JH, DeSoignie RC (1984) Rabbit proximal colon: a distinct transport epithelium. Am J Physiol 246:G603–G610

Sellin JH, Orayzabal H, Cragoe EJ, Potter GD (1989) Phenamil inhibits electrogenic Na absorption in rabbit ileum. Gastroenterology 96:997–1003

Semrad CE, Chang EB (1987) Calcium mediated cyclic AMP inhibition of Na-H exchange in small intestine. Am J Physiol 252:C315–C322

Semrad CE, Cragoe EJ, Chang EB (1991) Inhibition of Na/H exchange in avian intestine by atrial natriuretic factor. J Clin Invest (in press)

Sharkey KA, Gall DG, MacNaughton WK (1991) Distribution and function of brain natriuretic peptide in the stomach and small intestine of the rat. Regul Pept 34:61–70

Sheldon RJ, Riviere PJM, Malarchik ME, Mosberg HI, Burks TF, Porreca F (1990) Opioid regulation of mucosal ion transport in the mouse isolated jejunum. J Pharmacol Exp Ther 253:144–151

Shorofsky SR, Field M, Fozzard HA (1983) Electrophysiology of Cl secretion in canine trachea. J Membr Biol 72:105–115

Smith PL, Cascairo MA, Sullivan SK (1985) Sodium dependence of luminal alkalinization by rabbit ileal mucosa. Am J Physiol 249:G358–G368

Spring K (1991) Fluid transport across epithelia. In: Schultz SG, Field M, Frizzell RA (eds) The gastrointestinal system IV. American Physiological Society, New York, p 195 (Handbook of physiology)

Stewart CP, Turnberg LA (1989) A microelectrode study of responses to secretagogues by epithelial cells on villus and crypt of rat small intestine. Am J Physiol 257:G334–G343

Stoll R, Stern H, Ruppin H, Domschke W (1987) Effect of inhibitors on sodium and chloride transport in brush border vesicles from human jejunum and ileum. Digestion 37:228–237

Sudoh T, Kangawa K, Minamino N, Matsuo H (1988) A new natriuretic peptide in porcine brain. Nature 332:78–81

Sullivan SK, Field M (1991) Ion transport across mammalian small intestine. In: Schultz SG, Field M, Frizzell RA (eds) The gastrointestinal system IV. American Physiological Society, New York, p 287 (Handbook of physiology)

Sullivan SK, Smith PL (1986) Active potassium secretion by rabbit proximal colon. Am J Physiol 250:G475–G483

Suzuki Y, Kaneko K (1987) Acid secretion in isolated guinea pig colon. Am J Physiol 253:G155–G164

Tabcharani JA, Jensen TJ, Riordan JR, Hanrahan JW (1989) Bicarbonate permeability of the outwardly rectifying anion channel. J Membr Biol 112:109–122

Tasco M, Orsenigo MN, Eposito G, Faelli A (1988) Na/H exchange mechanism in the basolateral membrane of the rat enterocyte. Biochem Biophys Acta 944:473–476

Tatemoto K (1982) Isolation and characterization of peptide YY (PYY): a candidate gut hormone that inhibits pancreatic exocrine secretion. Proc Natl Acad Sci USA 79:2514–2518

Tatemoto K, Lundberg JM, Jornvall H, Mutt V (1985) Neurokinin K: isolation, structure and biological activities of a novel brain peptide. Biochem Biophys Res Commun 128:947–953

Townsley MC, Machen TE (1989) Na/HCO_3 cotransport in rabbit parietal cells. Am J Physiol 257:G350–G356

Traynor TR, O'Grady SM (1991) Brain natriuretic peptide stimulates K and Cl secretion across porcine distal colon epithelium. Am J Physiol 260:C750–C755

Trimble ER, Burzzone R, Biden TJ, Farese RV (1986) Secretin induces rapid increases in inositol triphosphate, cytosolic Ca^{+2} and diacylglycerol as well as cAMP in rat pancreatic acini. Biochem J 239:257–261

Turnberg LA (1971) Potassium transport in human small bowel. Gut 12:811–818

Turnberg LA (1983) Antisecretory activity of opiates in vivo and in vitro. Scand J Gastroenterol 18:79–83

Turnberg LA, Bieberdorf F, Morawski S, Fordtran J (1970a) Interrelationships of chloride, bicarbonate, sodium and hydrogen transport in the human ileum. J Clin Invest 49:557–567

Turnberg LA, Fordtran J, Carter N, Rector F (1970b) Mechanism of bicarbonate absorption and its relationship to sodium transport in human jejunum. J Clin Invest 49:548–556

Turner JT, Franklin CC, Bollinger DW, Kim HD (1990a) Vasoactive intestinal peptide stimulates active K transport and Na-K-Cl cotransport in HT29 cells. Am J Physiol 258:C266–C273

Turner JT, James-Kracke MR, Camden JM (1990b) Regulation of the neurotensin receptor and intracellular calcium mobilization in HT 29 cells. J Pharmacol Exp Ther 253:1049–1056

Vaandrager AB, DeJonge HR (1988) A sensitive technique for the determination of anion exchange activities in brush border membrane vesicles; evidence for two exchangers with different sensitivities for HCO_3 and SITS in rat intestinal epithelium. Biochem Biophys Acta 939:305–314

Vollmar AM, Friedrich A, Sinowatz F, Schultz R (1988) Presence of atrial natriuretic peptide-like material in guinea pig intestine. Peptides 9:965–971

Vuolteenaho O, Arjamaa O, Vakkuri T, Maksniemi L, Nikkila K, Kangus J, Puuruuen J, Ruskoaho H, Leppaluoto S (1988) Atrial natriuretic peptide (ANP) in rat gastrointestinal tract. FEBS Lett 233:79–82

Wakelam MJO, Murphy GJ, Hruby VJ, Houslay MD (1986) Activation of two signal transduction mechanisms in hepatocytes by glucagon. Nature 323:68–71

Waldman DB, Gardner JD , Zfass AM, Makhlouf GM (1977) Effects of vasoactive intestinal polypeptide secretin and related peptides on rat colonic transport and adenylate cyclase activity. Gastroenterology 73:518–523

Walling MW, Brasitus TA, Kimberg DV (1977) Effects of calcitonin and substance P on the transport of Ca, Na and Cl across rat ileum in vitro. Gastroenterology 73:89–94

Warhurst G, Smith G, Tonge A, Turnberg LA (1983) Effects of morphine on net water absorption, adenylate cyclase activity and PGE_2 metabolism in rat intestine. Eur J Pharmacol 86:77–82

Welsh MJ (1983) Intracellular chloride activities in canine tracheal epithelium. Direct evidence for Na-coupled intracellular chloride accumulation in a chloride-secreting epithelium. J Clin Invest 74:262–268

Welsh MJ, Smith PL, Fromm M, Frizzell RA (1982) Crypts are the site of intestinal fluid and electrolyte secretion. Science 218:1219–1221

Wiener H, van Os CH (1989) Rabbit distal colon epithelium. II. Characterization of (Na,K,Cl) cotransport and [^{3}H]bumetanide binding. J Membr Biol 110:163–174

Wills NK, Alles WP, Sandle GI, Binder HJ (1984) Apical membrane properties and amiloride binding kinetics of the human descending colon. Am J Physiol 247:G749–G757

White JF (1989) Conductive pathways for HCO_3 in basolateral membrane of salamander intestinal cells. Am J Physiol 257:C252–C260

Woll PJ, Rozengurt E (1989) Neuropeptides as growth regulators. Br Med Bull 45:492–505

Wu ZAC, O'Dorisio TM, Cataland C, Mekhijian HS, Gaginella TS (1979) Effects of pancreatic polypeptide and vasoactive intestinal polypeptide on rat ileal and colonic water and electrolyte transport. Dig Dis Sci 24:625–630

Yada T, Okada Y (1984) Electrical activity of an intestinal epithelial cell line: hyperpolarizing responses to intestinal secretagogues. J Membr Biol 77:33–44

Yada T, Oiki S, Ueda S, Okada Y (1989) Intestinal secretagogues increase cytosolic free Ca^{2+} concentration and K conductance in a human epithelial cell line. J Membr Biol 112:159–167

Zweibaum A, Laburthe M, Grasset E, Louvard D (1991) Use of cultured cell lines in studies of intestinal cell differentiation and function. In: Schultz SG, Field M, Frizzell RA (eds) The gastrointestinal system IV. American Physiological Society, New York, p 223 (Handbook of physiology)

CHAPTER 11

Peptidergic Regulation of Gastrointestinal Blood Flow

C.C. Chou and A. Alemayehu

A. Introduction

Gastrointestinal (GI) peptides can increase or decrease blood flow in addition to their actions on digestive functions, such as GI motility, intestinal absorption, and gastric and pancreatic secretions. Their vasoactive potency (i.e., the dose required to produce vasoactivity) may differ from their potency to induce changes in digestive functions and, therefore, the respective potencies should be compared. The relative potency of some chemicals on GI motility and blood flow, and the relationship between these two variables, have been summarized (Chou 1989). Administration of a peptide often produces concomitant changes in a digestive function, local oxidative metabolism, and local blood flow. A change in the tissue activity and metabolism can cause changes in local blood flow, and a change in blood flow can influence tissue activity. One should determine which aspect is the cause and which is the result. Another issue to be considered is whether or not the digestive and vascular actions of a peptide are physiological. Physiologically, the GI peptides are released following a meal to regulate digestive functions and local blood flow. In order to be a physiological peptide, the postprandial tissue or blood concentrations must be sufficient to produce vasoactivity. This topic has been discussed previously by comparing the minimal local blood concentration required to produce vasodilation with that achieved following a meal (Chou and Kvietys 1981; Chou et al. 1984). These two articles described and presented in tabular format the relative vasoactivities of these peptides and other circulating or locally released chemicals in the stomach, small intestine, colon, liver, and pancreas. Because of space limitations in the present chapter, these two articles are given as references both for discussion based on their data and also data cited in the bibliographies from these articles. Utilizing the studies published after 1983, the present chapter will reevaluate the vascular functions of cholecystokinin (CCK), secretin, gastrin, gastric inhibitory peptide (GIP), neurotensin (NT), glucagon, vasoactive intestinal polypeptide (VIP), substance P (SP), and somatostatin (SS), and descriptions will be given of the vasoactivities of novel peptides, calcitonin gene-related peptide (CGRP), peptide YY (PYY), and neuromedin U (NMU), and the vascular functions of capsaicin-sensitive sensory nerves in the GI tract. SP, CCK,

VIP and CGRP have been proposed to be neuropeptides of these nerves. The possible role of each peptide in regulation of blood flow under physiological and pathophysiological conditions, and the relationship between their actions on digestive functions and local blood flow, will be discussed. These peptides may play a physiological role as endocrine, paracrine, or neurocrine mediators.

B. Gastrointestinal Peptides

I. Cholecystokinin

The vasodilatory action of CCK is organ and tissue specific. It produces vasodilation in the small intestine and pancreas, but does not have a significant vascular effect in the colon and nondigestive organs, such as heart, kidney, forelimb, spleen, skin, and muscle. In the stomach and small intestine, it primarily increases blood flow to the mucosa (Chou 1989; Chou and Kvietys 1981; Chou et al. 1984). The selective action of CCK mimics the hemodynamic changes observed during nutrient absorption, i.e., an increase in blood flow primarily in the digestive organs (e.g., pancreas) and tissues (e.g., the mucosa) actively engaged in digestive activities (Chou 1983). Long-chain fatty acids (oleic acid being the most common dietary fatty acid) are not only the most potent stimuli for CCK release, but are also much more potent intestinal vasodilators than glucose or amino acids, when each nutrient is placed into the gut lumen (Chou and Kvietys 1981). In cross-perfusion studies, Fara et al. (1972) showed that intraduodenal instillation of corn oil increases intestinal blood flow in both the animal receiving corn oil and in an animal perfused by the blood from the test animal. In addition to its vasodilatory action, CCK increases intestinal oxygen uptake, capillary filtration coefficient, and lymph flow (Chou et al. 1984). These events also occur during nutrient absorption. Based on the above findings, CCK has been proposed to play a role as a circulating hormone in postprandial intestinal hyperemia. However, recent studies using synthetic analogs of CCK rather than tissue extracts have challenged this hypothesis (Gallavan and Chou 1985). Since tissue extracts are contaminated by other vasoactive agents, their vasodilatory potency does not reflect true CCK action. Synthetic CCK-8 in intraarterial infusions at 0.01 and 0.5 μg/min increases the blood flow through the superior mesenteric artery (SMA) by 10% and 100% respectively (Chou et al. 1984). Plasma levels of total CCK (CCK-8 and CCK-33), CCK-8, and CCK-like immunoreactivity have been shown to be 30 pmol/l (Maton et al. 1982), 7 pmol/l (Liddle et al. 1985), and 6–7 pmol/l (Eysselein et al. 1990) following a fat or mixed meal. These values were used as references for recent studies.

Cholecystokinin, secretin, GIP, gastrin, and NT have been proposed to play a role as circulating hormones in postprandial intestinal hyperemia.

Two recent studies reevaluated this possibility. PREMEN et al. (1984b) found that i.a. infusions of CCK-8 (34 pmol/l), gastrin (76 pmol/l), secretin (16 pmol/l), or GIP (191 pmol/l) at a rate that raised local arterial blood concentrations near the postprandial levels (as indicated in the parentheses) did not alter the vascular resistance of the canine ileum. The ileal wall tension was decreased by secretin and GIP, increased by CCK-8, and unaltered by gastrin. A subsequent study by PREMEN et al. (1985) further showed that, at levels close to postprandial arterial levels, CCK-8 (60 pmol/l), secretin (33 pmol/l), or NT (350 pmol/l) also did not significantly alter jejunal blood flow in dogs. At concentrations 100 times greater than the above levels, secretin, NT, and CCK-8 increased jejunal blood flow by 34%, 31%, and 24%, respectively. It is well known that secretin and CCK potentiate each other in stimulating pancreatic secretion. In order to determine if the above three peptides will potentiate one another in producing vasodilation, all three peptides were infused simultaneously. Combined infusion at the postprandial levels did not alter jejunal blood flow. A 47% increase in blood flow was observed when the doses were increased 100-fold. It thus appears that none of these five peptides contribute significantly as circulating hormones, individually or in combination, to the hyperemia occurring in the jejunum or ileum following a meal.

The possibility that CCK and NT may act as paracrine mediators of jejunal blood flow during lipid absorption has been studied by GALLAVAN et al. (1985, 1986). Placement of a solution containing bile and oleate into the jejunal lumen produced a sustained increase in jejunal blood flow by (21%–25%), but the increases in CCK and NT releases into the local venous blood were transient and on the order of 1 ng/min/100 g. Furthermore, the concentration of CCK or NT in the jejunal venous blood observed during oleate placement was much less than that required to produce jejunal vasodilation. Two more studies used CCK antagonists to address this same issue. PAWLIK et al. (1989) showed that a CCK-receptor antagonist, L-364,718, abolished the vascular and motor responses to exogenous CCK and significantly reduced the increases in intestinal blood flow and oxygen uptake observed during luminal placement of oleic acid in dogs. Luminal placement of glucose also increased blood flow and oxygen uptake, but these responses were unaffected by this antagonist. It is well known that CCK release is stimulated by oleic acid, but not by glucose. The above study, therefore, indicates that the released CCK may participate in the oleate-induced intestinal hyperemia as a paracrine. However, as will be described later (Sect. C), ROZSA and JACOBSON (1989) found that CCK-8 antiserum did not inhibit oleate-induced intestinal hyperemia in rats. The variation in results could be due to species differences. In contrast to humans, dogs, and pigs, in rats intact protein is the major stimulus for CCK release.

II. Secretin

Secretin is an established circulating hormone, but unlike CCK does not appear to be a neurotransmitter. The vasoactivity of secretin also differs from that of CCK. Earlier studies which utilized tissue extracts have shown that secretin is a universal vasodilator, producing a similar degree of vasodilation in the small intestine, as well as in nondigestive organs such as kidney, heart, spleen, forelimb, muscle, and skin (Chou et al. 1977). The vasoactivity of the extracts differs among different sources, but an extract from Karolinska Institute can increase intestinal blood flow by 54% and 70% at intravenous (i.v.) doses of 4.7 and 11 U/kg per hour, respectively. In addition to its vasodilatory action, secretin also increases intestinal oxygen uptake and the capillary filtration coefficient (Chou and Kvietys 1981; Chou et al. 1984).

Recent studies utilizing synthetic secretin have shown that local i.a. infusions at doses that increased local arterial blood concentrations of 1.5 nmol/l and 3 nmol/l significantly decreased ileal vascular resistance by 11% and increased jejunal blood flow by 34%, respectively (Premen et al. 1984b, 1985). As described in Sect. B.I, these two studies also showed that secretin does not alter jejunal or ileal blood flow at physiological postprandial plasma concentrations (16–33 pmol/l) alone, or in combination with other peptides. In the pancreas, secretin-induced near-maximal pancreatic secretion can occur in the absence of an increase in pancreatic blood flow. Therefore, secretin does not appear to play any significant role in regulation of intestinal or pancreatic blood flow as a circulating hormone under physiological conditions.

III. Gastrin

Gastrin is a classic circulating GI hormone, and is a physiological stimulator of gastric acid secretion. Pentagastrin increases blood flows to the stomach, pancreas, and small intestine, but decreases colonic blood flow (Chou and Kvietys 1981; Chou et al. 1984). In dogs, pentagastrin (1 and 10 μg/min) does not significantly alter total intestinal blood flow but depresses "absorptive-site blood flow" in fasted conditions, which is associated with a net decrease in the absorption of Na and water. However, absorptive-site flow is unchanged by pentagastrin in fed dogs. At higher i.a. infusion rates (3 μg/kg per hour, or 50 μg/min), it produces a 35%–50% increase in intestinal blood flow. Pentagastrin also increases intestinal motility and oxygen uptake at high infusion rates and depresses capillary filtration coefficient at a dose that does not alter blood flow (Chou et al. 1984). Inasmuch as the doses required to produce significant vasodilation are high, circulating gastrin does not appear to play a significant role in postprandial regulation of intestinal blood flow. Gastrin, however, may play a physiological role in the regulation of gastric mucosal blood flow.

Pentagastrin, at 1.98 and 19.8 µg/kg per hour i.v., increases gastric acid output by 70% and 280%, respectively, and blood flow to the gastric corpus by 8% and 41%, respectively (LEUNG et al. 1984). A 60% increase in pentagastrin-stimulated gastric secretion is accompanied by a 47% increase in gastric mucosal blood flow (HOLM-RUTILE and BERGLINDH 1986). Furthermore, LEUNG et al. (1986) and PIQUE et al. (1988) have found a linear correlation between pentagastrin-stimulated gastric acid output and gastric corpus mucosal blood flow. In the former study, the gastric mucosal blood flow was decreased by hemorrhage, which was accompanied by proportional decreases in the acid output. The latter study, on the other hand, increased the blood flow by i.v. infusions of graded doses of pentagastrin from 0 to 40 µg/kg per hour. The dose-related increase in acid output was linearly correlated with the dose-related increase in mucosal blood flow. Cimetidine or omeprazole inhibited both pentagastrin-induced increases in acid secretion and mucosal blood flow. The correlation between blood flow and acid output, however, is not valid evidence that the increased flow is the cause of the enhanced secretion. It has been shown that an increase in blood flow *per se* does not induce gastric secretion, and that secretory effects of some gastric secretagogues are independent of their vascular actions (CHOU and KVIETYS 1981).

The gastric mucosal hyperemia may result from the direct vasodilatory action of pentagastrin on vascular smooth muscle or secondary to an increase in local oxidative metabolism. Adenosine has been shown to be a metabolic mediator of the hyperemia that accompanies an increase in oxidative metabolism in various organs, including the small intestine (SAWMILLER and CHOU 1988). Utilizing an adenosine receptor blocker, 8-phenyltheophylline (8-PT), GERBER and GUTH (1989) determined whether adenosine also plays a similar role in the stomach. Pentagastrin 80 µg/kg per hour i.v. caused a 60% increase in gastric mucosal blood flow and this increase was reduced to 20% upon the addition of 8-PT. The pentagastrin-stimulated acid secretion, however, was enhanced from 2.06 to 2.84 mEq/min by the addition of 8-PT. Adenosine, therefore, may mediate the pentagastrin-induced mucosal hyperemia. This study also shows that the enhancement of acid output by 8-PT accompanies a decrease rather than an increase in the mucosal blood flow.

IV. Gastric Inhibitory Polypeptide

Gastric inhibitory polypeptide (GIP) has been renamed glucose-dependent insulin-releasing peptide. Earlier studies which utilized tissue extracts showed that GIP is a vasodilator. Although it has no significant vascular action at 10 ng/kg per minute i.v., it produces 64%, 45%, and 45% increases in blood flows to the duodenum, jejunum, and superior mesenteric artery (SMA), respectively, at an infusion rate of 50 ng/kg per minute (CHOU et al. 1984). A synthetic GIP produced a 12% decrease in the vascular resistance of the ileum when local arterial concentration as raised to 1.9 nmol/l (Premen et al. 1984b). Kogire et al. (1988) showed that injection of a synthetic human GIP at 3, 50, and 800 pmol/kg increased blood flow to the SMA by 9%, 43%, and 139%; decreased pancreatic blood flow by 11%–17% at the two higher doses; and had no significant effect on blood flow in the celiac artery. Thus, GIP vasoactivity appears to be organ selective. The vasodilatory action of GIP, however, is not physiological as described in Sect. B.I.

V. Neurotensin

Neurotensin (NT) may be a circulating GI hormone, but does not appear to be a neurotransmitter of perivascular nerves. It is a vasodilator in the small intestine and colon, but the vasodilatory action is confined to the muscularis layer, particularly that of the ileum. In addition to its vasodilator action, NT also increases oxygen uptake and motility of the ileum. Because of these findings NT has been proposed to play a role in postprandial regulation of blood flow in the ileum (Chou et al. 1984). Harper et al. (1984) have provided evidence to support this hypothesis by showing that, at 182 pmol/l local arterial plasma concentration, NT increased blood flow (by 28%) and capillary permeability of the terminal ileum. However, as described in Sect. B.I, recent studies have shown that NT does not play a role in postprandial jejunal hyperemia (Gallavan et al. 1986; Premen et al. 1985). In addition to its vascular action, NT has been shown to increase fluid secretion in the ileum, but the action appears to be pharmacological. MacKay et al. (1990) showed that the basal and peak postprandial plasma NT concentrations were 3–27 and 15–82 pmol/l, respectively, in humans, and that an increase in NT plasma levels to 16–108 pmol/l by i.v. infusions did not alter ileal secretion.

VI. Glucagon

Glucagon is secreted by the A cells of the pancreatic islets and is primarily involved in the regulation of plasma glucose concentration. A glucagon-like peptide (enteroglucagon) is present in intestinal epithelial cells (L-cell), but its function is unclear. This review, therefore, deals only with pancreatic glucagon, which is a potent vasodilator. It increases intestinal blood flow to as much as 250% of the control value. In addition to its vasodilatory action, glucagon enhances intestinal oxygen uptake, lymph flow, fluid secretion, and capillary pressure and filtration coefficient, but decreases intestinal motility. All these actions, however, appear to be pharmacological (Chou et al. 1984).

Recent studies have shown that circulating glucagon may play a role in portal hypertension. Plasma glucagon levels and intestinal blood flow in portal hypertensive rats (450 pg/ml and 200 ml/min/100 g) are much higher than those in normal rats (167 pg/ml and 125 ml/min/100 g) (Benoit et al. 1984). It has been estimated, after studying its vascular potency in the intestine, that glucagon may account for 40% of the increased intestinal blood flow observed in chronic portal hypertension. Indeed, infusion of glucagon antiserum to portal hypertensive rats produced a 22%–27% decrease in blood flow to the stomach, duodenum, jejunum, ileum, cecum, colon, pancreas, spleen and liver (BENOIT et al. 1986). Similarly, KRAVETZ et al. (1988) showed that a 29% decrease in portal venous blood flow induced by SS was accompanied by a 47% decrease in plasma glucagon concentration. The vascular actions of somatostatin are described in Sect. B.IX.

VII. Vasoactive Intestinal Polypeptide

Vasoactive intestinal polypeptide (VIP) is a potent intestinal vasodilator. The minimal i.a. infusion rate required to produce intestinal vasodilation is 17 pmol/min, which raises the calculated local arterial plasma VIP to 4.9 nmol/l. At plasma levels of 0.15–0.45 µmol/l, VIP produces a 40%–50% increase in intestinal blood flow (CHOU et al. 1984). In addition to its vasodilatory action, VIP decreases net intestinal fluid absorption, and increases ileal oxygen uptake and secretion. However, its action on transmucosal fluid transport appears to be greater than its vasodilatory action, because at 2.0 nmol/l VIP does not alter blood flow but still decreases net fluid absorption. VIP also increases pancreatic blood flow and secretion, and its vasoactivity is occurs at lower doses than its secretory action. At 12.5 and 200 pmol/kg i.v. bolus injections, VIP increased pancreatic blood flow by 33% and 170%, respectively, but a significant increase in pancreatic bicarbonate secretion occurred only at doses above 50 pmol/kg (KONTUREK et al. 1989).

Significant increases in plasma VIP levels have not been observed after normal feeding (CHOU et al. 1984). HOLM et al. (1988) also found in humans that a 30%–60% increase in portal blood flow, following a meal, was not accompanied by a rise in portal venous VIP concentration, the fasting level being 14.1 ± 2.7 pmol/l. However, portal VIP transport (a product of blood flow and VIP concentration) increased from a basal level of 8.3 to 12.5 pmol/min at 15 min, but returned to baseline value 30 min following the meal. This indicates that the digestive organs do release VIP into the portal vein following a meal. If VIP plays a role in the regulation of GI blood flow following a meal, it must act as a neurocrine or paracrine mediator but not as a hormone.

GALLAVAN et al. (1985) showed that luminal placement of a solution containing bile and oleic acid produced 21% and 118% increases in jejunal blood flow and VIP release, respectively. The close temporal relationship between increases in jejunal VIP release and blood flow suggests that VIP may play a role as a neurocrine or paracrine mediator in the oleate-induced jejunal hyperemia. ROZSA and JACOBSON (1989) further showed that a 64%–118% increase in intestinal blood flow during luminal placement of the bile-oleate solution is significantly attenuated by an antiserum to VIP.

The role of VIP as a neuropeptide of the capsaicin-sensitive afferent nerves is described in Sect. C.

Vasoactive intestinal peptide has been also proposed to play a role as a neurotransmitter in the following physiological events, which are mediated by noncholinergic and nonadrenergic enteric nerves, i.e., gastric receptive relaxation, mechanical stimulation of the mucosal surface of the small intestine or rectum, and colonic contraction induced by pelvic nerve stimulation. Thus, FAHRENKRUG et al. (1978) showed that the vasodilation occurring during the above events and that induced by electrical or serotonin-induced stimulation of the intramural nerves are accompanied by a significant increase in the VIP concentration of the venous effluent from the organs studied. Local i.a. infusions of VIP to the stomach and colon produced vascular and motility responses similar to those described above (EKLUND et al. 1979). Apamin, a component of bee venom, abolished or markedly reduced both the release of VIP and the vasodilation described above (JODAL et al. 1983; SJÖQVIST et al. 1983). Apamin, however, did not affect the vasodilatory action of exogenous VIP in vivo and in vitro. The authors, therefore, suggested that apamin acts to block the release of VIP via the presynaptic VIP receptor controlling VIP release. Two recent studies further support the involvement of VIP in the above events. A VIP antagonist, $[N\text{-Ac-Tyr}^1,\text{D-Phe}^2]\text{GRF}_{1-29}\text{NH}_2$, reduced the vasodilation induced by pelvic nerve stimulation or i.a. injection of VIP without affecting the nerve-stimulated increase in VIP release into the portal vein (BLANK et al. 1990). ITO et al. (1988) also showed that i.a. injections of VIP produced gastric relaxation and vasodilation. Stimulation of the central vagal nerve reproduced these VIP effects, and increased the gastric venous plasma VIP concentration to 40–50 pmol/l without altering the arterial VIP levels, via a vagovagal reflex. The gastric motor and vascular responses as well as VIP release were abolished by the ganglionic blocker hexamethonium.

VIII. Substance P

Substance P (SP) is a potent vasodilator and an established perivascular neuropeptide. Its vascular action results from its direct action on the vascular smooth muscle, because tetrodotoxin and various neurotransmitter receptor blockers do not influence the vasodilation evoked by SP. At 0.3 and 14.4 ng/kg per minute i.v., SP increases SMA blood flow in dogs as much as 102% and 171%, respectively. In pigs, i.v. SP produces dose-dependent increases in the SMA flow at infusion rates above 0.6 ng/kg per minute. These actions, however, appear not to be physiologically relevant, because the plasma SP levels before and after a meal are 22 pg/ml (16 pmol/l) and 37 pg/ml (27 pmol/l) (CHOU et al. 1984). At local arterial concentrations near these levels, SP produces vasodilation in the ileum (PREMEN et al. 1984a) and the muscularis layer of the ileum and gastric fundus without affecting blood flows to other sections of the GI tract (YEO et al. 1984). Thus, SP may play a role in postprandial regulation of blood flow in these two tissues. SP is released from the feline duodenum and jejunum into the gut lumen under basal conditions, and the release is enhanced by vagal stimulation. Instillation of SP through the jejunal lumen at 1 pmol/ml per minute increases blood flow to the mucosa/submucosa, without altering flow to the muscularis layer (GRONSTAD et al. 1986). Therefore, SP may act as a luminal hormone regulating jejunal mucosal blood flow. The role of neuronal SP in regulation of intestinal blood flow is described in Sect. C.

IX. Somatostatin

In the gut, somatostatin (SS) is both a hormone and neurotransmitter, and also a vasoconstrictor. At 0.2 µg/kg per minute i.v. it decreases blood flows through the SMA and pancreaticoduodenal arteries by 15%–25%, without affecting flows to the left gastric artery, portal vein, or hepatic artery. At a higher rate (2 µg/kg per minute), SS decreases flows through all of the above blood vessels by 25%–35%. These actions, however, are pharmacological (CHOU et al. 1984). In addition to acting as a gastric vasoconstrictor, SS also decreases pentagastrin-stimulated acid secretion in humans (IVARSSON et al. 1982).

The vascular action of SS may be secondary to a reduction in the secretion of other GI peptides. PRICE et al. (1985) have shown that in conscious dogs the SS-induced decreases in blood flows to the gastric antrum (53%), fundus (28%), and duodenum (30%), jejunum (22%), and pancreas (35%) were accompanied by decreases in plasma levels of insulin (87%), glucagon (78%), pancreatic polypeptide (75%), and gastrin (33%). The latter three peptides are vasodilators, and reduction in their releases could contribute to the SS vasoconstrictor effect, in addition to its own vasoconstrictor action. The above SS actions were produced by raising SS plasma levels to 1.25 nmol/l, which is supraphysiological. The portal and femoral venous plasma concentrations have been shown to increase from basal levels of 30 and 15 pmol/l to 69 and 33 pmol/l, respectively, following a meal (CHAYVIALLE et al. 1980).

Somatostatin has been used in the treatment of bleeding from esophageal varices, because of its vasoconstrictor action on the splanchnic circulation. KRAVETZ et al. (1988) showed that a bolus i.v. injection followed by a continuous infusion of SS produced a significant reduction in GI blood flow and a 29% reduction in portal venous inflow in portal hypertensive rats. A long-acting analog of SS (octreotide) has been developed to overcome the short half-life of SS (see Chap. 15). This analog decreases portal blood flow and pressure in various experimental portal hypertension models (CERINI et al. 1988). In human patients with liver cirrhosis and portal hypertension, this analog decreases intravariceal pressure by 42% (JENKINS et al. 1985).

X. Calcitonin Gene-Related Peptide

Calcitonin gene-related peptide (CGRP) is a potent vasodilator and may act as a neuropeptide (see Sect. C) and circulating hormone. DIPETTE et al. (1989) found that i.v. bolus injections of CGRP decreased systemic arterial blood pressure and total peripheral vascular resistance, while increasing heart rate, and without changing cardiac output at doses above 65 pmol. At doses above 200 pmol, it increased gastric and hepatic blood flows. The vasodilatory action of CGRP is organ selective, and most potent in the stomach. Injection of rat α-CGRP, 0.5 nmol i.v., increased celiac artery blood flow, but decreased the SMA blood flow in conscious rats (GARDINER et al. 1990a). The celiac artery supplies blood to the stomach, liver, and pancreas, whereas the SMA primarily supplies blood to the small intestine. In rabbits, human α-CGRP at 0.1 μg/kg per minute i.v. increased gastric and duodenal blood flow by 200% and 50%, respectively. On the other hand, human β-CGRP at the same dosage increased gastric and pancreatic blood flows by 225% and 86%, respectively, without altering duodenal blood flow (BAUERFEIND et al. 1989). ANDO et al. (1990) further showed that i.v. injections of human CGRP at 220 and 650 pmol in rats produced 53% and 80% decreases in gastric vascular resistance, respectively, but the vascular resistances of the heart, brain, and liver were only moderately reduced. The reasons for the selective action of CGRP and its physiological importance require further investigation. The plasma concentration of CGRP in humans ranges between 9.7 and 71 pmol/l (GIRGIS et al. 1985).

XI. Peptide YY

Peptide YY (PYY), a vasoconstrictor, is a 36-amino-acid peptide found in endocrine cells of the mucosa, and is particularly abundant in the ileum and colon (LUNDBERG et al. 1982). PYY at 25–150 pmol/kg per minute i.a. produced a dose-dependent transient decrease in intestinal blood flow and a sustained inhibition of jejunal and colonic motility in normal as well as guanethidine- and phentolamine-treated cats (LUNDBERG et al. 1982). BUELL and HARDING (1989), however, obtained different results, which might be due to differences in animals or dosages used. They found that PYY at 25 pmol/kg per minute i.v. decreased blood flow to the muscularis layer of the jejunum and ileum without significantly altering the net jejunal or ileal blood flow, indicating that PYY redistributed blood flow from the muscularis to the mucosa/submucosa. The significant decrease in muscularis blood flow was accompanied by an increase in the amplitude of contractions in these regions. Thus, PYY can increase or decrease jejunal motility, which might indirectly affect the direct vasoconstrictor action of PYY. A change in intestinal motility can increase or decrease local blood flow, depending on the types and strength of the contractions (CHOU 1989).

Peptide YY also affects gastric and pancreatic blood flows and secretions. It is unclear whether or not the PYY-induced decrease in gastric mucosal blood flow (TEPPERMAN and WHITTLE 1991) is related to its inhibitory action on gastric secretion (PAPPAS et al. 1985). However, PYY does produce a concomitant decrease in pancreatic blood flow and secretion (INOUE et al. 1988). KONTUREK et al. (1988) also showed that a 30% decrease in pancreatic blood flow was accompanied by 18% decreases in pancreatic bicarbonate and protein secretions. The decreased secretion may be in part due to the decreased blood flow, because a 30% decrease in pancreatic blood flow by partial occlusion of the local artery produced an 18% decrease in the bicarbonate secretion without affecting the protein secretion. KONTUREK et al. (1988) also showed that the fasting and postprandial plasma PYY levels were 63 and 135 pmol/l, respectively. Perfusion of oleate solution at 16 nmol/h through the ileal lumen in the fed dogs increased the plasma PYY level to about 295 pmol/l, and inhibited pancreatic secretion. Intravenous infusion of PYY to raise its plasma concentration to this same level also inhibited pancreatic secretion and decreased pancreatic blood flow. PYY, therefore, may play a role as a circulating hormone in the inhibition of pancreatic secretion induced by the presence of lipids in the distal intestine (PAPPAS et al. 1985). The physiological significance of this event may become apparent in persons who have malabsorption of fat in the upper small intestine, resulting in the presence of a substantial amount of fat in the distal intestine to significantly raise plasma PYY concentration.

The vasoconstrictor action of PYY may be due to its direct action on vascular smooth muscle or may be secondary to releases of other chemicals, such as catecholamines. KONTUREK et al. (1988) showed that the PYY-induced decreases in pancreatic blood flow and secretion were changed to increases after administration of phentolamine and propranolol. Furthermore, these two adrenergic antagonists reversed the oleate-induced decrease in pancreatic secretion without affecting the oleate-induced increase in plasma PYY levels. LUNDBERG et al. (1982), however, found that guanethidine and phentolamine did not alter the PYY action on intestinal blood flow and motility. The differing findings may be due, at least in part, to PYY actions on intestinal motility and pancreatic secretion, which could influence local blood flow. In general, an increase or decrease in tissue activities is accompanied by a corresponding change in local blood flow, as a result of changes in local oxidative metabolism.

XII. Neuromedin U

Neuromedin U (NMU)-like immunoreactivity is found in nerve terminals of the small intestine, but not in endocrine cells (AUGOOD et al. 1988). At 320 and 32 pmol/kg i.v. bolus injection, NMU produced a 48% and 42% decrease in SMA and portal venous blood flows, respectively, without significantly affecting pancreatic and axillary arterial blood flows and systemic arterial pressure in dogs (SUMI et al. 1987). GARDINER et al. (1990a,b) compared the vascular effects of porcine NMU-8, porcine NMU-25, and rat

NMU in conscious rats, and found that the action of the NMU-25 is more potent than NMU-8, but NMU-25 and rat NMU are comparable in potency of action. NMU-25 decreased SMA flow by 23% and 60% at 0.01 and 1 nmol i.v. bolus, but increased blood flow through the celiac artery by 30% at 1 nmol, without significantly affecting the flows through the kidney and distal abdominal aorta. The selective vasoconstriction or vasodilation of NMU in different organs, and its physiological significance, remain to be investigated.

C. Capsaicin-Sensitive Afferent Nerves

Recent studies have shown that primary afferent nerves (C fibers) in the GI tract not only serve a sensory role but also play a local effector role in regulating local blood flow by releasing a variety of neuropeptides from their peripheral endings (Holzer 1988). As described below, these nerves can be activated by mechanical, chemical, thermal or ischemic-hypoxic stimulation to increase local blood flow in the GI tract. The vasodilatory action is mediated by neuropeptides such as CCK, SP, VIP and CGRP. In addition, SS may reside in these nerves. All studies on this subject utilized capsaicin, which is given to neonatal rats to produce permanent loss of these nerve fibers, given to adult animals by injection into a local artery, or placed into the gut lumen. Local acute application of capsaicin in adult animals stimulates the release of neuropeptides from sensory primary afferent nerve endings.

Rozsa et al. (1984) showed that injections of small doses of capsaicin (0.1–7 μg/kg) into the SMA dose-dependently increased intestinal blood flow in dogs. The increased flow was not affected by hexamethonium or adrenergic antagonists, but was attenuated or inhibited by prior administration of atropine or SS. Rozsa et al. (1985) further showed that antiserum to SP, CCK, or VIP inhibits not only the vascular action of the individual peptide but also the capsaicin-induced increase in SMA blood flow. The SS antibody, however, enhances capsaicin-induced vasodilation. As described in Sect. B.IX, SS is a vasoconstrictor, and its vasoconstrictor action, direct or indirect, could counteract the vasodilatory actions of the other three peptides. The capsaicin-induced vasodilation was reduced by 80% after combined administration of SP, CCK, and VIP antisera, and was completely abolished by simultaneous administration of atropine and the antibodies. Since the CCK antiserum used was found to cross-react with CGRP, CGRP may also be involved as a neuropeptide. Thus, stimulation of intestinal capsaicin-sensitive afferent nerves can increase local blood flow, and the response is predominantly mediated by the release of SP, VIP, and CCK-like or CGRP peptides from perivascular C fibers of the gut with involvement of cholinergic mechanisms. The sensory afferent and enteric neurons in the gut, however, have several neuropeptide mediators in common. It is therefore possible that the capsaicin-induced vasodilation may involve both the afferent and other enteric neurons in the enteric nervous system.

It is well known that mucosal sensory receptors in the GI tract play a significant role in regulation of secretion, absorption, and motor activity. Mechanical or chemical stimulation of the mucosal surface has also been shown to increase local blood flow (CHOU and KVIETYS 1981; CHOU et al. 1984). A recent study further showed that application of capsaicin (5 mg) on the jejunal or ileal mucosa produced an immediate increase (by 30% in 28 of 35 applications) or decrease (by 34% in the remaining 7 applications) in SMA blood flow in dogs (ROZSA et al. 1986). The changes in blood flow induced by mucosal application of capsaicin were prevented after application of a local anesthetic, lidocaine, to the mucosal surface, or after i.a. injection of SS. Thus, mucosal application of capsaicin can produce either intestinal vasodilation or vasoconstriction via an axon reflex arrangement of mucosal afferent nerves. The physiological significance of this study is that capsaicin is the pungent ingredient of hot peppers, intake of which in sufficient amount can affect intestinal blood flow.

The mechanisms involved in postprandial intestinal hypermia are complex and multifactorial (GALLAVAN and CHOU 1985). ROZSA and JACOBSON (1989) provided evidence for adding capsaicin-sensitive afferent nerves to this complex list. Placement of a solution containing oleic acid and bile into the jejunal lumen produced an abrupt and transient increase in SMA blood flow in rats. This hyperemia was prevented by treating the jejunal lumen with lidocaine or capsaicin, and was absent in rats treated with capsaicin during their neonatal life. VIP antiserum attenuated this hyperemia by 69%, but SP or CCK-8 antiserum, as well as hexamethonium, atropine, or reserpine, failed to affect the hyperemia. If enteric nerves are involved in the oleate-induced hyperemia, the nerves involved must be resistant to tetrodotoxin, because this neurotoxin does not block oleate-induced jejunal hyperemia in dogs (NYHOF and CHOU 1983). Reactive hyperemia (i.e., the transient increase in blood flow following arterial occlusion for 1–2 min) as well as autoregulatory escape (an escape from vasoconstriction induced by sympathetic neural stimulation) in the small intestine may also involve capsaicin-sensitive afferrent nerves (ROZSA and JACOBSON 1987; REMAK et al. 1990). As in the nutrient-induced hyperemia, several local mechanisms, such as vasodilator metabolites (e.g., adenosine), have been proposed to play a role in both reactive hyperemia and autoregulatory escape.

The cardiovascular changes induced by warming the mucosal or serosal surfaces of the stomach and small intestine to 45°C may also involve SP and capsaicin-sensitive nerves (ROZSA et al. 1988). The thermal application with warm normal saline evoked tachycardia, and decreased systemic arterial pressure and SMA blood flow (by 50%), which occurred within 3–5 s of application and lasted for 2–4 min. These responses were abolished or attenuated after treating the mucosal or serosal surfaces with 1% lidocaine or capsaicin, or by SP antiserum, SS or hexamethonium, but were unaltered by atropine, propranolol, or vagotomy. Reserpine or splanchnic ganglionectomy abolished the cardiac responses, but reversed the decrease in SMA blood flow into a vasodilator response. The vasodilatory response in reserpinized rats was prevented after treating the rats with SP antiserum. Thus, SP may be a major neurotransmitter of the gastrointestinal thermoreflex, and the resulting decrease in SMA blood flow is mediated by an adrenergic mechanism.

In summary, stimulation of capsaicin-sensitive nerves of the small intestine appears to produce local vasodilation or vasoconstriction, which is mediated by VIP, SP, CCK, SS, and cholinergic mechanisms. It also appears that different stimuli effect releases of different types of neuropeptides, e.g., VIP in the oleate-induced hyperemia, and SP in the thermally induced reflex. Inasmuch as stimulation of the submucosal nerve plexus in the guinea pig small intestine causes both cholinergic (Neild et al. 1990) and noncholinergic (Galligan et al. 1990) vasodilation of submucosal arterioles, the vascular responses to stimulation of capsaicin-sensitive afferent nerves may also involve the enteric nervous system.

In addition to their physiological role, capsaicin-sensitive sensory neurons may play a role in protecting the gastrointestinal mucosa in disease states. Ablation of these afferent neurons aggravates gastric and intestinal lesion formation in a variety of experimental ulcer models, whereas activation of these neurons affords protection against ulcerogenic factors (Holzer 1988). The underlying mechanism for this protection is unknown, but an increase in mucosal blood flow could be a significant factor. Thus, activation of these neurons by intragastric perfusion of a capsaicin solution ($160\,\mu M$) not only prevents mucosal damage produced by ethanol (Holzer and Lippe 1988; Holzer et al. 1990) or acidified aspirin (Holzer et al. 1989), but also increases gastric mucosal blood flow (Hlzer et al. 1990; Lippe et al. 1989). A subsequent study (Holzer et al. 1991) further showed that intragastric perfusion of solutions containing both capsaicin (at concentrations between 10 and $640\,\mu M$) and an injurious concentration of ethanol (25%) dose dependently increased gastric mucosal blood flow. The dose-dependent mucosal vasodilation was significantly correlated with a dose-dependent reduction of gross ethanol-induced mucosal damage. Capsaicin prevented deep, but not superficial, mucosal damage, and its vasodilator and protective effects were blocked by functional ablation of capsaicin-sensitive sensory neurons or tetrodotoxin. Therefore, the conclusion was drawn that these capsaicin effects are mediated by a local neural reflex (an axon reflex) initiated by the presence of capsaicin in the gastric lumen. The enhancement of gastric mucosal blood flow may be the major mechanism for the sensory nerve-mediated prevention of gastric mucosal injury. The stimuli that activate capsaicin-sensitive nerves in pathological conditions are unclear. The stimuli may be ischemic hypoxia per se or the local chemicals released in response to ischemic hypoxia (Longhurst and Dittman 1987).

References

Ando K, Pegram BL, Frohich ED (1990) Hemodynamic effects of calcitonin gene-related peptide in spontaneously hypertensive rats. Am J Physiol 258:R425–R429

Augood SJ, Keast JR, Emson PC (1988) Distribution and characterization of neuromedin U-like immunoreactivity in rat brain and intestine and in guinea pig intestine. Regul Pept 20:281–291

Bauerfeind P, Hof R, Hof A, Cucala M, Siegrist S, von Ritter C, Fischer JA, Blum AL (1989) Effects of hCGRP I and II on gastric blood flow and acid secretion in anesthetized rabbits. Am J Physiol 256:G145–G149

Benoit JN, Barrowman JA, Harper SL, Kvietys PR, Granger DN (1984) Role of humoral factors in the intestinal hyperemia associated with chronic portal hypertension. Am J Physiol 247:G486–G493

Benoit JN, Zimmerman B, Premen AJ, Liang V, Go W, Granger DN (1986) Role of glucagon in splanchnic hyperemia of chronic portal hypertension. Am J Physiol 251:G674–G677

Blank MA, Kimura K, Fuortes M, Jaffe BM (1990) VIP antagonist [*N*-Ac-Tyr^1, D-Phe^2]-GRF-(1–29)-NH_2: an inhibitor of vasodilation in feline colon. Am J Physiol 259:G252–G257

Buell MG, Harding RK (1989) Effects of peptide YY on intestinal blood flow distribution and motility in the dog. Regul Pept 24:195–208

Cerini R, Lee SS, Hadengue A, Koshy A, Girod C, Lebrec D (1988) Circulatory effects of somatostatin analogue in two conscious rat models of portal hypertension. Gastroenterology 94:703–708

Chayvialle JA, Miyata M, Rayford PL, Thompson JC (1980) Effects of test-meal, intragastric nutrients, and introduodenal bile on plasma concentrations of immunoreactive somatostatin and vasoactive intestinal peptide in dogs. Gastroenterology 79:844–853

Chou CC (1983) Splanchnic and overall cardiovascular hemodynamics during eating and digestion. Fed Proc 42:1658–1661

Chou CC (1989) Gastrointestinal circulation and motor function. In: Wood JD (ed) Handbook of physiology. Am Physiological Society, Bethesda, pp 1475–1518

Chou CC, Kvietys PR (1981) Physiological and pharmacological alterations in gastrointestinal blood flow. In: Granger DN, Bulkley GB (eds) Measurement of splanchnic blood flow. Williams and Wilkins, Baltimore, pp 475–509

Chou CC, Hsieh CP, Dabney JM (1977) Comparison of vascular effects of gastrointestinal hormones on various organs. Am J Physiol 232:H103–H109

Chou CC, Mangino MJ, Sawmiller DR (1984) Gastrointestinal hormones and intestinal blood flow. In: Shepherd AP, Granger DN (eds) Physiology of the intestinal circulation. Raven, New York, pp 121–130

Dipette DJ, Schwazenberger K, Kerr N, Holland OB (1989) Dose-dependent systemic and regional hemodynamic effects of calcitonin gene-related peptide. Am J Med Sci 297:65–70

Eklund S, Jodal M, Lundgren O, Sjoqvist A (1979) Effects of vasoactive intestinal polypeptide on blood flow, motility and fluid transport in the gastrointestinal tract of the cat. Acta Physiol Scand 105:461–468

Eysselein VE, Eberlein GA, Hesse WH, Schaeffer M, Grandt D, Williams R, Goebell H, Reeve JR Jr (1990) Molecular variants of cholecystokinin after endogenous stimulation in humans: a time study. Am J Physiol 258:G951–G957

Fahrenkrug J, Haglund V, Jodal M, Lundgren O, Olbe L, Schaffalitzky de Muchadell OB (1978) Nervous release of vasoactive intestinal polypeptide in the gastrointestinal tract of cats: possible physiological implications. J Physiol (Lond) 284:291–303

Fara JW, Rubinstein EH, Sonnenschein RR (1972) Intestinal hormones in mesenteric vasodilation after intraduodenal agents. Am J Physiol 223:1058–1067

Gallavan RH, Chou CC (1985) Possible mechanisms for the initiation and maintenance of postprandial intestinal hyperemia. Am J Physiol 249:G301–G308

Gallavan RH, Chen MH, Joffe SN, Jacobson ED (1985) Vasoactive intestinal polypeptide, cholecystokinin, glucagon, and bile-oleate-induced jejunal hyperemia. Am J Physiol 248:G208–G215

Gallavan RH Jr, Shaw C, Murphy RF, Buchanan KD, Joffe SN, Jacobson ED (1986) Effects of micellar oleic acid on canine jejunal blood flow and neurotensin release. Am J Physiol 251:G649–655

Galligan JJ, Jiang M-M, Shen K-Z, Surprenant A (1990) Substance P mediates neurogenic vasodilatation in extrinsically denervated guinea-pig submucosal arterioles. J Physiol (Lond) 420:267–280

Gardiner SM, Compton AM, Bennett T (1990a) Differential effects of neuropeptides on coeliac and superior mesenteric blood flows in conscious rats. Regul Pept 29:215–227

Gardiner SM, Compton AM, Bennett T, Domin J, Bloom SR (1990b) Regional hemodynamic effect of neuromedin U in conscious rats. Am J Physiol 258:R32–R38

Gerber JG, Guth PH (1989) Role of adenosine in the gastric blood flow response to pentagastrin in the rat. J Pharmacol Exp Ther 251:550–556

Girgis SI, Stevenson JC, Lynch C, Self CH, Macdonald DWR, Bevis PJR, Wimolawansa SJ, Morris HR, Macintyre I (1985) Calcitonin gene-related peptide: potent product of calcitonin gene. Lancet 2:14–16

Gronstad KO, Dahlstrom A, Jaffe BM, Zinner MJ, Ahlman H (1986) Studies on the mucosal hyperemia of the feline small intestine observed at endoluminal perfusion with substance P. Acta Physiol Scand 128:97–108

Harper SL, Barrowman JA, Kvietys PR, Granger DN (1984) Effect of neurotensin on intestinal capillary permeability and blood flow. Am J Physiol 247:G161–G166

Holm C, Biber B, Gustavsson B, Milsom I, Winso O, Fahrenkrug J (1988) Postprandial portal venous blood flow and portal plasma vasoactive intestinal peptide in man. In: Said SI, Mutt V (eds) Vasoactive intestinal peptide and related peptides. New York Academy of Sciences, New York, p 591

Holm-Rutile L, Berglindh T (1986) Pentagastrin and gastric mucosal blood flow. Am J Physiol 250:G575–G580

Holzer P (1988) Local effector functions of capsaicin-sensitive sensory nerve endings: involvement of tachykinins, calcitonin gene-related peptide and other neuropeptides. Neuroscience 24:739–768

Holzer P, Lippe IT (1988) Stimulation of afferent nerve endings by intragastric capsaicin protects against ethanol-induced damage of gastric musosa. Neuroscience 27:981–987

Holzer P, Pabst MA, Lippe IT (1989) Intragastric capsaicin protects against aspirin-induced lesion formation and bleeding in the rat gastric mucosa. Gastroenterology 96:1425–1433

Holzer P, Pabst MA, Lippe IT, Peskar BM, Peskar BA, Livingston EH, Guth PH (1990) Afferent nerve-mediated protection against deep mucosal damage in the rat stomach. Gastroenterology 98:838–848

Holzer P, Livingston EH, Saria A, Guth PH (1991) Sensory neurons mediate protective vasodilation in rat gastric mucosa. Am J Physiol 260:G363–G370

Inoue K, Hosotani R, Tatemoto K, Yajima H, Tobe T (1988) Effect of natural peptide YY on blood flow and exocrine secretion of pancreas in dogs. Dig Dis Sci 33:828–832

Ito S, Ohga A, Ohta T (1988) Gastric relaxation and vasoactive intestinal peptide output in response to reflex vagal stimulation in the dog. J Physiol (Lond) 404:683–693

Ivarsson LE, Darle N, Lundgren O (1982) The effects of somatostatin on blood flow in the secretory part of the human stomach. Surg Gastroenterol 1:29–34

Jenkins SA, Baxter JN, Corbett WA, Shields R (1985) Effects of a somatostatin analogue SMS 201-995 on hepatic hemodynamics in the pig and on intravariceal pressure in man. Br J Surg 72:1009–1012

Jodal M, Lundgren O, Sjoqvist A (1983) The effect of apamin on non-adrenergic, non-cholinergic vasodilator mechanisms in the intestines in the cat. J Physiol (Lond) 338:207–219

Kogire M, Inoue K, Sumi S, Doi R, Takaori K, Yun M, Fujii N, Yajima H, Tobe T (1988) Effects of synthetic human gastric inhibitory polypeptide on splanchnic circulation in dogs. Gastroenterology 95:1636–1640

Konturek JS, Bilski J, Pawkik W, Tasler J, Domschke W (1988) Adrenergic pathway in the inhibition of pancreatic secretion by peptide YY in dogs. Gastroenterology 94:266–273

Konturek SJ, Yanaihara N, Pawlik W, Jaworek J, Szewczyk K (1989) Comparison of helodermin, VIP and PHI in pancreatic secretion and blood flow in dogs. Regul Pept 24:155–166

Kravetz D, Bosch J, Arderiu MT, Pizcueta MP, Casamitjana R, Rivera F, Rodes J (1988) Effects of somatostatin on splanchnic hemodynamics and plasma glucagon in portal hypertensive rats. Am J Physiol 254:G322–G328

Leung FW, Guth PH, Scremin OV, Golanska EM, Kauffman GL (1984) Regional gastric mucosal blood flow measurements by hydrogen gas clearance in the anesthetized rat and rabbit. Gastroenterology 87:28–36

Leung FW, Kauffman GC, Washington J, Scremin OU, Guth PH (1986) Blood flow limitation of stimulated gastric acid secretion in the rat. Am J Physiol 250:G794–799

Liddle RA, Goldfine ID, Rosen MS, Taplitz RA, Williams JA (1985) Cholecystokinin bioactivity in human plasma. J Clin Invest 75:1144–1152

Lippe IT, Pabst MA, Holzer P (1989) Intragastric capsaicin enhances rat gastric acid elimination and mucosal blood flow by afferent nerve stimulation. Br J Pharmacol 96:91–100

Longhurst JC, Dittman LE (1987) Hypoxia, bradykinin, and prostaglandins stimulate ischemically sensitive visceral afferents. Am J Physiol 253:H556–H567

Lundberg JM, Tatemoto K, Terenius L, Hellstrom PM, Muff V, Hokfelt T, Hamberger B (1982) Localization of peptide YY (PYY) in gastrointestinal endocrine cells and effects on intestinal blood flow and motility. Proc Natl Acad Sci USA 79:4471–4475

MacKay AD, Jones HW, Gough A, Peart SW, Freeman TC, Springer CJ, Calam J (1990) Do normal plasma neurotensin concentrations increase ileal output? Regul Pept 27:299–306

Maton PN, Seldon AC, Chadwick VS (1982) Large and small forms of cholecystokinin in human plasma: measurement using high pressure liquid chromatography and radioimmunoassay. Regul Pept 4:251–260

Neild TO, Shen KZ, Surprenant A (1990) Vasodilatation of arterioles by acetylcholine released from single neurones in the guinea-pig submucosal plexus. J Physiol (Lond) 420:247–265

Nyhof RA, Chou CC (1983) Evidence against local neural mechanism for intestinal postprandial hyperemia. Am J Physiol 245:H437–H446

Pappas TN, Debas HT, Goto Y, Taylor IL (1985) Peptide YY inhibits meal-stimulated pancreatic and gastric secretion. Am J Physiol 248:G118–G123

Pawlik WW, Gustaw P, Sendur R, Czarnobilski K, Konturek SJ (1989) Role of CCK receptors in the postprandial intestinal hyperemia. Proc Int Union Physiol Sci 17:503

Pique JM, Leung FW, Tan HW, Livingston E, Scremin OU, Guth PH (1988) Gastric mucosal blood flow response to stimulation and inhibition of gastric acid secretion. Gastroenterology 95:642–650

Premen AJ, Dobbins DE, Soika CY, Dabney JM (1984a) Relationship between substance P, intestinal wall compliance and vascular resistance in the canine ileum. Regul Pept 9:119–127

Premen AJ, Soika CY, Dabney JM, Dobbins DE (1984b) Effects of gastrointestinal hormones on ileal vascular and visceral smooth muscle. Am J Physiol 246:G1–G7

Premen AJ, Kvietys PR, Granger DN (1985) Postprandial regulation of intestinal blood flow; role of gastrointestinal hormones. Am J Physiol 249:G250–G255

Price BA, Jaffe BM, Zinner MJ (1985) Effect of exogenous somatostatin infusion on gastrointestinal blood flow and hormones in the conscious dog. Gastroenterology 88:80–85

Remak G, Hottenstein OD, Jacobson ED (1990) Sensory nerves mediate neurogenic escape in rat gut. Am J Physiol 258:H778–H786

Rozsa Z, Jacobson ED (1987) Postocclusive intestinal vasodilation is mediated by sensory substance P. Gastroenterology 92:1604

Rosza Z, Jacobson ED (1989) Capsaicin-sensitive nerves are involved in bile-oleate induced intestinal hyperemia. Am J Physiol 256:G476–G481

Rozsa Z, Jancso G, Varro V (1984) Possible involvement of capsaicin-sensitive sensory nerves in the regulation of intestinal blood flow in the dog. Naunyn Schmeidebergs Arch Pharmacol 326:352–356

Rozsa Z, Varro V, Jancso G (1985) Use of immunoblockade to study the involvement of peptidergic afferent nerves in the intestinal vasodilatory response to capsaicin in the dog. Eur J Pharmacol 115:59–64

Rozsa Z, Sharkey KA, Jancso G, Varro V (1986) Evidence for a role of capsaicin-sensitive mucosal afferent nerves in the regulation of mesenteric blood flow in the dog. Gastroenterology 90:906–910

Rozsa Z, Mattila J, Jacobson ED (1988) Substance P mediates a gastrointestinal thermoreflex in rats. Gastroenterology 95:265–276

Sawmiller DR, Chou CC (1988) Adenosine plays a role in food induced jujunal hyperema. Am J Physiol 255:G168–G174

Sjoqvist A, Fahrenkrug J, Jodal M, Lundgren O (1983) Effect of apamin on release of vasoactive intestinal polypeptide (VIP) from cat intestines. Acta Physiol Scand 119:69–76

Sumi S, Inoue K, Kogire M, Doi R, Takaori K, Suzuki T, Yajima H, Tobe T (1987) Effect of synthetic neuromedin U-8 and U-25, novel peptides identified in porcine spinal cord, on splanchnic circulation in dogs. Life Sci 41:1585–1590

Tepperman BL, Whittle BJR (1991) Comparison of the effects of neuropeptide Y and noradrenaline on rat gastric mucosal blood flow and integrity. Br J Pharmacol 102:95–100

Yeo CJ, Jaffe BM, Zinner MJ (1984) The effects of intravenous substance P infusion on hemodynamics and regional blood flow in conscious dogs. Surgery 95:175–182

CHAPTER 12

Peptidergic Regulation of Cell Proliferation Through Multiple Signaling Pathways

I. Zachary and E. Rozengurt

A. Introduction

The cells of many tissues and organs in vivo are maintained in a nonproliferating state (the G_o/G_1 phase of the cell cycle), but can be stimulated to resume DNA synthesis and cell division in response to external stimuli such as hormones, antigens, or growth factors. In this manner the growth of individual cells is regulated according to the requirements of the whole organism. The regulation of cell proliferation is therefore central to many physiological and pathological processes, including embryogenesis, growth and development, selective cell survival, hemopoiesis, tissue repair, immune responses, atherosclerosis, and neoplasia (Evered et al. 1985). For these reasons the identification of the extracellular factors which modulate cell proliferation and the elucidation of the molecular mechanisms involved have emerged as a fundamental problem in biology.

In both respects (i.e., the identification of growth factors and the understanding of their mechanism of action) studies in cultured fibroblasts, and in particular the Swiss 3T3 cell model system, have provided a successful approach which avoids the complexities involved in research on intact tissues and organisms. These cells cease to proliferate when they deplete the medium of its growth-promoting activity and can be stimulated to reinitiate DNA synthesis and cell division either by replenishing the medium with fresh serum or by the addition of defined combinations of growth factors or pharmacological agents in serum-free medium. An important feature of mitogenic signaling that has emerged from these studies is that cell proliferation can be stimulated through multiple, independent signal-transduction pathways which act in a synergistic and combinatorial fashion.

In recent years an increasing number of small regulatory peptides[1] have been discovered in the neural and neuroendocrine cells of the gastro-

---

[1] In what follows, we shall refer throughout to regulatory or neuropeptides rather than gastrointestinal peptides, though to a very large extent these are overlapping terms. This will avoid ambiguities of meaning which may arise because some peptides under consideration (e.g., vasopressin) are *not* gastrointestinal, but are included because of their relevance to a discussion of peptidergic regulation of cell growth. It should be pointed out, however, that almost all of the peptides discussed are found not only in the gut but also in other tissues, notably the brain and certain tumors.

intestinal (GI) tract and central nervous system (WALSH 1987). Defining the physiological functions of GI peptides is complicated, however, by the various modes of action displayed by these molecules (DOCKRAY 1987). Some are localized in neurons and act as neurotrasmitters in the central or peripheral nervous system, while others are released by endocrine cells and have effects both as systemic hormones circulating through the blood stream and by acting locally either on adjacent cells (paracrine communication) or on the same cells that released them (autocrine effects). Moreover, a number of peptides are found in both neuronal and endocrine cells, and a major effect of some regulatory peptides in vivo, e.g., bombesin/gastrin-releasing peptide (GRP), is to stimulate the release of other GI peptides. The classical role of these peptides as fast-acting neurohumoral signalers has recently been challenged by the discovery that they also stimulate cell proliferation (ZACHARY et al. 1987; WOLL and ROZENGURT 1989; ROZENGURT 1991). Furthermore, indirect evidence is accumulating that the mitogenic effects of neuropeptides may be physiologically relevant for a number of normal and abnormal biological processes, and in particular tumorigenesis.

This chapter will review the direct evidence for the growth-promoting activity of GI peptides, with particular emphasis on the biochemical events and pathways initiated by these factors in quiescent Swiss 3T3 cells. Evidence supporting a role for neuropeptide regulation of cell growth in vivo will also be discussed.

B. Growth-Promoting Activities of Neuropeptides in Cultured Cells

Neuropeptides are increasingly implicated in the control of cell proliferation and the mechanisms involved are attracting intense interest. The effects of neuropeptides or small regulatory peptides on the proliferation of cultured cells are summarized in Table 1. The finding that the neurohypophyseal nonapeptide vasopressin is a mitogen for quiescent cultures of these cells provided the first unambiguous evidence for a direct growth-promoting action of a neuropeptide in a homogeneous cell population (Table 1). Subsequently, bombesin, a 14-amino acid peptide first isolated from the skin of the European frog *Bombina bombina*, was found to be a potent mitogen for Swiss 3T3 cells (ZACHARY et al. 1987). Structurally related peptides found in mammals (gastrin-releasing peptide or GRP, neuromedin B, neuromedin C) have similar mitogenic effects (ZACHARY et al. 1987).

The list of small peptides which can act as mitogens in these cells has now grown considerably (Table 1) and bradykinin, vasoactive intestinal peptide (VIP), endothelin and the endothelin-related peptide vasoactive intestinal contractor (VIC) have all been shown to stimulate DNA synthesis in Swiss 3T3 cells in synergistic combinations with other growth factors. Evidence for the direct growth-promoting activities of several other peptides,

Table 1. Growth-promoting effects of neuropeptides in cultured cells[a]

Neuropeptide	Cell type
Bombesin	Swiss 3T3
GRP, neuromedin B	Swiss 3T3
Bombesin, GRP	Bronchial epithelial
Bombesin, GRP	SCLC
Vasopressin	Rat chondrocytes
	Swiss 3T3
Bradykinin	Swiss 3T3
	SCLC
CCK	SCLC
Neurotensin	SCLC
Galanin	SCLC
SP and SK	Human skin fibroblasts and rat smooth muscle cells
SP	Human rheumatoid synoviocytes
β-Endorphin	Rat spleen lymphocytes
Gastrin	Rat gastric carcinoma cells

[a] Primary references to the effects described in this table and in Table 2 can be found in ZACHARY et al. (1987) and WOLL and ROZENGURT (1989).

including the gastrointestinal peptides cholecystokinin (CCK) and neurotensin, has recently come from work in another cultured cell model system, namely cell lines established from small cell lung carcinoma (SCLC) (SETHI and ROZENGURT 1991). This work will be discussed more fully in a later section.

The tachykinin family of peptides, including substance P (SP) and neurokinin A, also known as substance K (SK), have no receptors in Swiss 3T3 cells, but direct growth-promoting effects of these peptides have been reported in smooth muscle cells and human skin fibroblasts. SP also enhances the proliferation of human blood T-lymphocytes, an effect apparently mediated by specific receptors for this peptide. Recently, it was also reported that SP stimulates release of prostaglandin E_2 (PGE_2) and proliferation in rheumatoid synoviocytes. These findings are in accord with other evidence which indirectly suggests that the release of tachykinins from sensory nerves in the skin, joints and other peripheral tissues might function as mediators of local inflammatory and wound healing responses and in the pathogenesis of rheumatoid arthritis (see WOLL and ROZENGURT 1989 for ref.).

In this chapter we shall focus on the mitogenic actions of bombesin and related peptides in Swiss 3T3 cells. There are a number of considerations which make bombesin attractive as a model peptide with which to investigate the mechanisms underlying peptidergic regulation of cell growth. In serum-free medium it stimulates DNA synthesis and cell division in the absence of other growth-promoting agents. The ability of bombesin, like platelet-derived growth factor (PDGF), to act as a sole mitogen for these cells contrasts with other peptide growth factors which are active only in synergistic combinations (ROZENGURT 1985, 1986). The mitogenic effects of bombesin are markedly potentiated by insulin, which both increases the maximal response and reduces the bombesin concentration required for half-maximal effect. Furthermore, receptors for bombesin-like peptides have been well characterized at the molecular level.

I. Mitogenic Action of Bombesin in Swiss 3T3 Cells: A Paradigm for Peptidergic Regulation of Cell Proliferation

To determine how bombesin/GRP stimulates mitogenesis, we sought specific cell-surface receptors on Swiss 3T3 fibroblasts, using radiolabeled [^{125}I]GRP. Binding measurements and chemical cross-linking experiments using this ligand show that bombesin-like peptides interact with specific, high-affinity receptors located on the cell surface. [^{125}I]GRP binding is inhibited by various bombesin-like peptides in proportion to their ability to stimulate DNA synthesis but not by structurally unrelated mitogens (ZACHARY et al. 1987). Bombesin antagonists inhibit both GRP binding and bombesin/GRP-stimulated mitogenesis (WOLL et al. 1992). Thus, bombesin and related peptides interact with receptors that are distinct from those for other mitogens in Swiss 3T3 cells.

[^{125}I]GRP binding to membrane preparations of Swiss 3T3 cells, i.e., in the absence of ligand internalization and degradation, is specific, saturable and reversible (ROZENGURT et al. 1990). Scatchard analysis indicates the presence of a single class of high-affinity binding sites with a K_d (2×10^{-10} M) which is in excellent agreement with the equilibrium constant derived from rate constants. These results are consistent with the existence of a homogeneous population of bombesin/GRP-binding sites in membranes of Swiss 3T3 cells.

The physical properties of the bombesin/GRP receptor have been investigated using an affinity-labeling method. Analysis of extracts of cells which have been preincubated with [^{125}I]GRP and then treated with disuccinimidyl cross-linking agents reveals the presence of a major band migrating with an apparent M_r of 75000–85000. Furthermore, [^{125}I]GRP can be cross-linked to an M_r 75000–85000 glycoprotein of Swiss 3T3 membranes but not to membranes from cells lacking bombesin receptors. The radiolabeled M_r 75000–85000 protein binds to wheat germ lectin-Sepharose columns and can be eluted with *N*-acetyl-D-glucosamine, suggesting that it is a glycoprotein. In addition, treatment with endo-β-*N*-acetylglycosaminidase F reduces the apparent molecular weight of the affinity-labeled band from 75000–85000 to 43000, indicating the presence of N-linked oligosaccharide groups. Thus, the bombesin/GRP receptor is a glycoprotein of apparent M_r 75 000–85 000 with N-linked carbohydrate side chains and a polypeptide core of M_r 43 000 (reviewed in ROZENGURT et al. 1990; ROZENGURT 1991).

A decrease in ligand affinity for receptors produced by added guanine nucleotides is characteristic of a receptor-guanine nucleotide binding (G) protein interaction (BOURNE et al. 1988). The nonhydrolyzable guanosine triphosphate (GTP) analog GTP-γ-S causes a specific and concentration-dependent inhibition of [^{125}I]GRP binding and cross-linking to 3T3 cell membranes. The effect is due primarily to an increase in the equilibrium dissociation constant (K_d) rather than to a decrease in the number of receptors. This modulation of ligand affinity by guanine nucleotides strongly suggests that a G protein couples the mitogenic bombesin receptor with intracellular effector systems.

Binding of various hormonal peptides to their corresponding membrane receptors is known to induce tight association between the receptor and their respective G proteins (BOURNE et al. 1988). [^{125}I]GRP-receptor complexes were solubilized from Swiss 3T3 cell membranes by using the detergents taurodeoxycholate or deoxycholate (COFFER et al. 1990). To determine whether the solubilized [^{125}I]GRP receptor complex is functionally coupled to a G protein(s) we tested whether the ligand-receptor complex isolated by gel filtration retains the ability to be regulated by guanine nucleotides. GTP-γ-S caused a specific and dose-dependent decrease in the level of the [^{125}I]GRP receptor complex (ROZENGURT et al. 1990). These results provided the first functional evidence for a bombesin/GRP receptor-G protein interaction.

II. Purification and Molecular Cloning

The solubilization of the bombesin receptor in an active form proved an important step for attempting its purification (FELDMAN et al. 1990; COFFER et al. 1990) and reconstitution into phospholipid vesicles (ROZENGURT et al. 1990; ROZENGURT 1991). Recently, COFFER et al. (1990) described a novel procedure to partially purify bombesin receptors using biotinylated [Lys^3]bombesin, a ligand that retains specific receptor binding and biological activity. ^{125}I-GRP binding activity that exhibited cellular and ligand specificity and saturability was demonstrated when affinity-purified receptor preparations were reconstituted into phospholipid vesicles (COFFER et al. 1990). The reconstitution of the affinity-purified bombesin receptor may provide an approach to elucidate the molecular nature of the G protein(s) that participate in the transduction of the mitogenic signal.

The bombesin receptor from Swiss 3T3 cells was also purified using a different approach, namely sequential lectin and ligand affinity chromatography. The purified receptor displayed an apparent M_r of 75 000–95 000, which was markedly reduced by removing the N-linked oligosaccharide moieties (FELDMAN et al. 1990) in agreement with previous results with the affinity-labeled receptor. Recently, the bombesin receptor from Swiss 3T3 cells has been cloned (BATTEY et al. 1990; SPINDEL et al. 1990) and the deduced amino acid sequence predicts a polypeptide core of M_r 43 000 and demonstrates that it belongs to the superfamily of G protein-linked receptors characterized by seven hydrophobic domains thought to traverse the cytoplasmic membrane. These results are therefore in excellent agreement with the biochemical data described above. Other receptors for the regulatory peptides substance P (YOKOTA et al. 1989; HERSHEY and KRAUSE 1990), substance K (MASU et al. 1987), angiotensin (SASAKI et al. 1991; MURPHY et al. 1991), the endothelins (SAKURAI et al. 1990; ARAI et al. 1990), VIP (SREEDHARAN et al. 1991), and bradykinin (MCEACHERN et al. 1991) also belong to this superfamily (reviewed in STROSBERG 1991; and see also Chap. 5, this volume). As yet, however, the elucidation of the structure of these receptors does not provide information concerning the class (i.e., G_s, G_i, G_o, etc.) or number of G proteins to which they are coupled.

C. Early Signaling Events

The binding of growth factors to their receptors promotes the generation of early signals in the membrane, cytosol, and nucleus which lead to cell proliferation (ROZENGURT 1985, 1986, 1991). Since the initiation of DNA synthesis occurs 12–15 h after the addition of the mitogens, it is expected that knowledge of the early events will provide fundamental insights into mechanisms underlying the proliferative response.

The early cellular and molecular responses elicited by bombesin and structurally related peptides have been elucidated in detail. The cause-effect relationships and temporal organization of these early signals and molecular events provide a paradigm for the study of other growth factors and mitogenic neuropeptides and illustrate the activation and interaction of a variety of signaling pathways.

I. Inositol Phospholipid Turnover and Ca^{2+} Mobilization

One of the earliest events to occur after the binding of bombesin to its specific receptor is a rapid mobilization of Ca^{2+} from intracellular stores, which leads to a transient increase in the concentration of cytosolic Ca^{2+} and subsequently to a decrease in the Ca^{2+} content of the cells (ROZENGURT 1991 and references cited therein). The mobilization of Ca^{2+} by bombesin is mediated by inositol 1,4,5-trisphosphate [Ins(1,4,5)P_3], which acts as a second messenger in the action of many ligands that stimulate inositol lipid turnover and Ca^{2+} efflux (BERRIDGE and IRVINE 1990). Bombesin causes a rapid increase in Ins(1,4,5)P_3, which coincided with the increase in cytosolic Ca^{2+}. Ins(1,4,5)P_3 is formed as a result of phospholipase C catalyzed hydrolysis of phosphatidylinositol 4,5-bisphosphate in the plasma membrane, a process that also generates 1,2-diacylglycerol (DAG). DAG can also be generated from other sources such as phosphatidylcholine hydrolysis and acts as second messenger in the activation of protein kinase C (PKC) by bombesin.

The Ca^{2+}-mobilizing action of bombesin can be distinguished from that of PDGF in terms of kinetics and sensitivity to PKC-medicated feedback inhibition (ROZENGURT 1991). It is likely that these differences reflect fundamental differences in signal transduction i.e., the PDGF receptor directly phosphorylates phospholipase C γ, whereas bombesin does not increase the tyrosine phosphorylation of this enzyme (CANTLEY et al. 1991; ZACHARY et al. 1991). Furthermore, bombesin, unlike PDGF, did not induce an enhanced inositol phosphate response in cells overexpressing phospholipase C γ (CUADRADO and MOLLOY 1990). Hence, it is likely that bombesin stimulates phosphoinositide breakdown and DAG formation via a phospholipase C isoform different from phospholipase C γ.

II. Protein Kinase C and Initiation of DNA Synthesis

Protein kinase C activated by diacylglycerols and phorbol esters is a Ca^{2+}- and phospholipid-dependent serine/threonine kinase, comprising multiple subspecies (NISHIZUKA 1988), and has been implicated in the signal transduction of many short-term cellular responses including secretion and contraction (KIKKAWA and NISHIZUKA 1986). PKC is directly activated in vivo by membrane-permeable diacylglycerols and by the plant-derived potent tumor promoters, phorbol esters. Of particular interest here is the evidence implicating PKC in mediating long-term responses. Phorbol esters stimulate DNA synthesis and cell division in synergy with insulin and other growth-promoting factors (ROZENGURT 1986). The mitogenic effect is mediated by high-affinity binding sites which were identified as PKC (NISHIZUKA 1988; ROZENGURT 1991). Furthermore, addition of the synthetic diacylglycerol 1-oleoyl-2-acetylglycerol (OAG) mimics the action of phorbol esters in stimulating reinitiation of DNA synthesis and cell division (ROZENGURT 1991).

Another approach to testing the role of PKC in the production of biological responses is to exploit the selective removal of this enzyme caused by a prolonged pretreatment of the cells with phorbol ester. Chronic exposure to phorbol esters leads to the disappearance of measurable PKC activity in cell-free preparations (ROZENGURT 1986). In parallel with this downregulation of PKC activity, the cells become desensitized to the mitogenic effects elicited by phorbol esters or OAG. Hence, activation of PKC is a potential pathway leading to mitogenesis.

1. Activation of Protein Kinase C in Intact Fibroblasts

Since activation of PKC may play a role in eliciting mitogenesis, it was of importance to test directly whether growth factors lead to activation of this enzyme in intact, quiescent cells. A rapid increase in the phosphorylation of an acidic cellular protein that migrates with an apparent M_r of 80000 (termed 80K) has provided a specific signal for activation of PKC in intact fibroblastic cells (reviewed in ROZENGURT 1991). The phosphorylation of this protein is stimulated by addition of biologically active phorbol diesters and synthetic diacylglycerols. Downregulation of PKC activity prevents the increase in 80K phosphorylation by subsequent addition of these agents. Furthermore, the same 80K protein is phosphorylated in cell-free systems either by activation of endogenous PKC or by addition of the purified enzyme. Recently, this prominent PKC substrate has been purified to homogeneity from extracts of Swiss 3T3 cells and the cDNA encoding this protein has been cloned and sequenced revealing specific protein kinase C phosphorylation sites (BROOKS et al. 1991). Thus, it is well established that an increase in the phosphorylation of 80K in intact cells provides a specific marker for PKC activation.

One of the most striking events induced by bombesin and structurally related peptides is a rapid increase in the phosphorylation of 80K which, as described above, is a prominent substrate of protein kinase C in Swiss 3T3 cells. Addition of bombesin causes a rapid increase in 80K phosphorylation, which is rapidly reversed upon removal of bombesin (ZACHARY et al. 1987; ROZENGURT 1991). In addition, bombesin promotes PKC translocation as judged by its ability to induce tight association of PKC to cellular membranes (STADDON et al. 1990). These findings strongly suggest that bombesin rapidly stimulates PKC activity in intact and quiescent Swiss 3T3 cells.

The phosphorylation of 80K has been characterized in digitonin permeabilized Swiss 3T3 cells and this technique was used to study the mechanism of bombesin-induced activation of protein kinase C (ERUSALIMSKY et al. 1988). A salient feature of the results is that the GDP analog GDP-β-*S* inhibited the stimulation of 80K phosphorylation by bombesin in a selective manner. GDP-β-*S* is known to prevent the activation of G proteins by inhibiting the binding of GTP. The fact that GTP can reverse the inhibitory effect of GDP-β-*S* is consistent with this notion. These findings indicate that guanine nucleotides modulate the transduction of the signal from the bombesin receptor and imply that a G protein couples the bombesin receptor to the generation of DAG through phospholipase C.

2. Cross-Talk Between Protein Kinase C, Cyclic AMP, and Epidermal Growth Factor Receptor Affinity

It is now recognized that a sustained increase in the cellular level of adenosine 3′,5′ cyclic monophosphate (cAMP) constitutes a growth-promoting signal for Swiss 3T3 cells (ROZENGURT 1991), and a variety of other cell types including thyroid and liver epithelial cells (DUMONT et al. 1989). As shown in Fig. 1, a variety of agents that promote cAMP accumulation in Swiss 3T3 cells, including PGE_1, the adenosine agonist 5′-*N*-

ethylcarboxamide-adenosine (NECA), cholera toxin, and permeable cAMP analogs, stimulate DNA synthesis by acting synergistically with insulin, phorbol esters, and other factors (ROZENGURT 1986). Furthermore, PDGF, one of the most potent mitogens for fibroblasts, induces a striking accumulation of cellular cAMP mediated, at least in part, by increased synthesis of E-type prostaglandins, which in turn leave the cell and stimulate cAMP synthesis through their own receptor. The neuropeptide vasoactive intestinal peptide (VIP), which stimulates DNA synthesis in Swiss 3T3 cells in the presence of insulin, also increases intracellular cAMP probably through G_s-coupled activation of adenylyl cyclase (ROZENGURT 1991). Together these findings indicate that a sustained increase in cellular cAMP acts as a mitogenic signal for Swiss 3T3 cells. While the cAMP and PKC signal transduction pathways are functionally separated from the cell surface to the nucleus (ROZENGURT 1991), other results indicate the existence of interactions between these major transmembrane-signaling systems. Specifically, activation of PKC by either phorbol esters or diacylglycerols markedly enhances the accumulation of cAMP in response to forskolin or cholera toxin, while downregulation of PKC blocks this enhancing effect (ROZENGURT et al. 1991).

A further example of cAMP/PKC signal pathway interactions has come from studies with the neuropeptides of the bombesin family. Bombesin causes a marked enhancement of cAMP accumulation (in the presence of forskolin); this increase is partially diminished both by downregulation of PKC and by the cyclooxygenase inhibitor indomethacin (MILLAR and ROZENGURT 1988). The inhibitory effects are additive in nature, suggesting the existence of two mechanisms by which bombesin can enhance cAMP accumulation.

One of the most intriguing areas of PKC/cAMP "cross-talk" involves the molecular basis for these pathway interactions. Further studies using phorbol esters, cAMP-increasing agents, and bombesin have shown that this "cross-talk" is abolished by treatment with pertussis toxin (PTX) in a time- and dose-dependent fashion (ROZENGURT 1991; MILLAR and ROZENGURT 1988). Since PTX does not itself promote cAMP accumulation in Swiss 3T3, it is unlikely to act by removing a tonic inhibitory influence on the adenylate cyclase via G_i. An attractive possibility is that a PTX substrate mediates the "cross-talk" between the PKC and the cAMP pathways.

Bombesin and other growth factors cause a rapid decrease in the apparent affinity of the epidermal growth factor (EGF) receptor population for EGF in Swiss 3T3 cells (reviewed in ZACHARY and ROZENGURT 1985). Considerable evidence implicates PKC in the mediation of this effect (ZACHARY and ROZENGURT 1985; ZACHARY et al. 1987). In particular, the inhibition of EGF binding induced by phorbol esters, DAG, or bombesin is prevented by downregulation of PKC (ZACHARY et al. 1987). The EGF receptor is phosphorylated by PKC at specific sites both in vitro and in vivo (ULLRICH and SCHLESSINGER 1990). Thus, transmodulation of the EGF receptor may result from the covalent modification of the EGF receptor catalyzed by PKC, though other mechanisms are not excluded.

III. Monovalent Ion Fluxes

The stimulation of the monovalent K^+, H^+, and Na^+ ion fluxes is a general early response seen in most types of quiescent cells stimulated to proliferate by multiple combinations of growth-promoting factors (ROZENGURT and MENDOZA 1986). This ubiquity suggests a possible role for enhanced ion fluxes in the mitogenic response. Addition of bombesin to quiescent 3T3 cells also causes a rapid increase in the activity of the ouabain-sensitive Na^+/K^+ pump (ZACHARY et al. 1987). The activity of the Na^+/K^+ pump in intact fibroblasts is limited and regulated by the supply of Na^+ (ROZENGURT and MENDOZA 1986). Growth-promoting agents including bombesin stimulate the pump by increasing Na^+ entry into the cells (ROZENGURT and MENDOZA 1986). This translocation of Na^+ across the plasma membrane is mediated in part by an amiloride-sensitive Na^+/H^+ antiport system that is driven by the Na^+ electrochemical gradient. Studies to determine the mechanisms by which bombesin stimulates these monovalent ion fluxes have shown the existence of PKC-dependent and -independent pathways of activation (ROZENGURT and MENDOZA 1986).

IV. Neuropeptide Stimulation of Tyrosine Kinase Activity

The occurrence of tyrosine phosphorylation in the action of bombesin has been controversial (ZACHARY et al. 1991). As indicated, however, it has been unequivocally established that the receptor for peptides of the bombesin family is coupled to a G-protein(s) and does not possess intrinsic tyrosine kinase activity. Recently, however, we reported that bombesin, vasopressin, and endothelin rapidly increase tyrosine and serine phosphorylation of multiple substrates in intact quiescent Swiss 3T3 cells (ZACHARY et al. 1991) including a major band of M_r 115000 (designated p115). The substrates for neuropeptide tyrosine phosphorylation in these cells appear to be unrelated to known targets for the PDGF receptor (ZACHARY et al. 1991), including GTPase-activating protein, phosphatidylinositol 3′ kinase, and phospholipase C-γ (reviewed in CANTLEY et al. 1991). Moreover, this effect is not mediated either by PKC activation or mobilization of Ca^{2+} from intracellular stores (ZACHARY et al. 1991).

Studies from other laboratories have also demonstrated that neuropeptides can stimulate tyrosine phosphorylation in intact liver epithelial (HUCKLE et al. 1990) and renal mesangial cells (FORCE et al. 1991). Similar to our previous findings, bombesin and bradykinin have also been shown to stimulate tyrosine phosphorylation of a group of 120-kDa proteins in Swiss 3T3 cells (LEEB-LUNDBERG and SONG 1991). Collectively these results strongly support the conclusion that neuropeptides acting through receptors linked to G-proteins can increase tyrosine phosphorylation of protein substrates in intact cells. It remains unclear, however, whether the effect of neuropeptides on tyrosine phosphorylation is due to activation of a cellular tyrosine kinase. The further elucidation of this intriguing and potentially exciting finding will require the measurement of neuropeptide-stimulated tyrosine kinase activity in a cell-free system. Nevertheless, given the importance of tyrosine phosphorylation in the action of growth factors and nonreceptor oncogenes, it is likely that this novel event plays a role in neuropeptide mitogenic signaling.

V. Arachidonic Acid Release and Prostaglandin Synthesis: Differential Effects of Bombesin and Vasopressin

While bombesin and structurally related mammalian peptides stimulate DNA synthesis in the absence of other factors, vasopressin is mitogenic for Swiss 3T3 cells only in synergistic combination with insulin (ZACHARY et al. 1987). Despite their differential ability to stimulate DNA synthesis, bombesin and vasopressin appear to trigger a set of similar early signaling events. Binding of vasopressin to its distinct receptor on quiescent cultures of Swiss 3T3 cells causes a rapid production of Ins(1,4,5)P_3, mobilization of Ca^{2+} from intracellular stores, and sustained activation of PKC via a G-protein-linked transduction pathway (reviewed in ZACHARY et al. 1987; ROZENGURT 1991). In addition, vasopressin potentiates cAMP accumulation via PKC, stimulates monovalent ionic fluxes and induces EGF receptor transmodulation (ZACHARY et al. 1987). Since the initiation of DNA synthesis is triggered by independent signal-transduction pathways that act synergistically in mitogenic stimulation (ROZENGURT 1991), the ability of bombesin to act as a sole mitogen could be due to activation of an additional but distinct signaling pathway not stimulated by vasopressin.

Recently, MILLAR and ROZENGURT (1990a,b) demonstrated that bombesin, but not vasopressin, induces a marked release of arachidonic acid and its cyclooxygenase metabolite PGE_2 into the medium. These results showed a clear difference in the pattern of early signals induced by the neuropeptides bombesin and vasopressin in Swiss 3T3 cells. If the stimulation of arachidonic acid release by bombesin constitutes a synergistic signal that contributes to bombesin-induced mitogenesis, externally applied arachidonic acid should potentiate mitogenesis induced by agents that stimulate polyphosphoinositide breakdown but not arachidonic acid release, e.g., vasopressin. In line with this prediction, arachidonic acid has been shown to act synergistically with vasopressin to stimulate DNA synthesis (MILLAR and ROZENGURT 1990a).

Arachidonic acid released by bombesin is converted into E-type prostaglandins. We have previously shown that bombesin enhances cAMP accumulation via a cyclooxygenase-dependent pathway (MILLAR and ROZENGURT 1988). Since elevated cAMP levels constitute a mitogenic signal for Swiss 3T3 cells (ROZENGURT 1986, 1991), at least one consequence of arachidonic acid release may be the modulation of intracellular cAMP levels. However, other arachidonic acid metabolites may also play a role in mitogenic signal transduction by bombesin. It is noteworthy that the only other growth factors known to induce a large and sustained release of arachidonic acid in Swiss 3T3 cells are the PDGF AA and PDGF BB homodimers, which, like bombesin, stimulate DNA synthesis in the absence of other synergistic agents (ROZENGURT 1991). We conclude that the liberation of arachidonic acid is an early signal that contributes to bombesin-mediated mitogenesis.

VI. Bombesin Induction of the Proto-Oncogenes c-*fos* and c-*myc*

In addition to the events in the membrane and cytosol described above, bombesin rapidly and transiently induces the expression of the cellular oncogenes c-*fos* and c-*myc* in quiescent fibroblasts (reviewed in ROZENGURT and SINNETT-SMITH 1988). Since these cellular oncogenes encode nuclear proteins it is plausible that their transient expression may play a role in the transduction of the mitogenic signal in the nucleus (CURRAN 1988). The demonstration that the product of the protooncogene c-*jun*, identified as a major component of the *trans*-acting factor AP-1, forms a tight complex with *fos* protein is consistent with a role for c-*fos* in the regulation of gene transcription (CURRAN 1988; ROZENGURT and SINNETT-SMITH 1988).

There has been considerable interest in elucidating the signal transduction pathways involved in c-*fos* induction. There is increasing evidence implicating PKC activation in the sequence of events linking receptor occupancy and protooncogene induction (reviewed in ROZENGURT and SINNETT-SMITH 1988). Accordingly, bombesin-induced oncogene expression is markedly reduced by downregulation of PKC. Interestingly, it has recently been shown that PKC activation causes a decrease in phosphorylation and a concomitant increase in the DNA-binding activity of c-*jun* (BOYLE et al. 1991). However, neither direct activation of PKC by phorbol esters nor addition of vasopressin evoke a maximal increase in c-*fos* mRNA levels. It is likely that the induction of c-*fos* by bombesin is mediated by the coordinated effects of PKC activation, Ca^{2+} mobilization, and an additional pathway dependent on arachidonic acid release (ROZENGURT and SINNETT-SMITH 1988; ROZENGURT 1991).

VII. Regulation of Cellular Responsiveness to Bombesin-Stimulated Mitogenesis

Exposure of cells to many peptide hormones or neurotransmitters decreases the subsequent response of target cells to further challenge with the same ligand (homologous desensitization) or with a structurally unrelated ligand which elicits reponses through a separate receptor (heterologous desensitization). Desensitization has been well documented for hormones that elicit short-term metabolic responses such as those mediated through adenylate cyclase-coupled receptors. However, little is known about desensitization of long-term responses such as cellular growth and differentiation. Since Swiss 3T3 cells have been extensively used to analyze the mechanisms of mitogenic stimulation by several mitogenic neuropeptides, these cells provide an ideal model system for investigating the role of cellular desensitization in the control of cell proliferation.

Recent work from this laboratory has demonstrated that the mitogenic response induced by bombesin in Swiss 3T3 cells is sensitive to at least two distinct desensitization mechanisms (MILLAR and ROZENGURT 1989, 1990a,b). Prolonged (40 h) treatment of the cells with bombesin or structurally related peptides causes homologous desensitization of both mitogenic stimulation and generation of early signals by progressive downregulation of cell surface receptors (MILLAR and ROZENGURT 1990b). In contrast, prolonged pretreatment with vasopressin induces a selective and reversible heterologous desensitization of the mitogenic activity of bombesin and GRP (MILLAR and ROZENGURT 1989). The block to bombesin-stimulated mitogenesis occurs at a postreceptor locus and may involve an uncoupling of ligand-bound bombesin receptor from the generation of its early signals.

The heterologous mitogenic desensitization to bombesin caused by vasopressin pretreatment was not due to functional uncoupling of phospholipase C from the bombesin/GRP receptor: bombesin stimulation of inositol phosphate accumulation, Ca^{2+} mobilization, DAG formation, and EGF receptor transmodulation was unimpaired in vasopressin-pretreated cells (MILLAR and ROZENGURT 1990a). In contrast, prolonged exposure to vasopressin completely blocked bombesin-induced release of arachidonic acid and its cyclooxygenase metabolite PGE_2 into the medium (MILLAR and ROZENGURT 1990a). These results demonstrate that the liberation of arachidonic acid represents a novel target for heterologous desensitization and provide further support to the proposition that this early event plays an important role in mitogenic signaling by bombesin. The existence of homologous and heterologous mitogenic desensitization suggests that the control of cell proliferation by neuropeptides may result from a delicate interplay between growth-stimulatory and growth-inhibitory signals.

D. Evidence for Growth-Promoting Effects of Neuropeptides In Vivo

The mitogenic effects of neuropeptides in cultured cells assume a wider relevance in view of evidence suggesting that these molecules are involved in the regulation of cell growth in vivo. Indeed, neuropeptide modulation of cellular proliferation may be important for several biological processes in both the normal and pathological states, including regulation of the immune

Table 2. Evidence for putative trophic effects of neuropeptides in vivo

Neuropeptide	Species	Tissue	Observed effect
Bombesin/GRP	Rat	Pancreas and gastrointestinal tract	Hyperplasia
	Mouse	SCLC	Antibodies inhibit growth
CCK	Rat	Pancreas	Increase in DNA, RNA, and protein content
			Incorporation of [^{3}H] thymidine
Gastrin	Rat and man	Duodenal and gastric mucosa	Cell proliferation
	Rat	Gastric mucosa	Reverses atrophy induced by antrectomy
	Human	Gastric and colonic carcinoma	Increase in tumor growth
	Mouse	Colonic carcinoma	Increase in tumor growth
SP	Rat	Skin and cornea	Capsaicin depletes SP and causes epithelial lesions
	Rat	Skin	Wounding depletes SP
Vasopressin	Rat	Bone marrow	Increase in mitotic index
	Rat	Liver	Stimulation of regeneration
β-Endorphin	Newt	–	Amphibian limb regeneration

system, tissue regeneration, development, and tumorigenesis (summarized in Table 2).

I. Trophic Effects of Gastrointestinal Peptides in the Gut

Gastrin is a 17 or 34 amino acid peptide localized primarily in the G cells of the gastric antral mucosa, and in vivo plays a major role in the regulation of gastric acid secretion (Walsh 1987). There is considerable indirect evidence that gastrin may also be an important regulator of normal and abnormal gastric mucosal growth in vivo (Table 2). In spite of the potential significance of its trophic effects in whole animals, further progress is hampered by the lack of an appropriate model cell system in which to perform more detailed studies of the proliferative response to this peptide.

Cholecystokinin is found in the mucosa of the small intestine as well as neurons associated with both the intestine and pancreas (Walsh 1987). CCK has been shown to produce striking trophic effects in the pancreas, as judged by increases in DNA, RNA, and protein content following repeated injections in whole animals or by incorporation of [^{3}H]thymidine in pancreatic acini (Table 2). Interestingly, one of the most potent effects of bombesin in vivo is release of gastrin and CCK. Injection of bombesin in rats also causes pancreatic hyperplasia. Thus, peptidergic regulation of cell growth in the gastrointestinal tract may involve complex interactions involving multiple regulatory peptides.

II. Development, Tissue Repair, and Tumorigenesis

There is some evidence which suggests that bombesin-like peptides play a role in lung development (reviewed in ZACHARY et al. 1987). Bombesin-like peptides are present in high levels in fetal and neonatal lung, but significant amounts of the peptide are absent from the adult respiratory tract. The changes which occur in pulmonary bombesin levels during gestation reflect alterations in the number of pulmonary neuroendocrine cells. Neonates suffering from the acute respiratory distress syndrome have immature respiratory tracts and in their lungs bombesin-like immunoreactivity was markedly reduced. The finding that bombesin and GRP are mitogenic for normal bronchial epithelial cells is consistent with this evidence. Whether this association between lung development and levels of bombesin-like peptides is due to a direct trophic or maturation effect of this family of peptides requires much further experimental investigation.

Neuropeptides have also been implicated in amphibian limb regeneration. This process results from the proliferation of a population of undifferentiated progenitor cells at the site of amputation, called the blastema, and is dependent upon innervation of the limb stump. Treatment of blastema explants with low (nanomolar) concentrations of SP caused an increase in the mitotic index of these cells which is suppressed by SP antiserum. SP and other neuropeptides are present in afferent sensory nerves which innervate the basal layers of the epidermis and the surrounding connective tissue of the skin. It is an intriguing and attractive idea that locally released peptide factors acting in a paracrine fashion may contribute to the normal regulation of keratinocyte and connective tissues cell growth and/or the proliferative response of these cells to wounding. A depletion of SP, somatostatin, and calcitonin-gene related peptide has been observed in rat skin in the region of experimentally induced wounds. It is unknown, however, whether this represents increased release of the peptides from nerve terminals or simply loss of the peptidergic nerve endings.

The dramatic proliferation of hepatocytes which occurs as a result of partial hepatectomy is a striking example of the ability of cells in the adult animal to divide in response to an appropriate stimulus. The neurohypophyseal peptide vasopressin, which is a potent mitogen for cultured Swiss 3T3 cells, has been reported to promote liver regeneration. This peptide has also been implicated in the development of the brain. Thus, vasopressin-deficient Brattleboro rats exhibit impaired brain development which is corrected only by prenatal treatment with vasopressin, suggesting an early role for this peptide in neurogenesis.

In addition to tumors which secrete polypeptide growth factors, a variety of tumors mainly of neural and neuroendocrine origin also produce large amounts of neuropeptides (SETHI and ROZENGURT 1991). It is known that GRP, vasopressin, CCK, and neurotensin are secreted by some SCLC tumors (reviewed in SETHI and ROZENGURT 1991; ROZENGURT 1991). Other peptides may be released by a variety of normal cells in the lung or, like bradykinin, produced extracellularly as a result of the proteolytic cleavage of plasma precursors in the damaged tissue surrounding tumors. Recent work from this laboratory has demonstrated that multiple neuropeptides can stimulate both early biochemical responses and clonal growth in SCLC cell lines. At optimal concentrations, bradykinin, neurotensin, vasopressin, CCK, galanin, and GRP induce comparable increases of SCLC clonal growth in responsive cell lines (SETHI and ROZENGURT 1991). Collectively, these findings support the hypothesis that SCLC growth is sustained by an extensive network of autocrine and paracrine interactions involving multiple neuropeptides. Approaches designed to block SCLC growth must take into account this mitogenic complexity.

Gastrin has been implicated in the growth of colorectal and gastric carcinomas in vivo, as well as the growth of cell lines derived from these tumors (Table 2; WATSON et al. 1989). In the absence of unambiguous evidence for the direct growth-promoting activity of this peptide, the precise role of gastrin in the growth of these tumors remains to be determined.

E. Conclusions

Recent findings have established that small regulatory peptides can stimulate cultured cells to reinitiate DNA synthesis and cell division. This work is supported by indirect evidence in vivo which suggests that neuropeptides can modulate animal cell growth acting in an autocrine or paracrine fashion in the local microenvironment. This is a particularly attractive idea, since it provides at least one important way in which the structurally diverse neuropeptides would be able to integrate different biological systems, e.g., the nervous and endocrine systems. Studies with neuropeptides in Swiss 3T3 cells have been particularly useful in elucidating the signaling processes important for the initiation of the proliferative response (Fig. 1).

As emphasized in the present article, cell proliferation is stimulated by multiple, independent signal-transduction pathways that act in a synergistic and combinatorial manner. Although neuropeptide receptors linked to phosphoinositide breakdown and Ca^{2+} mobilization have been considered to induce similar cellular responses, a comparative study of the early signals induced by bombesin, vasopressin, and bradykinin has demonstrated that the intensity, duration, and even the occurrence of early signals can differ substantially (ROZENGURT 1991). Thus, the bombesin receptor is coupled both to phospholipase C and to arachidonic acid release possibly

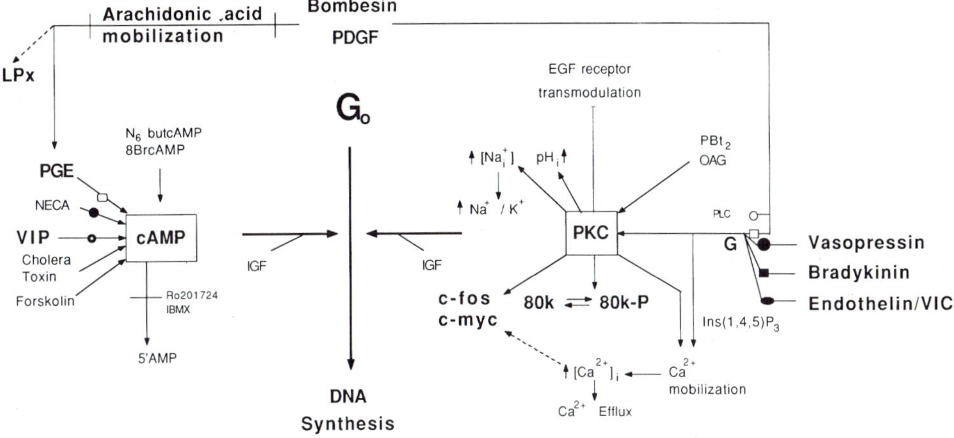

Fig. 1. Initiation of cell proliferation in 3T3 cells is stimulated by multiple signal transduction pathways that act in a synergistic and combinatorial fashion. The interactions between the PKC and cAMP pathways have been well defined in these cells and provide crucial experimental evidence for a model that invokes multiple pathways (see text). The mechanisms of action of bombesin, vasopressin, bradykinin, and endothelin are explained within the framework of this model. Abbreviations used: *LPx*, lipooxygenase pathway; *NECA*, 5′*N*-ethylcarboxamide adenosine; *PGE*, prostaglandin E_1 and E_2; *Ro 20 1724*, 4-(3-butoxy-4-methoxybenzyl)-2-imidazolidine; *IBMX*, 1-methyl-3-isobutylxanthine; *PBt*$_2$, phorbol 12,13-dibutyrate; *OAG*, 1-oleoyl-2-acetyl glycerol; *DAG*, diacylglycerol; *PLC*, phospholipase C; *G*, guanine nucleotide binding protein. All other details are explained in the text

via phospholipase A_2 (Fig. 1). In contrast, vasopressin does not induce a sustained increase in the release of arachidonic acid. Similarly, bradykinin unlike either vasopressin or bombesin causes only a transient activation of PKC as judged by transient increases in DAG formation and in 80K phosphorylation and also failed to induce EGF receptor transmodulation and to potentiate cAMP accumulation (Rozengurt 1991). Thus, the binding of bombesin, vasopressin, and bradykinin to their distinct receptors causes rapid Ins(1,4,5)P_3-mediated Ca^{2+} mobilization, but subsequently the pattern of molecular responses is strikingly different. It is likely that these differences in signaling provide an important mechanism for the "fine-tuning" of cellular regulation by neuropeptide receptors.

Recent findings are also beginning to call into question the sharp distinction generally drawn between polypeptide growth factors which act through receptor tyrosine kinases on the one hand, and G protein-transducing neuropeptides on the other. The stimulation of tyrosine kinase activity by neuropeptides is perhaps the most striking illustration that the signal transduction pathways initiated by these two classes of mitogen may be more closely related than hitherto realized (Zachary et al. 1991). It is anticipated that studies with neuropeptide growth factors will continue to bring fresh insights to our understanding of cellular growth regulation.

References

Arai H, Hori S, Aramoi I, Ohkubo H, Nakanishi S (1990) Cloning and expression of a cDNA encoding an endothelin receptor. Nature 348:730–732

Battey JF, Way JM, Corjay MH, Shapira H, Kusano K, Harkins R, Wu JM, Slattery T, Mann E, Feldman RI (1990) Molecular cloning of the bombesin/GRP receptor from Swiss 3T3 cells. Proc Natl Acad Sci USA 88:395–399

Berridge MJ, Irvine RF (1990) Inositol phosphates and cell signalling. Nature 341: 197–205

Bourne HR, Masters SB, Miller RT, Sullivan KA, Heideman W (1988) Mutations probe structure and function of G-protein alpha chains. Cold Spring Harbor Symp Quant Biol 53:221–228

Boyle WJ, Smeal T, Defize LHK, Angel P, Woodgett JR, Karin M, Hunter T (1991) Activation of protein kinase C decreases phosphorylation of c-*jun* at sites that negatively regulate its DNA binding activity. Cell 64:573–584

Brooks SF, Herget T, Erusalimsky JD, Rozengurt E (1991) Protein kinase C activation potently down-regulates the expression of its major substrate, 80K, in Swiss 3T3 cells. EMBO J 10:2497–2505

Cantley LC, Auger KR, Carpenter C, Duckworth B, Graziani A, Kapeller R, Soltoff S (1991) Oncogenes and signal transduction. Cell 64:281–302

Coffer A, Sinnett-Smith J, Rozengurt E (1990) Bombesin receptor from Swiss 3T3 cells: affinity chromatography and reconstitution into phospholipid vesicles. FEBS Lett 275:159–164

Cuadrado A, Molloy CJ (1990) Overexpression of phospholipase C-γ in NIH 3T3 fibroblasts results in increased phosphatidylinositol hydrolysis in response to platelet-derived growth factor and basic fibroblast growth factor. Mol Cell Biol 10:6069–6072

Curran T (1988) The *fos* oncogene. In: Reddy EP, Skalka AM, Curran T (eds) The oncogene handbook. Elsevier, Amsterdam, pp 307–325

Dockray GT (1987) Physiology of enteric neuropeptides. In: Johnson LR (ed) Physiology of the gastrointestinal tract, 2nd end. Raven, New York, p 41

Dumont JE, Jauniaux JC, Roger PP (1989) The cyclic AMP-mediated stimulation of cell proliferation. Trends Biochem Sci 14:67–71

Erusalimsky JD, Friedberg I, Rozengurt E (1988) Bombesin, diacylglycerols and phorbol esters rapidly stimulate the phosphorylation of an Mr = 80,000 protein kinase C substrate in permeabilized 3T3 cells: effect of guanine nucleotides. J Biol Chem 263:19188–19194

Evered D, Nugent J, Whelen J (eds) (1985) Growth factors in biology and medicine. Ciba Foundation Symp 116

Feldman RI, Wu JM, Jenson JC, Mann E (1990) Purification and characterization of the bombesin/gastrin-releasing peptide receptor from Swiss 3T3 cells. J Biol Chem 265:17364–17372

Force T, Kyriakis JM, Avruch J, Bonventre JV (1991) Endothelin, vasopressin and angiotensin II enhance tyrosine phosphorylation by protein kinase C-dependent and -independent pathways in glomerular mesanglial cells. J Biol Chem 266: 6650–6656

Hershey AD, Krause JE (1990) Molecular characterization of a functional cDNA encoding the rat substance P receptor. Science 247:958–962

Huckle WR, Prokop CA, Dy RC, Herman B, Earp S (1990) Angiotensin II stimulates protein-tyrosine phosphorylation in a calcium-dependent manner, Mol Cell Biol 10:6290–6298

Kikkawa U, Nishizuka Y (1986) The role of protein kinase C in transmembrane signalling. Annu Rev Cell Biol 2:148–154

Leeb-Lundberg LMF, Song X-H (1991) Bradykinin and bombesin rapidly stimulate tyrosine phosphorylation of a 120 kDa group of proteins in Swiss 3T3 cells. J Biol Chem 266:7746–7749

Masu Y, Nakayama K, Tamaki H, Harada Y, Kuno M, Nakanishi S (1987) cDNA cloning of bovine substance K receptor through oocyte expression system. Nature 32:836–837

McEachern AE, Shelton ER, Bhakta S, Obernolte R, Bach C, Zuppan P, Fujisaki J, Aldrich RW, Jarnagin K (1991) Expression cloning of a rat B_2 bradykinin receptor. Proc Natl Acad Sci USA 88:7724–7728

Millar JBA, Rozengurt E (1988) Bombesin enhancement of cAMP accumulation in Swiss 3T3 cells: evidence of a dual mechanism of action. J Cell Physiol 137: 214–222

Millar JBA, Rozengurt E (1989) Heterologous desensitization of bombesin-induced mitogenesis by prolonged exposure to vasopression: a post-receptor signal transduction block. Proc Natl Acad Sci USA 86:3204–3208

Millar JBA, Rozengurt E (1990a) Arachidonic acid release by bombesin: a novel post-receptor target for heterologous mitogenic desensitization. J Biol Chem 265:19973–19979

Millar JBA, Rozengurt E (1990b) Chronic desensitization to bombesin by progressive down-regulation of bombesin receptors in Swiss 3T3 cells: distinction from acute desensitization. J Biol Chem 265:12052–12058

Murphy TJ, Alexander RW, Griendling KK, Runge MS, Bernstein KE (1991) Isolation of a cDNA encoding the vascular type-1 angiotensin II receptor. Nature 351:233–236

Nishizuka Y (1988) The molecular heterogene of protein kinase C and its implications for cellular recognition. Nature 334:661–665

Rozengurt E (1985) The mitogenic response of cultured 3T3 cells: integration of early signals and synergistic effects in a unified framework. In: Cohen P, Houslay M (eds) Molecular mechanisms of transmembrane signalling. Elsevier, Amsterdam, pp 429–452

Rozengurt E (1986) Early signals in the mitogenic response. Science 234:161–166

Rozengurt E (1991) Neuropeptides as cellular growth factors: role of multiple signalling pathways. Eur J Clin Invest 21:123–134

Rozengurt E, Mendoza SA (1986) Early stimulation of Na^+/H^+ antiport, Na^+/K^+ pump activity and Ca^{2+} fluxes in fibroblast mitogenesis. Curr Top Membr Transport 27:163–191

Rozengurt E, Sinnett-Smith J (1988) Early signals underlying the induction of the c-*fos* and c-*myc* genes in quiescent fibroblasts: studies with bombesin and other growth factors. Prog Nucleic Acid Res Mol Biol 35:261–295

Rozengurt E, Fabregat I, Coffer, A, Gil J, Sinnett-Smith J (1990) Mitogenic signalling through the bombesin receptor: role of a guanine nucleotide regulatory protein. J Cell Sci Suppl 13:43–56

Sakurai T, Yanagisawa M, Takuwa Y, Miyazaki H, Kimura S, Goto K, Masaki T (1990) Cloning of a cDNA encoding a non-isopeptide-selective subtype of the endothelin receptor. Nature 348:732–735

Sasaki K, Yamano Y, Bardhan S, Imai N, Murray JJ, Hasegawa M, Matsuda Y, Inagami T (1991) Cloning and expression of a complementary DNA encoding a bovine adrenal angiotensin II type-1 receptor. Nature 351:230–233

Sethi T, Rozengurt E (1991) Multiple neuropeptides stimulate clonal growth of small cell lung cancer: effects of bradykinin, vasopressin, cholecystokinin, galanin and neurotensin. Cancer Res 51:3621–3623

Spindel ER, Giladi E, Brehm P, Goodman RH, Segerson TP (1990) Cloning and functional characterization of a complementary DNA encoding the murine fibroblast bombesin/GRP receptor. Mol Endocrinol 4:1956–1963

Sreedharan SP, Robichon A, Peterson KE, Goetzl EJ (1991) Cloning and expression of the human vasoactive intestinal peptide receptor. Proc Natl Acad Sci USA 88:4986–4990

Staddon JM, Chanter N, Lax AJ, Higgins TE, Rozengurt E (1990) *Pasteurella multocida* toxin, a potent mitogen, stimulates protein kinase C-dependent and

-independent protein phosphorylation in Swiss 3T3 cells. J Biol Chem 265: 11841–11848

Strosberg AD (1991) Structure/function relationships of proteins belonging to the family of receptors coupled to GTP-binding proteins. Eur J Biochem 196:1–10

Ullrich A, Schlessinger J (1990) Signal transduction by receptors with tyrosine kinase activity. Cell 61:203–212

Walsh JH (1987) Gastrointestinal hormones. In: Johnson LR (ed) Physiology of the gastrointestinal tract, 2nd edn. Raven, New York, p 181

Watson S, Durrant L, Morris D (1989) Gastrin: growth enhancing effects on human gastric and colonic tumour cells. Br J Cancer 59:554–558

Woll PJ, Rozengurt E (1989) Neuropeptides as growth regulators. Br Med Bull 45:492–505

Woll PJ, Sethi T, Rozengurt E (1992) Neuropeptide growth factors and antagonists. In: Hickman J, Tritton T (eds) Frontiers in pharmacology and therapeutics. Blackwell, Oxford (in press)

Yokota Y, Sasai Y, Tanaka K, Fujiwara T, Tsuchida K, Shigemoto R, Kazizuka A, Ohkubo H, Nakanishi S (1989) Molecular characterization of a functional cDNA for rat substance P receptor. J Biol Chem 264:17649–17652

Zachary I, Rozengurt E (1985) Modulation of the epidermal growth factor receptor by mitogenic ligands: effects of bombesin and role of protein kinase C. Cancer Surv 4:729–765

Zachary I, Woll P, Rozengurt E (1987) A role for neuropeptides in the control of cell proliferation. Dev Biol 124:295–308

Zachary I, Gil J, Lehmann W, Sinnett-Smith J, Rozengurt E (1991) Bombesin, vasopressin and endothelin rapidly stimulate tyrosine phosphorylation in intact Swiss 3T3 cells. Proc Natl Acad Sci USA 88:4577–4581

CHAPTER 13

Peptidergic Regulation of Mucosal Immune Function

FENG CHEN and M.S. O'DORISIO

A. Introduction

The mucosa of the gastrointestinal (GI) system is the body's main defense against external antigens such as bacteria, viruses and parasites which enter via oral-fecal routes as well as against antigens in food. The mucosal immune system also defends against abnormal internal antigens such as developing gastrointestinal malignancies. The GI tract is richly supplied with peptidergic neurons, endocrine cells and immune effector cells concentrated in gut-associated lymphoid tissue (GALT); GALT comprises one of the largest immune tissues in the body and accounts for nearly 40% of all immune effector cells. Thus GALT is endowed with the necessary components for a neuroendocrine-immune axis.

Recent evidence suggests that mucosal immune responses to both external and internal antigens are subject to neuroendocrine regulation. An expanding family of peptides and their receptors have been identified on neurons, endocrine cells and immune cells in GALT. Interactions among neuronal, endocrine and immune cells were first recognized in phenomenological studies of the effects of brain lesions on antibody production (KORNEAV and KHAI 1964). More recently, investigators have recognized that gut peptides may play a critical role in the neuroendocrine-immune network (O'DORISIO and PANERAI 1990). The question of whether these peptides modulate immune function in vivo has become the subject of intense investigation. This review will focus on gut hormone regulation of mucosal immune function and will begin with a brief overview of mucosal immune responses.

B. Overview of Immune Function in Gut-Associated Lymphoid Tissue

The GALT includes spleen, Peyer's patches, appendix, mesenteric lymph nodes, lamina propria lymphoid nodules, and intraepithelial lymphocytes. Just as bronchus-associated lymphatic tissue is the first line of defense against antigens which enter via the respiratory tract, GALT is the first line of defense against pathogens transmitted by the oral-fecal route. Immune effector cells in GALT express both humoral and cellular immunity, which are modulated by gut hormones (FREIER 1990). Abnormal GALT function may lead to food allergies, infectious disease, inflammatory bowel disease, or even neuroendocrine tumors unique to the gut (O'DORISIO and O'DORISIO 1986).

Peyer's patches are clusters of lymphoid nodules in the intestine. They consist of dense lymphatic tissue with a large lymphopoietic reaction center surrounded by a layer of dense reticular fibers. They differ from lymph nodes in that they are predominantly selective for the production of certain classes of immunoglobulin, namely, immunoglobulin A (IgA) and IgM. The intracellular space in Peyer's patches is innervated by nonmyelinated neurons. Axons and synaptic-like contacts are associated with the reticular cells, while the lymphocytes have close and restricted access to both nerve terminals and synaptic vesicles (FREIER 1990).

The appendix is composed of serous, muscular, submucosal, and mucosal layers. An abundance of lymphoid tissue that defends the gut against the invasion of pathogens is found within the submucosa. B and T lymphocytes are segregated into different regions of the lymphoid areas of the appendix. An extensive neuronal network traverses T-cell-dependent lymphoid areas and also extends between lymph nodules (FREIER 1990), suggesting that neurotrophic factors participate in localized immune functions in gut.

The lamina propria, a lining structure of the GI tract, includes the muscularis mucosa, the vasculature and connective tissue. Antigen-presenting cells, B and T lymphocytes and plasma cells are the principal immune effector cells in lamina propria. One of the most striking features of the lamina propria is that secretory IgA produced in this area (sIgA) is in the dimeric form and is secreted into the lumen of the GI tract, while IgA in plasma is in a monomeric form (HANSON and BRANDTZAEG 1989).

Gut-associated lymphoid tissue contains a large contingent of antigen-specific immune effector cells including antigen-presenting cells (APC), T cells, B cells, and plasma cells. Antigen-independent immune effector cells including natural killer (NK) cells, neutrophils, eosinophils, basophils, and mast cells are also abundant in GALT.

Antigen recognition is important in both cellular and humoral immune functions. APC present in GALT include macrophages, the dendritic cells of spleen, specialized intestinal epithelial cells called M cells which overlie the domes of Peyer's patches and villous enterocytes (HANSON and BRANDTZAEG 1989). Villous enterocytes of the small intestine express MHC class II molecules and MHC-like CD1 molecules (MAYER and SHLIEN 1987; BLAND and WARREN 1986; BLUMBERG et al. 1991). T-cell-dependent antigens, such as large globular proteins of microorganisms, undergo endocytosis, unfolding and proteinase degradation to peptide fragments in macrophages (UNANUE and ALLEN 1987). Enterocytes do not appear to process protein antigens to small peptides as macrophages do. Nevertheless, they are capable of presenting antigen to T lymphocytes (BLAND and WARREN 1986).

When T cells encounter antigen-bearing APCs, antigen-reactive T cells bind to APCs via the T-cell receptor (TcR) (ASHWELL and KLAUSNER 1990). T-cell recognition of antigen on enterocytes or other intestinal APCs induces proliferation and differentiation of antigen-specific cells. T cells enter the circulation in large numbers and migrate to distinct sites of the body, including mucosal surfaces and sites of inflammation.

The TcR is composed of two subunits, $\alpha\beta$ or $\delta\gamma$, linked by an interchain disulfide bridge near the transmembrane region (O'DORISIO and PANERAI 1990). The $\delta\gamma$ TcR is more highly expressed in the immune cells in mucosa, where it appears to play a role in bacterial recognition (GUY-GRAND et al. 1991), while T cells in peripheral blood express predominantly $\alpha\beta$ TcR. Accessory adhesion molecules, such as CD4, CD8, and CD2, help to stabilize binding of the TcR to antigen and MHC on APC, and may be involved in the activation signal (LITTMAN 1987). CD8 is expressed on most cytotoxic/suppressor T cells ($T_{c/s}$). $T_{c/s}$ lymphocytes directly lyse or destroy antigen-bearing target cells through direct cell contact. $CD8^+$ cells also release a variety of lymphokines which control the differentiation of hematopoietic cells. Some $CD8^+$ cells appear to down-regulate the number of Thelper (T_h) cells available to regulate B-cell activation and proliferation, thereby regulating immunoglobulin production. $CD4^+$ T_h cells provide positive signals which induce B cells to proliferate and differentiate into antibody-secreting plasma cells.

The activation of B cells is assisted by activated T_h cells, which release lymphokines (Table 1) to regulate B-cell proliferation and differentiation. T_h1 cells synthesize interleukin (IL)-2, IL-3, IL-9, interferon (IFN)-γ, granulocyte macrophage colony-stimulating factor (GM-CSF), and lymphotoxin whereas T_h2 cells synthesize IL-3, IL-4, IL-5, IL-6, IL-10, and GM-CSF (MOSMANN and COFFMAN 1989). Activation of B cells can be divided into inductive, proliferative, and effector stages. The inductive stage covers all events between antigenic entry and T_h contact with B-cell antigen receptors (sIg); the proliferative stage begins with antigen binding to Ig on the surface of antigen-specific B cells, followed by direct T_h-B cell contact via either surface MHC or surface Ig molecules; the effector stage is characterized by synthesis and secretion of Ig from immunoglobulin-secreting plasma cells. Recent evidence suggests that selective release of lymphokines by T_h1 and T_h2 lymphocytes may be able to modulate both the amount and subtype of Ig produced by plasma cells in this stage. The net result is that primed B cells can proliferate into long-lived memory cells, undergo clonal expansion to daughter antigen-sensitive cells, or differentiate into immunoglobulin-secreting plasma cells. Long-lived memory B cells will become reactivated upon reexposure to antigen, thus giving rise to the secondary Ig response. B cells are also activated directly by soluble antigen

Table 1. Cellular source and function of interleukins

Interleukin	Source	Function	References
IL-1	Monocytes, macrophages, T cells, fibroblasts, endothelial cells, epithelial cells, Langerhan's cells	T-cell activation, NK-cell function, B/T-cell differentiation, synthesis of IL-2, IL-6, synthesis of NGF, synthesis of substance P, synthesis of somatostatin	O'DORISIO and PANERAI (1990) O'DORISIO and PANERAI (1990) O'DORISIO and PANERAI (1990) LINDHOLM et al. (1987) JONAKAIT and SCHOTLAND (1990)
IL-2	Resting B cells, T_h1 lymphocytes, Neurite outgrowth	B/T-cell proliferation, NK-cell activation	O'DORISIO and PANERAI (1990) HAUGEN et al. (1990)
IL-3	Bone marrow stroma, pre-B, leukemia, T_h1 lymphocytes, T_h2 lymphocytes	Hematopoietic growth, macrophage proliferation, mast cell proliferation	O'DORISIO and PANERAI (1990)
IL-4	Bone marrow stroma, activated B cells, T_h2 lymphocytes	MHC class II expression, B-cell proliferation, IgG_1, and IgE production, Fc receptor expression, mast-cell proliferation	O'DORISIO and PANERAI (1990)
IL-5	Activated B cells, T_h2 lymphocytes	B-cell differentiation, IgM, IgG_1 and IgA production; $T_{c/s}$ differentiation; eosinophil growth	O'DORISIO and PANERAI (1990)
IL-6	B cells, T_h2 lymphocytes, macrophages, monocytes, mast cells, fibroblast, epithelial cells, endothelial cells	B-cell differentiation, Ig production, $T_{c/s}$ proliferation, fibrinogen secretion, hepatocyte growth	O'DORISIO and PANERAI (1990)
IL-7	Bone marrow stroma	Mature B-cell proliferation, mature T-cell proliferation	O'DORISIO and PANERAI (1990)
IL-8	Lymphocytes, macrophages, epithelial cells, fibroblasts	Neutrophil chemotaxis, T-cell chemotaxis, neutrophil activation	O'DORISIO and PANERAI (1990)
IL-9	T_h1 lymphocytes	T-cell growth, hematopoiesis, mast cell growth and activation	KELLEHER et al. (1991)
IL-10	B cell lymphoma B cells, T_h2 lymphocytes	B-cell activation, T-cell growth, mast cell activation	MACNEIL et al. (1990)
IL-11	Bone marrow stroma	Hematopoiesis and lymphopoiesis, B-cell activation	MUSASHI et al. (1991)

binding to B-cell surface Ig molecules. This noncognate response is independent of APC or T_h regulation (SAXON and STIEHM 1989).

In GALT, immunoglobulin-secreting plasma cells synthesize IgM, IgG, IgA, IgE, and IgD (FREIER 1990). However, sIgA is the predominant antibody in intestinal secretions and absence of sIgA in mucosal tissues may impair survival. sIgA exists in polymeric form coupled to J chain and secretory component (SC). IgA and J chain are produced by plasma cells in the interstitial space of the mucosal lamina propria (MESTECKY and MCGHEE 1987); SC is synthesized by specialized intestinal epithelial cells. The production of sIgA is one of the most important functions of the mucosal immune system; neuropeptide regulation of IgA synthesis occurs at all levels of sIgA production, including antigen presentation, T_h and $T_{c/s}$ function, B-cell synthesis of immunoglobulin, and synthesis of SC.

This review will focus on immune regulation by most of the key gut regulatory peptides including vasoactive intestinal peptide, somatostatin, substance P, opioid peptides, neuropeptide Y, calcitonin gene-related peptide, and cholecystokinin. Each of these peptides has been demonstrated to modulate some aspect of mucosal immune function.

C. Peptide Effects on Immune Function

I. Vasoactive Intestinal Peptide

Vasoactive intestinal peptide (VIP), a 28-amino-acid peptide, is a member of the secretin-glucagon family. VIP was originally described as a vasodilatory substance in rabbit lung and subsequently isolated from porcine intestine by SAID and MUTT (1972). The peptide was later localized to central and peripheral nervous tissue and described as a neurotransmitter because of its location (HÖKFELT et al. 1982). In mammalian intestine, a high density of VIP-containing nerves has been identified as ending in close proximity to neuronal, endocrine, and immune cells, in addition to the well-described endings juxtaposed to epithelial, vascular, and smooth muscle cells of the gut. VIP has also been located in the submucosal plexus of the intestine (O'DORISIO 1990). Each of these cell types has been demonstrated to recognize and respond functionally to VIP. The clinical significance of VIP was first recognized when investigators observed that VIP-secreting tumors caused watery diarrhea (SAID and FALOONA 1975). The peptide is now known to be a major regulator of water and electrolyte secretion in the gut. VIP regulates exocrine secretion in pancreatic acinar cells and secretion of prolactin from pituitary (YAJIMA et al. 1986). The role of VIP in immunomodulation is less clear and is an area of intense investigation.

Functional VIP receptors were first identified on human peripheral blood lymphocytes by demonstration of VIP-mediated activation of adenylate cyclase in lymphocyte membranes (O'DORISIO et al. 1981). GUERRERO et al.

(1981) and DANEK et al. (1983) subsequently identified high-affinity binding sites for VIP on human peripheral blood lymphocytes by classical radioligand binding techniques. This finding has since been confirmed by several investigators using murine lymphocytes (KRCO et al. 1986; OTTAWAY and GREENBERG 1984). The binding kinetics are remarkably similar in all of these studies. Affinity constants for the high-affinity binding site are in the nanomolar range for both human and murine lymphocytes (Table 2). The nature of the low-affinity site observed by GUERRERO et al. (1981) remains to be explained. Monocytes do exhibit chemotaxis to VIP (SACERDOTE et al. 1987) and appear to bind VIP; however, VIP does not activate adenylate cyclase in monocytes (O'DORISIO et al. 1981). Thus, the low-affinity binding sites observed in peripheral blood mononuclear cells are conceivably monocyte chemotactic receptors which are not linked to adenylate cyclase. The estimated number of high-affinity VIP binding sites per cell varies from 1500 in unseparated mononuclear cells to 5000 in enriched T-cell populations (KRCO et al. 1986; OTTAWAY and GREENBERG 1984). This is fewer binding sites than have been observed in intestinal epithelial cells, colon carcinoma cells, or pancreatic acinar cells. The human lymphoblastic cell line, Molt 4b, has an estimated 15 000 binding sites per cell, which suggests the possibility that only selected lymphocyte subpopulations express the VIP receptor (O'DORISIO and PANERAI 1990).

Using bifunctional reagents to covalently cross-link radiolabeled VIP to intact cells or to a plasma membrane fraction, the VIP receptor on Molt 4b lymphoblasts has been characterized as a 47-kilodalton (kDa) protein without disulfide bonds (WOOD and O'DORISIO 1985). Using similar techniques, a receptor protein of the same molecular weight has been identified in lymphoblasts, frontal cortex, pancreas, and pituitary GH_3 cells (O'DORISIO and PANERAI 1990), as well as in lung, colon carcinoma cells, and intestinal epithelial cells (BATTARI et al. 1987; LABURTHE et al. 1984; PAUL and SAID 1987). In each of these tissues a higher molecular weight form (67–87 kDa) of the receptor is observed which can be converted to 47 kDa by treatment with glycohydrolase, suggesting that the ligand-binding component is a 47-kDa protein and that various amounts of carbohydrate account for the molecular weight differences in these tissues. The molecular weight of VIP receptor has recently been confirmed by SREEDHARAN et al. (1991), who have cloned the VIP receptor from a transformed human B-cell line and a human colon carcinoma cell line. The VIP receptor in these cells is a 362 amino acid protein (42 kDa) with 7 transmembrane domains and the corresponding extracellular and cytoplasmic domains which characterize the G-protein family of receptors including several possible sites for attachment of carbohydrate on the extracellular NH_2-terminal domain and several phosphorylation sites on the cytoplasmic domains.

The VIP receptor on lymphoblasts is linked to the adenylate cyclase catalytic unit via interaction with a stimulatory guanine nucleotide binding protein, G_s (CHRISTOPHE et al. 1990). The ability of VIP to stimulate

Table 2. Vasoactive intestinal peptide, somatostatin, and substance P receptors on immune cells

Peptide		Cell type		K_D (nM)	References
VIP	Human	PBMC		0.24	OTTAWAY et al. (1983)
		Monocytes		0.25	WIIK et al. (1985)
		T cells		0.47	DANEK et al. (1983)
		T-cell lines:	Jurkat	5.2	FINCH et al. (1989)
			Molt4b	7.3	BEED et al. (1983)
			SUP-TI	15.0	ROBBERECHT et al. (1988)
		B-cell lines:	U266	7.6	O'DORISIO et al. (1989)
			Nalm-6	12.6	O'DORISIO et al. (1989)
			Dakiki	9.1	FINCH et al. (1989)
	Murine	MLN T cell		0.26	OTTAWAY (1984)
		GALT T cell		1.9–2.4	OTTAWAY et al. (1984)
SS	Human	LPMC		2.1	FAIS et al. (1991)
		PBMC		910.0	FAIS et al. (1991)
		Monocytes		500.0	PAYAN and GOETZL (1985)
		Lymphocytes		500.0	PAYAN and GOETZL (1985)
		T-cell lines:	Molt4b	0.22	NAKAMURA et al. (1987)
			MT-2	0.64	NAKAMURA et al. (1987)
			IM-9	16.0	STEAD et al. (1987a)
			Jurkat	3.0; 66.0	SREEDHARAN et al. (1989)
		B-cell lines:	Isk	1.1	NAKAMURA et al. (1987)
			U266	0.005, 100	SREEDHARAN et al. (1989)
	Murine	GALT T and B cells		CB	SCICCHITANO et al. (1987)
		MOPC-315		1.6	SCICCHITANO et al. (1988)
SP	Human	T cells		180.0	PAYAN and GOETZL (1985)
		B cells		CB	PAYAN et al. (1984b)
		Monocytes		CB	PAYAN et al. (1984b)
		Neutrophils		CB	PAYAN et al. (1984b)
		Platelet		CB	PAYAN et al. (1984b)
		T-cell lines:	Jurkat	CB	PAYAN et al. (1984b)
			Molt4b	CB	PAYAN et al. (1984b)
			Hut78	CB	PAYAN et al. (1984b)
			IM9	0.65	PAYAN et al. (1984a)
		Monocyte lines:	HL-60	CB	PAYAN et al. (1984a)
			U937	CB	PAYAN et al. (1984a)
	Murine	GALT	T cells	0.75	STANISZ et al. (1987)
			B cells	0.90	STANISZ et al. (1987)
	Guinea pig	Macrophage		19.0	HARTUNG et al. (1986)

CB, cytofluorimetric binding; PBMC, peripheral blood mononuclear cell; LPMC, lamina propria mononuclear cell; MLN, mesenteric lymph node.

adenylate cyclase is concentration dependent in the range of 0.1 nM to 1 μM (O'DORISIO et al. 1981). The effects of VIP and guanine nucleotide (GppNHp) are synergistic, supporting the concept that the VIP-receptor complex interacts with a G_s-adenylate cyclase complex. The effects of VIP and prostaglandin E_2 (PGE_2) are simply additive, indicating that VIP and PGE_2 act on separate and distinct receptors. VIP-or forskolin-mediated activation of lymphoblast adenylate cyclase results in phosphorylation of a 41-kDa protein. The identical protein is phosphorylated by both VIP and forskolin in pituitary and colonic cells, indicating the 41-kDa protein is a physiological substrate for adenosine-3′,5′-cyclic monophosphate (cAMP)-dependent protein kinase in multiple cell types (O'DORISIO and CAMOLITO 1989).

Immunomodulation by VIP includes inhibitory effects on lymphocyte migration and proliferation; synthesis of Ig and lymphokines; and NK cell activity (O'DORISIO and PANERAI 1990). VIP and cAMP modulate migration of T cells in intestine of mice and sheep (MOORE 1984; OTTAWAY 1984). Preincubation of lymphocytes with VIP induces a time-and dose-dependent downregulation of VIP receptors and reduces the ability of lymphocytes to migrate into mesenteric lymph nodes and Peyer's patches of recipient animals; migration to spleen was not affected (OTTAWAY and GREENBERG 1984). This is consistent with the hypothesis that VIP regulates entry of T cells into lymphoid tissues in the gastrointestinal tract. Several investigators have demonstrated VIP-mediated inhibition of lymphocyte proliferation in response to the T-cell mitogens concanavalin A (Con A) and phytohemagglutinin (PHA), but failed to show any response of lymphocytes to the B-cell mitogen lipopolysaccharide (LPS; KRCO et al. 1986; OTTAWAY and GREENBERG 1984; STANISZ et al. 1986). The poor response of B cells to VIP suggested that peripheral blood B cells might not express VIP receptors. However, STANISZ et al. (1986) demonstrated that VIP inhibited secretion of IgA in culture. The VIP inhibition of Ig synthesis and secretion may be indirect, as a result of inhibition of T_h cells or a direct effect on IgA-secreting plasma cells. Recent studies have identified high-affinity VIP receptors on IgA-secreting plasma cells and on pre-B cells (FINCH et al. 1989; O'DORISIO et al. 1989). VIP inhibits IgA synthesis but enhances IgM synthesis in lymphocytes from Peyer's patches. The differential effects of VIP on IgA versus IgM/IgG synthesis suggests that immunomodulation by VIP may be more important in mucosal immune tissues than in the peripheral circulation.

To examine the immunomodulatory effects of VIP in vivo, several parameters of peripheral blood lymphocyte function were studied in a patient with a VIP-secreting pancreatic endocrine tumor (ANNIBALE et al. 1990). The circulating lymphocyte subpopulations had a normal ratio of $T_h/T_{c/s}$ cells and proliferation of those lymphocytes was normal in response to mitogens in vitro. However, serum IgM levels and the number of IL-2 receptors expressed on these cells were significantly increased. Comparable increases in both IgM production and IL-2 receptor expression were examined in normal peripheral lymphocyte cultures with similar concentrations of VIP. These results demonstrate that a modulatory effect of VIP on lymphocyte activation and IgA synthesis may be operating in vivo.

Vasoactive intestinal peptide inhibits NK cell activity at concentrations as low as 0.1 n*M*, though preincubation with VIP could increase NK cell activity (Rola-Pleszczynski et al. 1985). Integrity of the 14–28 C-terminal portion of VIP is necessary for the activity of VIP on NK cell function. Sirinanni et al. (1990) observed that, in addition to inhibition of NK lysis, VIP inhibited microtubule polarization required for NK cell conjugation with target cells. Microtubule polarization is thought to be important in the transfer of lytic enzymes from NK cells to target cells (Bonavida and Wright 1986).

II. Somatostatin

Somatostatin (SS) is a tetradecapeptide which was originally isolated from hypothalamus, where it was found to inhibit release of growth hormone (Reichlin 1986). SS is concentrated in the central nervous system (CNS), the endocrine system, and the GI system, where it is synthesized in both neural and endocrine cells (McGuigan 1989). The intestinal sources include neurons in the lamina propria, the submucosal plexus, gut endocrine cells, and basophils (Keast et al. 1984). SS is thought to be one of several peptides released from peripheral nerve endings in quantities capable of regulating immune function.

High-affinity SS receptors have been characterized in the brain, pituitary, adrenal cortex, pancreas, and gastric parietal cells. Bhathena et al. (1981) first identified SS-binding sites on peripheral blood lymphocytes. Subsequently, a fluorescent analog of SS has been used to characterize the distribution of SS receptors on both T and B cells isolated from spleen and Peyer's patches (Scicchitano et al. 1987). In spleen, 35% of T cells and 37% of B cells appear to express SS receptors. An even larger proportion of Peyer's patch lymphocytes express the receptor: 49% and 52% for T and B cells, respectively (Scicchitano et al. 1987). SS-binding sites were also found in IM-9 lymphoblast cells (Stead et al. 1987a). Sreedharan and colleagues observed two subclasses of SS-binding sites on the Jurkat line of human leukemic T cells and on U266 IgE-producing human myeloma cells (Sreedharan et al. 1989). Specific, saturable, time- and temperature-dependent binding sites for SS have been identified on human lymphocytes from the lamina propria (Fais et al. 1991; Pallone et al. 1990), suggesting that SS regulates human immune cells in GALT in vivo.

As discussed in Chap. 5, SS inhibits agonist-induced activation of adenylate cyclase in a number of cell types, including intestinal epithelial cells (Carter et al. 1978), gastric parietal cells (Park et al. 1987), anterior pituitary cells (Spada et al. 1984), and human lymphocytes (O'Dorisio et al. 1981). The mechanism has been characterized in both pituitary GH_3 and S49 lymphoma cells (Jakobs et al. 1983). In both GH_3 and S49 cell lines, the effect of SS appears to be mediated by an inhibitory guanine nucleotide-binding protein G_i which forms a complex with adenylate cyclase and inhibits cAMP accumulation; this hypothesis is supported by the observation that SS effects can be prevented by pretreatment of cells with pertussis toxin. The inhibitory effects of SS on substance P mediated enhancement of prolactin release occur by a cAMP-independent mechanism which probably also involves SS interaction with an inhibitory G-protein (YAJIMA et al. 1986). This pertussis toxin-sensitive G protein appears to mediate somatostatin effects on a Ca^{2+} channel (KLEUSS et al. 1991).

In both the neuroendocrine system and the immune system, SS appears to be an inhibitory factor. SS inhibits hormone release from gut tissues, e.g., insulin, gastrin, secretin, motilin, enteroglucagon, and VIP (REICHLIN 1986). SS appears to have inhibitory actions on immune cells in gut tissue as well, including the inhibition of immunoglobulin synthesis and the proliferation of lymphocytes from Peyer's patches, mesenteric lymph nodes, and spleen (STANISZ et al. 1987).

Somatostatin has been shown to inhibit the release of IgA, but not IgM and IgG, by plasma cells isolated from murine spleen and Peyer's patches (STANISZ et al. 1986). SS levels are elevated to picomolar-nanomolar concentrations in tissues during immediate hypersensitivity and inflammatory reactions (PAYAN et al. 1986). This is similar to the concentration of SS required in vitro to inhibit IgE-dependent stimulation of basophils (GOETZL and PAYAN 1984), though the mechanism of the inhibition remains an open question.

Crude extracts of rat hypothalami inhibit the proliferation of mouse splenic lymphocytes in vitro (PAWLIKOWSKI et al. 1985), an effect mimicked by SS, but not by other known neuropeptides. The proliferation of T and B cells derived from murine spleen and Peyer's patches was also inhibited by SS (STANISZ et al. 1986). PAWLIKOWSKI et al. (1985) observed that low concentrations of SS inhibited mouse splenocyte proliferation in vitro, as measured by [^{3}H] thymidine incorporation, whereas the higher concentrations of SS exerted a stimulatory effect (GHOSE et al. 1978). The mechanism of SS inhibition of lymphocyte proliferation is not clear. SS appears to block both RNA and DNA synthesis (PAYAN et al. 1984c), which is similar to its inhibition of DNA synthesis in cultures of intestinal tissue. Conceivably, SS effects on DNA synthesis may be secondary to its inhibition of release of growth factors and/or lymphokines.

III. Substance P

Substance P (SP) was identified as an 11-amino-acid peptide member of the tachykinin family of peptides which share a common C-terminal sequence Phe-X-Gly-Leu-Met-NH_2. The tachykinin family includes SP, substance K (also known as neurokinin A), and neuromedin K (also known as neurokinin B) in mammals, and eledosin, kassinin, and physalaemin in non-mammalian vertebrates. SP is found in high concentrations in both mammalian brain and intestine. In CNS, SP has been identified in basal ganglia and hypothalamus; in the peripheral nervous system, it is localized in descending neurons, especially in the dorsal horn of the spinal cord, where SP is involved in sensory and nociceptive pathways with serotonin. SP was also identified in peripheral sensory neurons throughout the body, including the skin, vascular and gastrointestinal tissue, mucosa tissue and around joints (McGILLIS et al. 1990). In the GI tract, SP immunoreactivity has been detected in perivascular nerve plexuses, in myenteric and submucosal plexuses of the gut wall, and in endocrine cells of the mucosa, suggesting an important but as yet undefined role of SP in regulaton of gut function (McGUIGAN 1989). SP receptors have also been identified within the GI tract, including epithelium, lymphoid nodules, vasculature, and smooth muscle (GATES et al. 1988). During inflammation, SP is released from sensory nerve endings in the skin where it causes vasodilation and plasma exudation. Immunomodulatory roles of SP include its ability to act as a chemoattractant, to stimulate cellular proliferation, and to influence immune cell metabolism.

Functional studies with lymphocytes and macrophages suggest that SP binds to specific high-affinity receptors on lymphocytes and macrophages (PAYAN et al. 1986) since many immune effects can be blocked by SP antagonists (STANISZ et al. 1987; HARTUNG et al. 1986). This hypothesis has been further validated by the demonstration of receptors for SP on human lymphocytes using a fluoresceinated bioactive SP analog DTAF-SP (PAYAN et al. 1984). Using this technique, 10% of human circulating $T_{c/s}$ cells and 18% of T_h cells appear to show SP-specific binding with approximately 7000 binding sites/cell (PAYAN et al. 1984). A single class of high-affinity SP receptors, with a K_D of 0.87 nM and 23000 receptors/cell, has also been identified on cultured human IM-9 lymphoblasts, a pre-B-like lymphoblastic cell line which expresses surface IgG (PAYAN et al. 1984c). Similar studies have demonstrated SP receptors on a lymphoblastic cell line (PAYAN et al. 1984), murine T and B lymphocytes (STANISZ et al. 1987), and guinea pig macrophages (HARTUNG et al. 1986). A single class of SP-binding sites has been characterized on guinea pig macrophages using [^{3}H] SP as a ligand with a K_D of 19 nM (HARTUNG et al. 1986). Human macrophages may express a novel neurokinin-1-like receptor or another tachykinin receptor since effects of SP on macrophages are not blocked by SP antagonist and neurokinin A was as effective as SP in stimulating cytokine release (McGILLIS et al. 1990). Three tachykinin-binding sites have been identified by bioassay and radioreceptor assay. Among those subtypes of neurokinin receptor, the type 2 receptor has been cloned and characterized as a member of the G-protein-associated membrane receptor family (MASU et al. 1987). Competition binding with other tachykinins indicates that the SP receptor on lymphocytes is of the neurokinin type 1, though some of the effects may be mediated by activation of other neurokinin receptors (McGILLIS et al. 1990). SP receptors are coupled to the phosphatidyl inositol pathway, thus inducing the activation of protein kinase C and calcium mobilization (PAYAN and GOETZL 1987).

Immunomodulation by SP has been well documented. In general, SP appears to have a stimulatory effect on immune responses both in vivo and in vitro. The first evidence demonstrating a direct stimulatory effect of SP on lymphocytes was that SP could stimulate human peripheral blood lymphocyte proliferation in response to the T-cell mitogens PHA and ConA (PAYAN et al. 1983). SP also enhances the proliferation of murine splenocytes and immunoglobulin secretion stimulated with Con A, but has no effect on LPS-induced proliferation (STANISZ et al. 1986). The observation that SP had no effect on LPS-induced lymphocyte proliferation coupled with the demonstration of SP receptors on T cells implies that the stimulation of immunoglobulin production by SP was indirectly mediated by an effect on T_h cells (STANISZ et al. 1987). The proliferative response of lymphocytes from spleen, mesenteric lymph nodes, and Peyer's patches was also enhanced by SP (O'DORISIO 1990). In GALT, the high concentration of SP in intestinal nerve endings and expression of a specific SP receptor on T and B cells in Peyer's patches suggest that SP may act as a trophic factor, a homing factor, or a differentiating factor for IgA-secreting plasma cells (STANISZ et al. 1987). SP stimulates the secretion of IgA and IgM in the spleen; in the gut, however, SP enhances only IgA production.

Substance P stimulates several macrophage functions, which are characteristic of activation including downregulation of membrane-associated 5′-nucleotidase, enhancement of lysosomal enzyme synthesis, and enhancement of phagocytosis (BAR-SHAVIT and GOLDMAN 1990). Production of arachidonic acid metabolites, prostaglandin E_2, thromboxane B_2, leukotriene C_4, and reactive oxygen species such as O_2 and H_2O_2 is also enhanced by SP (LOTZ et al. 1987; HARTUNG and TOYKA 1983).

Substance P may contribute to some chronic inflammatory conditions, such as rheumatoid arthritis (RA), by stimulating macrophages or other macrophage-like cells. Early studies have shown that SP is elevated in synovial fluid (COLPAERT et al. 1987) and that SP innervation is involved in joint inflammation in an animal model for rheumatoid arthritis (RA; LEVINE et al. 1986). SP has a potent chemotactic effect on human monocytes that is dose dependent and can be blocked by SP antagonists while responses to the chemotactic peptide formyl Met-Leu-Phe (fMLP) are not affected by SP antagonists (RUFF et al. 1985). SP regulates the production of cytokines from macrophages (LOTZ et al. 1988); nanomolar concentrations of SP stimulate the production of IL-1, tumor necrosis factor-α (TNF-α), and IL-6, all of which are important in the initiation of an immune response and inflammation (LOTZ et al. 1988).

IV. Opioid Peptides

Clinical studies have demonstrated a correlation between abuse of addicting drugs such as morphine, heroin, or alcohol and susceptibility to infection

(BHARGAVA 1990). While inadequate nutrition and poor sanitation contribute to immune suppression, opiate receptor agonists such as morphine also cause immunosuppression in both animals and humans. Reversal of morphine-induced immune suppression by the opioid antagonist naloxone implies that endogenous opioid peptides may be involved in immunomodulation (GILMORE et al. 1990). Opiates constitute a family of related peptides that were originally isolated from CNS (STEFANO 1989) and are derived from at least three distinct genes which code for large peptide precursors: proopiomelanocortin (POMC); proenkephalin (proenkephalin A, PEA); and prodynorphin (proenkephalin B, PEB) (GILMORE et al. 1990). The N-terminal sequence Tyr-Gly-Gly-Phe-(Met or Leu) is shared by all opioid peptides cleaved from the precursor molecules. To date, at least five subtypes of opioid receptors have been thought to exist; these include the δ-, ε-, κ-, μ-, and σ-subtypes, which appear to bind certain opioid peptides or synthetic opiates preferentially: enkephalin (δ), endorphin (ε), dynorphin (κ), morphine (μ and σ), and SKF 10047 (σ). Only two major classes of opioid peptides, enkephalins (ENKs) and endorphins (ENDs), will be discussed since these peptides have been identified in both the GI tract and immune cells (BHARGAVA 1990; GILMORE et al. 1990) and have effects on the immune system (TESCHEMACHER et al. 1990).

Proenkephalin A mRNA and its products, Met-ENK and Leu-ENK, are widely located in both the neuroendocrine system and the GI tract (GILMORE et al. 1990). ENK-positive fibers have been detected throughout the GI tract, especially in the myenteric plexus of the small intestine (McGUIGAN 1989), and in the spleen (FELTEN et al. 1985). PEA mRNA and its products have also been identified in immune cells, including T cells (ROTH et al. 1989), B cells (ROSEN et al. 1989), macrophages, and mast cells (MARTIN et al. 1987).

WYBRAN et al. (1979) demonstrated that addition of Met-ENK increased the percentage of T-rosette-forming cells in peripheral blood lymphocytes while morphine decreased the number of T rosettes. Both effects could be antagonized by naloxone. These results suggested that subtypes of opioid receptors, either enkephalin-preferring (δ) or morphine-preferring (μ) receptors, might be expressed on immune cells. This study was subsequently supported by two other observations. JOHNSON et al. (1982) demonstrated a specific, high-affinity ($K_D = 0.59\,nM$) opioid-binding site for Met-ENK on mouse spleen cells; MEHRISHI and MILLS (1983) demonstrated that [^{3}H] naloxone binding to human peripheral blood lymphocytes was displaced up to 57% by morphine. Recently, δ- and μ-opioid receptors have been characterized as glycoproteins of 53–58 Da and 65 Da, respectively (GILMORE et al. 1990), and the complementary DNA (cDNA) for an opiate-binding protein has been cloned from bovine brain (SCHOFIELD et al. 1989). The δ-opiate åeceptor is linked to a G-protein and inhibits adenylate cyclase; the evidence for adenylate cyclase inhibition via μ- and κ-opiate receptors is less convincing (LOH and SMITH 1990).

Enkephalins have effects on mitogen-induced proliferation of lymphocytes. Both Met- and Leu-ENK enhance proliferation of human and murine lymphocytes in response to PHA (Kharkevich and Kadagidze 1989). In contrast to these reports, Li and Fraker (1989) showed that Met-ENK inhibited the murine splenocyte response to Con A and LPS while enhancing the response to PWM. DALARGIN, a synthetic analog of Leu-ENK (Zozulia et al. 1987), has been shown to inhibit the proliferation of human peripheral blood lymphocytes.

Several studies have shown Met-ENK or Leu-ENK enhancement of NK-cell or mononuclear cell activity in vitro (Faith et al. 1984, 1987; Van Epps and Saland 1984). However, Shavit et al. (1984) showed suppression of NK activity by opioid peptides and morphine, and proposed that opioid peptide release during stress may cause inhibition of NK cell activity and allow tumor development.

The heterogeneity of opioid peptides together with their multiple receptor subtypes can easily lead to confusion regarding the role of opiate peptides in immunomodulation. Definitive studies are needed to clarify the role of each endogenous opioid peptide. These experiments will require systematic evaluation of the interaction of each peptide with the various opioid receptor subtypes, the delineation of the signal transduction pathway involved, and confirmation of the putative peptide-receptor interaction by use of agonists and antagonists specific for the μ-, ε-, δ-, or κ-opiate receptor subtypes. Such studies have been performed in the nervous system (Stefano 1989; Pierre et al. 1990; Stevens et al. 1991), but have not yet been accomplished for immune effector cells. Thus, the role of opioid peptides in immunomodulation appears to be a fruitful area for further investigation.

V. Neuropeptide Y

Neuropeptide Y (NPY) is a member of the family of tyrosine-rich neuropeptides. NPY consists of 36 amino acids with a C-terminal α-amide group and is one of the most abundant and widely distributed neuropeptides of the mammalian nervous system. NPY is identified in sympathetic nerves, where it is colocalized with norepinephrine (McGuigan 1989). NPY has also been identified in the myenteric and submucosal plexuses of the GI tract, where it is colocalized with somatostatin and cholecystokinin. Nerve fibers containing NPY in parenchyma of mesenteric lymph nodes end in close proximity to lymphocytes (McGuigan 1989), but NPY receptors on immune cells have not yet been demonstrated using radioligand-binding techniques.

Reports of NPY effects on immune function have been somewhat controversial. Jones et al. (1990) found variable NPY effects on ConA-induced proliferation in human peripheral blood T cells. Further studies have suggested that the effect of NPY might be altered by the physiological state of the cell donor. However, early observations by Soder and Hellstrom

(1987) demonstrated NPY-mediated inhibition of both the spontaneous and the mitogen-induced proliferative responses of guinea pig lymph node cells. The immunomodulatory role of NPY remains an open-ended question.

VI. Calcitonin Gene-Related Peptide

Calcitonin gene-related peptide (CGRP), a 37 amino acid polypeptide, is encoded along with calcitonin and katacalcin in one gene complex (ZAIDI et al. 1987). CGRP is widely distributed in both the central and peripheral nervous systems (STERNINI and BRECHA 1986) and in thyroid tissue (MACINTYRE 1989). Similarities exist between CGRP and SP in sensory function. Both peptides are found in high concentrations in the dorsal horn of spinal cord, where they are stored in the same vesicles and appear to be released simultaneously (ZAIDI et al. 1987). Both peptides are potent vasodilators (BRAIN et al. 1985). All of these similarities suggest that, like SP, CGRP may be involved in inflammatory events and may also play a role in regulation of immune function. In the GI tract, CGRP has been identified with a subepithelial plexus in the mucosa of the esophagus. STEAD et al. (1987b) have reported finding CGRP-containing neurons in contact with immune cells in the lamina propria. Both high- and low-affinity CGRP-binding sites have been characterized on murine lymphocytes with dissociation constants of $0.35\,\mathrm{n}M$ and $48\,\mathrm{n}M$, respectively. In these cells the EC_{50} values for CGRP increasing cAMP and the IC_{50} in inhibiting mitogen-induced T-cell proliferation are similar to the K_D values of CGRP for its high-affinity binding sites (UMEDA et al. 1988).

Calcitonin gene-related peptide profoundly inhibits the ability of macrophages to produce H_2O_2 in response to interferon and to act as antigen-presenting cells (NONG et al. 1989). These results were supported by the observations of ABELLO et al. (1990), who demonstrated high-affinity binding of CGRP functionally linked to adenylate cyclase in a macrophage-like P388D1 cell line. Eosinophil chemotactic activity is enhanced following treatment with CGRP (MANLEY and asHaynes 1989), suggesting that CGRP can also interact with the chemotactic peptide receptor. Taken together, these findings suggest that CGRP is an active immunomodulator in mucosal immune function.

VII. Cholecystokinin

Cholecystokinin (CCK) is a member of the family of gut hormones which share a C-terminal pentapeptide with gastrin (EYSSELEIN et al. 1984). CCK was originally isolated from porcine small intestine and characterized as a 33 amino acid peptide (MCGUIGAN 1989). Subsequently, CCK was observed to function in several molecular forms, including CCK-39 and CCK-58 (EYSSELEIN et al. 1983). A much smaller form, CCK-8, consisting of the carboxy-terminal octapeptide of CCK-33, is the predominant form localized in the small intestine, blood, and CNS, and is equally potent with CCK-33 (McGuigan 1989). CCK-8 is identified in the cerebral cortex, hypothalamus, and posterior pituitary (McGuigan 1989). CCK-8 is synthesized by neurons or endocrine cells in the lamina propria (Freier and Lebenthal 1990). In spleen, CCK-8 is identified in the white pulp, where it appears to surround cell clusters (Felten et al. 1985). The biological effect of CCK has been well documented (McGuigan 1989). Generally speaking, CCKs appear to be stimulatory factors affecting motility, immunoglobulin secretion, and hormone release in the gastroenteropancreatic axis.

Although specific, high-affinity CCK receptors have not yet been unequivocally identified on lymphocytes, injection of CCK stimulates immunoglobulin secretion in enteric fluid of rats and humans. The intravenous injection of CCK-8 results in a significant rise of IgA antibody activity within 2.5 min after injection; a rise in IgG, IgM, and IgE antibody activity is also observed; the antibody activity of IgA and IgG remained elevated for 12.5 min and 22.5 min, respectively (Freier 1990). The stimulation of Ig antibody activity may be via a specific CCK receptor since the effect of CCK was reduced by the CCK antagonist proglumide (Freier et al. 1987). Another study which supports the above assumption is that the increase of IgA and IgG antibody activity, caused by protein feeding, was inhibited when proglumide was administrated before the injection of CCK (Gershon and Erde 1981). The mechanism of CCK simulation of Ig secretion is not certain. It seems not to be a direct effect of CCK on Ig synthesis by plasma cells because the time interval between the injection of CCK and the rise in Ig is quite short. The rise in Ig may be the result of CCK stimulation of the secretory activity of enterocytes (Freier and Lebenthal 1990).

Cholecystokinin treatment of human peripheral blood mononuclear cells resulted in increased proliferation of cells stimulated with either suboptimal doses of PHA or anti-CD3 antibody. CCK stimulated the release of intracellular Ca^{2+} from human peripheral blood mononuclear cells. This action of CCK may be accomplished via a specific receptor on human peripheral blood mononuclear cells since CCK stimulation could be reduced or abolished by pretreatment of CCK antagonist L-364,718 (McMillen et al. 1990).

D. Conclusion

The mucosal immune system of the GI tract is the first line of defense against oral antigens derived from both microorganisms and foods as well as an important surveillance system against GI malignancies. Nonspecific and antigen-specific immune mechanisms are integral to this defense system. Recent evidence supports an important role for several gut peptides in modulation of mucosal immunity. Three gut peptides, VIP, substance P, and somatostatin, have well-characterized effects mediated via high-affinity receptors which utilize classical signal transduction pathways. Several other gut peptides including endorphins, enkephalins, cholecystokinin, CGRP, and neuropeptide Y appear to modulate the function of immune cells, but the mechanism of immunomodulation by these peptides remains to be defined. Delineation of the specific role of each of these gut peptides in the regulation of mucosal immunity presents a scientific challenge which has therapeutic implications for the treatment of infections, immunodeficiencies, and food allergies.

References

Abello J, Kaiserlian-Nicolas D, Cuber JC, Revillard JP, Chayvialle JA (1990) Identification of high affinity calcitonin gene-related peptide receptors on a murine macrophage-like cell line. Ann NY Acad Sci 594:364–366

Annibale B, Fais S, Boirivant M, delle Fave G, Pallone R (1990) Effects of high in vivo levels of vasoactive intestinal polypeptide on function of circulating lymphocytes in humans. Gastroenterology 98:1693–1698

Ashwell JD, Klausner RD (1990) Genetic and mutational analysis of the T-cell antigen receptor. Annu Rev Immunol 8:139–167

Bar-Shavit Z, Goldman R (1990) Modulation of phagocyte activity by substance P and neurotensin. In: Freier S (ed) The neuroendocrine-immune network. CRC, Boca Raton, pp 177–186

Bathena SJ, Louie J, Schecter GPP, Redman RS, Wahl L, Recant L (1981) Identification of human mononuclear leukocytes bearing receptors for somatostatin and glucagon. Diabetes 30:127–131

Battari E, Luis J, Martin JM, Fantani J, Muller JM, Marvaldi J, Pichon J (1987) The vasoactive intestinal peptide receptor on intact human colonic adenocarcinoma cells (HT29-D4). Biochem J 242:185–191

Beed EA, O'Dorisio MS, O'Dorisio TM, Gaginella TS (1983) Demonstration of a functional receptor for vasoactive intestinal polypeptide on Molt 4b T lymphoblasts. Regul Pept 6:1–12

Bhargava HN (1990) Opioid peptides, receptors and immune function. NIDA Res Monogr 96:220–233

Bland PW, Warren LG (1986) Antigen presentation by epithelial cells of the rat small intestine. Kinetics, antigens, specificity and blocking by anti-Ia antisera. Immunology 58:1

Blumberg RS, Terhorst C, Bleicher P, McDermott FV, Allan CH, Landau SB, Trier JS, Balk SP (1991) Expression of a nonpolymorphic MHC class I-like molecule, CDlD, by human intestinal epithelial cells. J Immunol 147:2518–2524

Bonavida, B, Wright SC (1986) Role of natural killer cytotoxic factors in the mechanism of target-cell killing by natural killer cells. J Clin Immunol 6:1–8

Brain SD, Williams TJ, Tippins JR (1985) Calcitonin gene-related peptide is a potent vasodilator. Nature 313:54–56

Carter RF, Bitar KN, Zfass AM, Makhlouf GM (1978) Inhibition of VIP-stimulated intestinal secretion and cyclic AMP production by somatostatin. Gastroenterology 74:726–730

Christophe J, Cauvin A, Vervisch E, Buscail L, Damine C, Abello J, Gourlet P, Robberecht P (1990) VIP receptors in human SU-T1 lymphoblasts. Digestion 46 Suppl 2:148–155

Colpaert DH, Donnerer J, Lembeck F (1987) Effects of capsaicin on inflammation and on the substance P content of nervous tissues in rats with adjuvant arthritis. Life Sci 32:1827

Danek A, O'Dorisio MS, O'Dorisio TM, George JM (1983) Specific binding sites for vasoactive intestinal polypeptide on nonadherent peripheral blood lymphocytes. J Immunol 131:1173–1177

Eysselein V, Deveney C, Sankaran H, Reeve J, Jr, Walsh J (1983) Biological activity of canine intestinal cholecystokinin-58. Am J Physiol 245:G313

Eysselein V, Reeve J, Shiveley J, Miller C, Walsh J (1984) Isolation of a large cholecystokinin precursor from canine brain. Proc Natl Acad Sci USA 81:6565

Fais S, Annibale B, Boirivant M, Santoro A, Pallone F, delle Fave G (1991) Effects of somatostatin on human intestinal lamina propria lymphocytes. J Neuroimmunol 31:211–219

Faith RE, Liang HJ, Murgo AJ, Plotnikoff NP (1984) Neuroimmunomodulation with enkephalins: enhancement of human natural killer cell activity in vitro. Clin Immunol Immunopathol 31:412

Faith RE, Liang HJ, Plotnikoff NP, Murgo AJ, Nimeh NF (1987) Neuroimmunomodulation with enkephalins: in vitro enhancement of natural killer cell activity in peripheral blood lymphocytes from cancer patients. Nat Immun Cell Growth Regul 6:88–98

Felten DL, Felten SY, Carlson SL, Olschowka JA, Livnat S (1985) Noradrenergic and peptidergic innervation of lymphoid tissue. J Immunol 135:755s–765s

Finch RJ, Sreedharan SP, Goetzl EJ (1989) High-affinity receptors for vasoactive intestinal peptide on human myeloma cells. J Immunol 142:1977–1981

Freier S (ed) (1990) The neuroendocrine-immune network. CRC, Boca Raton

Freier S, Lebenthal E (1990) The neuroendocrine-immune network in the gastrointestinal tract. In: Freier S (ed) The neuroendocrine-immune network. CRC, Boca Raton, pp 239–255

Freier S, Eran M, Faber J (1987) Effect of cholecystokinin and of its antagonist, of atropine and of food on the release of immunoglobulin A and immunoglobulin G antibodies in rat intestine. Gastroenterology 93:1242

Gates TS, Zimmerman RP, Mantyh CR, et al. (1988) Substance P and substance K receptor binding sites in the human gastrointestinal tract: localization by autoradiography. Peptides 9:1207–1219

Gershon MD, Erde SM (1981) The nervous system of the gut. Gastroenterology 80:1571

Ghose LH, Schnagl RD, Holmes IH (1978) Comparison of an enzyme-linked immunosorbant assay for quantitation of rotavirus antibodies with complement fixation in an epidemiological survey. J Clin Microbiol 8:268–276

Gilmore W, Moloney M, Weiner LP (1990) The role of opioid peptides in immunomodulation. Ann NY Acad Sci 597:252–263

Goetzl EJ, Payan DG (1984) Inhibition by somatostatin of the release of mediators from human basophils and rat leukemic basophils. J Immunol 133:3255–3259

Guerrero JM, Prieto JC, Elorza FL, Ramirez R, Goberna R (1981) Interaction of vasoactive intestinal peptide with human blood mononuclear cells. Mol Cell Endocrinol 21:151–160

Guy-Grand D, Cert-Bensussan N, Malissen B, et al. (1991) Two gut intraepithelial CD8+ lymphocyte populations with different T cell receptors: a role for the gut epithelium in T cell differentiation. J Exp Med 173:471–481

Hanson LA, Brandtzaeg P (1989) The mucosal defense system. In: Stiehm ER (ed) Immunologic disorders in infants and children, vol. 3. Saunders, Philadelphia, pp 116–155

Hartung HP, Toyka KV (1983) Activation of macrophages by substance P: induction of oxidative burst and thromboxane release. Eur J Pharmacol 89:301–305

Hartung H-P, Wolters K, Toyka KV (1986) Substance P: binding properties and studies on cellular responses in guinea pig macrophages. J Immunol 136: 3856–3863

Haugen, Keely P, Letourneau PC (1990) Interleukin-2 enhances chick and rat sympathetic, but not sensory neurite outgrowth. J Neurosci Res 25:443–452

Hökfelt T, Schultzbert M, Lundberg JM, Fuxe K, Mutt V, Fahrenkrug J, Said SI (1982) Distribution of vasoactive intestinal polypeptide in the central and peripheral nervous systems as revealed by immunochemistry. In: Said SI (ed)

Vasoactive intestinal peptide. Raven, New York, p 65 (Advances in peptide hormone research series)

Jakobs KH, Aktories K, Schultz G (1983) A nucleotide regulatory site for somatostatin inhibition of adenylate cyclase in S49 lymphoma cells. Nature 303:177

Johnson HM, Smith EM, Torres BA, Blalock JE (1982) Neuroendocrine hormone regulation of in vitro antibody production. Proc Natl Acad Sci USA 79:4171

Jonakait GM, Schotland S (1990) Conditioned medium from activated splenocytes increases substance P in sympathetic ganglia. J Neurosci Res 26:24–30

Jones JV, James H, Mansour M (1990) Neuropeptide Y modulates T cell proliferation in vitro: a new interaction of the immune and neuroendocrine systems. Ann NY Acad Sci 594:388–390

Keast JR, Furness JB, Costa M (1984) Somatostatin in human enteric nerves. Distribution and characterization. Cell Tissue Res 237:299

Kelleher K, Bean K, Clark SC, Leung W-Y, Yang-Feng TL, Chen JW, Lin P-F, Luo W, Yang Y-C (1991) Human interleukin-9: genomic sequence, chromosomal location, and sequences essential for its expression in human T-cell leukemia virus (HTLV)-I-transformed human T cells. Blood 77:1436–1441

Kharkevich DD, Kadagidze ZG (1989) The effect of the stimulation of opioid receptors on lymphocyte functional activity in vivo. Biull Eksp Biol Med 108: 315–316

Kleuss C, Hescheler J, Ewel C, Rosenthal W, Schultz G, Wittig B (1991) Assignment of G-protein subtypes to specific receptors inducing inhibition of calcium currents. Nature 353:43–48

Korneav EA, Khai LM (1964) Effect of destruction of hypothalmic areas on immunogenesis. Fed Proc 23:88–92

Krco CJ, Gores A, Go VLW (1986) Gastrointestinal regulatory peptides modulate in vitro immune reactions of mouse lymphoid cells. Clin Immunol Immunopathol 39:308–318

Laburthe M, Breant B, Rouyer-Fessard C (1984) Molecular identification of receptors for vasoactive intestinal peptide in rat intestinal epithelium by covalent cross-linking. Eur J Biochem 139:181–187

Levine JD, Dardick SJ, Roizen MF, Helms C, Basbaum AI (1986) Contribution of sensory afferents and sympathetic efferents to joint injury in experimental arthritis. J Neurosci 6:3423–3429

Li G, Fraker PJ (1989) Methionine-enkephalin alteration of mitogenic and mixed lymphocyte culture responses in zinc-deficient mice. Chung Kuo Yao Li Hsueh Pao 10:216–221

Lindholm D, Heumann R, Meyer M, Thoenen H (1987) Interleukin-1 regulates synthesis of nerve growth factor in non-neuronal cells of rat sciatic nerve. Nature 330:658–659

Littman DR (1987) The structure of CD4 and CD8 genes. Annu Rev Immunol 5:561

Loh HH, Smith AP (1990) Molecular characterization of opioid receptors. Annu Rev Pharmacol Toxicol 30:123–147

Lotz M, Carson DA, Vaughan JH (1987) Substance P activation of rheumatoid synoviocytes: neural pathway in pathogenesis of arthritis. Science 235:893–895

Lotz M, Vaughan JH, Carson DA (1988) Effect of neuropeptides on production of inflammatory cytokines by human monocytes. Science 241:1218–1221

Macintyre I (1989) Calcitonin: physiology, biosynthesis, secretion, metabolism, and mode of action. In: DeGroot LJ, Besser GM, Cahill GF (eds) Endocrinology. Saunders, Philadelphia, pp 892–901

MacNeil IA, Suda T, Moore KW, Mosmann TR, Zlotnik A (1990) IL-10, a novel growth cofactor for mature and immature T cells. J Immunol 145:4167–4173

Manley HC, Haynes LW (1989) Eosinophil chemotactic response to rat CGRP-1 is increased after exposure to trypsin or guinea-pig lung particulate fraction. Neuropeptides 13:29–34

Martin J, Prystowsky MB, Angeletti RH (1987) Preproenkephalin mRNA in T-cells, macrophages, and mast cells. J Neurosci Res 18:82–87
Masu Y, Nakayama H, Tamaki Y, Harada M, Kuno M, Nakanishi S. (1987) cDNA cloning of bovine substance K-receptor through oocyte expression system. Nature 329:836–838
Mayer L, Shlien R (1987) Evidence for function of Ia molecules on gut epithelial cells in man. J Exp Med 166:1417–1483
McGillis JP, Mitsuhashi M, Payan DG (1990) Immunomodulation by tachykinin neuropeptides. Ann NY Acad Sci 594:85–94
McGuigan JE (1989) Hormones of the gastrointestinal tract. In: DeGroot LJ, Besser GM, Cahill GF (eds) Endocrinology. Saunders, Philadelphia, pp 2741–2768
McMillen MA, Ferrara A, Schaefer HC, Goldenring JR, Zucker KA, Modlin IM (1990) Cholecystokinin mediates a calcium signal in human peripheral blood mononuclear cells and is a co-mitogen. Ann NY Acad Sci 594:399–402
Mehrishi JN, Mills IH (1983) Opiate receptors on lymphocytes and platelets in man. Clin Immunol Immunopathol 27:240–249
Mestecky J, McGhee JR (1987) Immunoglobulin A (IgA): molecular and cellular interactions involved in IgA biosynthesis and immune response. Adv Immunol 40:153
Moore TC (1984) Modification of lymphocyte traffic by vasoactive neurotransmitter substances. Immunology 52:511–518
Mosmann TR, Coffman RL (1989) TH1 and TH2 cells: different patterns of lymphokine secretion lead to different functional properties. Annu Rev Immunol 7:145–173
Musashi M, Yang YC, Paul SR, Clark SC, Sudo T, Ogawa M (1991) Direct and synergistic effects of interleukin 11 on murine hemopoiesis in culture. Proc Natl Acad Sci USA 88:765–769
Nakamura H, Koike T, Hiruma K, Sato T, Tomioka H, Yoshida S (1987) Identification of lymphoid cell lines bearing receptors for somatostatin. Immunology 62:655–658
Nong YH, Titus RG, Ribeiro JM, Remold HG (1989) Peptides encoded by the calcitonin gene inhibit macrophage function. J Immunol 143:45–49
O'Dorisio MS (1990) The role of substance P, somatostatin and vasoactive intestinal peptide in modulation of mucosal immunity. In: Baker M (ed) Neuroendocrine-immune network. CRC, Boca Raton, pp 187–198
O'Dorisio MS, Campolito LB (1989) Comparison of vasoactive intestinal peptide-mediated protein phosphorylation in human lymphoblasts and colonic epithelial cells. Mol Immunol 26:583–590
O'Dorisio MS, Panerai AE (1990) Neuropeptides and immunopeptides: messengers in a neuroimmune axis. New York Academy of Sciences, New York
O'Dorisio MS, Hermina N, Balcerzak SP, O'Dorisio TM (1981) Vasoactive intestinal polypeptide stimulation of adenylate cyclase in purified human leukocytes. J Immunol 127:2551–2554
O'Dorisio MS, Shannon BT, Fleshman DJ, Campolito LB (1989) Identification of high affinity receptors for vasoactive intestinal peptide on human lymphocytes of B cell lineage. J Immunol 142:3533–3536
O'Dorisio TM, O'Dorisio MS (1986) Endocrine tumors of the gastroentero-pancreatic (GEP) axis. In: Massaferri EL (ed) Endocrinology, vol 3. Medical Exam, New York, pp 764–817
Ottaway CA (1984) In vitro alteration of receptors for VIP changes the in vivo localization of mouse T cells. J Exp Med 160:1054–1069
Ottaway CA, Greenberg GR (1984) Interaction of VIP with mouse lymphocytes: specific binding and the modulation of mitogen responses. J Immunol 132: 417–423
Ottaway CA, Bernaerts C, Chan B, Greenberg GR (1983) Specific binding of VIP to human circulating mononuclear cells. Can J Physiol Pharmacol 61:664–671

Pallone F, Fais S, Annibale B, Boirivant M, Morace S, delle Fave G (1990) Modulatory effects of somatostatin and vasoactive intestinal peptide on human intestinal lymphocytes. Ann NY Acad Sci 594:408–410

Park, J., Chiba T, Yamada T (1987) Mechanisms for direct inhibition of canine gastric parietal cells by somatostatin. J Biol Chem 262:14190–14196

Paul S, Said SI (1987) Characterization of receptors for vasoactive intestinal peptide solubilized from the lung. J Biol Chem 262:158–162

Pawlikowski M, Stepien H, Kunert-Radek J, Schally AV (1985) Effect of somatostatin on the proliferation of mouse spleen lymphocytes in vitro. Biochem Biophys Res Commun 129:52–55

Payan DG, Goetzl EJ (1985) Modulation of lymphocyte function by sensory neuropeptides. J Immunol 135:783s–786s

Payan DG, Goetzl EJ (1987) Substance P receptor-dependent responses of leukocytes in pulmonary inflammation. Am Rev Respir Dis 136:s39–s43

Payan DG, Brewster DR, Goetzl EJ (1983) Specific stimulation of human T lymphocytes by substance P. J Immunology 131:1613–1617

Payan DG, Brewster DR, Goetzl EJ (1984a) Stereospecific receptors for substance P on cultured human IM-9 lymphoblast. J Immunol 133:3260–3265

Payan DG, Brewster DR, Missirian-Bastian A, Goetzl EJ (1984b) Substance P recognition by a subset of human T lymphocytes. J Clin Invest 74:1532–1539

Payan DG, Hess CA, Goetzl EJ (1984c) Inhibition by somatostatin of the proliferation of T-lymphocytes and MOLT-4 lymphoblasts. Cell Immunol 84:433–438

Payan DG, McGillis JP, Goetzl EJ (1986) Neuroimmunology. Adv Immunol 39: 299–323

Pierre J, Vaysse J, Zukin SR, Fields KL, Kessler JA, (1990) Characterization of opioid receptors in cultured neurons. Neurochem 55:624–631

Reichlin S (1986) Somatostatin: historical aspects. Scand J Gastroenterol 21:1–10

Robberecht P, Waelbroeck M, DeNeef P, Tastenoy M, Gourlet P, Cogniaux J, Christophe J (1988) A new type of functional VIP receptor has an affinity for helodermin in human SUP-T1 lymphoblasts. FEBS Lett 228:351–355

Rola-Pleszczynski M, Bolduc D, St Pierre S (1985) The effects of VIP on NK cell function. J Immunol 135:2569–2573

Rosen H, Behar O, Abramsky O, Ovadia H (1989) Regulated expression of proenkephalin in a normal lymphocytes. J Immunol 143:3703–3707

Roth KA, Lorenz RA, Unanue RA, Weaver CT (1989) Nonopiate active proenkephalin-derived peptides are secreted by T helper cells. FASEB J 3: 2401–2407

Ruff MR, Wahl SM, Pert CB (1985) Substance P receptor-mediated chemotaxis of human monocytes. Peptides 2 Suppl 6:107–111

Sacerdoce P, Ruff MR, Pert CB (1987) Vasoactive intestinal peptide 1–12: a ligand for the CD4(T4)/human immunodeficiency virus receptor. J of Neurosci Res 18:102–107

Said SI, Faloona GR (1975) Elevated plasma and tissue levels of vasoactive intestinal polypeptide in the watery-diarrhea syndrome due to pancreatic, bronchogenic and other tumors. N Engl J Med 293:155–160

Said SI, Mutt V (1972) Isolation from porcine-intestinal wall of a vasoactive octacosapeptide related to secretin and to glucagon. Eur J Biochem 28:199–204

Saxon A, Stiehm ER (1989) The B-Lymphocyte system. In: Stiehm ER (ed) Immunologic Disorders in Infants and Children, vol 3rd. Saunders Co WB, Philadelphia. pp. 40–67

Schofield PR, McFarland KC, Hayflick JS, Wilcox JN, Cho, TM, et al. (1989) Molecular characterization of a new immunoglobin superfamily protein with potential roles in opioid binding and cell contact. EMBO J 8:489–85

Scicchitano R, Dazin P, Bienenstock J, Payan DG, Stanisz AM (1987) Distribution of somatostatin receptors on murine spleen and Peyer's patch T and B lymphocytes. Brain Behav Immun 1:173–184

Scicchitano R, Dazin P, Bienenstock J, Payan DG, Stanisz AM (1988) The murine IgA-secreting plasmacytoma MOPC-315 expresses somatostatin receptors. J Immunol 141:937–941

Shavit Y, Lewis JW, Terman GW, Gale RP, Liebeskind JC (1984) Opioid peptides mediate the suppressive effect of stress on natural killer cell cytotoxicity. Science 223:188–190

Sirianni MC, Tagliaferri F, Fais S, Annibale B, Pallone F, delle Fave G (1990) Effect of vasoactive intestinal peptide neuropeptide on natural killer activity from peripheral blood lymphocytes of normal subjects. Ann NY Acad Sci 594: 421–422

Soder O, Hellstrom PM (1987) Neuropeptide regulation of human thymocyte, guinea pig T lymphocyte and rat B lymphocyte mitogenesis. Int Arch Allergy Appl Immunol 84:205–211

Spada A, Vallar L, Giannattasio G (1984) Presence of an adenylate cyclase dually regulated by somatostatin and human pancreatic growth hormone (GH)-releasing ractor in GH-secreting cells. Endocrinology 115:1203–1209

Sreedharan SP, Kodama KT, Peterson KE, Goetzl EJ (1989) Distinct subsets of somatostatin receptors on cultured human lymphocytes. J Biol Chem 264: 949–952

Sreedharan SP, Robichon A, Peterson KE, Goetzl E (1991) Cloning and expression of the human vasoactive intestinal peptide receptor. Proc Natl Acad Sci USA 88:4986–4990

Stanisz AM (1987) In vitro studies of immunoregulation by substance P and somatostatin. Ann NY Acad Sci 496:217–225

Stanisz AM, Befus D, Bienenstock J (1986) Differential effects of vasoactive intestinal peptide, substance P, and somatostatin on immunoglobulin synthesis and proliferations by lymphocytes from Peyer's patches, mesenteric lymph nodes, and spleen. J Immunol 136:152–156

Stanisz AM, Scicchitano R, Dazin P, Bienenstock J, Payan DG (1987) Distribution of substance P receptors on murine spleen and Peyer's patch T and B cells. J Immunol 139:749–754

Stead RH, Bienenstock J, Stanisz AM (1987a) Neuropeptide regulation of mucosal immunity. Immunol Rev 100:333

Stead RH, Tomioka M, Quinonez G, Simon GT, Felten SY, Bienenstock J. (1987b) Intestinal mucosal mast cells in normal and nematode-infected rat intestines are in intimate contact with peptidergic nerves. Proc Natl Acad Sci USA 8490: 2975–2979

Stefano GB (1989) Role of opioid neuropeptides in immunoregulation. Prog Neurobiol 33:149–159

Sternini C, Brecha N (1986) Immunocytochemical identification of islet cells and nerve fibers containing calcitonin gene-related peptide-like immunoreactivity in the rat pancreas. Gastroenterology 90:1155–1163

Stevens CW, Lacey CB, Miller KE, Elde RP, Seybold VS (1991) Biochemical characterization and regional quantification of Mu, delta and kappa opioid binding sites in rat spinal cord. Brain Res 550:77–85

Teschemacher H, Koch G, Scheffler H, Hildebrand A, Brantl V (1990) Opioid peptides, immunological significance? Ann NY Acad Sci 594:66–77

Umeda Y, Takamiya M, Yoshizaki H, Arisawa M (1988) Inhibition of mitogen-stimulated T lymphocyte proliferation by calcitonin gene-related peptide. Biochem Biophys Res Commun 154:227–235

Unanue ER, Allen PM (1987) The basis for the immunoregulatory role of macrophages and other accessory cells. Science 236:551-557

Van Epps DE, Saland L (1984) Beta-endorphin and met-enkephalin stimulate human peripheral blood mononuclear cell chemotaxis. J Immunol 132:3046

Wiik P, Opstad PK, Boyum A (1985) Binding of vasoactive intestinal peptide (VIP) by human blood monocytes: demonstration of specific binding sites. Regul Pept 12:145–153

Wood CL, O'Dorisio MS (1985) Covalent cross-linking of vasoactive intestinal polypeptide to its receptors on intact human lymphoblasts. J Biol Chem 260: 1243–1247

Wybran J, Appelbloom T, Famaey JP, Govaerts A (1979) Suggestive evidence for morphine and met-enkephalin receptors on normal blood T lymphocytes. J Immunol 123:1070

Yajima Y, Akita Y, Saito T (1986) Pertussis toxin blocks the inhibitory effects of somatostatin on cAMP-dependent vasoactive intestinal peptide and cAMP-independent thyrotropin releasing hormone-stimulated prolactin secretion of GH3 cells. J Biol Chem 261:2684–2689

Zaidi M, Breimer LH, MacIntyre I (1987) Biology of peptides from calcitonin genes. Q J Exp QJ Physiol 72:371

Zozulia AA, Voronkova TL, Patsakova E, Ivanushkin AM (1987) Importance of opiate receptor ligands in regulating lymphocyte proliferative activity. Farmakol Toksikol 50:58–59

CHAPTER 14

Pathophysiological Aspects of Gut Peptide Hormones

R.J. PLAYFORD, J. CALAM, and S.R. BLOOM

A. Introduction

This chapter deals with the role of regulatory peptides in human disease states. The gastrointestinal tract is discussed starting at the esophagus and working anatomically downwards. A brief description of the normal anatomy and endocrine elements in each region is included. A more detailed description of the location of gastrointestinal peptides is given in Chap. 1.

B. Esophagus

I. Anatomy

The esophagus is about 40 cm long and lined with a stratified squamous epithelium. It has upper and lower sphincters to prevent easy entry of air from above and gastric contents from below. The esophagus is supplied by the vagus nerve and also has sympathetic innervation. The importance of sympathetic innervation is unclear because sympathectomy does not appear to alter esophageal motility in man (INGELFINGER 1958). There is no well-defined anatomical lower esophageal sphincter although manometry studies undoubtedly show a physiological sphincter exists. The possibility that regulatory peptides are involved in esophageal disease has been considered in relation to achalasia and gastroesophageal reflux disease.

II. Achalasia

In achalasia the lower esophageal sphincter does not relax normally so that food and drink do not pass through normally. The patient may be aware that what is swallowed "sticks" in the chest (dysphagia) and that it regurgitates into the mouth. The primary cause of achalasia is still unknown but abnormalities of muscle and nerve components have been identified. There is a reduction in the number of ganglion cells (CASSELLA et al. 1965). The circular muscle of the lower esophagus is abnormally thick but this is likely to be secondary to its abnormal innervation. These patients have an exaggerated contraction in response to acetylcholine analogs (KRAMER and

INGELFINGER 1951), probably due to denervation hypersensitivity. They also have a paradoxical rise in esophageal pressure in response to cholecystokinin (CCK) (DODDS et al. 1981), which may represent loss of inhibitory neurons which are normally stimulated by CCK. A high density of vasoactive intestinal peptide (VIP)-containing fibers is found at the lower esophageal sphincter (UDDMAN et al. 1978). Studies using immunoblockade of VIP receptors in animals suggest that VIP may mediate normal relaxation of the lower esophageal sphincter (GOYAL et al. 1980) and it is possible that loss of VIP-containing neurons is important in achalasia.

III. Reflux Esophagitis

In this condition diminished contraction of the lower esophageal sphincter (LES) contributes to the reflux of gastric contents into the esophagus leading to inflammation known as esophagitis. Lower esophageal sphincter tone is influenced by a variety of circulating gut hormones including gastrin, CCK, motilin and secretin (LIPSCHUTZ and COHEN 1971), in addition to the VIP-containing neurons (GOYAL et al. 1980). Patients with esophageal reflux do have lower esophageal sphincter pressure compared with normal subjects (STREMPLE 1974). It was considered that circulating gastrin levels were the major determinant of lower esophageal sphincter closure force (COHEN 1973) but several studies have failed to show abnormalities of circulating gastrin in these patients (STREMPLE 1974). In a review of the literature, STURDEVANT (1974) concluded that gastrin is not a major determinant of normal lower esophageal pressure and that the diminished lower esophageal sphincter pressure seen in patients with esophageal reflux is not due to altered circulating gastrin concentrations.

C. Stomach

I. Anatomy

The human stomach is lined throughout by a simple columnar epithelium consisting of surface mucus cells with numerous invaginations called gastric pits. The gastric glands, which may be single or multiple tubular structures, empty into the lumen via the pits. The glands contain several different types of specialized cells: the oxyntic (or parietal) cells secrete hydrochloric acid into the gastric lumen; the chief cells secrete the proteolytic pro-enzyme pepsinogen and "intrinsic factor" required for absorption of dietary vitamin B_{12}. In addition, there are mucus-secreting "neck" cells, undifferentiated cells and endocrine cells. Although the endocrine cells only constitute a small proportion of the total gastric cells, they are of importance in maintaining the normal function of the stomach. The endocrine cells are commonly called "enterochromaffin" cells on the basis that they are stained by chromic acid. The advent of radioimmunoassay and immunocytochemical staining has allowed the identification of the hormones present.

The stomach is normally considered anatomically as three distinct regions, the corpus (or body), which contributes the vast majority of acid and pepsinogen secretion, the fundus, which contains these plus mucus-secreting cells, and the antrum, which consists largely of mucus-secreting cells but is the major gastric source of gastrin. Gastrin (G-) cells are present throughout the stomach but are considerably more abundant in the antrum. These cells contain mainly gastrin-17. Somatostatin cells (D) are present throughout the stomach. Other endocrine cells present in the stomach whose function has not been as clearly defined include glucagon, pancreatic polypeptide (PP) and bombesin-containing cells. There are enterochromaffin-like (ECL) cells which are found exclusively in the stomach and constitute the majority of gastric endocrine cells. The function of ECL cells is not fully understood but they contain large intracellular granules. It is likely that they release histamine and they may be involved in the control of acid secretion.

1. Nerves of the Epithelium

In addition to the classical sympathetic (norepinephrine) and parasympathetic (acetylcholine) systems, the stomach is innervated by peptidergic neurons containing enkephalin, dynorphin, glucose-dependent insulinotropic peptide (GIP), gastrin-releasing peptide (GRP; mammalian bombesin), substance P, vasoactive intestinal peptide (VIP), calcitonin gene-related peptide (CGRP) and neuropeptide Y (NPY). (For details of localization of peptides, see Chap. 1.)

2. Interactions

The interactions of peptide-containing endocrine cells and neurons occur in the neuron → endocrine cell direction (e.g., GRP released from local neurons stimulates gastrin release), endocrine cell → endocrine cell (e.g., release of somatostatin from D cells decreases gastrin release) and endocrine cell → neuron direction. The control of acid secretion and gastrin release is an example of the complex interactions of endocrine cells and final target organ: somatostatin has been shown to decrease acid secretion in the stomach. This is probably due to a combination of its inhibiting gastrin release, decreasing histamine release in response to gastrin, and decreasing the response of the parietal cell itself to histamine (MICHELANGELI et al. 1988). Stimulation of the vagus nerve in the pig results in increased release of both GRP and gastrin with an increase in secretion of acid and pepsin. Gastrin release together with an increased secretion of acid/pepsin was seen when GRP was infused directly. These findings support the idea that the vagus might affect gastrin release and thus the secretion of acid/pepsin by leading to release of GRP in the gastric antrum (SKAK-NIELSON et al. 1988). In addition to this complex picture somatostatin release increases in response to infusion of GRP, histamine or pentagastrin, suggesting that somatostatin is involved in restraint of stimulated acid secretion. As well as responding to these peptidergic factors, the G cell also releases gastrin directly in response to amino acids (DELVALLE and YAMADA 1990) and is also influenced by γ-amino butyric acid (GABA) and β-adrenergic agonists (SAKAMOTO et al. 1990). These experiments demonstrate the intricate mechanisms which control gastric physiology. (For details of control of gastric acid secretion see Chap. 7.)

II. Gastric Ulcer Disease

Gastric ulcer disease is common and is to be distinguished from duodenal ulceration. The presence of ulcers within the stomach is almost invariably associated with gastritis (GEAR et al. 1971) and the diseased gastric epithelium secretes less acid than normal (JOHNSON et al. 1964). Patients with gastric ulcer disease have higher plasma gastrin levels than normal individuals and this is thought to be due to diminished acid inhibition of gastrin release. There is no doubt that gastric acid secretion, although diminished, contributes to gastric ulceration, because pharmacological inhibition of gastric acid secretion accelerates healing of gastric ulcers and also prevents their recurrence (WALEN et al. 1989). It seems likely that gastrin release contributes to acid secretion in these patients. Reflux of duodenal contents into the stomach occurs more frequently in patients with gastric ulcer disease than in normal individuals (RHODES et al. 1969) and there is an association between duodenogastric reflux and increased gastrin release (CALAM and TRACY 1980). Gastric ulcers are more common in patients with rheumatoid arthritis who are taking nonsteroid antiinflammatory drugs (NSAIDs) and plasma gastrin levels are increased in these patients. However, ingestion of NSAIDs does not appear to affect gastrin release (CALDARA et al. 1981; ORME et al. 1979), and patients with rheumatoid arthritis probably have increased gastrin release because their secretion of gastric acid is diminished (DE WHITTE et al. 1979). In view of the evidence that gastritis itself may increase gastrin release, it is conceivable that a similar process affects gastrin release in patients with gastric ulcer (GU) disease.

III. Gastric Cancer

Studies have shown that patients with gastric cancer tend to have increased plasma gastrin concentrations. This is probably because, like patients with GU disease, they have gastritis which diminishes acid secretion. At present there is little evidence that gastrin stimulates the growth of ordinary gastric cancer cells although gastrin receptors are present on some tumors. However, the incidence of gastric cancer is raised in subjects with chronically diminished gastric acid secretion, whether due to previous gastric surgery or to atrophic gastritis, as in patients with pernicious anemia (SVENDSEN et al. 1986). Thus gastrin could be involved in the pathogenesis of gastric adenocarcinoma. However, the evidence for this is weaker than that implicating gastrin in ECL-cell tumors, and mechanisms based on the presence of luminal carcinogens have been put forward to explain the association between diminished acid secretion and gastric adenocarcinoma. At present there is no evidence that normal clinical doses of acid-suppressing drugs such as ranitidine and omeprazole affect the incidence of gastric adenocarcinoma, either by their effect on circulating gastrin or by any other mechanism (LA VECCHIA 1990).

IV. Pernicious Anemia

Pernicious anemia is a disease which results from severe atrophy of the gastric epithelium, often associated with metaplasia to intestinal-type mucosa. Auto-antibodies which cross-react with the H^+, K^+-ATPase, the gastric proton pump, are present. The chief cells fail to release intrinsic factor leading to malabsorption of vitamin B_{12}. These patients also have complete achlorhydria. Plasma gastrin levels are elevated by up to 30 times in some patients; much higher than those in patients taking normal doses of acid-suppressing medication. The patients who have achlorhydria with low gastrin levels generally have gastritis involving the antrum, and their lower gastrin concentrations are probably due to loss of antral G cells. It has been discovered that gastrin is trophic for enterochromaffin (ECL) cells, which are the most abundant endocrine cells in the gastric body. Chronic marked hypergastrinemia, whether associated with pernicious anemia, or with pharmacological suppression of gastric acid secretion in rats, leads to hyperplasia of ECL cells, eventually leading to gastric carcinoid tumors. These tumors should not be confused with gastric adenocarcinomas, which are the common form of human gastric cancer. The role of gastrin in the etiology of ECL-cell hyperplasia and tumors is strongly supported by the effect of antrectomy: rats given omeprazole develop gastric carcinoid tumors, but this does not occur if the antrum has been removed (H. LARSSON et al. 1988). In one patient with pernicious anemia, antrectomy led to regression of gastric carcinoid tumors (RICHARDS et al. 1987). It appears that in man the degree of hypergastrinemia determines whether gastric carcinoid tumors develop or not. On the basis of present evidence, suppression of acid secretion to the extent produced by omeprazole 20 or 40 mg/day for several years does not lead to hyperplasia of ECL cells (LAMBERTS et al. 1988).

V. Drugs and Hypergastrinemia

Omeprazole an H^+, K^+-ATPase inhibitor, and histamine H_2-receptor blockers produce an elevation of plasma gastrin levels to about three or four times normal (LANZON-MILLAR 1987). This rise in gastrin is likely to be due to diminished inhibition of gastrin release by low intragastric pH. In addition, there is some evidence that overgrowth of bacteria within the neutralized stomach might contribute to this effect: when conventional and germ-free rats were given the irreversible H_2-histamine antagonist loxtidine in doses sufficient to completely inhibit acid secretion, the rise in plasma gastrin levels was diminished in the germ-free rats (CALAM et al. 1991). This might have been due to a bacterial product stimulating gastrin release. Alternatively the bacteria might have altered gastrin release indirectly, for example by causing inflammation: loxtidine produced a dose-dependent rise in mucosal eosinophils in conventional, but not in germ-free, rats. Certain inflammatory mediators, such as interleukin 2, can increase gastrin release (TEICHMAN et al. 1986).

D. Duodenum

The duodenum contains a variety of endocrine cells which release secretin, motilin, CCK, GIP and gastrin. Duodenal G cells release more gastrin-34 than gastrin-17, whereas in the stomach the reverse is the case. In addition there are nerve fibers containing somatostatin, VIP, NPY, substance P, bombesin and enkephalin. (See Chap. 1 for details of localization of hormones.)

I. Duodenal Ulcers

There are breaches in the epithelium of the proximal duodenum which are distinct from gastric ulcers. Subjects with duodenal ulcers (DU) tend to have increased gastric acid secretion (BLAIR et al. 1987) and increased circulating gastrin (WALSH et al. 1975), but the reason why this occurs has been unclear. Understanding of the etiology of the disease and the associated hypergastrinemia has been greatly enhanced by the identification and culture of the organism *Helicobacter pylori* from the stomachs of patients. Evidence that *H. pylori* is involved in the etiology of duodenal ulcer disease includes:

1. Almost all patients with duodenal ulcer disease (about 95%) have *H. pylori* present in their stomach compared with only a minority of controls (DOOLEY and COHEN 1988).
2. Once *H. pylori* is eradicated from the stomach, duodenal ulcers generally do not recur until *H. pylori* returns (MARSHALL et al. 1988).

Helicobacter pylori does not colonize *normal* duodenal mucosa and it has been proposed that *H. pylori* causes duodenal ulcers in one or both of two ways:

1. The organism colonizes areas of the duodenum which have undergone histological change to resemble the gastric mucosa, i.e., gastric metaplasia. They then cause local injury either by toxic or immunological mechanisms. *H. pylori* has been shown to colonize areas of gastric

metaplasia which are present in these patients (GOODWIN 1988). How *H. pylori* in the duodenum causes ulcers remains open to speculation. Some strains of *H. pylori* produce a toxin which produces vacuoles in cell culture lines (LEUNK et al. 1988). There is evidence that toxin-producing strains are more prevalent in patients with DU disease than in patients with *H. pylori* but no ulcer (FIGURA et al. 1989). It is also possible that *H. pylori*-related inflammation leads to local damage.

2. Alternatively it has been suggested that *H. pylori* might exert its damaging effect on the duodenum by increasing antral gastrin release and thus increasing gastric acid secretion. It might then be the increased acid load that damages the duodenum. Evidence in support of this idea is that patients with DU disease generally have gastric acid secretion greater than normal individuals (BLAIR et al. 1987). In addition they tend to have higher peak postprandial gastrin concentrations than normal individuals and the raised gastrin falls when the organism is eradicated (LEVI et al. 1989b). The increased gastrin in duodenal ulcer disease is generally regarded as inappropriate because intragastric acid normally inhibits gastrin release, by a mechanism which may involve local release of somatostatin (BRAND and STONE 1988). Thus one might expect patients with duodenal ulcer disease to release less gastrin than normal individuals. This raises the possibility that the normal inhibition of gastrin release by luminal acid is diminished or that gastrin release has been upregulated. WALSH et al. (1975) used the technique of intragastric titration to show that the normal inhibition of gastrin release by a low intragastric pH is indeed impaired in duodenal ulcer disease.

How does *H. pylori* infection alter the normal feedback control? *H. pylori* organisms produce large amounts of a potent urease which cleaves urea normally found within the gastric juice to produce alkaline ammonia. So much urease is produced that incubating antral biopsies in urea and watching for a alkali pH change is a commonly used test for the presence of *H. pylori*. We speculated that the alkali produced by the organism might impair the normal inhibition of normal gastrin release by a low intragastric pH. In our initial study, we compared DU patients with a positive and negative biopsy urease test for *H. pylori* and found that the presence of a positive test was associated with both elevated postprandial plasma gastrin concentrations and higher maximal acid output (see Fig. 1) (LEVI et al. 1989a). Subsequently, we and other groups showed that elimination of *H. pylori* was associated with normalization of plasma gastrin concentrations (LEVI et al. 1989a; MCCOLL et al. 1989). Therefore, the relationship between *H. pylori* infection and increased plasma gastrin seems clear, but the mechanism by which this occurs is not resolved.

The two questions which are presently being addressed are:

1. Is *H. pylori*'s enzyme urease involved in the increase in circulating gastrin?

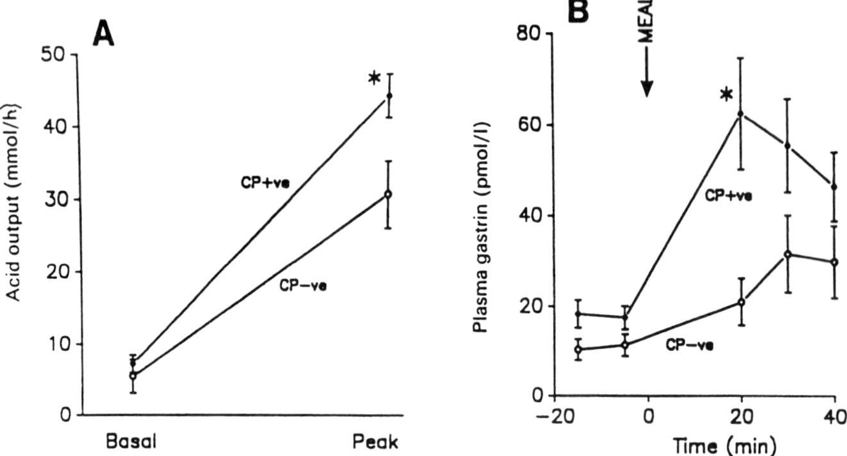

Fig. 1A,B. Mean basal and pentagastrin-stimulated acid secretion (**A**) and basal and meal-stimulated gastrin release (**B**) in CP + ve and CP − ve patients with duodenal ulcers. Vertical bars indicate SEM; $^*P < 0.05$. (From LEVI et al. 1989a) CP = *Campylobacter* (*Melicobacter*) *pylori*.

2. Is it some other product secreted by the organism itself, or the inflammatory response to the organism in the antrum which increases gastrin release?

The role of the urease has been addressed by studies involving short-term administration of urease inhibitors. Acetohydroxamic acid did not alter plasma gastrin concentrations (El-NUJUMI et al. 1991). The bismuth-containing ulcer-healing drug DeNol also inhibits urease and, particularly when given with antibiotics, can eradicate the organism. After administration of DeNol with metronidazole for 24 h urease was more or less completely inhibited but no change in plasma gastrin concentrations could be detected (McCOLL et al. 1990). Neither regime altered gastric acidity. TEICHMAN et al. (1986) have shown that certain cytokines released during inflammation can increase gastrin release and WYATT et al. (1989) showed increased basal gastrin concentrations in two patients with antral gastritis not due to *H. pylori*. This has led to the idea that the inflammation present within the antral epithelium itself, rather than the urease enzyme, might be responsible for the increased gastrin release in patients with DU disease. Leukocytes attracted to the antrum and activated by the presence of *H. pylori* would be expected to release a host of bioactive factors. In vitro, *H. pylori* increases the release of the interleukins-1 and -6 and tumor necrosis factor by leukocytes and can itself secrete platelet-activating factor (CRABTREE et al. 1990; DENIZOT et al. 1990; MAI et al. 1990).

A minority of DU patients have greatly increased antral gastrin release, which is familial and is associated with a high acid output and elevated serum pepsinogen I levels. In these patients gastrin release was inhibited normally by a low intragastric pH but the antrum was abnormally sensitive to a weak luminal stimulant of gastrin release (COOPER et al. 1985). This condition might be associated with increased fasting duodeno-gastric reflux (CALAM and TRACY 1980).

McCOLL et al. (1989) showed that eradication of *H. pylori* produced a significantly more alkaline postprandial pH within the stomach, although the change was not marked and subsequent studies by the same group have not shown significant changes in intragastric acidity on eradication of the organism. The failure of gastric acidity to fall at a time when gastrin release is diminished might be due to a number of factors. For example, it might be that acidity was measured under conditions in which endogenous gastrin was not predominantly responsible for the stimulation of acid secretion. Alternatively it might be that the eradication of *H. pylori* produces opposite forces which tend to increase gastric acid secretion; CAVE and VARGAS (1989) have demonstrated that *H. pylori* produces a protein which suppresses gastric acid secretion and it is possible that its elimination might tend to increase acid secretion.

Although the link between *H. pylori* and gastrin seems clear, the complex regulation of gastrin release would allow this link to be indirect, for example: secretin inhibits gastric acid secretion, especially during stimulation with gastrin (HUBEL 1972), but, interestingly, PRIMROSE et al. (1985) showed that an infusion of secretin did not affect plasma gastrin levels in normals but produced an increase in gastrin in patients with duodenal ulcers.

E. Pancreas

I. Anatomy

The majority of the pancreas is composed of acinar cells which secrete digestive enzymes via the pancreatic duct into the duodenum. Pancreatic islets consist of several different types of endocrine cells. The majority (about 75%) are B cells which produce insulin. The other endocrine-containing cells are the A cells which produce glucagon and the D cells which produce somatostatin, pancreatic polypeptide (PP) and, in fetal life, gastrin. All of these regulatory peptides can influence pancreatic exocrine secretion (WILLIAMS and GOLDFINE 1985). In addition, the pancreas is penetrated by sympathetic (noradrenergic), parasympathetic (cholinergic) and peptidergic (VIP and enkephalin) nerve fibers.

Pancreatic endocrine tumors are generally classified according to the hormone that they secrete, e.g., insulinoma, or the clinical syndrome which they produce, e.g., the watery diarrhea achlorhydria syndrome. Tumors secreting gastrin, insulin, VIP or glucagon generally cause characteristic clinical syndromes, whereas tumors releasing somatostatin or PP tend to have less obvious presentations.

II. Gastrinoma and the Zollinger-Ellison Syndrome

The pancreas is the commonest site of gastrin-secreting tumors even though G cells are not found in the normal adult pancreas. Pancreatic gastrinomas are believed to arise from D_1 cells of the islets of Langerhans, or fetal rest cells (VASSALO et al. 1972). About 60% are malignant (ELLISON 1964) and 50% recur even if total surgical excision was thought to have been achieved (ZOLLINGER 1975). Some patients have multiple pancreatic microadenomas rather than a single tumor. In about 10% of cases the gastrin-releasing tumor is located in the duodenum (GERBER 1960). About 30% of gastrinomas occur as part of the syndrome of multiple endocrine neoplasia type 1 (MEN-1). Duodenal gastrinomas may be particularly common in these patients (PIPELEERS-MARICHAL et al. 1990).

Peptic ulceration, due to excessive gastric secretion of acid and pepsin, is the commonest clinical feature of the Zollinger-Ellison (ZE) syndrome. Over 90% of patients develop ulceration at some time of their disease. Ulceration often resembles that seen in "ordinary" DU disease, but the presence of ulcers in the second or third parts of the duodenum is a clue to the diagnosis. Diarrhea is present in about one-third of patients, and is the presenting symptom in about 7% of cases (ELLISON 1964). The diarrhea disappears if gastric juice is aspirated and is thought to be due to the increased volume load entering the duodenum, plus the injurious effects of a low pH on pancreatic enzymes and the absorptive epithelium of the small intestine. Steatorrhea occurs in some patients. It may be due to effects of low pH on luminal factors such as lipase and bile salts and the absorptive epithelium as well as obstruction of the pancreatic duct by tumor.

The ZE syndrome should be considered in patients presenting with ulcers which are multiple, distal to the duodenal bulb, difficult to heal, which recur particularly promptly or are unusually aggressive. The ulcer may appear unremarkable. The combination of duodenal ulcer with diarrhea suggests the diagnosis. Once suspected the diagnosis can usually be made by measurement of the fasting plasma gastrin concentration, which is generally >70 pmol/l and often >500 pmol/1. The presence of peptic ulceration usually eliminates the possibility that hypergastrinemia is due to low acid secretion, so long as anti-secretory drugs were stopped before the blood sample was taken. However, difficulty can arise if the plasma gastrin concentrations are only slightly elevated. Other investigations may be helpful, although their results may also overlap with the normal range. Most patients with the ZE syndrome have a basal acid output of >15 mEq/h. Most, but not all, patients with the ZE syndrome have a ratio of basal: stimulated (by pentagastrin or histamine) acid secretion rates >2/3. Intravenous administration of secretin diminishes or does not change plasma gastrin concentrations in normals or patients with DU whereas patients with ZE syndrome respond with a rise in plasma gastrin (McGuigan and Wolfe 1980). Intravenous infusion of calcium causes a rise in plasma gastrin concentrations in patients with the ZE syndrome, but this may also occur in patients with "ordinary" DU disease. The administration of a meal leads to a marked rise in plasma gastrin concentrations if hypergastrinemia is of antral origin as in G-cell hyperfunction, but not in the ZE syndrome. Imaging of the pancreas and duodenum by the newer techniques may assist diagnosis by allowing the tumor to be demonstrated, although this is not always the case.

Chey et al. (1989) observed that 11 out of 30 patients with gastric hypersecretion and peptic ulcers had near normal plasma gastrin concentrations, even though 10 of these had pancreatic tumors. Plasma from these patients contained a peptide which did not cross-react in gastrin radioimmunoassays but stimulated acid secretion in rats and dogs. Further work to identify this peptide is required.

1. Treatment

The treatment of the ZE syndrome has been revolutionized by the development of first the histamine H_2-receptor antagonists and more recently the H^+, K^+ ATPase inhibitor omeprazole. High doses are often required and it is best to titrate the dose to bring the basal acid output under control. Omeprazole is preferred because of its longer duration of action and more marked effect on acid secretion. Before these drugs became available, peptic ulceration could only be controlled by total gastrectomy. This should still be performed in noncompliant patients because of the danger of ulcer-related death.

The treatment of choice is surgical resection but this is often impossible to perform because the tumors are often difficult to find, and may be multiple. They are generally slow growing and do usually not threaten life by direct effects. If acid secretion is controlled, 5-year survival is at least 50% (Fox et al. 1974). However, advances in medical therapy, imaging and pancreatic surgery are improving the chances for removal of tumors. Preoperative and intraoperative endoscopic ultrasound scanning of the pancreas may be particularly useful. Trans-hepatic venous sampling to localize gastrin gradients has also been useful in some cases but is technically demanding. A recent technique for localization of pancreatic tumors has employed systemic injection of radiolabeled somatostatin which is taken up by the tumor, but this technique is not generally used at present (Harris 1990).

III. VIPomas and the Verner-Morrison or WDHA Syndrome

Vasoactive intestinal peptide secreting tumors produce watery diarrhea, hypokalemia and achlorhydria (WDHA). The severe diarrhea was known as pancreatic cholera because patients can die of dehydration and electrolyte imbalance which are more dangerous than local effects of the tumor itself. In fact, the majority of cases present with symptoms related to the abnormal circulating hormones, rather than a direct effect of tumor bulk itself. Most VIPomas are located in the pancreas but ganglioneuroblastomas can secrete VIP and produce the syndrome more often in children (LONG et al. 1981). The diagnosis is made by measuring plasma VIP concentrations which are >60 pmol/1. Smaller rises in VIP are also seen in patients with chronic renal failure, and during prolonged fasting (KOCH et al. 1991). The association between elevated circulating VIP and the WDHA syndrome has been established. However, the VIP gene also contains a related peptide with similar biological activity: peptide histidine methionine (PHM) and its C-terminally extended variant, peptide histidine valine (PHV). These are also secreted by VIPomas and often circulate at higher concentrations than VIP (FAHRENKRUG and PEDERSEN 1986). All three peptides have been shown to stimulate secretion of water and electrolytes from rat intestine. However, CALAM et al. (1990) found that VIP had by far the greatest effect on human ileal output, when the three peptides were infused at rates which were similar and reproduced their respective plasma concentrations in patients with the WDHA syndrome.

Treatment of the WDHA syndrome is normally palliative as surgical resection of the tumor(s) is not generally feasible. We have found good responses to hepatic embolization which reduce the bulk of liver secondaries, and it is particularly useful in cases with portal vein compression. An alternative therapy which we have also found useful is treatment with the cytotoxic drug streptozotocin. The synthetic somatostatin analog octreotide is very effective in controlling the diarrhea and in controlling symptoms before and during embolization therapy. Somatostatin acts by diminishing both the release of VIP and the response of the intestine to it.

IV. Glucagonomas

The main clinical features of excessive circulating glucagon are weight loss, which is probably due to increased catabolism, normochromic normocytic anemia, angular stomatitis and a sore, red tongue. Many patients have psychiatric disturbances. In addition patients may present with a characteristic migratory erythematous rash or venous thrombosis. One-third of patients have glucose intolerance. Stimulation of gluconeogenesis by glucagon results in low plasma amino acids (ALLISON 1978), which may be of importance in the etiology of the rash. However, patients are also deficient in zinc, and zinc supplements will usually improve the rash. Treatment is with zinc, a high-protein diet, insulin (if diabetic) and somatostatin (MATON et al. 1989).

V. Somatostatinomas

Despite the numerous inhibitory effects of somatostatin on pituitary and pancreatic function, these patients have little metabolic disturbance. They present with diabetes and gall-stones and have decreased gastric acid secretion. It is possible that somatostatin normally acts as a paracrine mediator so that high local levels are required for maximal effects. Some of these patients have diarrhea. This is perhaps surprising because somatostatin is used in the treatment of secretory diarrhea. Acute administration of somatostatin in man does not affect jejunal absorption (KREJS et al. 1980) but increases ileal absorption (RUSKONE et al. 1982). Diarrhea in these patients may be due to maldigestion due to diminished secretion of digestive fluids such as pancreatic juice.

VI. Multiple Hormone Elevations

Immunocytochemical staining has shown that many pancreatic tumors contain multiple gastrointestinal peptides (CREUTZFELDT et al. 1975). In a study by CHIANG et al. (1990), 62% of patients with the ZE syndrome had an increased plasma level of a second regulatory peptide at the time of presentation. In another study, WYNICK et al. (1988) showed that patients with tumors secreting one hormone often start to secrete another, producing a change in the clinical picture. This is an important observation as four of their patients died from perforated duodenal ulcer due to the development of a second tumor which secreted gastrin. In view of these findings they recommended regular screening of patients to identify such changes so that treatment may be adjusted accordingly. Multiple endocrine neoplasia (MEN) syndromes are inherited and predispose to hyperplasia, adenoma or carcinoma in several endocrine organs. MEN type 1 consists of primary hyperparathyroidism, pituitary hyperplasia and endocrine pancreatic endocrine tumor. MEN type 2 consists of medullary carcinoma of the thyroid and pheochromocytoma. The treatment and prognosis of patients with MEN is different from that of patients with isolated endocrine lesions and a family history should always be taken. Some initial results suggested that increased circulating PP was a marker for MEN type 1 but this does not appear to be true (CHIANG et al. 1990). The gene responsible for MEN type 1 has been found to be located on chromosome 11 (G. LARSSON et al. 1988).

VII. Chronic Pancreatitis

Circulating cholecystokinin (CCK) plays a major role in the stimulation of pancreatic exocrine secretion (ANDERSON and DOCKRAY 1988). There is good evidence that CCK release is inhibited by intraduodenal proteases in rats. A similar effect in man has been demonstrated in some studies using phenylalanine as a stimulant (OYWANG et al. 1986) but not a study by CANTOR et al. (1988), who used fat as a stimulant. Studies of the mechanism of this effect in the rat have shown the presence of trypsin-labile CCK-releasing factors(s) derived from the duodenum (WIDER et al. 1988) and pancreas (IWAI et al. 1987). It is envisaged that food proteins and protease inhibitors normally increase CCK release by occupying the active sites of proteases, thus elevating intraduodenal concentrations of the CCK-releasing factors. Patients with chronic pancreatitis who have diminished secretion of pancreatic enzymes might be expected to have increased release of CCK. A number of studies have compared basal and postprandial plasma CCK concentrations between patients with chronic pancreatic exocrine insufficiency with normal individuals. The results are conflicting with increased levels in some studies (KOIDE et al. 1989) and decreased levels in others (FUNAKOSHI et al. 1990). Diminished secretion of the putative CCK-releasing factor by the diseased pancreas could contribute to these results. Also, CCK cells might adapt to a chronic lack of enzymes. Abdominal pain is a major clinical problem in patients with chronic pancreatitis. Studies suggest that this can be decreased by administering oral pancreatic enzyme supplements, which may act by decreasing CCK-driven stimulation of the pancreas (RAMO et al. 1989). CCK-receptor blockers might be expected to diminish pain in chronic pancreatitis, but this has not been tested experimentally. Studies in rats have shown that CCK receptor blockade can prevent experimentally induced acute pancreatitis (SHINYA and FUJIMURA 1989). Chronic pancreatic insufficiency results in malabsorption, particularly of fat, causing steatorrhea. The increased fat load reaching the terminal ileum and colon is presumably responsible for the increased plasma levels of peptide YY (PYY), neurotensin and enteroglucagon (e.g., ADRIAN et al. 1986) found in these patients. PYY is known to decrease pancreatic exocrine secretion (TATEMOTO 1982), which might contribute to malabsorption, although the effect of PYY is predominantly on secretion of water and bicarbonate.

VIII. Pancreatic Cancer

The incidence of pancreatic cancer in the United Kingdom and the United States has increased at least twofold over the last 50 years (GORDIS and GOLD 1984). This rise is almost certainly due to an environmental change. Cigarette smoking is a known risk factor but dietary changes may also have contributed. Epidemiological studies have shown a correlation between dietary fat intake and pancreatic cancer in different populations (WYNDER 1975). Studies in rats have shown that CCK can promote pancreatic growth and neoplasia (FOLSCH et al. 1978) and that pancreatic carcinogenesis is also promoted by diets which increase plasma CCK such as a high-fat diet (ROEBUCK et al. 1981) or raw soya flour (DOUGLAS et al. 1989). The CCK antagonist CR1409 diminished the cancer-promoting effect of raw soya flour, confirming the role of CCK in its effect (DOUGLAS et al. 1989). At present there is no direct evidence that increased release of CCK also causes

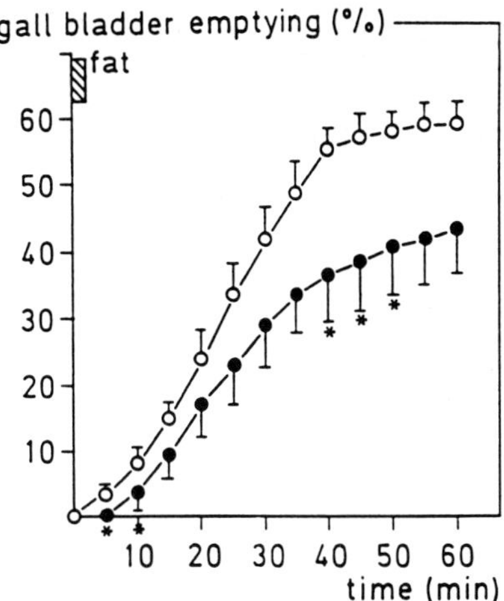

Fig. 2. Effect of intraduodenal fat on gallbladder emptying (mean ± SEM) in 20 control subjects (*open circles*) and 20 gallstone patients (*closed circles*). *Asterisks* denote significant differences between control subjects and gallstone patients. $^{*}P < 0.05$. (Reproduced from MASCLEE et al. 1989)

human pancreatic cancer, though this remains a possibility. BEARDSHALL et al. (1989) showed that in humans the CCK-releasing effect of fats is greater if they are less saturated. Interestingly the pancreatic cancer-promoting effect of fat in the rat is greatest if the fat is unsaturated (ROEBUCK et al. 1981). Thus it is conceivable that increased ingestion of polyunsaturated fatty acids might have contributed to the increase in pancreatic cancer, although this remains highly speculative.

F. Gallstones

Gallbladder (GB) motility is affected by more than one control system. For example, stimulation of GB contraction by the sight or smell of food is mediated by vagal cholinergic fibers because it is inhibited by atropine (HOPMAN et al. 1987) or vagotomy (FICHER et al. 1986). However, CCK seems to predominate because, when normal subjects eat during CCK-A receptor blockade, the GB becomes larger rather than smaller (LIDDLE et al. 1989; MEYER et al. 1989). This may be because CCK receptor blockade may prevent relaxation of the sphincter of Oddi, causing the GB to become distended with bile secreted by the liver in response to food.

Many patients who present to hospital with gallstones have diminished GB contraction (FORGACS et al. 1984; MASCLEE et al. 1989; see Fig. 2) but it is not clear whether this is a cause or a result of the stones. Perhaps the most popular view is that stones form because the bile is abnormal and that

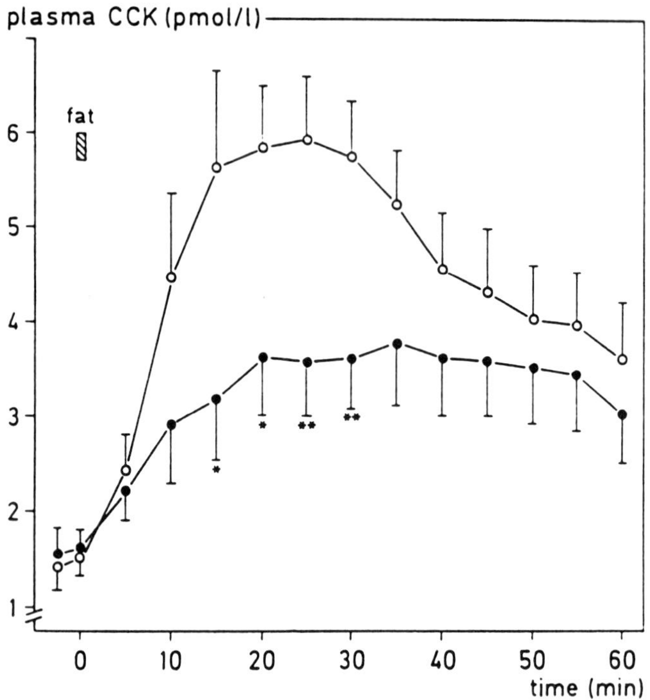

Fig. 3. Plasma CCK concentrations before and after intraduodenal administration of fat in 20 control subjects (*open circles*) and 20 gallstone patients (*closed circles*). *Asterisks* indicate significant differences between the two groups: $^{*}P < 0.05$; $^{**}P < 0.01$. (Reproduced from MASCLEE et al. 1989)

abnormal motility is the result of chronic inflammation of the gallbladder muscle. However, diminished gallbladder contraction might play a primary role in gallstone formation. Less contraction might allow crystals to form as well as diminishing the chance that they will be ejected. Diminished contraction of the gallbladder in pregnancy (EVERSON et al. 1982), probably due to progesterone, may contribute to gallstones in multiparous women. Diminished gallbladder contraction in diabetics may contribute to gallstones in this group. Diabetics may develop autonomic neuropathy, affecting gallbladder contraction (STONE et al. 1988) but also have hypercholesterolemia and it has been suggested that cholesterol itself might diminish gallbladder contraction because cholesterosis of the gallbladder and cholesterol gallstones are both associated with weaker isometric contraction of gallbladder muscle strips in vitro compared with the presence of pigment stones (BEHAR et al. 1989). Against this background, studies of the role of cholecystokinin have investigated whether its release or its effect on the gallbladder are diminished. The rise in plasma CCK produced by infusion of fat into the duodenum was indeed diminished in patients with uncomplicated gallstones in one study (MASCLEE et al. 1989; see Fig. 3). However, in another study from the same group the rise in CCK after fat was eaten was equal to that in controls (VAN ERPECUM et al. 1990). Studies of the contraction of the gallbladder in response to exogenous CCK have shown, surprisingly, that the sensitivity of the gallbladder to CCK, expressed as the dose required to produce half maximal contraction, is actually increased in some gallstone patients (MASCLEE et al. 1989). One possible explanation is that inflammatory factors are potentiating the effect of cholecystokinin.

Therefore, the question "Is the diminished gallbladder contraction seen in patients with gallstones a primary or secondary phenomenon?" is still not answered. Diminished release of cholecystokinin is likely to be responsible for noncontraction and accumulation of GB sludge during total parenteral nutrition (MESSING et al. 1983). Primates given CCK receptor blockers for long periods did develop gallstones.

Cholesterol gallstones can be made to dissolve by the administration of bile salts for long periods by mouth. Bile salts can inhibit CCK release (KOOP et al. 1989) and might thus diminish gallbladder contraction. However, results have been conflicting. In one study bile acid dissolution therapy increased the resting GB volume, perhaps by increasing the volume of bile, but postprandial plasma CCK levels and GB contraction after a fatty meal were normal (VAN-ERPECUM et al. 1990). However, according to another study, CCK-stimulated contraction of the gallbladder is diminished during bile salt therapy (SYLWESTROWICZ and SHAFFER 1988). Lithotripsy has no apparent effect on GB function (SPENGLER et al. 1989).

G. Small Bowel

I. Anatomy

The human small intestine is about 20 feet in length and is usually considered in three parts: duodenum, jejunum and ileum. The diameter of the lumen gradually decreases as one proceeds distally. Intestinal endocrine cells are flask shaped and are distinguished by the presence of secretory granules. Cells containing secretin, CCK, gastrin inhibitory peptide (GIP), somatostatin, motilin and endorphin are mainly found in the duodenum and jejunum. Gastrin-containing cells are restricted to the duodenum (and stomach). Somatostatin cells are found in the ileum and jejunum. Further down the small intestine, neurotensin cells are found in the ileum and cells containing both PYY and enteroglucagon are found in the ileum and colon. In addition to endocrine cells, the intestine is innervated with neurons containing VIP, CCK, gastrin-releasing peptide (GRP), substance P and somatostatin. For details of distribution see Chap. 1.

In general, studies have shown that diseases diminish local release of hormones, but the release of hormones from bowel distal to the disease is often increased. For example, celiac disease is a sensitivity to dietary gluten seen in adults and children which affects the upper part of the small intestine. Subjects have been studied before and after treatment with a gluten-free diet. Gastrin and PP are not affected, but there is decreased release of CCK, secretin and GIP from the upper small intestine (e.g., CALAM et al. 1982). However, there is markedly increased release of hormones derived from further down the bowel, such as neurotensin, enteroglucagon and PYY (BESTERMAN et al. 1978), their release presumably due to the increased nutrient load reaching the distal bowel.

II. Effects of Gastric Surgery

A few patients with peptic ulcers still require operations to control their disease, but most patients with symptoms resulting from peptic ulcer surgery had their operations prior to the recent improvements in medical therapy and surgical techniques.

Recent operations are based on reducing acid secretion by vagotomy and this leads to increased circulating gastrin. The vagal trunks were often divided at the cardia; this delayed gastric emptying and was combined with a gastric "drainage" procedure: gastroenterostomy or pyloroplasty. Patients who have had "vagotomy and drainage" may develop diarrhea, due to a combination of rapid gastric emptying and the effects of vagotomy on upper gastrointestinal function. This problem and the need for "drainage" are now usually avoided by selectively dividing the branches of the vagus which supply the region of the stomach which secretes acid, leaving the innervation of antral/pyloric motility intact.

Another surgical approach was to diminish gastrin release by removing the gastric antrum, including the pylorus. These operations often produced rapid emptying, "dumping" of gastric contents into the small intestine and the "dumping syndrome." Symptoms include nausea, faintness and flushing after meals. Some patients have reactive hypoglycemia resulting from rapid absorption of sugar and excessive insulin release. Hypotension probably results from the rapid shift of water into the gut and vasodilation which may be due to increased release of regulatory peptides from intestine. There is increased release of VIP (ADRIAN et al. 1985) and neurotensin, both of which increase intestinal secretion (see Fig. 4) and are vasodilators. Movement of fluid from the plasma volume into the duodenum following a meal may also be driven by osmosis. Release of enteroglucagon and PYY are also increased in the dumping syndrome (ADRIAN et al. 1985; SAGOR et al. 1981; BLACKBURN et al. 1980).

It is sometimes necessary to resect parts of the small intestine because they are affected by diseases such as tumors, Crohn's disease, and intestinal infarction. BESTERMAN et al. (1982) found that resection of part of the ileum resulted in rises in enteroglucagon, PYY, motilin, PP and gastrin. The cause

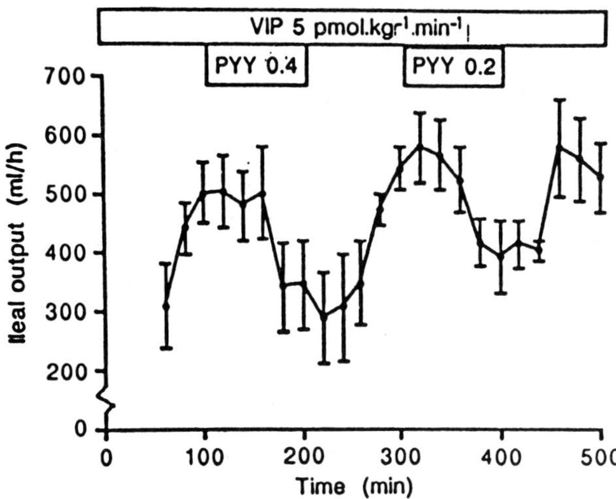

Fig. 4. Mean net ileal secretion, as assessed by dilution of polyethylene glycol, during stimulation with VIP in six patients. Unit for PYY infusion = pmol $kg^{-1} min^{-1}$. (Reproduced from PLAYFORD et al. 1990)

of these changes is not known precisely, but may be due to altered stimulation of endocrine cells from the lumen. For example, resection of the ileum where bile salts are absorbed might lead to depletion of the bile salt pool, diminishing proximal absorption of fat which then stimulates release of enteroglucagon and PYY from cells located distally. Increased release of peptides from cells proximal to the resection might reflect loss of an inhibitory signal from the ileum. The human intestine is capable of adaption following partial removal by hypertrophy of the remaining intestine and increased absorption. The changes in plasma regulatory peptides may well explain some of these effects. For example, enteroglucagon is trophic to the small intestine, and motilin and PYY can both delay intestinal transit, (e.g., PLAYFORD et al. 1990), allowing more time for absorption to occur.

Some patients with extreme obesity were treated by jejunoileal bypass: an anastomosis is made between the jejunum and ileum to cause malabsorption and weight loss. These patients have diminished release of jejunal peptides and increased release of peptides from the terminal ileum and colon. They have diminished postprandial insulin release (BLOOM 1980), which was unexpected because other obese patients have increased release of insulin. This might be explained by decreased jejunoileal release of GIP, which has an "insulinotropic" effect, increasing insulin release.

III. Gastrointestinal Hormones During Diarrhea

Severe diarrhea can be life threatening, usually because of the dehydration which it causes. Different regulatory peptides may be involved in the pathogenesis of diarrhea, and also in minimizing its effects. VIP is a potent secretagogue in man, and excess circulating VIP in patients with VIPomas results in severe secretory diarrhea (CALAM et al. 1990). There is also some evidence that two potent infective causes of secretory diarrhea in man, *Vibrio cholerae* and *Escherichia coli*, may cause diarrhea by causing a local release of VIP in the intestine. The toxin produced by *Vibrio cholerae*, in addition to a direct effect on intestinal cells, stimulates the release of VIP from VIP-containing nerves of the small intestine (CASSUTO et al. 1981). Also, EKLUND et al. (1985) showed that the diarrhea induced by *E. coli* toxin is decreased in animals if they are given tetrodotoxin which destroys mesenteric neurons, suggesting that they are involved in the pathogenesis of the diarrhea. Thus, local release of VIP might mediate the excessive secretion seen in infective diarrhea (see Chap. 10). Patients with infectious diarrhea have increased circulating PYY, which may be due to unabsorbed fat reaching the ileum as high levels of PYY are seen in patients with steatorrhea (ADRIAN et al. 1986). Recent work suggests that elevated PYY release might act to minimize fluid loss during a diarrheal illness. Infusion of PYY into patients with ileostomies at rates which reproduce postprandial plasma concentrations diminished VIP-stimulated ileostomy output and delayed small intestinal transit time (PLAYFORD et al. 1990; see Fig. 4). We also recently showed that plasma levels of PYY seen during a diarrheal illness result in generalized vasoconstriction which may help to maintain blood pressure during dehydration. Vasoconstriction of the mesenteric vessels may be one of the mechanisms by which intestinal secretion is diminished (BENITO-ORFILA et al. 1990). Both VIP and cholera toxin stimulate intestinal secretion by increasing cellular adenosine-3′-5′-cyclic monophosphate (cAMP) and there is evidence that PYY has the opposite effect on cAMP (SERVIN et al. 1989; see also Chaps. 5, 10).

IV. Inflammatory Bowel Disease

Crohn's disease is a chronic relapsing inflammatory disease of unknown etiology which can affect all layers and any part of the gastrointestinal tract but most often affects the terminal ileum. Circulating PP, GIP, motilin, and enteroglucagon are elevated (BESTERMAN et al. 1983a). These changes are thought to be due to the stimulatory effect of unabsorbed nutrients. An interesting but unexplained histological finding in Crohn's disease is that VIP-containing nerves in affected tissue are greatly swollen and irregular in shape (BISHOP et al. 1980; POLAK and BLOOM 1982). This change does not occur in the other major inflammatory bowel disease, ulcerative colitis, which only affects the colonic mucosa (see below) but is also associated with abnormal circulating concentrations of regulatory peptides including raised gastrin, pancreatic polypeptide and motilin levels (BESTERMAN et al. 1983b). The explanation for these changes is still unclear.

H. Colon

I. Anatomy

A number of regulatory peptides have been identified in the epithelium of the human colon. Endocrine cells containing enteroglucagon, peptide YY and somatostatin are present in the mucosa as are nerve fibers containing vasoactive intestinal peptide and substance P. Major functions of the colon include fluid and electrolyte absorption and the storage of feces prior to defecation. These are regional differences in the tissue concentrations of regulatory peptides in different parts of the colon. For example, enteroglucagon, PYY and somatostatin are higher in the distal colon than the cecum and in fact PYY levels were three times higher in the rectum of normal subjects than in the cecum. However, the reverse was true for the neuroendocrine hormones VIP and PHM, where levels in the rectum were about half of those in the cecum (CALAM et al. 1989).

II. Colon and Malabsorption

Diseases of the small intestine such as celiac disease interfere with the normal absorptive function of the small bowel. This results in an increased nutrient load reaching the colon. This probably explains the increases in circulating colonic hormones such as PYY in these diseases (see preceding sections; BESTERMAN et al. 1978).

III. Polyp and Cancers

Several studies have examined the tissue concentrations of regulatory peptides in human adenomas (polyps) and cancers of the colon. Some tumors have approximately normal amounts of neuropeptides. Numbers of endocrine cells and the peptides that they contain vary from normal to diminished, reflecting less differentiation within the polyp or tumor. However, it remains possible that these hormones stimulate growth by paracrine (local) effects. Of particular interest are findings which suggest that gastrin plays a trophic role in colonic cancer. SMITH et al. (1989) showed that patients with colonic carcinoma or adenomatous polyps have elevated plasma gastrin levels. Also, gastrin levels fall following resection of the tumor (WONG et al. 1990). Gastrin receptors have been identified on several human and animal colonic cancer cell lines (e.g., CHICONE et al. 1989). HOOSEIN et al. (1990) found the growth of two human colonic carcinoma cell lines in vivo to be inhibited by a gastrin antagonist, suggesting an autocrine stimulation by a gastrin-like peptide. NEMETH et al. (1992) have shown the presence of progastrins in some colonic tumors. In one study there was an association between colonization of the stomach with *H. pylori* (which can also raise plasma gastrin) and colonic cancer (TALLY et al. 1990). Thus, is it possible that, in addition to an autocrine stimulation of growth, increased plasma gastrin concentrations might stimulate growth of some cancers.

IV. Abnormalities of Gut Peptidergic Innervation in the Colon

Two diseases affecting the colon, Hirschsprung's disease and Chagas' disease, are associated with severe chronic constipation. Chagas' disease is caused by infestation with *Trypanosoma cruzi*, whereas Hirschsprung's disease is of unknown etiology. In Hirschsprung's disease there is a congenital aganglionic segment of distal colon. In this segment there are fewer VIP and substance P containing nerve fibers and diminished tissue concentrations of these peptides (FREUND et al. 1979). This loss of peptidergic innervation is interesting because the number of cholinergic and adrenergic nerves is increased (BENNETT et al. 1968). Also, the amount of substance P is increased in the colon just proximal to the aganglionic section compared to normal controls (EHRENPREIS and PERNOW 1953). The aganglionic section of bowel is chronically constricted, perhaps because the remaining excitatory nerves outweigh the inhibitory influence of inhibitory nerves, such as those releasing VIP. This suggests that the peptidergic neurons are involved in the inhibitory part of colonic control and propulsion. Chagas' disease causes destruction of intramural ganglion cells and also causes a reduction in substance P and VIP containing neurons (LONG et al. 1980). Mucosal levels of enteroglucagon and somatostatin cells were also decreased, raising the possibility that the innervation of mucosal endocrine cells contributes to their growth.

In contrast to the two diseases discussed above, Crohn's disease is associated with diarrhea and immune-staining has shown the diameter of VIP-containing nerves to be greatly increased, both in segments of the colon affected by the disease and other parts of the colon (BISHOP et al. 1980). This finding has been confirmed by others. It was suggested that increased secretion of VIP might contribute to diarrhea. The cause of the neural change remains speculative but might be due to effects of local inflammation. According to this idea, enlargement of VIP-containing nerve fibers is not seen in ulcerative colitis (UC) because it only involves the epithelium, which is superficial to the nerves, whereas Crohn's disease is transmural. Initial results suggested that tissue concentrations of VIP are elevated in Crohn's disease (POLAK and BLOOM 1982), but this has not been confirmed in later studies (CALAM et al. 1989).

V. Irritable Bowel Syndrome

Patients with this disease experience abdominal bloating and altered bowel habit in the absence of detectable organic pathology. They have been shown to have an abnormal intestinal motility which resembles that produced by psychological stress. However, HARVEY and READ (1973) showed that infusion of CCK into these subjects reproduced their symptoms while stimulating motility. This suggests that increased sensitivity to CCK may be involved. Studies on the value of CCK antagonists in this disease are under way (see Chap. 15).

J. Gastrointestinal Hormones and Cardiovascular Function

The distribution of VIP (and possibly CGRP) containing neurons suggests that they are important in the control of vasomotor tone. VIPergic neurons are found in subglandular, periglandular and subepithelial plexuses of the mucosa. They are often found in association with small blood vessels in the mucosa of the GI tract and many other vascular beds. VIP is a potent vasodilator and has positive inotropic and chronotropic actions. Eating normally increases cardiac output by about 60% (KELBAEK et al. 1989) and markedly increases superior mesenteric blood flow (KOONER et al. 1989). Release of VIP may be important in mediating this response because "stroking" the intestinal mucosa of the cat results in local release of VIP (EKLUND et al. 1980). In many species, VIP innervation is found mainly around vessels in the mucosa rather than the submucosa and this may be a mechanism to regulate regional blood flow. We recently showed that postprandial levels of PYY inhibit VIP-induced vasodilatation (BENITO-ORFILA et al. 1990), suggesting that PYY may be important in "turning off" postprandial vasodilatation. However, the importance of peptides in the control of regional blood flow is still unclear. The vascular effects of regulatory peptides may be important in disease states. Vasodilatation, tachycardia and flushing associated with VIPomas are probably due to the VIP released from the tumor. Also, the high PYY concentrations seen during a diarrheal illness may have a negative inotropic effect on the heart as well as a vasoconstrictor effect on blood vessels (BENITO-ORFILA et al. 1990). Generalized vasoconstriction during a diarrheal illness may help to maintain blood pressure but a decrease in cardiac output might be detrimental.

K. Conclusions

Many diseases of the gastrointestinal tract are associated with abnormalities of the release of regulatory peptides or altered responses to them. With a few exceptions, the reasons for their altered release and their role in the disease process are incompletely understood. Further work is required because of the marked effects that these peptides have on gastrointestinal function. The development of specific peptide antagonists is proving particularly useful in the elucidation of their roles in disease (see Chap. 15).

References

Adrian TE, Long RG, Fuessl HS, Bloom SR (1985) Plasma peptide YY (PYY) in dumping syndrome. Dig Dis Sci 30:1145–1148

Adrian TE, Savage AP, Bacarese-Hamilton AJ, Wolfe K, Bestermann HS, Bloom SR (1986) Peptide YY abnormalities in gastrointestinal diseases. Gastroenterology 90:379–384

Allison DJ (1978) Therapeutic embolisation. Br J Hosp Med 20:707–715

Anderson L, Dockray GJ (1988) The cholecystokinin antagonist L364, 718 inhibits the action of cholecystokinin but not bombesin on rat pancreatic secretion in vivo. Eur J Pharmacol 146:307–311

Behar J, Lee KY, Thompson WR, Biancani P (1989) Gall bladder contraction in patients with pigment and cholesterol stones. Gastroenterology 97:1479–1484

Beardshall K, Frost G, Morarji Y, Domin J, Bloom SR, Calam J (1989) Saturation of fat and cholecystokinin release: implications for pancreatic carcinogenesis. Lancet 2:1008–1010

Benito-Orfila MA, Playford RJ, Nandha K, Bloom SR (1990) Postprandial concentrations of peptide YY induce changes on VIP cardiovascular response. Brit. J Clin Pharmacol 31:613P

Bennett A, Garrett JR, Howard ER (1968) Adrenergic myenteric nerves in Hirschsprung's disease. Brit Med J 1:487–489

Besterman HS, Bloom SR, Sarson DL, Blackburn AM, Johnston DI, Patel HR, Stewart JS, Modigliani R, Guerin S, Mauison CN (1978) Gut-hormone profile in coeliac disease. Lancet 1:785–788

Besterman HS, Adrien TE, Mallinson CN, Christofides ND, Sarson DL, Pera A, Lombardo L, Modigliani R, Bloom SR (1982) Gut hormone release after intestinal resection. Gut 23:854–861

Besterman HS, Christofides ND, Welby PD, Adrian TE, Sarson DL, Bloom SR (1983a) Gut hormones in acute diarrhoea. Gut 24:665–671

Besterman HS, Mallinson CN, Modigliani R, Christofides ND, Pera A, Pontt V, Sarson DL, Bloom SR (1983b) Gut hormones in inflammatory bowel disease. Scand J Gastroenterol 18:845–852

Bishop AE, Polak JM, Bryant MG, Bloom SR, Hamilton S (1980) Abnormalities of VIP containing nerves in Crohn's disease. Gastroenterology 79:853–860

Blackburn AM, Christofides ND, Ghatei M, Sarson DL, Ebeid FH, Ralphs DN, Bloom SR (1980) Elevation of plasma neurotensin in the dumping syndrome. Clin Sci 59:237–243

Blair AJ, Feldman M, Barnett C, Walsh JM, Richardson CT (1987) Detailed comparison of basal and food stimulated gastric acid secretion rates and serum gastrin concentrations in duodenal ulcer patients and normal subjects. J Clin Invest 79:582–587

Bloom SR (1980) Hormonal changes after jejuno ileal bypass and their physiological significance. In: Maxwell JD, Gazet JC, Pilkington TR (eds) Surgical management of obesity. Academic, London, pp 115–123

Brand SJ, Stone D (1988) Reciprocal regulation of antral gastrin and somatostatin gene expression by omeprazole-induced achlorhydria. J Clin Invest 82: 1059–1066

Calam J, Tracy HJ (1980) Pyloric reflux and gastrin cell hyperfunction. Lancet 2:918

Calam J, Ellis A, Dockray G (1982) Identification and measurement of molecular variants of CCK in duodenal mucosa and plasma. J Clin Invest 69:218–225

Calam J, Ghatei MA, Domin J, Adrian TE, Myszor M, Gupta S, Tait C, Bloom SR (1989) Regional differences in concentrations of regulatory peptides in human colonic mucosal biopsies. Dig Dis Sci 34:1193–1198

Calam J, Yiangou Y, Nikou GC, Chrysanthou BJ, Beacheman JL, Bloom SR (1990) Effects of preprovasoactive intestinal polypeptide derived peptides on ileal output. Gastroenterology 98:505–508

Calam J, Goodlad RA, Lee CY et al. (1991) Achlorhydria-induced hypergastrinaemia: the role of bacteria. Clin Sci 80:281–284

Caldara R, Grimaldi MG, Barbieri C, Grimaldi MG, Barbieri C, Ferrari C (1981) Serum gastrin concentration in patients with rheumatoid arthritis: effect of long-term immunosuppressive or antidopaminergic treatment. Horm Metab Res 13:615–617

Cantor P, Olsen O, Schaffalitzky de Muckadell OB, Jodal M (1988) Effect of trypsin on the hormonal regulation of fat stimulated exocrine secretion. Biomed Res Suppl 1:A25

Cassella RR, Ellis FH, Brown AL (1965) Fine structure changes in achalasia of the eosophagus. I. Vagus nerves. Am J Pathol 46:279–288

Cassuto J, Fahrenkrug J, Jodal M, Tuttle R, Lundgren O (1981) Release of vasoactive intestinal polypeptide from the cat small intestine exposed to cholera toxin. Gut 22:958–963

Cave DR, Vargas M (1989) Effect of a *Campylobacter pylori* protein on acid secretion by parietal cells. Lancet 2:187–189

Chey WY, Chang TM, Lee KY, Sun G, Kim CK, You Ch, Hamilton DL, Shah A, Rhee JC, Mutt V, (1989) Ulcerogenic tumour syndrome of the pancreas associated with a non-gastrin secretogogue. Ann Surg 210:140–149

Chian HCV, O'Dorisio TM, Huang SC, Maton PN, Gardner JD, Jensen RT (1990) Multiple hormone elevations in Zollinger Ellison syndrome. Gastroenterology 99:1565–1575

Chicone L, Narayan S, Townsend CM, Singh P (1989) The presence of a 33–40 KDa gastrin binding protein on human and mouse colon cancer. Biochem Biophys Res Commun 164:512–519

Cohen S (1973) Hypogastrinaemia and sphincter incompetence. N Engl J Med 289:215–217

Cooper RG, Dockray GJ, Calam J, Walker R (1985) Acid and gastrin responses during intragastric titration in normal and DU patients with G-cell hyperfunction. Gut 26:232–236

Crabtree JE, Shallcross TM, Wyatt JI, Meatley RV (1990) Tumour necrosis factor alpha secretion by *Helicobacter pylori* colonised gastric mucosa. Gut 31: A600–A601

Creutzfeldt W, Arnold R, Creutzfeldt C, Track NS (1975) Pathomorphological aspects, biochemical and diagnostic aspects of gastrinomas. Hum Pathol 6:47–76

DelValle J, Yamada T (1990) Amino acids and amines stimulate gastrin release from canine antral G cells via different pathways. Clin Invest 85:139–143

Denizot Y, Sobhani I, Rambaud JC, Lewin M, Thomas Y, Beneviste J (1990) PAF aceter synthesis by *Helicobacter pylori*. Gut 31:1242–1245

De Witte TJ, Geerdink PJ, Lamers CB, Boerbooms AM, Vander Karst JK (1979) Hypochlorhydria and hypergastrinaemia in rheumatoid arthritis. Ann Rheum Dis 38:14–17

Dodds WJ, Dent J, Hogan WJ, Patel GK, Toouli J, Amdorfer RC (1981) Paradoxical lower oesophageal sphincter contraction induced by CCK-8 in patients with achalasia. Gastroenterology 80:327–333

Dooley CP, Cohen H (1988) The clinical significance of *Campylobacter pylori*. Ann Intern Med 108:70–79

Douglas BR, Wouterson RA, Jansen JM, de Jong AJ, Rovati LC, Lamers CB (1989) Modulation by CR-1409 a cholecystokinin antagonist of trypsin inhibitor enhanced growth of azaserine induced putative preneoplastic lesions in rat pancreas. Cancer Res 498:2438–2441

Ehrenpreis T, Pernow B (1953) On the occurrence of substance P in the rectosigmoid junction in Hirschsprungs disease. Acta Physiol Scand 27:380–388

Eklund S, Farenkrug F, Jodal M, Lundgren O, Schaffalitzky de Muckadell OB, Sjöqvist A (1980) Vasoactive intestinal polypeptide, 5-hydroxytryptamine and reflex hyperaemia in the small intestine of the cat. J Physiol (Lond) 302:549–557

Eklund S, Jodal M, Lundgren O (1985) The enteric nervous system participates in the secretory response to the heat stable enterotoxins of *Escherichia coli* in rats and cats. Neuroscience 14:673–681
Ellison EH (1964) The ZE syndrome: reappraised and evaluation of 260 registered cases. Ann Surg 160:512–530
El-Nujumi AM, Dorian CA, Chittajalla RS, Neithercut WD, McColl KE (1991) Effect of inhibition of *H. pylori* urease activity by acetohydroxamic acid on serum gastrin in duodenal ulcer subjects. Gut 32:866–870
Everson GT, McKinley C, Lawson M, Johnson M, Kern F Jr (1982) Gallbladder function in the human female: effect of the ovulatory cycle, pregnancy, and contraceptive steroids. Gastroenterology 82:711–719
Fahrenkrug J, Pedersen JH (1986) Cosecretion of PHM and VIP in patients with VIP producing tumours. Peptides 7:717–721
Figura N, Guglielmetti P, Rossolini A, Barberi A, Cusi G, Musmanno RA, Russi M, Quaranta S (1989) Cytotoxin production by *Campylobacter pylori* strains isolated from patients with peptic ulcers and patients with gastritis only. J Clin Microbiol 27:225–226
Fisher RS, Rock E, Malmud LS (1986) Gallbladder emptying response to sham feeding in humans. Gastroenterology 90:1854–1857
Folsch UR, Winkler K, Wormsley KG (1978) Influence of repeated administration of CCK and secretin on the pancreas of the rat. Scand J Gastroenterol 13:663–671
Forgacs IC, Maisey MN, Murphy GM, Dowling RH (1984) Influence of gallstones and ursodeoxycholic acid therapy on gallbladder emptying. Gastroenterology 87:299–307
Fox PS, Hofmann JW, Decosse JJ, Wilson SD (1974) The influence of total gastrectomy on survival in malignant Zollinger Ellison syndrome. Ann Surg 180:558–566
Freund HR, Humphrey CS, Fischer JE (1979) Reduced tissue content of VIP in aganglionic colon of Hirschsprung's disease. Gastroenterology 76:1135
Funakoshi A, Tateishi K, Shinozaki H, M. Yashaka K, Ito T, Wakasugi H (1990) Plasma pancreastatin responses after intrajejunal infusion of liquid meal in patients with chronic pancreatitis. Dig Dis Sci 35:721–725
Gear MWL, Truelove SC, Whitehead R (1971) Gastric ulcer and gastritis. Gut 12:639–645
Gerber BC (1960) Simultaneous duodenal carcinoma and non-beta cell tumours of the pancreas. Arch Surg 81:379–388
Goodwin CS (1988) Duodenal ulcer, *Campylobacter pylori* and the "leaking roof" concept. Lancet 2:1467–1469
Gordis, Gold EB (1984) Epidemiology of pancreatic cancer. World J Surg 8:808–821
Goyal RK, Ratten S, Said SI (1980) VIP as a possible neurotransmitter of non-cholinergic, non-adrenergic inhibitory neurones. Nature 288:378–380
Harris AG (1990) Future medical prospects for Sandostatin. Metabolism 39 Suppl 2:180–185
Harvey RF, Read AE (1973) Effect of CCK on colonic motility and symptoms in patients with the irritable bowel syndrome. Lancet 1:1
Hoosein NM, Keiner PA, Currey RC, Brattain MG (1990) Evidence for autocrine growth stimulation of cultured colon tumour cells by a gastrin/CCK-like peptide. Exp Cell Res 186:15–21
Hopman WP, Jansen JB, Rosenbusch G, Lamers CB (1987) Cephalic stimulation of gallbladder contraction in humans: role of cholecystokinin and the cholinergic system. Digestion 38:197–203
Hubel K (1972) Secretin: a long progress note. Gastroenterology 62:318–341
Ingelfinger FJ (1958) Esophageal motility. Physiol Rev 38:533–584
Iwai K, Fukuoka SI, Fushiki M, Tsujikawa M, Hirose M, Tsunasawa S, Sakiyama F (1987) Purification and sequencing of a trypsin sensitive cholecystokinin releasing peptide from rat pancreatic juice: its homology with pancreatic secretory trypsin inhibitor. J Biol Chem 262:8956–8959

Johnson HD, Love AHG, Rogers NC, Wyatt AP (1964) Gastric ulcers, blood groups and acid secretion. Gut 5:402–405

Kelbaek H, Munck O, Christensen NJ, Godtfredsen J (1989) Central haemodynamic effects after a meal. Br Heart J 61:506–509

Koch TR, Michener SR, Go VL (1991) Plasma VIP concentration determination in patients with diarrhoea. Gastroenterology 100:99–106

Koide M, Okabayashi Y, Hasegawa H, Tani S, Fujisawa T, Nakamura T, Fujii M, Otsuki M (1989) Plasma cholecystokinin concentrations in patients with chronic pancreatitis measured by bioassay. Nippon Shokakibyo Gakkai Zasshi 86: 2419–2424

Kooner JS, Peart WS, Mathias CJ (1989) The peptide release inhibitor, Octreotide (SMS 201–995), prevents the haemodynamic changes following food ingestion in normal human subjects. Q J Exp Physiol 74:569–572

Koop I, Koop H, Gerhardt C, Schafmaner A, Arnold R (1989) Do bile acids exert a negative feedback control of CCK release? Scand J Gastroenterol 24:315–320

Kramer P, Ingelfinger FJ (1951) Esophageal sensitivity to metholyl in cardiospasm. Gastroenterology 19:242

Krejs GJ, Browne R, Raskin P (1980) Effect of intravenous somatostatin on jejunal absorption of glucose, amino acids, water and electrolytes. Gastroenterology 78:26–31

Lamberts R, Creutzfeldt W, Stockman F, Jacubaschke V, Maas S, Brunner G (1988) Long term omeprazole treatment in man. Digestion 39:126–135

Lanzon-Millar S (1987) Twenty four hour intragastric acidity and plasma gastrin concentration before and during treatment with either ranidine or omeprazole. Aliment Pharmacol Ther 1:239–251

Larsson C, Skogseid B, Oberg K, Nakamura Y, Nordenskjold M (1988) Multiple endocrine neoplasia type 1 gene maps to chromosome 11 and is lost in insulinoma. Nature 332:85–87

Larsson H, Carlsson E, Hakanson R, Mattsson H, Nilsson-Seensalu R, Wallmark B, Sundler F (1988) Time course of development and, reversal of gastric endocrine cell hyperplasia after inhibition of acid secretion. Gastroenterology 95: 1477–1486

La Vecchia C (1990) Histamine-2-receptor antagonists and gastric cancer risk. Lancet 336:355–357

Leunk RD, Johnson PT, David BC, Kraft WG, Morgan DR (1988) Cytotoxic activity in broth culture filtrates of *Campylobacter pylori*. J Med Microbiol 26:93–99

Levi S, Beardshall K, Swift I, Foulkes W, Playford RJ, Gosh P, Calam J (1989a) Antral *Campylobacter pylori*, hypergastrinaemia and duodenal ulcers effect of eradicating the organism. Br Med J 299:1504–1505

Levi S, Beardshall K, Haddad G, Playford RJ, Gosh P, Calam J (1989b) *Campylobacter pylori* and duodenal ulcers: the gastrin link. Lancet 1:1167–1168

Liddle RA, Gertz BJ, Kanayama S, Beccaria L, Coker LD, Turnbull TA, Morita ET (1989) Effects of a novel cholecystokinin (CCK) receptor antagonist, MK-329, on gallbladder contraction and gastric emptying in humans. Implications for the physiology of CCK. J Clin Invest 84:1220–1225

Lipschutz WH, Cohen S (1971) Physiological determinants of lower esophagal sphincter function. Gastroenterology 61:16–24

Long RG, Bishop AE, Barnes AJ, Albuguergne RH, O'Shaughnessy DJ, McGregor GP, Baunister R, Polak JM, Bloom SR (1980) Neural and hormonal peptides in rectal biopsy specimens from patients with Chagas' disease. Lancet 1:559–562

Long RG, Bryant MG, Mitchell SJ, Adrian TE, Polak JM, Bloom SR (1981) Clinicopathological study of pancreatic and ganglioblastoma tumour-secreting VIP (VIPomas). Br Med J 282:1767–1771

Mai UEH, Perz GI, Wahl LM, Wahl SM, Blaser MJ, Smith PD (1990) Inflammatory and cytoprotective response by human monocytes are induced by *Helicobacter pylori*. Gastroenterology 98:A662

Marshal BJ, Goodwin CS, Warren JR, Murray R, Blincow ED, Blackboorn SJ, Phillips M, Waters TE, Sanderson CR (1988) Prospective double blind trial of duodenal ulcer relapse after eradication of *Campylobacter pylori*. Lancet 2: 1439–1441

Masclee AA, Jansen JB, Driessen WM, Geuskens LM, Lamers CB (1989) Plasma cholecystokinin and gallbladder responses to intraduodenal fat in gallstone patients. Dig Dis Sci 34:353–359

Maton PN, Gardner JD, Jensen RT (1989) Use of long-acting somatostatin analog SMS 201–995 in patients with pancreatic islet cell tumors. Dig Dis Sci 34 Suppl 3:28S–39S

McColl KE, Fullarton GM, El-Nujumi AM, Macdonald AM, Brown IL, Hilditch TE (1989) Lowered gastric acidity and gastrin after eradication of *Campylobacter pylori* in duodenal ulcer. Lancet 2:499–500

McColl KEL, Fullarton GM, El-Nujumi AM, MacDonald AMI, Dahil S, Hilditch TE (1990) Serum gastrin and gastric acid status one and seven months after eradication of *Helicobacter pylori* in duodenal ulcer patients. Gut 31:A601

McGuigan JE, Wolfe MM (1980) Secretin injection test in the diagnosis of gastrinoma. Gastroenterology 79:1324–1331

Messing B, Bories C, Kunstlinger F, Messing B, Bories C, Kunstlinger F, Bernier JJ (1983) Does total parenteral nutrition induce gallbladder sludge formation and lithiasis? Gastroenterology 84:1012–1019

Meyer BM, Werth BA, Beglinger C, Hildebrand P, Jansen JB, Zach D, Rovati LC, Stadler GA (1989) Role of cholecystokinin in regulation of gastrointestinal motor functions. Lancet 1:12–15

Michelangeli F, Ruiz MC, Rodriguez CL, Relacca A (1988) Somatostatin inhibition of secretagogue and forskolin-stimulated gastric acid secretion. Am J Physiol 254:G531–G537

Nemeth J, Taylor BA, Dockray GJ (1992) Human colonic cancers contain gastrin precursor. Gut (in press)

Orme ML, Dockray GJ, Baber N, Sibeon RG, Malliday LD, Littler TR (1979) The effects of indomethacin on serum gastrin concentrations (Letter). Ann Rheum Dis 38:578

Owyang C, May D, Louie DS (1986) Trypsin suppression of pancreatic enzyme secretion. Gastroenterology 91:637–643

Pipeleers-Marichal M, Somers G, Willems G, Foulis A, Jmrie C, Bishop AE, Polak JM, Hacki WH, Stamm B, Heitz PU (1990) Gastrinomas in the duodenum of patients with multiple endocrine neoplasia type 1 and the Zollinger Ellison syndrome. N Engl J Med 322:723–727

Playford RJ, Domin J, Beecham J, Parmar KB, Tatemoto K, Bloom SR, Calam J (1990) Peptide YY: a natural defence against diarrhoea. Lancet 335:1555–1557

Polak JM, Bloom SR (1982) Regulatory peptides in the gut and their relevance to disease. In: Crow TJ (ed) Disorders of neuroendocrine transmission. Academic, London, pp 83–104

Primrose N, Ratcliffe JG, Joffee SN (1985) Differences between peptic ulcer and control patients on the basis of response to secretin. Digestion 32:249–254

Ramo OJ, Puolakkainen PA, Seppala K, Schroder TM (1989) Self-administration of enzyme substitution in the treatment of exocrine pancreatic insufficiency. Scand J Gastroenterol 24:688–692

Rhodes J, Barnado DE, Sidney MB, Philips SF (1969) Increased reflux of bile into the stomach in patients with gastric ulcer. Gastroenterology 57:241–252

Richards AT, Hinder RA, Harrison AC (1987) Gastric carcinoid tumours associated with hypergastrinaemia and pernicious anaemia – regression of tumours by antrectomy. S Afr Med J 72:51–54

Roebuck BD, Yager JD, Longnecker DS, Wilpone SA (1981) Promotions by unsaturated fat of azaserine-induced pancreatic carcinogenesis in the rat. Cancer Res 41:3961–3966

Ruskone A, Rene E, Chayvialle JA, Bonin N, Pignal F, Kremer M, Bonfils S, Ramband JC (1982) Effects of somatostatin on diarrhoea and on small intestinal water and electrolyte transport. Dig Dis Sci 27:459

Sagor GR, Bryant MG, Ghatei MA, Kirk RM, Bloom SR (1981) Release of vasoactive intestinal peptide in the dumping syndrome. Br Med J [Clin Res] 282: 507–510

Sakamoto T, Miyata M, Hamaji M, Sakaguchi H, Tanaka Y, Hashimoto T, Kawashima Y (1990) Beta adrenergic regulation of gastrin release from gastrinoma cells. Surgery 107:282–288

Servin AL, Rouyer-Fessard C, Balasubramaniam A, Saint-Pierre S, Laburthe M (1989) Peptide-YY and neuropeptide-Y inhibit vasoactive intestinal peptide-stimulated adenosine 3′, 5′-monophosphate production in rat small intestine: structural requirements of peptides for interacting with peptide-YY-preferring receptors. Endocrinology 124:692–700

Shinya H, Fujimura M (1989) Possible role of cholecystokinin in the development of acute pancreatitis in rats. Nippon Geka Hokan 58:3–17

Skak-Nielson T, Holst JJ, Nielson OV (1988) Role of GRP in the neural control of pepsinogen secretion from the pig stomach. Gastroenterology 95:1216–1220

Smith JP, Wood JG, Solomen TE (1989) Elevated gastrin levels in patients with colon cancer or adenomatous polyps. Dig Dis Sci 34:171–174

Spengler U, Sackmann M, Sauerbruch T, Holl J, Paumgartner G (1989) Gallbladder motility before and after extracorporeal shock-wave lithotripsy. Gastroenterology 96:860–863

Stone BG, Gavaler JS, Belle SH, Shreiner DP, Peleman RR, Sarra RP, Yingvorapant N, Vanthiel DH (1988) Impairment of gallbladder emptying in diabetes mellitus. Gastroenterology 95:170–176

Stremple JF (1974) Serum gastrin levels in reflux oesophagitis. Gastroenterology 66:A131

Sturdevant RAL (1974) Is gastrin the major regulator of lower esophageal sphincter pressure? Gastroenterology 67:551–553

Svendsen JH, Dahl C, Svendsen LB, Christiansen PM (1986) Gastric cancer risk in achlorhydric patients. A long-term follow-up study. Scand J Gastroenterol 21: 16–20

Sylwestrowicz TA, Shaffer EA (1988) Gallbladder function during gallstone dissolution. Effect of bile acid therapy in patients with gallstones. Gastroenterology 95:740–748

Tally NJ, DiMagno E, Zinsmeister AR, Blaser M (1990) *Helicobacter pylori* and gastric cancer: a case controlled study. Enferm Dig 78 Suppl:O11

Tatemoto K (1982) Isolation and characterisation of peptide YY (PYY), a candidate gut hormone that inhibits pancreatic exocrine secretion. Proc Natl Acad Sci USA 79:2514–2518

Teichman K, Grab P, Hammer C, Brendel W (1986) Gastrin release by interleukin 2 and gamma interferon in vitro. Can J Physiol Pharmacol 64 Suppl:62

Uddman R, Alumets J, Edvinsson L, Hakanson R, Sundler F, (1978) Peptidergic (VIP) innervation of the eosophagus. Gastroenterology 75:5–8

Van Erpecum KJ, van Berge Henegouwen GP, Stolk MF, Hopman WP, Jansen JB, Lamers CB (1990) Effects of ursodeoxycholic acid on gallbladder contraction and cholecystokinin release in gallstone patients and normal subjects. Gastroenterology 99:836–842

Vassallo G, Solcia E, Bussolati G, Polak JM, Pearse AG (1972) Non G-cell gastrin producing tumours of the pancreas. Virchows Arch [B] 11:66–79

Walan A, Bader JP, Classen M, Lamers CB, Piper DW, Rutgersson K, Eriksson S (1989) Effect of omeprazole and ranitidine on ulcer healing and relapse rates in patients with benign gastric ulcers. N Engl J Med 320:69–73

Walsh JH, Richardson CT, Fordtran JS (1975) pH dependence of acid secretion and gastrin release in normal and ulcer subjects. J Clin Invest 55:462–468

Wider M, Lu L, Owyang C (1988) Extraction and characterisation of a CCK-releasing peptide mediating feedback regulation of pancreatic secretion (Abstr). Biomed Res 9 Suppl 1:54

Williams JA, Goldfine ID (1985) Insulin-pancreatic acinar axis. Diabetes 34:980–986

Wong K, Beardshall K, Calam J, Poston GJ (1990) Postprandial hypergastrinaemia in patients with colorectal cancer. Gut 31:A1166

Wyatt JI, Rathbone BJ, Green DM, Primrose J (1989) Raised fasting serum gastrin in chronic gastritis is independent of *Campylobacter pylori* status and duodenal ulcer. Gut 30:A1483

Wynder EL (1975) An epidemiological evaluation of the causes of cancer of the pancreas. Cancer Res 35:2228–2233

Wynick D, Williams SJ, Bloom SR (1988) Symptomatic secondary hormone syndromes in patients with established malignant pancreatic endocrine tumours. N Engl Med 319:605–609

Zollinger RM (1975) Islet cell tumours of the pancreas and alimentary tract. Am J Surg 129:102–110

CHAPTER 15

Gastrointestinal Peptides as Therapeutic Agents and Targets: Past, Present and Future

T.S. GAGINELLA

A. Introduction

Peptides of gastrointestinal (GI) origin, as discussed in detail throughout this volume, function in many capacities. Their endocrine, paracrine, neurocrine and autocrine actions control or modify the function of "distant" organs, other cells within the peptide-producing tissue, or through feedback upon the cells of origin. Among the disorders with a possible link to GI peptides (see Chap. 14) are hypergastrinemia, peptic ulcer disease, diarrheal syndromes (e.g., VIPoma syndrome), obesity, Hirshsprung's disease and irritable bowel syndrome (IBS) (DEBAS 1987; DEL VALLE and YAMADA 1990; GAGINELLA et al. 1990; SHULKES 1990; OHE et al. 1980; HARVEY and READ 1973).

The birth of gut endocrinology began with the discovery and initial characterization of the action of secretin (BAYLISS and STARLING 1902). The beginning of the therapeutic application of peptide hormones came with the dramatic demonstration of the efficacy of crude insulin in the treatment of diabetes mellitus (BANTING et al. 1922). Although insulin is not used to treat dysfunction of the GI tract, it does influence the activity of the gut. Insulin enhances the rate of gut maturation (HENNING 1987), indirectly activates vagal stimulation of gastric acid secretion (HIRSCHOWITZ and SACHS 1965), and exerts a choleraic effect on biliary secretion (BALDWIN 1966). The close proximity of A and D cells to the insulin-producing cells of the pancreas results in paracrine influence of glucagon and somatostatin on insulin secretion. This illustrates a potential pharmacologic complication of the use of peptides in therapy, whereby an administered peptide may initiate unintended consequences through changes in endogenous peptide levels.

B. Diagnostic Use of Gastrointestinal Peptides

An important value of GI peptides exists in their use as diagnostic aids (Tables 1, 2). The availability of specific and sensitive radioimmunoassays for peptides has been a key factor in diagnosis of several types of GI tumors. The availability of antisera permits circulating concentrations of the peptides to be evaluated and immunohistochemistry to be done on tissue sections or biopsies. The diagnosis of VIPoma, for example, is made by

Table 1. Radioimmunoassay in diagnosis of gut tumors

Peptide assayed	Condition associated with elevation of circulating concentration
Gastrin	Zollinger-Ellison syndrome (>500 pg/ml serum)
Cholecystokinin	Chronic pancreatitis[a]
Secretin	Secretinoma, Zollinger-Ellison syndrome
Vasoactive intestinal peptide (VIP)	VIPoma (watery diarrhea syndrome; Verner-Morrison syndrome; pancreatic cholera); (plasma level >170 pg/ml)

[a] Anomaly; elevated levels reported due to exaggerated CCK release in chronic pancreatitis (DEBAS 1987).

Table 2. Clinical use of GI peptides as agonists

Peptide	Clinical applications
Gastrin (pentagastrin)	• Test for gastric acid secretory capacity • Provocative test for carcinoid tumors
Cholecystokinin	• Diagnosis of chronic pancreatitis, based on pancreatic secretory capacity • Provocative test to evoke bile flow to determine origin of abnormal bile flow and abdominal pain
Secretin	• Diagnosis of Zollinger-Ellison syndrome; paradoxical elevation of gastrin (>100 pg/ml) • Enhance radiologic visualization of small pancreatic vessels
Glucagon	• Provokes release of catecholamines in pheochromocytoma • Reduction of gut tone (motility) to improve radiologic and endoscopic exams
Motilin (erythromycin analogs)	• Gastroparesis • Postoperative ileus (?)
Vasoactive intestinal peptide	• Bronchodilator in asthma
Somatostatin or octreotide	• ↓ GI bleeding • ↓ Release/action of other GI peptides • ↓ Pancreatic secretion • ↓ Secretory diarrhea (e.g., as in AIDS and VIPoma) and ileostomy • Acromegaly • Antiproliferation of GI tissue • Localization of tumor metastases

noting fasting plasma levels of vasoactive intestinal peptide (VIP) greater than 170 pg/ml; values are usually in excess of 225 pg/ml and as high as 1000 pg/ml (O'Dorisio and O'Dorisio 1985). Histochemical techniques have been used to identify VIP-containing nerve endings and endocrine cells in VIP-secreting neural tumors and pheochromocytomas (Solcia et al. 1982). Elevated plasma or serum levels of VIP have been reported in mastocytoma, bronchogenic carcinoma, medullary thyroid carcinoma, ganglioneuroma, pheochromocytoma, neuroblastoma and islet cell adenoma (Gaginella et al. 1989).

The availability of a radioimmunoassay for gastrin made it possible to confirm diagnosis of Zollinger-Ellison syndrome (gastrinoma). Exaggerated elevations of serum gastrin (in excess of 500 pg/ml) are observed in these patients. Besides its clinical applications in hypergastrinemia, the assay has better defined the physiology of gastrin release and action (Wiener et al. 1987).

Provocative tests have proved useful in the diagnosis of GI neoplasms and other pathologies. A common test for determining maximal gastric acid secretory capacity is to stimulate acid output with pentagastrin. This test is particularly useful after partial gastrectomy to determine residual gastric secretory capacity. Pentagastrin also has been used to diagnose gut carcinoid tumors.

A combination of secretin and cholecystokinin (CCK) is used to diagnose chronic pancreatitis. Both peptides are infused intravenously and pancreatic juice collected. The degree of secretory capacity can be assessed and compared with other indices of disease severity. Secretin is used in the diagnosis of Zollinger-Ellison syndrome. Although in the normal physiologic state secretin does not evoke the release of gastrin, intravenous administration of the peptide causes a rise in plasma gastrin of 100 pg/ml or greater in most patients with confirmed gastrinoma (Debas 1987).

The sections that follow are organized according to the therapeutic applicability of GI peptides based upon their agonist or antagonist activities. Consideration is given to the use of peptide agonists for conditions in which there is reduced release of a given peptide, and for antagonists (peptide and nonpeptide) where there is excessive release or hypersensitivity to a particular peptide. In some cases agonists that mimic the natural peptide will be discussed. Table 3 summarizes some applications for antagonists of GI peptides.

C. Peptides as Agonists in Therapeutics

I. Somatostatin

Natural somatostatin (SS) exists in at least two molecular forms. The tetradecapeptide and octaeicosapeptide forms of SS are considered to be of

Table 3. Potential applications of GI peptide antagonists

Antagonism of	Result
Gastrin	• Decrease gastric H^+ secretion
	• Reduce growth rate of gastrin-dependent cancers
Cholecystokinin	• Decrease intestinal hypermotility or inappropriate motility in IBS
	• Stimulate feeding behavior
Vasoactive intestinal peptide	• Inhibit secretory diarrhea
Substance P	• Mitigate pain/cramping in irritable bowel syndrome

greatest physiologic importance. About 70% of all the SS in the body is contained in the GI tract. Release into the circulation is stimulated by CCK, secretin, gastrin-17 and other hormones, intraluminal acid, bile, prostaglandins, circadian rhythms, motor activity of the gut and changes in autonomic nervous system tone (NEWMAN et al. 1987). Somatostatin has numerous inhibitory actions on the GI tract. Therefore, although an agonist, its activity is often expressed as a reduction in physiologic and pathologic events. These inhibitory effects, particularly on GI motility, acid secretion, hormone release and splanchnic blood flow, form the basis for somatostatin's clinical efficacy.

Clinical trials have been undertaken with SS in patients with upper GI bleeding (MAGNUSSON et al. 1985; KAYASSEH et al. 1980). Portal pressure and heptic blood flow in patients with cirrhosis is reduced by infusion of SS but efficacy in the treatment of acute variceal hemorrhage remains unproven (JENKINS et al. 1985). Another major use of SS is to control secretory diarrhea in post-pseudo-obstruction (MULVIHILL et al. 1984). Some success also has been obtained with the use of SS to induce temporary closure of pancreatic and intestinal fistulae (MULVIHILL et al. 1986). Beneficial effects were also achieved in the dumping syndrome (LONG et al. 1985) and the short bowel syndrome (MULVIHILL et al. 1986). Somatostatin may have additional applications in acromegaly and as an antiproliferative agent in tumor growth (NEWMAN et al. 1987).

Somatostatin is most readily cleaved at the tryptophan $(Trp)^8$ position. In addition, aminopeptidases can attack its N terminus (see Chap. 6). Upon the discovery of substitution of D for L isomers for several of the amino acids in the sequence, enhanced metabolic stability was noted. This information led several research laboratories to develop SS analogs with varying degrees of biological activity and enhanced metabolic stability (PLESS 1989).

Sandoz was successful in their development strategy and have marketed octreotide acetate as Sandostatin. This molecule is protected against metabolic degradation through the introduction of D-phenylalanine (Phe) at the N terminus, inclusion of threonine (ol) at the C terminus and substitution of

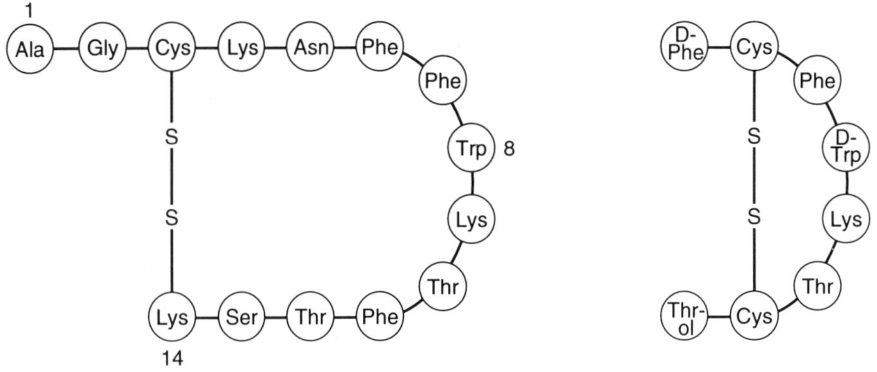

Fig. 1. Comparative structures of natural somatostatin and octreotide. (Modified from JENSEN and NORTON 1991)

D-Trp for L-Trp at position 8 (Fig. 1). This metabolically stable analog of SS can be administered intramuscularly three to four times daily, obviating the necessity for constant infusion. This represents a major accomplishment toward the development of peptides as therapeutic agents in the future; thus octreotide will be treated here in some detail.

Octreotide is nearly completely stable for up to 20h at 37°C when exposed to an ultrafiltrate of rat kidney homogenate, compared to less than 1h for native SS (PLESS 1989). This octapeptide is absorbed well and associated with good therapeutic activity after intramuscular or subcutaneous injection.

Receptor sites for SS have been identified throughout the GI tract, in the brain and in other tissues (NEWMAN et al. 1987; GAGINELLA 1990a). Virtually all of the pharmacological effects of somatostatin are mimicked by octreotide. This is probably due to the octapeptide sharing a four amino acid homology with native somatostatin; yet conclusive studies to determine by direct measurement the interaction of the synthetic peptide with receptors to native somatostatin have not been done.

The spectrum of activity of octreotide includes inhibition of the release of growth hormone, insulin, gastrin, gastric inhibitory peptide, glucagon, pancreatic polypeptide, motilin, secretin, neurotensin, and VIP (LONGNECKER 1988; JENSEN and NORTON 1991). The most important current use for octreotide is in the treatment of tumors of the gastroenteropancreatic axis (GAGINELLA et al. 1989) and for acquired immunodeficiency syndrome (AIDS) associated secretory diarrhea (GAGINELLA and O'DORISIO 1988; KATZ et al. 1988). In vivo radioisotopic labeling of tumor metastases also has been accomplished with octreotide (KRENNING et al. 1989).

Somatostatin-14 either enhances basal intestinal chloride ion absorption or reduces electrogenic anion secretion in a variety of in vitro and in vivo experimental paradigms (see also Chap. 10). Substitutions on the molecule have some effect on this activity compared to little or no action on the endocrine system, suggesting the possibility for subtypes of SS receptors in different tissues. Whether or not SS-14 or -28 play a role in the physiologic regulation of intestinal electrolyte transport is not known.

Like SS, octreotide exerts a net antisecretory effect in rabbit ileum and rat colon (GAGINELLA et al. 1990). It antagonizes the secretory effects of 5-HT, VIP, bethanechol, substance P, aminophylline and bradykinin (FASSLER et al. 1990). In these in vitro systems, in which short-circuit current was measured as an index of electrogenic secretion, the patterns of inhibition suggested that more than one mechanism for octreotide is operative. These are probably distal to the generation of intracellular effectors (e.g., cyclic nucleotides, calcium).

Clinically, octreotide reverses the secretory diarrhea and hypokalemia of VIPoma syndrome (SANTANGELO et al. 1985; MATON et al. 1985) and reduces the release of VIP from the tumor (ANDERSON and BLOOM 1986; MATON et al. 1985). This is accomplished with a dose of 50–100 μg administered twice daily subcutaneously. A similar dose regimen has been effective in treating the diarrhea of carcinoid syndrome (VINIK and MOATTARI 1989). Diarrheas associated with diabetes, amyloidosis and high-output ileostomy are responsive to octreotide therapy (GAGINELLA et al. 1990).

Additional studies with octreotide and other analogs (some of which have oral activity) will undoubtedly improve our approach to therapy of diseases of diverse pathophysiologies. Octreotide serves as an important model from which to develop other peptides such as mimics and peptide antagonists. It provides solid evidence that this is achievable.

II. Glucagon

In pharmacologic doses glucagon increases biliary flow but decreases gastric emptying, colonic motility and gallbladder pressure and relaxes the sphincter of Oddi in humans (ALWMARK and GREELEY 1987). Relaxation and dilation of the stomach and duodenum, which have been used to aid in small intestinal and pancreatic endoscopy (MILLER et al. 1982), are other effects of glucagon. The suppressive effect of glucagon on exocrine pancreatic secretion is useful in relieving the pain and improving the condition of patients with acute pancreatitis (DYCK et al. 1970; CONDON et al. 1973).

Another therapeutic action attributed to pharmacologic doses of glucagon is its inotropic effect on the heart in emergency situations (SALZBERG and GALLAGHER 1980). In addition, cardiac output is increased in dogs under experimental conditions; this is designed to preserve intestinal perfusion during acute hypovolemic shock (BOND and LEVITT 1980). Other actions of glucagon include its satiety effect (Geary et al. 1981; Smith 1980) and its ability to stimulate insulin release under fasting, hypoglycemic conditions (Alwmark and Greeley 1987).

III. Motilin

Motilin is a 22 amino acid peptide originally isolated from the duodenum (Brown 1967). It is also found in the CNS and in high concentrations in the gallbladder and the colon (Walker and Rayford 1987). The physiologic role of motilin is not completely understood but it appears to serve as an interdigestive regulator of phase III intestinal motility (see also Chap. 9); it also stimulates gastric motility and increases pancreatic bicarbonate secretion (Walker and Rayford 1987).

An important potential therapeutic application of motilin and agonists with motilin-like activity is in the treatment of gastroparesis and post-operative ileus. Decreased levels of motilin are reported in patients with ileus (Fujimoto et al. 1982). The condition subsides as plasma levels of motilin rise after surgery. Serum motilin is elevated in patients with irritable bowel syndrome (IBS), dumping syndrome and in acute diarrheal syndromes (Ohe et al. 1980). Thus, an antagonist of motilin would be desirable for these conditions.

The macrolide antibiotic erythromycin has motilin-like activity on the GI tract (see Chap. 9). This drug and its analogs are being investigated for conditions associated with motilin deficiency. Erythromycin normalizes impaired gastric emptying and evokes a gastrokinetic response in patients with diabetic gastroparesis (Janssens et al. 1990). Ten insulin-dependent diabetes patients were studied for emptying of liquids and solids, with and without intravenous administration of 200 mg erythromycin placebo. In this double-blind crossover, erythromycin shortened the emptying time for liquids and solids back to the normal range. Treatment with 250 mg three times daily for 4 weeks also improved emptying in these patients. These results are encouraging and may offer new hope for treatment alone or in combination with other drugs for severe gastroparesis.

Work continues on derivatives of erythromycin. These efforts have yielded some synthetic agents with greater GI motor-stimulating activity (in the dog), and virtual absence of antimicrobial activity. The analog *N*-propargyl-8,9-anhydroerythromycinA 6,9-hemiacetal bromide is almost 3000 times more potent than erythromycin A and three times more potent than motilin in stimulating canine intestinal motor activity (Sunazuka et al. 1989). This and other agents with motilin-like activity must survive the gauntlet of toxicological tests required to declare them devoid of unwanted activities. They might then be developed to enhance motor activity of specific areas of the gut as appropriate for ileus, gastroparesis or other dysmotilities.

D. Antagonists of Peptides in Therapy

I. Gastrin Antagonists

Gastrin was originally identified in an extract of the antral mucosa (EDKINS 1906). Different molecular forms of gastrin have been characterized. Among these are peptides comprising 17 (G-17), 34 (G-34) or 14 amino acids. The occurrence and distribution of different forms of gastrin depends upon the species, the tissue, and the presence of pathology.

Gastrin is released from antral and duodenal cells in response to neural, chemical, or mechanical stimuli. Its main function is the stimulation of gastric acid secretion (see Chap. 7). Other physiologic or pharmacologic effects of gastrin include stimulation of pepsin secretion and enhancement of gastric mucosal blood flow, contraction of the lower esophageal sphincter and trophic effects on GI tissues (WIENER et al. 1987). The trophic effect has been linked to stimulation of GI cancers (TOWNSEND et al. 1989; SIRINEK et al. 1985; MCGREGOR et al. 1982; WATSON et al. 1989; and see Chap. 12). Indeed, elevated plasma gastrin levels have been reported in patients with colon cancer (SMITH et al. 1989), and hypergastrinemia enhances chemical colorectal carcinogenesis in rats (KARLIN et al. 1985). Furthermore, gastrin receptors are present on human colon carcinoma cells (UPP et al. 1989; WEINSTOCK and BALDWIN 1988). Therefore, the principal applications of antagonists of gastrin are for inhibitory actions on gastric secretion and on the growth of gastrin-responsive cancers. The antagonists will have to be capable of interacting at what might be different forms of the gastrin receptor expressed on the carcinoma cells.

In this chapter the antagonists of gastrin and CCK will be treated separately. This is somewhat arbitrary, however, because the composition of the two peptides is so similar. The last five amino acids from the C terminal of both peptides are identical, thus there is crossover for antagonist activity against CCK and gastrin receptors. There is evidence that at least three subtypes of the gastrin/CCK receptor exist (FREIDINGER 1989). The gastrin receptor on parietal cells appears to be most closely related to the CCK receptor distributed in the central nervous system. Antagonists for these receptors have been derived from cyclic nucleotides, microbial sources, benzodiazepines and amino acids (Fig. 2).

Early attempts at the identification of antagonists to gastrin were unsuccessful, despite the synthesis of over 500 analogs of the native peptide (BASS 1974). In the late 1960s proglumide (benzoyl-DL-glutamic acid dipropylamide) was synthesized and identified as a weak gastrin antagonist. Concentrations generally higher than 1 m*M* are required in vitro for inhibition of gastrin (ROVATI 1976; FREIDINGER 1989). Doses in the range of 20–500 mg/kg are required for inhibition in vivo (ROVATI 1976). Benzotript (*p*-chlorobenzoyl-L-tryptophan) is another gastrin receptor competitive blocker (ROVATI 1987). The lack of selectivity is shown by proglumide and

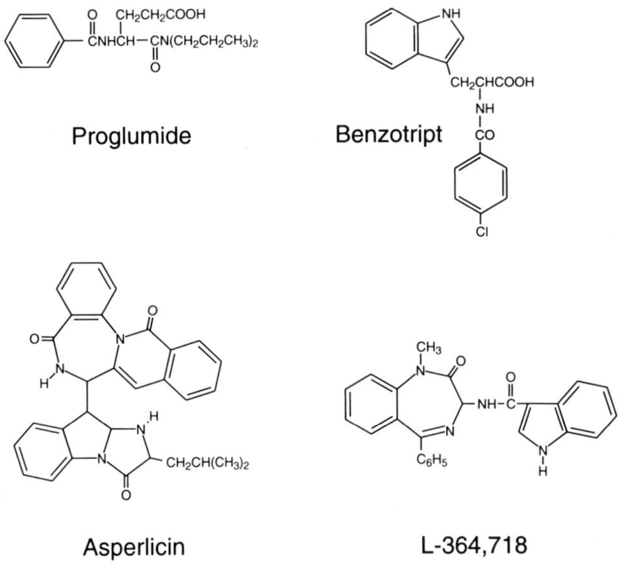

Proglumide Benzotript

Asperlicin L-364,718

Fig. 2. Chemical structures of CCK antagonists

benzotript inhibiting CCK binding and receptor activation (FREIDINGER 1989).

The trophic effects of pentagastrin on rat gastric oxyntic and the colonic mucosa were blunted by i.p. injection of 100–400 mg/kg proglumide (JOHNSON and GUTHRIE 1984). Pancreatic and duodenal growth were also inhibited. Growth of transplantable mouse cancer cells (MC-26) which contain gastrin receptors and are responsive to pentagastrin was inhibited in tumor-bearing mice by injections of proglumide (250 mg/kg three times daily for 7 days); tumor size, weight, DNA and RNA content were significantly lower and survival was increased in the treated compared to the control group (BEAUCHAMP et al. 1985). Subsequently it was demonstrated that proglumide can lower the binding affinity of gastrin receptors on MC-26 tumor cells, offering another explanation for how gastrin antagonists might reduce tumor growth (SINGH et al. 1987).

Proglumide and benzotript inhibit the growth of human colon carcinoma monolayers in culture, but at concentrations in the millimolar range (HOOSEIN et al. 1988). A variety of other human gut carcinoma cell lines are responsive to gastrin and to the inhibitory effects of proglumide (IMDAHL et al. 1989; TOWNSEND et al. 1989). In a randomized, controlled study in 110 patients with gastric carcinoma, HARRISON et al. (1990) found that proglumide (800 mg p.o. four times daily) did not significantly increase survival time. They attributed this to the low potency of proglumide in terms of its affinity for the gastrin receptor and possible partial agonist activity. Similarly, proglumide did not increase the survival of 41 patients with advanced colorectal cancer (MORRIS et al. 1990). We will have to await the further development of more specific and more potent gastrin antagonists and their performance in clinical trials of their efficacy in gastrin-responsive gastric and/or colon carcinomas.

Other gastrin antagonists have been synthesized. Most are derivatives of the C-terminal tetrapeptide of gastrin. The importance of the Phe residue and C-terminal carboxamides and binding through the aromatic ring were demonstrated by MARTINEZ et al. (1986). They also observed highly potent antagonist activity in the Boc-β-Ala-Trp-Leu-Asp phenethyl ester and the Boc-β-Ala-Trp-Leu-Asp *p*-fluorophenethyl ester. Later this group identified a smaller peptide, Boc-Trp-Leu-Asp-NH_2, which inhibited gastrin binding to isolated gastric mucosal cells with an IC_{50} of $50\,\mu M$, and inhibited pentagastrin-induced secretion in vivo (LAVEZZO et al. 1986). Removal of the α-amino group of Trp and conversion of the indole to a phenyl ring, with methylation of the amide bonds, conserved antisecretory activity in gastric fistula dogs (YASUI et al. 1990).

The synthetic agent L-365,260 binds competitively to gastrin receptors in the guinea pig stomach (LOTTI and CHANG 1989) and pancreatic acini (HUANG et al. 1989). This molecule also binds to CCK receptors in the guinea pig brain with a K_i of $2.0\,nM$, virtually identical to the value determined for affinity to the gastrin receptor in the stomach (LOTTI and CHANG 1989). The K_i of this compound for binding to gastrin receptors on pancreatic acini is reported to be $7.3 \pm 0.8\,nM$ (HUANG et al. 1989). L-365,260 did not affect basal acid secretion or that stimulated by histamine or carbachol, but was effective in antagonizing gastrin-stimulated acid secretion in mice (LOTTI and CHANG 1989). Another analog, L-364,718, was found to have properties similar to L-365,260 but is reported to be 125-fold higher in affinity for pancreatic CCK versus gastrin receptors (HUANG et al. 1989). Thus, these two agents are useful as models with which to develop other drugs with greater selectivity for peptide receptors of the gastrin/CCK class, and they should help define physiologic processes under the primary control of each of these peptides.

Another potential approach to blocking the actions of gastrin is through immunoneutralization. Animals have been actively immunized against crude extracts of gastric mucosa, which causes inhibition of gastric acid and pepsin secretion (BASS 1974). KOVACS et al. (1987) demonstrated that a specific monoclonal antibody (28.2), which interacts with the C-terminal region of gastrin and CCK, inhibited the acid secretory response to exogenous gastrin when the antibody was intravenously administered to dogs. The antibody also inhibited the response to a peptone meal while not affecting basal or histamine-induced acid secretion. Chronic immunoneutralization of gastrin is reported to significantly reduce basal acid secretion in addition to blocking the response to exogenously administered gastrin-17 (even 48 h after the last dose) with no blockade of the histamine response (OHNING et al. 1991). Clinically, HUGHES et al. (1976) demonstrated in a patient with

Zollinger-Ellison syndrome that gastrin antiserum inhibited gastric secretion without interfering with betazole-stimulated acid secretion.

II. Cholecystokinin Receptor Blockers

Like gastrin, CCK possesses many activities of physiologic importance. As stated previously, the C-terminal pentapeptide is identical to that of gastrin. And like gastrin, CCK exists in several molecular forms, with chain lengths from 5 to 58 amino acids (Marx et al. 1987). There is some controversy as to whether the predominant physiologically important form of the peptide is CCK-8 or CCK-33; the octapeptide is thought to be two or more times more potent than the longer sequence.

Antagonists to CCK may be useful to reduce gallbladder contraction, inhibit pancreatic bicarbonate secretion, reduce intestinal motility, tighten the lower esophageal sphincter or inhibit the growth of tissues responsive to the trophic action of CCK. In addition, it is conceivable that there may be applications to block the satiety effect of CCK (Reidelberger and O'Rourke 1989).

Another potential application of CCK antagonists is in irritable bowel syndrome (IBS). CCK is speculated to play a role in IBS (Harvey and Read 1973; Chasen et al. 1982). Besides direct effects of CCK gut smooth muscle, a central and/or peripheral neuromodulatory action could account for dysmotilities, cramping and the lowered visceral pain threshold noted in IBS. CCK is present in neurons of the brain and the autonomic nervous system (Alwmark and Greeley 1987). CCK-8 is excitatory for neurons in inferior mesenteric ganglia (Schumann and Kreulen 1986), and evokes [^{3}H] acetylcholine release from myenteric neurons (Zelles et al. 1990). Its actions on enteric neurons are described in Chap. 8. These findings support a role for CCK in the integration and/or disruption of gut myoelectric and contractile activities. Future studies may confirm the involvement of CCK in IBS and provide insight into the selective central or peripheral targeting of new therapeutic agents.

There are several classes of CCK receptors and antagonists with relative selectivity for each. A distinctly identifiable class of receptors exists in the brain (CCK-B) and another with different characteristics exists principally in the gallbladder, pancreas, vagus nerve and intestinal tract (CCK-A). A third type of receptor, closely related to the CCK-B receptor, is thought to be identical to the receptor for gastrin in the stomach (Freidinger 1989). CCK antagonists are classified as derivatives of cyclic nucleotides or amino acids, as well as partial sequences of the C-terminal portion of CCK, and benzodiazepines. A benzodiazepine-like CCK antagonist was discovered as a fermentation product from *Aspergillus alliaceus*; the substance was given the name asperlicin (Chang and Lotti 1986). Asperlicin was reported to be at least 300 times more potent than proglumide in CCK binding and functional assays in vitro (Silverman et al. 1987). This natural product is a

Table 4. Comparative binding IC_{50} (nM) values for CCK-8 and gastrin antagonists.[a] (Modified from CHANG and LOTTI 1986)

	[^{125}I]CCK-8		[^{125}I]gastrin
	Bovine gallbladder	Guinea pig brain	Guinea pig gastric gland
CCK-8	0.06 ± 0.01	0.39 ± 0.06	0.88 ± 0.21
L-364, 718	0.05 ± 0.003	245 ± 97	300 ± 72
Asperlicin	220 ± 11	>100 000	>100 000
Proglumide	200 000 ± 43 000	800 000 ± 26 000	900 000 ± 20 000

[a] $n = 3$ or more.

nonpeptide with no agonist activity which shows preferential selectivity for the CCK-B type receptors. Using this substance as a lead compound, medicinal chemists successfully synthesized L-364,718 (now known as MK-329) and other orally active competitive antagonists (EVANS et al. 1986; FREIDINGER 1989). Table 4 compares the competitive binding of CCK antagonists at peripheral (CCK-A), central (CCK-B) and gastrin-like CCK receptors. L-364,718 is most selective for CCK-A type receptors. This is consistent with its action to block CCK-induced release of [^{3}H]-acetylcholine from neurons of the myenteric plexus of the gut (ZELLES et al. 1990). The inhibitory effects of L-364,718 on rat and human gastric emptying (GREEN et al. 1988; LIDDLE et al. 1989), canine pancreatic secretion (HOSOTANI et al. 1989) and human gallbladder contraction (LIDDLE et al. 1989) have helped confirm a physiologic role for CCK in feedback mechanisms and direct regulation of GI function.

Loxiglumide (previously named CR1505) is another proglumide-related CCK antagonist. It is reported to block gallbladder emptying induced by a meal or the amphibian CCK analog cerulein (NIEDERAU et al. 1989a). This compound also competitively inhibits cerulein-stimulated pancreatic secretion in rats, but does not antagonize secretin-stimulated secretion (NIEDERAU et al. 1989b).

Current and other newly developed, more specific CCK antagonists will undoubtedly be valuable tools for physiologists and pharmacologists. They should help to more precisely define the physiologic role of CCK and provide new therapies for syndromes driven by CCK.

III. Vasoactive Intestinal Peptide Antagonists

Vasoactive intestinal peptide is widely distributed in and has numerous effects on mammals. Physiologically, VIP is thought to play paracrine and neurocrine roles in gut smooth muscle relaxation, and may be a neuro-modulator in the central nervous system and the myenteric plexus. Although the physiologic function of VIP is still not completely understood, it is a major mediator of the severe watery diarrhea associated VIPoma (pancreatic cholera, Verner-Morrison syndrome; GAGINELLA et al. 1990). Thus, while it may be undesirable to block the paracrine effects of low concentrations of VIP in normal physiological processes, it would be useful to have a pharmacological antagonist of VIP for application in pathophysiologic situations.

Peptide histidine isoleucine amide (PHI) is nearly identical in sequence to VIP except for the substitution of Lys^{12} for Arg^{12} and a C-terminal methionine amide in place of the isoleucine amide (TATEMOTO et al. 1981). No physiologic or pathologic role for PHI has been convincingly demonstrated. However, VIP and PHI are encoded in the same precursor molecule and appear to coexist in enteric nerves along the length of the mammalian intestinal tract (YIANGOU et al. 1985; SCHULTZBERG et al. 1980; SJÖLUND et al. 1983; FAHRENKRUG et al. 1985). VIP/PHI immunoreactive neurons project from the submucosa to areas of close proximity to crypt and villous epithelial cells (KEAST et al. 1985; COSTA et al. 1980; BISHOP et al. 1984; COSTA and FURNESS 1983), suggesting a possible role of PHI in regulating mucosal ion transport. It might be anticipated that antagonists of VIP would also be effective at blocking the effects of PHI.

As in the case of immunoneutralization of gastrin, antiserum to VIP has served as a "antagonist" in a number of studies aimed at demonstrating a physiologic role of the peptide (ANGEL et al. 1983; GRIDER et al. 1985; GOYAL et al. 1980; BIANCANI et al. 1984). However, therapeutic use of VIP antiserum is untenable for a number of reasons, including the requirement for large amounts to be available for use in patients.

The synthesis of analogs of VIP would seem to be the most logical approach to developing specific receptor antagonists to VIP. WAELBROECK et al. (1985) tested 14 growth hormone-releasing (GRF) analogs for their ability to compete for $[^{125}I]$i-VIP-binding sites and stimulation of adenylate cyclase activity in rat pancreatic plasma membranes. All of the analogs showed competition for the binding sites but they differed considerably in intrinsic activity (some were agonists with intrinsic activity equal to, and in some instances greater than, VIP itself). One analog, (N-Ac-Tyr^1, D-Phe^2)-GRF_{1-29}-NH_2, inhibited adenylate cyclase activity ($K_i = 1\,\mu M$) and had a binding IC_{50} of $1.2 \pm 0.40\,\mu M$. PANDOL et al. (1986) later demonstrated that [4-Cl-D-Phe^6,Leu^{17}]-VIP was a competitive antagonist of VIP in a rat pancreatic tissue preparation and a colonic cell line. This effect appeared to be specific; the antagonist did not inhibit the action of secretin, glucagon or GRF. Unfortunately, a high concentration ($30\,\mu M$) of this substance was required for notable inhibition of VIP action in pancreatic acini and hepatocytes. The 19 amino acid C-terminal fragment of VIP also acts as an antagonist at VIP-binding sites in membranes from HT29 colonic adenocarcinoma cells (TURNER et al. 1986). This peptide was without agonist activity but had an affinity of approximately 1000-fold less than VIP for VIP-binding sites and stimulation of adenylate cyclase activity in this preparation. Others have found compounds (based upon GRF; see Chap. 5), which demonstrate competitive antagonism of electrolyte secretion by the rat duodenal mucosa in vitro (Cox and CUTHBERT 1989).

Another approach has been to develop a neurotensin/VIP hybrid peptide. One such substance competes for VIP-binding sites on spinal cord and lymphoid cells (GOZES et al. 1991). Differences in affinity for binding sites in the two tissues suggest heterogeneity for receptors of immune and spinal cord cells. ROBBERECHT et al. (1990) have discussed their viewpoint for the existence of subtypes of VIP receptors. Their conclusions are based upon ligand binding, as well as biochemical data in normal and transformed cells. Direct evidence for the multiplicity of VIP receptors will require receptor purification and the application of molecular biological techniques to determine the sequence and to clone the receptor(s) in a particular cell type (see Chap. 5). From the viewpoint of the current rapid advancement of modern biological techniques it appears that we will soon have the answer to this question. This will clarify the physiologic role of VIP/PHI and related (newly identified?) peptides. More effective therapy for patients suffering from conditions due to excess VIP will be a further benefit of this avenue of research.

IV. Substance P Antagonists

Among the tachykinins, substance P is perhaps the most widely studied. Substance P is concentrated in neurons of the brain and gut (WALKER and THOMPSON 1987). The mammalian peptides neurokinin A, neurokinin B and neuropeptide K, and the nonmammalian peptides kassinin, eledoisin and physalaemin are other members of the tachykinin family with considerable cross-reactivity towards substance P receptors (DUTTA 1987). Named for their ability to produce rapid contraction of smooth muscle (as opposed to slow contraction produced by bradykinins), the physiologic functions of substance P and the other tachykinins are not yet defined (BROWN and MILLER 1991). The enteric and CNS distribution of substance P suggests a possible role in the regulation of gut motility and secretion (see Chaps. 9, 10) and as a mediator of visceral pain. Substance P may contribute to dysmotility and cramping in IBS. Binding sites for substance P (but not several other tachykinins) are increased in the submucosa of patients with ulcerative colitis and Crohn's disease (MANTYH et al. 1988), raising the possibility that this peptide is involved in the pathophysiology of inflammatory bowel disease. The fact that tachykinins are mediators in neurogenic inflammation (PAYAN 1989) supports their involvement in mucosal inflammation. Thus, substance P antagonists might be of use in the symptomatic control of IBS and inflammatory processes in the gut.

Early work identified peptide analogs of substance P as antagonists but their lack of selectivity, low potency and intrinsic agonist activity have been problems. Particularly important is the undesirable property of many potential substance P antagonists to release histamine (FEWTRELL et al. 1982; MAZUREK et al. 1981; PIOTROWSKI et al. 1984). Some of the peptide antagonists of substance P have indeed acted as analgesics and antinociceptive agents in several different experimental paradigms; none to date have been reported to block visceral pain (DUTTA 1987). Inflammation models other than those used in inflammatory bowel disease have responded to analogs of substance P (FERRELL and LAM 1989).

An intriguing example of a nonpeptide antagonist of substance P has recently been reported (SNIDER et al. 1991). This drug, designated as CP-96,345 or [(25,35S)-*cis*-2-(diphenylmethyl)-*N*-[(2-methoxyphenyl)-methyl]-1-azabicyclo[2.2.2]octan-3-amine], is a competitive antagonist of substance P binding to the neurokinin-1 subclass of tachykinin receptor in the bovine brain, and is without agonist activity in the systems in which it has been tested. Taken together with the discovery of nonpeptide antagonists of gastrin and CCK, the discovery of CP-96,345 serves to increase our confidence that selective agents for delineation of the function of tachykinins will be possible. New therapies will eventually arise if this approach is pursued tenaciously.

V. Opioid Peptides

The story of the discovery of endogenous peptides with opioid activity is fascinating and has led to an immense amount of research. Numerous agonists and antagonists, and a growing number of subtypes of opioid receptors, have now been identified (see BROWN and MILLER 1991). Space does not permit a detailed description of the distribution and function of the endorphin/enkephalin family of peptides and their receptors. Nevertheless, it is important to touch on the potential role of these peptides in GI function.

Following the ancient discovery of the euphoric and analgesic properties of opium, the antidiarrheal and constipating effects of opium-derived drugs were recognized. Contemporary research led to discovery of the endogenous opiate receptor ligands. Localization of two such peptides, methionine- and leucine-enkephalin, throughout the GI tract (within the myenteric neural network) has suggested a modulatory role on motility and secretory function of the gut. Considered from the agonist point of view, synthetic enkephalins might be useful in antidiarrheal therapy. Their effect can result from reduction in intraluminal flow (due to an effect on gut muscle) and enhancement of net electrolyte absorption by the mucosa. The influence on fluid absorption might be centrally mediated (BROWN and MILLER 1983) or through a local action on the gut (FOGEL and KAPLAN 1984; DOBBINS et al. 1980; KACHUR and MILLER 1980). Synthetic opiates also antagonize the electrolyte secretion induced by prostaglandins (GAGINELLA 1990b) and VIP (COUPAR 1983). Whether or not the local action of opioids is directly upon the epithelial cell is uncertain. Although specific receptors appear to mediate opiate effects on intestinal secretion (KACHUR and MILLER 1980; DOBBINS et al. 1980; COUPAR 1983), direct localization of receptors for opiates to epithelial cells has not been accomplished (BINDER et al. 1980; GAGINELLA et al. 1983). Tonic influences of enkephalins on mucosal transport may be regulated by interaction of the opioids with cholinergic (GAGINELLA and WU 1983) or other enteric neurons (see Chap. 10).

An abnormality of enhanced sensitivity to, or increased release of endogenous opioids, in certain GI disorders (e.g., IBS, constipation, ileus) might be responsive to an opioid antagonist. Naloxone is an effective antagonist at μ-type opioid receptors. This nonpeptide antagonist has been used to relieve severe constipation in two patients (KREEK et al. 1983). However, naloxone has not been reported to be effective in patients with constipation-predominant (or any other form of) IBS.

E. The Future

Important advances have been made in the development of peptides or peptide mimics as drugs, and as modulators of endogenous peptides in therapeutics. Future work will be directed toward identifying orally active peptides and new, more specific nonpeptide antagonists, especially with regard to CCK, VIP and substance P actions. Intensive work is underway to isolate, clone and characterize peptide receptors. Many such receptors are known, while others are being rapidly unmasked through the application of new techniques.

We will enrich our understanding of "known" peptides and, through molecular biology, gain insights as to receptor origins, specificity and manipulation. We will also learn more about GI physiology through new information on the paracrine, neurocrine, endocrine and autocrine functions of existing and newly identified peptides. This will result in more precise targeting of new drugs in the therapy of GI diseases and other disorders.

References

Anderson JV, Bloom SR (1986) Neuroendocrine tumors of the gut: long-term therapy with the somatostatin analog SMS 201–995. Scand J Gastroenterol 2:115–128

Angel F, Go VLW, Schmalz PF, Szurszewski JH (1983) Vasoactive intestinal polypeptide: a putative transmitter in the canine gastric muscularis mucosa. J Physiol (Lond) 341:641–654

Alwmark A, Greeley GH Jr (1987) Glucagon. In: Thompson JC, Greeley GH Jr, Rayford PL, Townsend CM Jr (eds) Gastrointestinal endocrinology. McGraw-Hill, New York, p 234

Baldwin J, Heer FW, Albo R, Peloso O, Ruby L, Silen W (1966) Effect of vagus nerve stimulation on hepatic secretion of bile in human subjects. Am J Surg 11:66–69

Banting FG, Best CH, Collip JB, Campbell WR, Fletcher AA (1922) Pancreatic extracts in the treatment of diabetes mellitus. Can Med Assoc 12:141–146

Bass P (1974) Gastric antisecretory and antiulcer agents. In: Harper NJ, Simmonds AB (eds) Advances in drug research. Academic, London, p 205

Bayliss WM, Starling EH (1902) The mechanisms of pancreatic secretion. J Physiol (Lond) 28:325–353

Beauchamp RD, Townsend CM, Singh P, Glass EJ, Thompson JC (1985) Proglumide, a gastrin receptor antagonist, inhibits growth of colon cancer and enhances survival of mice. Ann Surg 202:303–309

Biancani P, Walsh JH, Behar J (1984) Vasoactive intestinal polypeptide. A neurotransmitter for lower esophageal sphincter relaxation. J Clin Invest 73:963–967

Binder HJ, Lemp GF, Gardner JD (1980) Receptors for vasoactive intestinal peptide and secretin on small intestinal epithelial cells. Am J Physiol 238:G190–G196

Bishop AE, Polak JM, Yiangou Y, Christofides ND (1984) The distributions of PHI and VIP in porcine gut and their colocalization to a proportion of intrinsic ganglion cells. Peptides 5:255–259

Bond JH, Levitt MD (1980) Effects of glucagon on gastrointestinal blood flow of dogs in hypovolemic shock. Am J Physiol 238:G434–G439

Brown DR, Miller RJ (1983) CNS involvement in the anti-secretory action of [Met^5] enkephalinamide on the rat intestine. Eur J Pharmacol 90:441–444

Brown DR, Miller RJ (1991) Neurohormonal control in fluid and electrolyte transport in intestinal mucosa. In: Schultz SG, Field M, Frizzell RA (eds) Intestinal absorption and secretion. American Physiological Society, Bethesda, MD, p 527 (Handbook of Physiology, sect 6: The gastrointestinal system, vol 4)

Brown JC (1967) Presence of a gastric motor-stimulating property in duodenal extracts. Gastroenterology 52:225–229

Chang RSL, Lotti VJ (1986) Biochemical and pharmacological characterization of an extremely potent and selective nonpeptide cholecystokinin antagonist. Proc Natl Acad Sci USA 183:4923–4926

Chasen R, Tucker H, Palmer D, Whitehead W, Schuster M (1982) Colonic motility in irritable bowel syndrome and diverticular disease. Gastroenterology 82:1031

Condon JR, Knight M, Day JL (1973) Glucagon therapy in acute pancreatitis. Br J Surg 60:509–511

Costa M, Furness JB (1983) The origins, pathways, and terminations of neurons with VIP-like immunoreactivity in the guinea-pig small intestine. Neuroscience 8:665–676

Costa M, Furness JB, Buffa R, Said SI (1980) Distribution of enteric nerve cell bodies and axons showing immunoreactivity for vasoactive intestinal polypeptide in the guinea-pig intestine. Neuroscience 5:587–596

Coupar IM (1983) Characterization of the opiate receptor population mediating inhibition of VIP-induced secretion from the small intestine of the rat. Br J Pharmacol 80:371–376

Cox HM, Cuthbert AW (1989) Secretory actions of vasoactive intestinal polypeptide, peptide histidine isoleucine and helodermin in rat small intestine: the effects of putative VIP antagonist upon VIP-induced ion secretion. Regul Pept 26:127–135

Debas HT (1987) Clinical significance of gastrointestinal hormones. Adv Surg 21:157–188

Del Valle J, Yamada T (1990) The gut as an endocrine organ. Annu Rev Med 41:447–455

Dobbins J, Racusen L, Binder JH (1980) Effect of D-alanine methionine enkephalin amide on ion transport in rabbit ileum. J Clin Invest 66:19–28

Dutta AS (1987) Agonists and antagonists of substance P. Drugs Future 12:781–792

Dyck WP, Texter EC Jr, Lasater JM (1970) Influence of glucagon on pancreatic exocrine secretion in man. Gastroenterology 58:532–539

Edkins JS (1906) On the chemical mechanism of gastric secretion. J Physiol (Lond) 34:133–144

Evans BE, Bock MG, Rittle KE, DiPardo RM, Whitter WL, Veber DF, Anderson PS, Freidinger RM (1986) Design of potent, orally effective, nonpeptidal

antagonists of peptide hormone cholecystokinin. Proc Natl Acad Sci USA 83:4918–4922
Fahrenkrug J, Bek T, Lundberg JM, Hokfelt T (1985) VIP and PHI in cat neurons: colocalization but variable tissue content possible due to differential processing. Regul Pept 12:21–34
Fassler JE, O'Dorisio TM, Mekhjian HS, Gaginella TS (1990) Octreotide inhibits increases in short-circuit current induced in rat colon by VIP, substance P, serotonin and aminophylline. Regul Pept 29:189–197
Ferrell WR, Lam FY (1989) Inhibition of carrageenan induced inflammation in the rat knee joint by substance P antagonist. Ann Rheum Dis 48:928–932
Fewtrell CMS, Foreman JC, Jordan CC, Oehme P, Renner H, Stewart JM (1982) The effects of SP on histamine and 5HT release in the rat. J Physiol (Lond) 330:393–411
Fogel R, Kaplan RB (1984) Role of enkephalins in regulation of basal intestinal water and ion absorption in the rat. Am J Physiol 246:G386–G392
Freidinger RM (1989) Cholecystokinin and gastrin antagonists. Med Res Rev 9:272–290
Fujimoto S, Miyazaki M, Ishigami H et al. (1982) Plasma motilin levels in patients with abdominal surgery. Acta Chir Scand 148:33–37
Gaginella TS (1990a) Neurohumoral receptor pharmacology of intestinal secretion. In: Lebenthal E, Duffey M (eds) Textbook of secretory diarrhea. Raven, New York, p 163
Gaginella TS (1990b) Inhibition of eicosanoid-induced secretion. In: Lebenthal E, Duffey M (eds) Textbook of secretory diarrhea. Raven, New York, p 395
Gaginella TS, O'Dorisio TM (1988) Octreotide: entering the new era of peptidomimetic therapy. Drug Intell Clin Pharm 22:154–155
Gaginella TS, Wu Z (1983) [D-Ala^2, D-Met^{5}NH$_2$]-enkephalin inhibits acetylcholine release from the submucosal plexus of rat colon. J Pharm Pharmacol 35:823–825
Gaginella TS, Rimele TJ, Wietecha M (1983) Studies on rat intestinal epithelial cell receptors for serotonin and opiates. J Physiol (Lond) 335:101–111
Gaginella TS, O'Dorisio TM, Mekhjian HS, O'Dorisio MS, Woltering EA (1989) Tumors of the gastroenteropancreatic axis. In: O'Dorisio TM (ed) Sandostatin in the treatment of GEP endocrine tumors. Springer, Berlin Heidelberg New York, p 23
Gaginella TS, O'Dorisio TM, Fassler JE, Mekhjian HS (1990) Treatment of endocrine and nonendocrine secretory diarrheal states with sandostatin. Metabolism 39 Suppl 2: 172–175
Geary N, Langhans W, Scharrer E (1981) Metabolic concomitants of glucagon-induced suppression of feeding in the rat. Am J Physiol 241:R330–R335
Goyal RK, Rattan S, Said SI (1980) VIP as a possible neurotransmitter of non-cholinergic, non-adrenergic inhibitory neurons. Nature 288:378–380
Gozes Y, Brenneman DE, Fridkin M, Asofsky R, Gozes I (1991) A VIP antagonist distinguishes VIP receptors on spinal cord cells and lymphocytes. Brain Res 540:319–321
Green T, Dimaline R, Peikin S, Dockray GJ (1988) Action of the cholecystokinin antagonist L364,718 on gastric emptying in the rat. Am J Physiol 255:G685–G689
Grider JR. Cable MB, Said SI, Makhlouf GM (1985) Vasoactive intestinal peptide as a neural mediator of gastric relaxation. Am J Physiol 248:G73–G78
Harrison JD, Jones JA, Morris DL (1990) The effect of the gastrin receptor antagonist proglumide on survival in gastric carcinoma. Cancer 66:1449–1452
Harvey RF, Read AE (1973) Effect of cholecystokinin on colonic motility and symptoms in patients with the irritable bowel syndrome. Lancet 1:1–3
Henning SJ (1987) Functional development of the gastrointestinal tract. In: Johnson LR (ed) Physiology of the gastrointestinal tract. Raven, New York, p 85

Hirschowitz B, Sachs G (1965) Vagal gastric secretory stimulation by 2-deoxy-D-glucose. Am J Physiol 209:452–460
Hoosein NM, Kiener PA, Curry RC, Rovati LC, McGilbra DK, Brattain MG (1988) Antiproliferative effects of gastrin receptor antagonists and antibodies to gastrin on human colon carcinoma cell lines. Cancer Res 48:7179–7183
Hosotani R, Chowdhury P, Rayford PL (1989) L-364,718, a new CCK antagonist, inhibits postprandial pancreatic secretion and PP release in dogs. Dig Dis Sci 34:462–467
Huang SC, Zhang L, Chiang HCV, Wank SA, Maton PN, Gardner JD, Jensen RT (1989) Benzodiazepine analogues L365,260 and L364,718 as gastrin and pancreatic CCK receptor antagonists. Am J Physiol 257:G169–G174
Hughes WS, Hughes JB, Ritzmann SE (1976) Therapy with gastrin antibody in the Zollinger-Ellison syndrome. Dig Dis Sci 21:201–204
Imdahl A, Eggstein S, Crone C, Farthmann EH (1989) Growth of colorectal carcinoma cells: regulation in vitro by gastrin, pentagastrin and the gastrin-receptor antagonist proglumide. J Cancer Res Clin Oncol 115:388–392
Janssens J, Peeters TL, Vantrappen G, Tack J, Urbain JL, de Roo M, Muls E, Bouillon R (1990) Improvement of gastric emptying in diabetic gastroparesis by erythromycin. N Engl J Med 322:1028–1031
Jenkins SA, Baxter JN, Corbett W et al. (1985) Efficacy of somatostatin and vasopressin in the control of acute variceal hemorrhage. Hepatology 5:344–345
Jensen RT, Norton JA (1991) Endocrine neoplasms of the pancreas. In: Yamada T, Alpers DH, Owyang C, Powell DW, Silverstein FE (eds) Textbook of gastroenterology, vol 2. Lippincott, Philadelphia, p 1912
Johnson LR, Guthrie PD (1984) Proglumide inhibition of trophic action of pentagastrin. Am J Physiol 9:G62–G66
Kachur JF, Miller RJ, Field M (1980) Control of guinea pig intestinal electrolyte secretion by a delta-opiate receptor. Proc Natl Acad Sci USA 77:2753–2756
Karlin DA, McBath M, Jones RD, Elwyn KE, Romsdahl MM (1985) Hypergastrinemia and colorectal carcinogenesis in the rat. Cancer Lett 29:73–78
Katz MD, Erstad BL, Rose C (1988) Treatment of severe cryptosporidium-related diarrhea with octreotide in a patient with AIDS. Drug Intell Clin Pharm 22:134–136
Kayasseh L, Keller U, Gyr K et al. (1980) Somatostatin and cimetidine in peptic-ulcer haemorrhage. Lancet 1:844–846
Keast JR, Furness JB, Costa M (1985) Distribution of certain peptide-containing nerve fibres and endocrine cells in the gastrointestinal mucosa in five mammalian species. J Comp Neurol 236:403–422
Kovacs TOG, Maxwell V, Walsh JH (1987) The effect of immunoneutralization with antigastrin monoclonal antibody on gastric acid secretion in dogs. Gastroenterology 92:508
Kreek MJ, Schaeffer RA, Hahn EF, Fishman J (1983) Naloxone, a specific opioid antagonist, reverses chronic idiopathic constipation. Lancet 1:261–262
Krenning EP, Bakker WH, Breeman WAP, Koper JW, Kooij PPM, Ausema L, Lameris JS, Reubi JC, Lamberts SWJ (1989) Localization of endocrine-related tumours with radioiodinated analogue of somatostatin. Lancet 1:242–244
Kusyk CJ, McNiel NO, Johnson LR (1986) Stimulation of growth of a colon cancer cell line by gastrin. Am J Physiol 251:G597–G601
Lavezzo A, Bali JP, Magous R, Lignon MF, Nisato D, Laur J, Castro B, Martinez J (1986) Pharmacological profile of a new peptidic gastrin antagonist. Regul Pept 15:111–119
Liddle RA, Gertz BJ, Kanayama S, Beccaria L, Coker LD, Turnbull TA, Morita ET (1989) Effects of a novel cholecystokinin (CCK) receptor antagonist, MK-329, on gallbladder contraction and gastric emptying in humans. J Clin Invest 84:1220–1225

Long RJ, Adrian TE, Bloom SR (1985) Somatostatin and the dumping syndrome. Br Med J 290:886–888

Longnecker SM (1988) Somatostatin and octreotide: literature review and description of therapeutic activity in pancreatic neoplasia. Drug Intell Clin Pharm 22:99–106

Lotti VJ, Chang RSL (1989) A new potent and selective non-peptide gastrin antagonist and brain cholecystokinin receptor (CCK-B) ligand: L-365-,260. Eur J Pharm acol 162:273–280

Mantyh CR, Gates TS, Zimmerman RP, Welton ML, Passaro EP Jr, Vigna SR, Maggio JE, Kruger L, Mantyh PW (1988) Receptor binding sites for substance P, but not substance K or neuromedin K, are expressed in high concentrations by arterioles, venules, and lymph nodules in surgical specimens obtained from patients with ulcerative colitis and Crohn disease. Proc Natl Acad Sci USA 85:3235–3239

Magnusson I, Ihre T, Johansson C et al. (1985) Randomised double blind trial of somatostatin in the treatment of massive upper gastrointestinal haemorrhage. Gut 26:221–226

Martinez J, Rodriguez M, Bali JP, Laur J (1986) Phenethyl ester derivative analogues of the C-terminal tetrapeptide of gastrin as potent gastrin antagonists. J Med Chem 29:2201–2206

Marx M, Gomez G, Lonovics J, Thompson JC (1987) Cholecystokinin. In: Thompson JC, Greeley GH Jr, Rayford PL, Townsend CM Jr (eds) Gastrointestinal endocrinology. McGraw-Hill, New York, p 213

Maton PN, O'Dorisio TM, Brent AH et al. (1985) Effect of long-acting somatostatin analogue (SMS 201–995) in a patient with pancreatic cholera. N Engl J Med 312:17–22

Mazurek N, Pecht I, Teichberg VI, Blumberg S (1981) The role of the N-terminal tetrapeptide in the histamine releasing action of SP. Neuropharmacology 20:1025–1027

McGregor DB, Jones RD, Karlin DA, Romsdahl MM (1982) Trophic effects of gastrin on colorectal neoplasms in the rat. Ann Surg 195:219–223

Miller RE, Chernish SM, Greenman GF (1982) Gastrointestinal response to minute doses of glucagon. Radiology 143:317–320

Morris DL, Charnley RM, Ballantyne KC, Jones J (1990) A pilot randomized control trial of proglumide (a gastrin receptor antagonist) in advanced colorectal cancer. Eur J Surg Oncol 16:423–425

Mulvihill SJ, Passaro E Jr, Debas HT et al. (1984) Severe diarrhea after colonic pseudoobstruction: treatment with somatostatin. N Engl J Med 310:467–468

Mulvihill SJ, Pappas TN, Passaro E et al. (1986) The use of somatostatin and its analogs in the treatment of surgical disorders. Surgery 100:457–476

Newman JB, Lluis F, Townsend CM Jr (1987) Somatostatin. In: Thompson JC, Greeley GH Jr, Rayford PL, Townsend CM Jr (eds) Gastrointestinal endocrinology. McGraw-Hill, New York, p 286

Niederau C, Heintges T, Rovati L, Strohmeyer G (1989a) Effects of loxiglumide on gallbladder emptying in healthy volunteers. Gastroenterology 97:1331–1336

Niederau M, Niederau C, Strohmeyer G, Grendell JH (1989b) Comparative effects of CCK receptor antagonists on rat pancreatic secretion in vivo. Am J Physiol 256:G150–G157

O'Dorisio TM, O'Dorisio MS (1985) Endocrine tumors of the gastroenteropancreat (GEP) axis. In: Mazzaferri E (ed) Endocrinology. Medical Examination, New York, pp 76–81

Ohe K, Sumii K, Sano K et al. (1980) Serum motilin in gastrointestinal diseases. Endocrinol Jpn 27:167–172

Ohning GV, Lloyd KCK, Walsh JH (1991) Chronic immunoneutralization with anti-gastrin monoclonal antibody blocks the acid secretory response to gastrin. Ceastroeuterology 100:658

Pandol SJ, Dharmsathaphorn K, Schoeffield MS, Vale W, Rivier J (1986) Vasoactive intestinal peptide receptor antagonist [4Cl-D-Phe^{6},Leu^{17}] VIP. Am J Physiol 250:G553–G557

Payan DG (1989) Neuropeptides and inflammation: the role of substance P. Annu Rev Med 40:341–352

Piotrowski W, Devoy MAB, Jordan CC, Foreman JC (1984) The SP receptor on rat mast cells and in human skin. Agents Actions 14:420–424

Pless J (1989) Chemical structure-pharmacological profile of sandostatin. In: O'Dorisio TM (ed) Sandostatin in the treatment of GEP endocrine tumors. Springer, Berlin Heidelberg New York, p 3

Reidelberger RD, O'Rourke MF (1989) Potent cholecystokinin antagonist L-364,718 stimulates food intake in rats. Am J Physiol 257:R1512–R1518

Robberecht P, Cauvin A, Gourlet P, Christophe J (1990) Heterogeneity of VIP receptors. Arch Int Pharmacodyn Ther 303:51–66

Rovati LA (1976) Inhibition of gastric secretion by anti-gastrinic and H_2-blocking agents. Scand J Gastroent erol 11:113–118

Rovati LA (1987) Views on possible therapeutic use of gastrin and cholecystokinin antagonists. In: Bali JP, Martinez J (eds) Gastrin and cholecystokinin. Chemistry, physiology and pharmacology. Elsevier North-Holland, Amsterdam, p 225

Salzberg MR, Gallagher EJ (1980) Propranolol overdose. Ann Emerg Med 9: 26–28

Santangelo WC, O'Dorisio TM, Kim JG (1985) Pancreatic cholera syndrome: effect of a synthetic somatostatin analog on intestinal water and ion transport. Ann Intern Med 103:363–367

Schultzberg M, Hokfelt T, Nilsson G, Terenius L, Rehfeld JF, Brown M, Elde R, Goldstein M, Said S (1980) Distribution of peptide- and catecholamine-containing neurons in the gastrointestinal tract of rat and guinea-pig: immunohistochemical studies with antisera to substance P, vasoactive intestinal polypeptide, enkephalins, somatostatin, gastrin/cholecystokinin, neurotensin, and dopamine β-hydroxylase. Neuroscience 5:689–744

Schumann MA, Kreulen DL (1986) Action of cholecystokinin octapeptide and CCK-related peptides on neurons in inferior mesenteric ganglion of guinea pig. J Pharmacol Exp Ther 239:618–625

Shulkes A (1990) Gastrointestinal hormones: from basic science to a clinical perspective. Aust N Z J Surg 60:575–578

Silverman MA, Greenberg RE, Bank S (1987) Cholecystokinin receptor antagonists: a review. Am J Gastroenterol 82:703–708

Singh P, Le S, Beauchamp RD, Townsend CM, Thompson JC (1987) Inhibition of pentagastrin-stimulated up-regulation of gastrin receptors and growth of mouse colon tumor in vivo by proglumide, a gastrin receptor antagonist. Cancer Res 47:5000–5004

Sirinek KR, Levine BA, Moyer MP (1985) Pentagastrin stimulates in vitro growth of normal and malignant human colon epithelial cells. Am J Surg 149:35–39

Sjölund K, Sande G, Hakanson R, Sundler F (1983) Endocrine cells in human intestine: an immunocytochemical study. Gastroenterology 85:1120–1130

Smith GP (1980) Satiety effect of gastrointestinal hormones. In: Beers RF, Bassett EG (eds) Polypeptide hormones. Raven, New York, p 413

Smith JP, Wood JG, Solomon TE (1989) Elevated gastrin levels in patients with colon cancer or adenomatous polyps. Dig Dis Sci 34:171–174

Snider RM, Constantine JW, Lowe JA, Longo KP, Lebel WS, Woody HA, Drozda SE, Desai MC, Vinick FJ, Spencer RW, Hess HJ (1991) A potent nonpeptide antagonist of the substance P (NK_1) receptor. Science 251:435–437

Solcia E, Capella C, Buffa R et al. (1982) Histology, histochemistry and ultrastructure of diarrheogenic VIP-secreting tumors (VIP-omas). In: Said SI (ed) Vasoactive intestinal peptide. Raven, New York, pp 495–602

Sunazuka T, Tsuzuki K, Marui S, Toyoda H, Omura S, Inatomi N, Itoh Z (1989) Motilides, macrolides with gastrointestinal motor stimulating activity. II. Quaternary N-substituted derivatives of 8,9-anhydroerythromycin A 6,9-hemiacetal and 9,9-dihydroerythromycin A 6,9-epoxide. Chem Pharm Bull (Tokyo) 37:2701–2709

Tatemoto K, Carlquist M, Mutt V (1981) Isolation and characterization of the intestinal peptide porcine PHI (PHI-27), a new member of the glucagon secretin family. Proc Natl Acad Sci USA 781:6603–6607

Townsend CM, Singh P, Thompson JC (1987) Possible role of gut hormones in cancer. In: Thompson JC, Greeley GH Jr, Rayford PL, Townsend CM Jr (eds) Gastrointestinal endocrinology. McGraw-Hill, New York, p 178

Townsend CM, Singh P, Thompson JC (1989) Effects of gastrointestinal peptides on gastrointestinal cancer growth. Med Clin North Am 18:777–791

Turner JT, Jones SB, Bylund DB (1986) A fragment of vasoactive intestinal peptide, VIP (10–28), is an antagonist of VIP in the colon carcinoma cell line, HT29. Peptides 7:849–854

Upp JR Jr, Singh P, Townsend CM Jr, Thompson JC (1989) Clinical significance of gastrin receptors in human colon cancers. Cancer Res 49:488–492

Vinik A, Moattari AR (1989) Use of somatostatin analog in management of carcinoid syndrome. Dig Dis Sci 34:14S–27S

Waelbroeck M, Robberecht P, Coy DH, Camus JC, de Neef P, Christophe J (1985) Interaction of growth hormone-releasing factor (GRF) and 14 GRF analogs with vasoactive intestinal peptide (VIP) receptors of rat pancreas. Discovery of (N-Ac-Tyr^1,D-Phe^2)-GRF(1–29)-NH_2 as a VIP antagonist. Endocrinology 116:2643–2649

Walker JP, Rayford PL (1987) Motilin. In: Thompson JC, Greeley GH Jr, Rayford PL, Townsend CM Jr (eds) Gastrointestinal endocrinology. McGraw-Hill, New York, p 311

Walker JP, Thompson JC (1987) Substance P. In: Thompson JC, Greeley GH Jr, Rayford PL, Townsend CM Jr (eds) Gastrointestinal endocrinology. McGraw-Hill, New York, p 317

Watson S, Durrant L, Morris D (1989) Gastrin: growth enhancing effects on human gastric and colonic tumour cells. Br J Cancer 59:554–558

Weinstock J, Baldwin GS (1988) Binding of gastrin$_{17}$ human gastric carcinoma cell lines. Cancer Res 48:932–937

Wiener I, Khalil T, Thompson JC, Rayford PL (1987) Gastrin-CCK family. In: Thompson JC, Greeley GH Jr, Rayford PL, Townsend CM Jr (eds) Gastrointestinal endocrinology. McGraw-Hill, New York, p 194

Yasui A, Douglas AJ, Walker B, Magee DF, Murphy RF (1990) Novel C-terminal gastrin antagonists. Int J Pept Protein Res 35:301–305

Yiangou Y, Christofides ND, Blank MA, Yanaihara N, Tatemoto K, Bishop AE, Polak JM, Bloom SR (1985) Molecular forms of peptide histidine isoleucine-like immunoreactivity in the gastrointestinal tract. Nonequimolar levels of peptide histidine isoleucine and vasoactive intestinal peptide in the stomach explained by the presence of a big peptide histidine isoleucine-like molecule. Gastroenterology 89:516–524

Zelles R, Harsing LG, Vizi ES (1990) Characterization of neuronal cholecystokinin receptor by L-364,718 in Auerbach's plexus. Eur J Pharmacol 178:101–104

Subject Index

Springer-Verlag and the Environment

We at Springer-Verlag firmly believe that an international science publisher has a special obligation to the environment, and our corporate policies consistently reflect this conviction.

We also expect our business partners – paper mills, printers, packaging manufacturers, etc. – to commit themselves to using environmentally friendly materials and production processes.

The paper in this book is made from low- or no-chlorine pulp and is acid free, in conformance with international standards for paper permanency.

Printing: Druckerei Kutschbach, Berlin
Binding: Buchbinderei Lüderitz & Bauer, Berlin